CHEMICAL PROCESS CONTROL - CPCIII

Proceedings of the Third International Conference on Chemical Process Control

Asilomar, California, January 12 - 17, 1986

Editors:

Manfred Morari
California Institute of Technology

Thomas J. McAvoy
University of Maryland

A **CACHE** Publication

ELSEVIER
Amsterdam - Oxford - New York - Tokyo
1986

Distributors for the United States and Canada:

ELSEVIER SCIENCE PUBLISHERS INC.
52 Vanderbilt Avenue
New York, NY 10017, U.S.A.

For all other areas:

ELSEVIER SCIENCE PUBLISHERS B.V.
Sara Burgerhartstraat 25
P.O. Box 211, 1000 AE Amsterdam, The Netherlands

Library of Congress Catalog Card Number 86-71565

ISBN 0-444-99532-3

This work is subject to copyright. All rights reserved, whether the whole or part of the material is concerned, specifically those of translation, reprinting, re-use of illustrations, broadcasting, reproduction by photocopying machine or similar means, and storage in data banks.

© CACHE
  Austin, Texas

PREFACE

Manfred Morari
California Institute of Technology

Thomas J. McAvoy
University of Maryland

This volume contains the proceedings of the Third International Conference on Chemical Process Control which was held at the Asilomar Conference Grounds in California from January 12 to 17, 1986. The first conference[1] of this kind took place ten years earlier at the same location, the second[2] in Sea Island, Georgia in 1981. The goal of all three conferences was to bring together participants from industry, government and universities for a week-long intensive meeting to discuss and critique the current state of process control. The 145 participants attending the third conference included 67 from universities and government, and 78 from industry. The international flavor was provided by 31 foreign attendees representing 15 different countries.

The first two conferences have come to be regarded as milestones in the process control field. They have not only served to highlight recent developments, but they have had a decisive impact on research efforts in succeeding years. They have helped to narrow the gap between process control theory and application. It is hoped that the third conference will continue this tradition.

The conference focused on advances that have taken place over the last five years. In particular, developments in control system robustness, process operability, adaptive control, and model predictive control were highlighted. In addition new results on on-line identification and optimization, reactor control, and actual industrial applications of advanced control were presented. Lastly, the potential impact of artificial intelligence and expert systems was assessed.

The presentations and discussions were complemented by a demonstration of software for the interactive design of control systems.

A comparison with the earlier two conferences suggests the following conclusions:

1. The interest in process control has increased dramatically. About 100 registrants could not be admitted in order to retain the conference format and to allow active participation by all attendees.
2. Academic research has shifted toward more practical problems. For example, the issue of model error and uncertainty which has hindered the application of some of the modern control techniques is now much better understood.
3. Future progress in process control will depend critically on intensive university/industry cooperation. In addition to industrial sponsorship of academic research, this cooperation should include, for example, sabbatical leaves for faculty members in industry and onsite experimental testing of new techniques by graduate students working jointly with control engineers.

[1]"Chemical Process Control", A. S. Foss & M. Denn, eds. AIChE Symposium Series, Number 159, Vol. 72 (1976).
[2]"Chemical Process Control 2", T. F. Edgar & D. E. Seborg, eds. United Engineering Trustees (1982), available from AIChE.

ACKNOWLEDGEMENTS

We wish to acknowledge the assistance and support of the following organizations and individuals in making this control conference possible.

— The industrial contributors facilitated participation by faculty members.
— The support from the National Science Foundation allowed the organizers to invite several speakers from abroad.
— Richard S. H. Mah, president of CACHE, coordinated all conference organization efforts.
— With the help of Janet Sandy, CACHE secretary, David M. Himmelblau took care of financial matters, site planning and publicity.
— Thomas J. Harris of E. I. DuPont handled all registrations.
— John Hale of E. I. DuPont helped establish and arrange funding for an electronic mail system which enhanced communication between the organizers.
— Vicky Jones and two graduate student assistants Norm Jerome and Evanghelos Zafiriou, handled all local arrangements and day-to-day problems.
— Bradley R. Holt and Ahmed Palozoglu organized the software demonstrations.
— Jan Owen helped in the preparation of the proceedings.

CONFERENCE ORGANIZATION

TECHNICAL PROGRAM COMMITTEE

Manfred Morari, Co-Chairman, California Institute of Technology
Thomas J. McAvoy, Co-Chairman, University of Maryland

COMMITTEE MEMBERS

Thomas F. Edgar, University of Texas, Austin
Jeffrey C. Kantor, Notre Dame University
W. Harmon Ray, University of Wisconsin
Irven H. Rinard, Halcon S. D.
Dale E. Seborg, University of California, Santa Barbara
George Stephanopoulos, MIT
Robert G. Wilhelm, Jr., AccuRay
B. Erik Ydstie, University of Massachusetts

SPONSORED BY

CAST Division, AIChE
National Science Foundation
CACHE Inc.

INDUSTRIAL CONTRIBUTORS

Bailey Controls
Chevron Research
Dow Chemical
E. I. DuPont
Exxon Chemical
Foxboro
IBM
Proctor & Gamble
Set Point
Shell Development

TABLE OF CONTENTS

PREFACE iii

CONTROL IN THE PRESENCE OF MODEL UNCERTAINTY
Session Summary
   J. C. Kantor 1

A Unifying Framework for Control System Design Under Uncertainty and its Implications for Chemical Process Control
   J. C. Doyle and M. Morari 5

Impact of Model Uncertainties and Nonlinearities on Modern Controller Design. Present Status and Future Goals
   R. Shinnar 53

AN INDUSTRIAL VIEW OF ADVANCED PROCESS CONTROL
Session Summary
   R. G. Wilhelm, Jr. 95

The Intgration of Rigorous Dynamic Modeling and Control System Synthesis for Distillation Columns: An Industrial Approach
   S. Roat, J. Downs, E. Vogel and J. Doss 99

Progress and Challenges in Batch Process Control
   M. R. Juba and J. W. Hamer 139

Multivariable Constraint Control Using a Frequency Domain Design Approach
   S. Treiber and D. W. Hoffman 185

MODEL PREDICTIVE CONTROL
Session Summary
   D. E. Seborg 231

Industrial Applications of IDCOM
   B. Froisy and J. Richalet 233

Advances in Industrial Model-Predictive Control
   C. E. Garcia and D. M. Prett 245

Coordinated Control
   L. Popiel, C. Brosilow and T. Matsko 295

PROCESS OPERABILITY
Session Summary
   I. H. Rinard 321

Dynamic Process Operability. Important Problems, Recent Results and New Challenges
   Y. Arkun 323

A Short-Cut Method for Process Control and Operability Analysis
T. E. Marlin, T. J. McAvoy, M. Marino-Galarraga and N. Kapoor 369

ADAPTIVE CONTROL
Session Summary
B. E. Ydstie 421

Adaptation, Auto-Tuning and Smart Controls
K. J. Åström 427

On the Use of Adaptive Control in the Process Industries
G. Dumont 467

Industrial Perspective of the Use of Adaptive Process Control
M. Fjeld 501

ON-LINE IDENTIFICATION AND OPTIMIZATION
Session Summary
T. F. Edgar 513

Digital Adaptive Control. A Robust Approach
G. C. Goodwin 517

Universal Dynamic Matrix Control
A. M. Morshedi 547

The Integration of Planning, Scheduling and Process Control
L. Lasdon and T. Baker 579

CONTROL OF CHEMICAL REACTORS
Session Summary
W. H. Ray 621

Control Problems in Microelectronics Processing
K. F. Jensen 623

Computer Control of Bioreactors—Present Limits and Challenges for the Future
M. Reuss 641

Some New Approaches for Controlling Complex Processes in Chemical Engineering
E. D. Gilles 689

A Review of Industrial Reactor Control: Difficult Problems and Workable Solutions
P. D. Schnelle, Jr. and J. R. Richards 749

EXPERT SYSTEMS IN PROCESS CONTROL
Session Summary
G. Stephanopoulos 803

Expert Systems for Computer-Aided Control Engineering
J. H. Taylor 807

Expert Systems in On-Line Process Control
R. L. Moore and M. A. Kramer 839

Process Control System Synthesis by an Expert System
K. Niida and T. Umeda 869

An Expert System for the Design of Distillation Controls
F. G. Shinskey 895

REFLECTIONS ON CPC-III
 F. G. Shinskey 913

LIST OF AUTHORS AND DISCUSSORS 929

INDEX 933

CONTROL IN THE PRESENCE OF MODEL UNCERTAINTY
Session Summary

Jeffrey C. Kantor
University of Notre Dame, Notre Dame, IN 46556

Much of the recent theoretical research in linear control design has centered on the issue of model uncertainty. This is clearly an appropriate topic for chemical process control where a dynamical model is usually unavailable, or at best, poorly characterized. While one goal for this line of research is to develop practical, robust methods for control design, the results have much deeper implications than just simply defining new tuning algorithms. We are now beginning to understand the fundamental tradeoffs that uncertainties pose for control stability and performance. After all, if perfect dynamical models were available, and if all disturbances and constraints were precisely known, then exact solutions to the usual control problems could be found by simply applying a sufficiently large computer. The extent to which this cannot be done is due to our inexact knowledge of a process and of the externally imposed disturbances. In fact, the intrinsic merit of feedback control is that it can suppress the effects of unknown disturbances, whether they arise from external sources or model uncertainty. Understanding of these issues has increased markedly in the last few years for the limited class of systems that can be described by linear, multivariable models. The purpose of this session at CPC-III was to review the status of the linear theory and its application to chemical process control.

The alteration of process dynamics through linear multivariable feedback control is, of course, a small part of a comprehensive approach to process control. The larger task is primarily one of steady-state optimization with explicit consideration given to process constraints, available control action, and to safety, alarms, data gathering, and the operator interface. Dynamics are often a secondary consideration.

But if this is the case, then what is the role of multivariable control design? Certainly it is to establish the effective low-level regulation of complex, interacting systems; to suppress the influence unknown disturbances; and to suppress the effects of modeling uncertainties. Without good regulation, the higher-level optimization strategies are doomed to fail. And that is precisely because high-level strategies assume a well-characterized and well-behaved process. Multivariable feedback control enables optimization by reducing the influence of disturbances.

One might also ask why additional theoretical research is required if controllers can already be tuned on-line to accomodate a particular situation. There are several points to be made, perhaps the most important is that so-called 'tuning' procedures are, in fact, de-tuning procedures which do not account for the stability/performance trade-off that is implicitly being made. What we seek from theory is both understanding and a method that permits us this make this essential trade-off in an optimal way. Moreover, an effective theory should answer such additional questions as the nature of the trade-offs between model complexity and control performance (e.g., what is the minimum information required to achieve integral control action?), or between controller complexity and performance. These additional questions cannot be addressed by simply looking for alternative, or even automatic, 'tuning' rules for a PID controller.

Paradoxically, most of the recent theoretical development for robust multivariable control has been accomplished using frequency domain analyses. This is because of the convenient way in which both stability and certain useful types of performance criteria can be cast into the frequency domain. Norms defined on complex matrix transfer functions provide a natural means for analysis. A decade ago, at about the time of CPC-I, it was thought that the multivariable frequency domain methods would lead to simple graphical design tools analogous to the classical single-variable methods. Since then, the complexity of the multivariable problem has become more evident, the results have become more rigorous, and unfortunately, also more resistant to simple graphical interpretation. Thus the new results have a decidely algebraic flavor, while the computational methods require a complex mixture frequency domain and time domain manipulations. Improved computational aids largely remove the objection to computational complexity so long as the problem statements are clear and applicable to real problems, the methods are reliable, and the results useful.

This session of CPC-III was intended to introduce both the problems posed by model uncertainty and review at least one significant approach linear multivariable control design. The paper by M. Morari and J. Doyle illustrates the role of model uncertainty in two examples: the classical Smith predictor for single-variable time-delay compensation; and in a 2-input, 2-output high-purity distillation column. They show that the acheivable control performance in both cases is limited by the presence of model uncertainty. They develop criteria for robust stability and performance in single-variable systems, comment on the multivarible problem, and provide a entry into the recent literature. As it happens, Internal Model Control is a natural basis for this discussion because it properly focuses on the real issues of feedback control, and transparently handles the problem of nominal closed-loop stability. The point is that the choosing a particular control structure does not solve the problem, it only serves as a more or less convenient basis for analysis. The resulting controllers can be implemented in number of equivalent ways.

The second paper of this session was presented by R. Shinnar. This paper also focused on the issue of model uncertainty, but from a more general and application oriented viewpoint. There remain many more problems to addressed, including the effects of nonlinearities, constraints, and obtaining better characterizations of the model uncertainties that are encountered in real chemical systems.

A UNIFYING FRAMEWORK FOR CONTROL SYSTEM DESIGN UNDER UNCERTAINTY AND ITS IMPLICATIONS FOR CHEMICAL PROCESS CONTROL

Manfred Morari
California Institute of Technology, Pasadena, CA 91125

John C. Doyle
California Institute of Technology, Pasadena, CA 91125
and
Honeywell, Inc., Minneapolis, MN 55440

Abstract. Model uncertainty is one of the major problems facing control system designers in practice. In this paper a general approach is outlined for the assessment of the effects of model uncertainty on control system performance. Also discussed are methods for the design of controllers which meet given performance specifications despite model uncertainty. The power of the new techniques as well as the failure of the traditional ones is illustrated on two examples widely studied in the process control literature: The design of Smith Predictor controllers and the design of 2-point composition controllers for high purity distillation columns.

Keywords. Robustness; Optimal Control; Distillation Control; Smith Predictor; Singular Value; Structured Singular Value; H_∞- Optimal Control; Model Error; Model Uncertainty

INTRODUCTION

On one side there is the **theory:** Given a linear process model, given information on the disturbances and the measurement noise (for example the spectra) we can choose from a number of techniques to design a feedback compensator: The "optimal" controller, of course, is obtained via the Linear Quadratic Gaussian (LQG) technique (e.g. Kwakernaak & Sivan, 1972). More "engineering" oriented approaches, for example the multivariable frequency domain techniques introduced by Rosenbrock (1974) and MacFarlane (1980), allow to include other design criteria like "decoupling" and "actuator failure tolerance" and can also lead to reasonable controller designs at times. And there are many more techniques which have been proposed over the last several decades.

On the other side there is the **practical reality:** There are no linear processes. Even if a process is linear in some narrow operating range, the model parameters are usually not known exactly and can vary in an unpredictable manner during the operation. Invariably a constraint on some manipulated variable is reached eventually and the process becomes nonlinear. Equally unrealistic is the assumption of fixed known disturbance and noise spectra. Controllers designed on the basis of the theory mentioned above have been shown to be prone to failure in this environment of nonlinearities and uncertainty.

Therefore it is not surprising that today the chemical process industries try to solve difficult control problems largely through simulation, case studies and ad hoc on-line adjustment of controller knobs. This can require enormous manpower. Because no systematic insight is obtained into the system inherent properties which give rise to these control difficulties and because the fundamental trade-offs which are involved cannot be understood in this manner, trial-and-error "solutions" might be largely inferior to what is possible with a more systematic procedure or no acceptable solution might be found at all. Design modifications which improve the control characteristics of the process have to be arrived at by random experimentations with the process model or even worse the process itself.

It has become increasingly clear during the last five years that the lack of use of the "modern" control techniques by industry was not caused by an "education-gap" but simply by the fact that the available theory did not address the practical problems. During the last decade several developments started independently which are all focused at bringing the theory closer to the practical problems and which have attracted widespread interest.

For years critics like Horowitz (1963) and Kestenbaum et al. (1976) have emphasized that the key issue in feedback design is **model uncertainty**. Finally the research community woke up to this fact and shifted its attention to the effects of model uncertainty on closed loop stability and performance. The terms **robust stability** and **robust performance** are now commonly used to indicate that a closed loop system is stable and meets the performance specifications even though the model does not adequately describe the plant.

The objective of this paper is to introduce the reader to the insights which have been gained during the last few years into the robust controller design problem and to outline their implications for process control. To motivate the discussion we will start with the description of two examples all of which have been the topic of dozens of publications during the last two to three decades and which are still subjects of controversy. The power of the new theory is that it ties them all together and explains and solves them through one unifying approach.

Example 1: Effect of Model Uncertainty on Smith-Predictor Controller

One possible representation of the Smith Predictor control structure is shown in Fig. 1. Here p is the physical process and $\tilde{p}$ the process model used to describe it. u, y, d and r are the process input, output, disturbances and reference or setpoint respectively. c is the controller. $\tilde{p}^*$ is the process model $\tilde{p}$ with the time delay removed, i.e.

$$\tilde{p} = \tilde{p}^* e^{-s\theta} \qquad (1)$$

where $\tilde{p}^*$ is rational. The objective of the two model blocks $\tilde{p}$ and $\tilde{p}^*$ is to improve the closed loop performance over what is possible with the controller c alone.

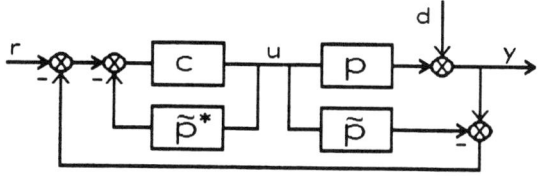

Fig. 1. Smith Predictor Controller

The most widely studied system is

$$\tilde{p} = \frac{ke^{-s\theta}}{\tau s+1} \qquad (2)$$

with the controller c chosen to be a PI controller. It can be shown (Cook & Price, 1978) that for (2) the Smith Predictor/PI controller with the integral time set equal to the process time constant τ is optimal in the sense of a quadratic performance objective. The optimality implies that the closed loop system has a gain margin of at least 2 and a phase margin of at least $60°$ for all positive controller gains (Palmor, 1980).

Several authors (e.g. Palmor, 1980; Ioannides et al., 1979; Owens, 1982) have noticed and explained in different ways that the performance improvements by the Smith Predictor controller can be very sensitive to errors in the model $\tilde{p}$. It was demonstrated that for high controller gains the Smith Predictor controller can be destabilized by arbitrarily small deadtime errors. The parametric studies underlying these observations prove that optimality in the classic sense as well as gain and phase margin have little to do with practically useful design criteria.

While these demonstrations of parametric sensitivity are illustrative for academics they are of limited value to the design engineer: Parametric studies are infeasible when more complex models than (2) are involved and when more than one or two parameters vary at the same time. Equally important, errors in the model structure (2nd order vs. 1st order, for example) cannot be studied at all in this manner. Finally, the design engineer is faced (at best) with errors in all three parameters of (2) and wants to know how to choose the PI parameters such as to guarantee at least stability but also acceptable performance despite the uncertainty: The design engineer is primarily interested in controller <u>synthesis</u> rather than <u>analysis</u>. Until recently a comprehensive theory for both analysis and synthesis in the presence of uncertainty was lacking.

Example 2: Control of High-Purity Distillation Column

For the distillation column specified below, the overhead composition is to be controlled at $y_D = 0.99$ and the bottom composition at $x_B = 0.01$, by manipulating the boilup V and the reflux L.

Feed:	1
Distillate:	0.5
Bottoms:	0.5
Feed Composition:	0.5
Relative Volatility:	$\alpha = 1.5$
Number of Trays:	40
Reflux Ratio:	5.4 (1.39 times minimum reflux)

Short cut calculations (Skogestad, 1986) yield the dynamic model

$$\begin{bmatrix} y_D \\ x_B \end{bmatrix} = \frac{1}{3.6s+1} \begin{bmatrix} 0.99 & 0.98 \\ 0.97 & 0.98 \end{bmatrix} \begin{bmatrix} L \\ -V \end{bmatrix} \qquad (3)$$

with a Relative Gain (Bristol, 1966) of 49.5 indicating "strong interactions". We will study three different disturbances and two different control configurations. The disturbances are represented by setpoint changes, which have equivalent steady state effects on the outputs.

(i) $r^T = (1 \quad 1)$ corresponds to a feed composition disturbance

(ii) $r^T = (1 \quad -1)$ corresponds to a setpoint change or measurement disturbance

(iii) $r^T = (0.463 \quad 0.517)$ corresponds to a feed flowrate disturbance

Controller A consists of two single loop PI controllers, Controller B employs in addition a steady state decoupler.

(A) $$C(s) = \frac{4(3.6s+1)}{s} \cdot I \qquad (4)$$

(B) $$C(s) = \frac{0.5(3.6s+1)}{s} G(0)^{-1} \qquad (5)$$

In theory Controller B gives rise to decoupled first order responses with time constant 2. In practice, the plant behavior will be different from the model (3). To introduce a flavor of reality we assumed for the simulation runs shown in Figs. 2-4 an error of 20% in the value of the manipulated variable. That is, any change in u computed by the controller differed by 20% from the change in u actually implemented on the plant.

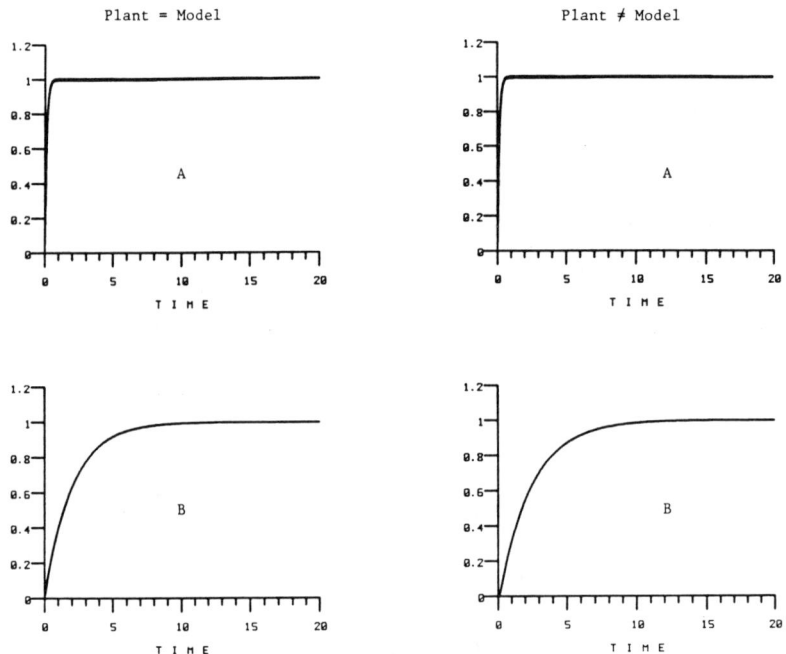

Fig. 2. Response to feed composition disturbance for controllers A and B with and without model error.

A Unifying Framework for Control System Design

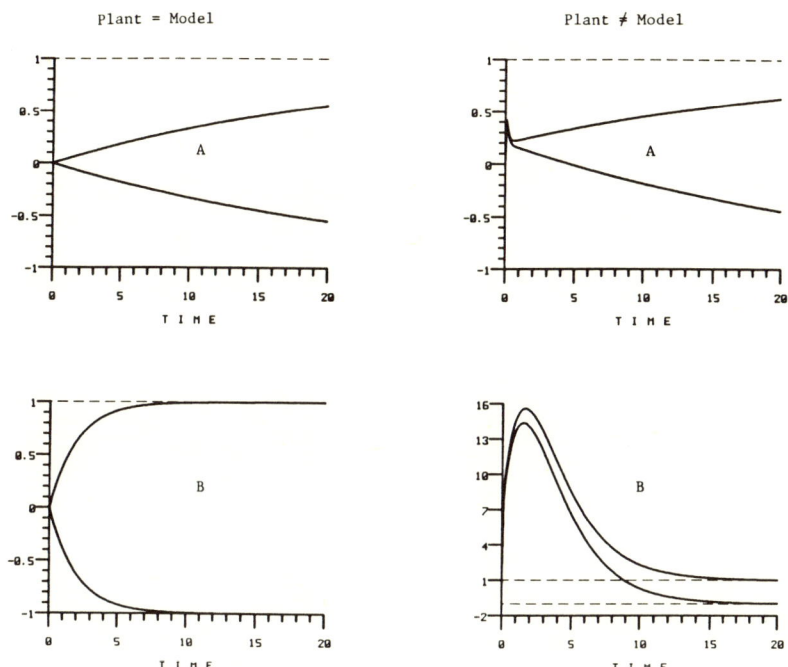

Fig. 3. Response to setpoint change for controllers A and B with and without model error.

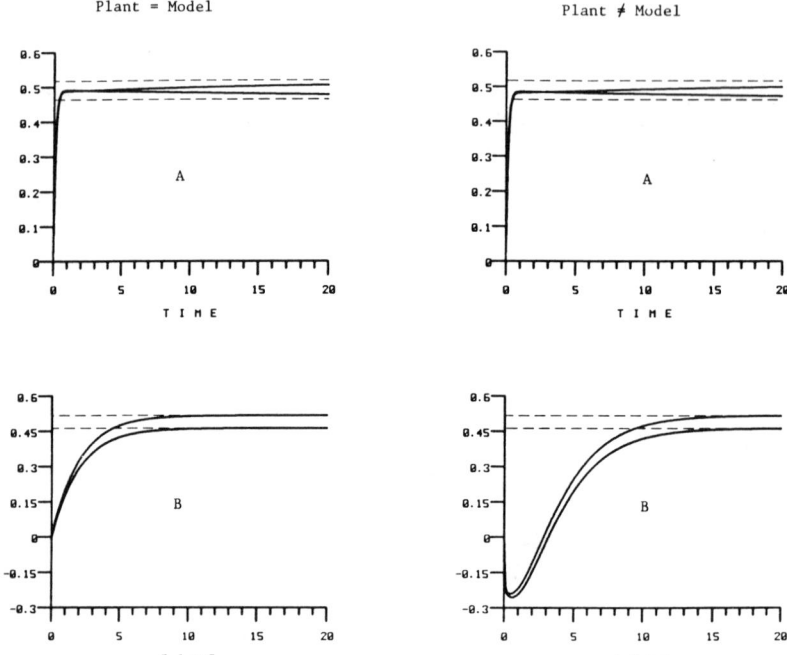

Fig. 4. Response to feed flowrate disturbance for controllers A and B with and without model error.

The responses in the presence of model error can differ drastically from those predicted on the basis of the model. The following observations can be made.

- The dynamic response is strongly input-dependent. While the rejection of the feed composition disturbance is very good (Fig. 2), the disturbance rejection for input (ii) is <u>very</u> sluggish (Fig. 3). The response to feedflow disturbances (Fig. 4) is a combination of the responses in Figs. 2 and 3: Fast initial response and subsequent sluggish approach to steady state.
- Removal of the interactions (steady state decoupling) leads to a significant deterioration of performance – even when the controller gain is decreased by a factor of 8. The responses to inputs (i) and (iii) are much more sluggish (compare Fig. 2 and 4). The 1500% overshoot for input (ii) is unacceptable (Fig. 3).

It should be emphasized that simple controller adjustments do not correct the demonstrated definiencies: By increasing both loop gains of (5) the response to input (i) can be speeded up (Fig. 2) but at the same time the overshoot for input (ii) (Fig. 3) will further increase. Choosing different gains in the two single loop PI controllers (5) makes all the responses worse.

Though input(ii) is not very important in practice, there are always some small disturbances in this direction. A decoupler would amplify these disturbances by a factor of 20 or more (Fig. 3); two single loop PI controllers are not very effective either (Fig. 3) but at least no amplification occurs.

It is quite evident that the experience gained on Single-Input Single-Output (SISO) Systems does not help to explain these phenomena. In particular, multivariable systems appear to exhibit a type of "directionality" which manifests itself in two ways: (1) The dynamic responses for different input directions are different. (2) A control system design which aims at changing these natural directions (e.g. decoupling) makes the dynamic behavior worse. The consequences of this directionality are not immediately evident from the system model. They become apparent mainly in closed loop and under model uncertainty. This makes it very difficult to account for them in the control system design.

There are very few papers in the current distillation control literature which mention and try to explain the unusual dynamic behavior of high purity distillation columns. No systematic guidelines are reported on how to design control systems for these columns, even less for general multivariable systems, in the presence of uncertainty.

Formulation of Control Problem

For any design procedure to yield a control algorithm which works satisfactorily in a real environment the following essentials have to be specified
- process model
- model uncertainty bounds
- types of inputs (i.e. setpoints and disturbances)
- performance objectives

Omission of any one of these four items invariably leads to controllers which fail in practice: Every design or tuning procedure is centered around a process model. Its complexity can vary. The experience of the operator who "knows" how the plant responds to certain inputs is the simplest kind of model. At the other end of the scale would be, for example, a coupled system of nonlinear partial differential equations. Neglecting model uncertainty leads to controllers which are too "tight" and which are likely to become unstable in the real operating environment. It is physically impossible to design controllers which work well for all types of setpoint changes and disturbances. The designer has to decide which inputs are most important and most frequent and accept inferior performance if the inputs encountered during the operation are not exactly equal to the inputs assumed for the design. Finally, the designer has to specify what is meant by "good performance" for the particular problem at hand. What is good in one case might be entirely unacceptable in another.

The ultimate objective of control system design is clearly that the controller works "well" when implemented on the real plant. This goal can be best understood if we decompose it into a series of subobjectives:

Because we assume that the model is at least an approximate description of the time plant it is reasonable to require stability when the controller is applied to the plant model. Thus the minimal requirement on the closed loop system is nominal stability. But the controller also has to work in the real world where the actual plant behavior can be quite different from that of the model. Thus we will require that the controller be designed such that the closed loop system is stable and meets the performance specifications for all members of a family of possible plants which we will define later. If a closed loop property holds for a family of plants we will refer to it as robust. Thus our controller design objectives are robust stability and robust performance.

In this paper we will distinguish analysis and synthesis. Analysis refers to the process of determining whether a system with a given controller has desired characteristics, whereas synthesis refers to the process of finding a controller that gives desired characteristics, usually expressed in terms of some analysis method. This is the fairly standard usage of these terms in the control community. It should be obvious that the question of analysis must be settled to some degree before a reasonable synthesis problem can be posed. The formal analysis and synthesis techniques discussed in this paper are only some of the methods that might make up the overall process of engineering design.

There are very few controller synthesis techniques available. The most successful techniques express the control objectives in the form of an optimization problem and determine the controller structure and parameters from its solution (Optimal Control). In the formulation practical relevance always has to be sacrificed for mathematical convenience, i.e. an optimization problem which can be solved. Thus the "optimality" of a controller does not imply its usefulness. If a controller determined from optimal control principles solves a specific practical problem depends to a large extent on how the control system designer has approximated reality to fit into the mathematical formalism. For several decades only the so-called Linear Quadratic (H_2-) Optimal Control was available. In the seventies the new paradigm of H_∞-Optimal Control emerged. Depending on the problem, the assumptions inherent in one or the other optimal control formalism can be more applicable and can allow a better approximation and thus a better solution.

FORMULATION OF CONTROL PROBLEM FOR SISO SYSTEMS
Process Model

Strictly speaking, most results stated in this paper are only applicable to systems described by finite dimensional linear time invariant models represented by rational transfer function. In a somewhat careless manner we will extrapolate our results to systems involving time delays. We are confident that this extrapolation is generally correct. In order to avoid technical arguments inappropriate here we only provide loose engineering justifications. In particular we emphasize that <u>all</u> models are inadequate at high frequencies and that over a finite frequency range a delay can be approximated arbitrarily well by a finite dimensional system. Furthermore, one key objective of this paper is to show how to account for model uncertainty in the controller design procedure: Therefore the difference between a delay and a system of high order can always be interpreted as "model uncertainty" and the control system can be designed accordingly.

Model Uncertainty Description

The linear time invariant models used throughout this paper describe the actual plant dynamics only approximately. At times this approximation can be quite poor: Any real system is nonlinear. Even if its behavior is essentially linear in some small operating range, high frequency dynamics (e.g. valve dynamics) are usually neglected in the model.

Uncertainty can be described in many different ways: Bounds on the parameters of a linear model, bound on nonlinearities, frequency domain bounds, etc. To account for model uncertainty we will assume that the dynamic behavior of a plant is described not by a single linear time invariant model but a <u>family</u> Π of linear time invariant models. This is a somewhat primitive uncertainty description especially when the effect of nonlinearities is to be captured, but it is the only feasible approach at present.

The family Π of plants will be defined in the frequency domain. We will assume that the transfer function magnitude and phase at a particular frequency ω can lie in a region $\pi(\omega)$ on the Nyquist plane. In general, the region can have a very complex shape and mathematical description and be very difficult to deal with in the context of control system design. Therefore we will assume the regions to be disk shaped with radius $\bar{\ell}_a(\omega)$ (Fig. 5). Any complex region can be approximated by a disk with more or less conservatism. Algebraically, the family Π of plants described by the disk is defined by

$$\Pi = \{p: |p(i\omega)-\tilde{p}(i\omega)| < \bar{\ell}_a(\omega)\} \tag{6}$$

Here $\tilde{p}(i\omega)$ is the nominal plant or the model defining the center of all the disk shaped regions.

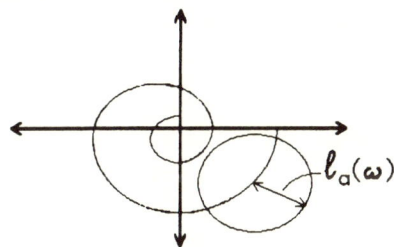

Fig. 5. Norm bounded additive uncertainty represented as disks on the Nyquist plot.

Any member of the family Π satisfies

$$p(i\omega) = \tilde{p}(i\omega) + \ell_a(i\omega) \tag{7}$$

with

$$|\ell_a(i\omega)| < \bar{\ell}_a(\omega) \tag{8}$$

(7) is referred to as an additive uncertainty description and (8) states a bound on the allowed additive uncertainty.

Equivalently we can define the multiplicative uncertainty

$$p(i\omega) = \tilde{p}(i\omega)(1+\ell_m(i\omega)) \tag{9}$$

with

$$|\ell_m(i\omega)| < \bar{\ell}_m(\omega) \tag{10}$$

A typical plot of $\bar{\ell}_m(\omega)$ is shown in Fig. 10. The uncertainty usually increases with frequency and eventually exceeds or becomes equal to unity. $\vec{\ell}_m(\omega) > 1$ implies that the disc shaped regions include the origin and that the phase of the plant is completely unknown. Fig. 6 shows a block diagram representation of the two types of uncertainties.

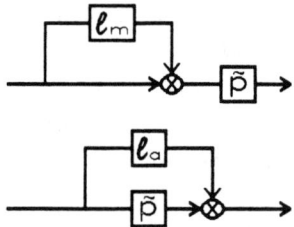

Fig. 6. Block diagram representation of additive and multiplicative uncertainty.

Nominal Performance

The most basic objective of a feedback controller is to keep the error between the plant output y and the reference r small when the overall system is affected by external signals r and d. In order to quantify "performance" a measure of "smallness" for the error has to be defined. Furthermore the set of external input signals has to be specified for which the error is to be made small. Control system design techniques differ in the way they measure "magnitude" and how they define the permissible set of external inputs. Two particularly popular approaches are outlined in this section.

Sensitivity and complementary sensitivity functions. The most important relationships between the inputs and outputs in Fig. 7 are

$$\frac{e}{r-d} = \frac{y}{d} = \frac{1}{1+pc(s)} \triangleq \varepsilon(s) \qquad (11)$$

$$\frac{y}{r} = -\frac{y}{n} = \frac{pc(s)}{1+pc} \triangleq \eta(s) \qquad (12)$$

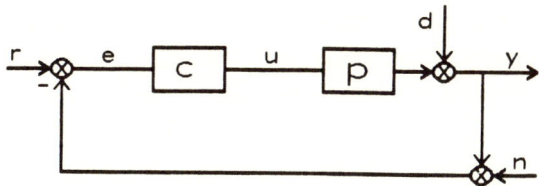

Fig. 7. Classic feedback control loop.

The <u>sensitivity function $\varepsilon(s)$</u> relates the external inputs r-d to the error e. It also expresses the effect of the disturbance d on the output y. $\varepsilon(s)$ is of primary importance in judging the performance of a feedback controller. It is desirable to make $\varepsilon(s)$ as "small" as possible.

The <u>complementary sensitivity function $\eta(s)$</u> derives its name from the equality

$$\varepsilon(s) + \eta(s) = 1 \qquad (13)$$

$\eta(s)$ relates the reference r to the output y (12). From this point of view $\eta(s)$ should be made as close to unity as possible.

$\eta(s)$ also expresses the effect of measurement noise n on y (12). From this point of view $\eta(s)$ should be made small. This illustrates one of the basic trade-offs in feedback design: Good reference following and disturbance rejection ($\varepsilon \cong 0$, $\eta \cong 1$) has to be traded off against suppression of measurement noise ($\eta \cong 0$, $\varepsilon \cong 1$). The optimal compromise can be found via stochastic optimal control theory. Experience has shown that this particular trade-off is usually irrelevant in process control. Measurement noise is often small and η resulting from stochastic optimal control considerations is close to unity over a wide frequency range. As we will learn later, model uncertainty also imposes an upper bound on the magnitude of η. With the poor models generally available for process control this bound is much more restrictive than the measurement noise constraint. Therefore measurement noise will be neglected here.

<u>Linear Quadratic (H_2-) Optimal Control</u>. The controller c is determined such that the integral square error

$$||e||_2^2 = \int_0^\infty e^2(t)dt \qquad (14)$$

is minimized for a particular input d. Referring back to the introductory discussion we note that the "measure of smallness" is the 2-norm of the error ($||e||_2$) and the "input-set" is just one specific input d. Using Parsevals Theorem we can state the optimization problem in the frequency domain

$$\text{Min}_c \frac{1}{2\pi} \int_{-\infty}^{\infty} |e(i\omega)|^2 \, d\omega \qquad (15)$$

or upon substitution for e from (15) with r = 0

$$\text{Min}_c ||\varepsilon d||_2^2 = \text{Min}_c \frac{1}{2\pi} \int_{-\infty}^{\infty} |\varepsilon(i\omega) d(i\omega)|^2 \, d\omega \qquad (16)$$

(16) can be interpreted as the <u>minimization of the 2-norm of the sensitivity function ε weighted by the input d</u>. More general objective functions than (14) including for example a term u^2, to penalize excessive input variations, are well known and are not discussed here. Entirely equivalent stochastic formulations are also available.

<u>H_∞- Optimal Control</u>. Here the objective is to design the controller to satisfy

$$|\varepsilon(i\omega)| < 1/|w(i\omega)| \quad \forall \omega \qquad (17)$$

or equivalently using the definition of the ∞-norm

$$\text{Max}_\omega |\varepsilon w(i\omega)| = ||\varepsilon w||_\infty < 1 \qquad (18)$$

The controller design objective

$$\text{Min}_c ||\varepsilon w||_\infty = \text{Min}_c \text{Max}_\omega |\varepsilon w(i\omega)| \qquad (19)$$

is the <u>minimization of the ∞-norm of the sensitivity function ε weighted by w</u> and to make it at least less than unity. Fig. 10 shows a typical bound. Expressing the control objective as a bound on the sensitivity function ε has much practical appeal. Often the designer wishes to specify a minimum bandwidth and to limit the maximum peak of the sensitivity function to avoid excessive disturbance amplification. In the H_∞- Optimal Control formulation this can be done explicitly. Most important, however, this new formalism allows to address the robust controller design problem in a consistent framework as will be discussed in the next section.

Note that for high frequencies $|\tilde{p}c|$ is small and therefore

$$|\varepsilon| = \left|\frac{1}{1+pc}\right| \cong 1 \quad \text{large } \omega \tag{20}$$

Thus tight performance specifications are only meaningful in the low frequency range, where $|\tilde{p}c|$ is large and $|\varepsilon| \cong |\tilde{p}c|^{-1}$. Then the performance specification (17) reduce to

$$|\tilde{p}c| > |w| \quad \text{small } \omega \tag{21}$$

The loop gain $|\tilde{p}c|$ has to be shaped to fall above the performance weight $|w|$.

An alternate interpretation of (17) can be found in Zames (1981). In short, it specifies that for a norm bounded set of input signals d the output signals e lie in a norm bounded set with a frequency dependent weight.

Robust Stability

We wish to derive conditions for the robust stability of the family of plants Π defined by (6). The following theorem follows easily from the Nyquist stability criterion.

Theorem 1 (Robust Stability): Assume that all plants p in the family Π have the same number of RHP poles and that a particular controller c stabilizes the nominal plant $\tilde{p}$. Then the system is robustly stable with the controller c if and only if

$$||\tilde{\eta}\bar{\ell}_m||_\infty \stackrel{\Delta}{=} \mathrm{Max}_\omega |\tilde{\eta}\bar{\ell}_m(\omega)| < 1 \tag{22}$$

This result is most instructive for several reasons. First note that robust stability imposes a bound on the ∞-norm of the complementary sensitivity function $\tilde{\eta}$ weighted by $\bar{\ell}_m$. Recall, on the other hand, that for the H_2- Optimal Control formulation the performance specifications were expressed in terms of the 2-norm of the sensitivity function $\tilde{\varepsilon}$ weighted by the input d. The different norms are an indication that performance, robust stability and robust performance cannot be treated "jointly" when the ISE is employed as a performance objective. This is one of the main motivations for introducing the H_∞- Optimal Control.

Furthermore, recall that measurement noise also tends to impose a bound on the magnitude of η. However, in process control the constraint imposed by model uncertainty tends to dominate.

Also not that for high frequencies $|\tilde{p}c|$ is small and therefore

$$|\tilde{\eta}| = \left| \frac{\tilde{p}c}{1+\tilde{p}c} \right| \cong |\tilde{p}c| \qquad (23)$$

Thus for large ω (22) reduces to

$$|\tilde{p}c| < \frac{1}{\bar{\ell}_m} \qquad \omega \text{ large} \qquad (24)$$

The design implication is that the controller gain at high frequency is limited by uncertainty. The loop gain $|\tilde{p}c|$ has to be shaped to fall below the uncertainty bound $1/\bar{\ell}_m$.

Finally note that in the H_∞-Optimal Control framework we want to minimize $||\tilde{\varepsilon}w||_\infty$ for performance (see (19)) and $||\tilde{\eta}\bar{\ell}_m||_\infty$ for robustness. A trade-off between performance and robustness arises from the fact that ε and η are not independent (see (13)) and making one small will automatically make the other one large. This problem is inherent in feedback control and cannot be removed by clever controller design. The objective of a design method for robust control systems is to reach the best compromise between the conflicting objectives of performance and robustness.

Robust Performance

Robust stability is the minimum requirement a control system has to satisfy to be useful in a practical environment where model uncertainty is an important issue. However, robust stability alone is not enough. If the bound (22) is satisfied for a family Π then there exists a particular plant $p \varepsilon \Pi$ for which the closed loop system is on the verge of instability and for which the performance is arbitrarily poor. Thus we also have to make sure that the performance specifications are met for all plants in the family Π. Nyquist plane arguments yield the following theorem:

Theorem 2 (Robust Performance): Assume that all plants in the family Π have the same number of RHP poles. Then the closed loop system will meet the performance specification

$$||\varepsilon w||_\infty = \text{Max}_\omega |\varepsilon w| < 1 \qquad \forall p \varepsilon \Pi$$

if and only if
$$|\tilde{\eta}\bar{\ell}_m| + |\tilde{\epsilon}w| < 1 \quad \forall \omega \quad (25)$$
Trivially, robust performance (25) implies robust stability (22) and nominal performance (18). The interdependence between $\tilde{\epsilon}$ and $\tilde{\eta}$ makes it a challenge to meet (25). Improving the nominal performance (decreasing $|\tilde{\epsilon}w|$) worsens the robustness (increases $|\tilde{\eta}\bar{\ell}_m|$) and pushes the system closer to the point of instability for some plant $p \in \Pi$.

If $\bar{\ell}_m(0) < 1$ the bound (25) can be satisfied at low frequencies by making $|\tilde{p}c|$ large (21). If $w < 1$ at high frequencies the bound (25) is met by choosing $|\tilde{p}c|$ small (24). By shaping the loop gain $|\tilde{p}c|$ to be large at low frequencies and small at high frequencies the robust performance specifications can be met assuming that they are not tight (w large) in a frequency range where the uncertainty $\bar{\ell}_m$ is large (Fig. 8).

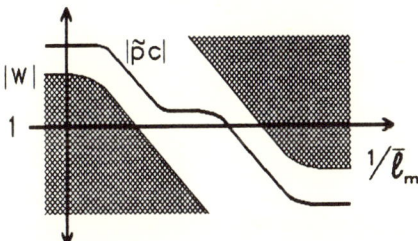

Fig. 8. For robust performance the loop gain $|\tilde{p}c|$ has to lie above $|w|$ for low and below $1/|\bar{\ell}_m|$ for high frequencies.

A controller which optimizes robust performance solves
$$\min_c \max_\omega (|\tilde{\eta}\bar{\ell}_m| + |\tilde{\epsilon}w|) \quad (26)$$
Compare problem (26) for optimizing robust performance with problem (19) for optimizing nominal performance. For <u>nominal performance</u> the weighted ∞-norm of the sensitivity function is to be minimized. For <u>robust performance</u> the ∞-norm ceases to be a suitable measure. As we will learn later the appropriate measure for robust performance is the Structured Singular Value (SSV) which will be discussed in the context of

multivariable systems. The LHS of (25) turns out to be the SSV or a particular matrix.

A controller which optimizes robust performance solves

$$\operatorname*{Min}_{c} \operatorname*{Max}_{\omega} (|\tilde{\mu}\tilde{\ell}_m| + |\tilde{\varepsilon}w|) \qquad (26)$$

Compare problem (26) for optimizing robust performance with problem (19) for optimizing nominal performance. For <u>nominal performance</u> the weighted ∞-norm of the sensitivity function is to be minimized. For <u>robust performance</u> the ∞-norm ceases to be a suitable measure. As we will learn later the appropriate measure for robust performance is the Structured Singular Value (SSV) which will be discussed in the context of multivariable systems. The LHS of (25) turns out to be the SSV of a particular matrix.

CONTROLLER DESIGN FOR SISO SYSTEMS

The development of algorithms to solve (26) is an active research area at present. No convenient and efficient method is available to date. In principle the designer could simply shape the loop gain $|\tilde{p}c|$ to satisfy (21) and (24) (Fig. 8). In practice this can be difficult because (21) and (24) imply nothing about the stability of the nominal system. A more convenient technique is Internal Model Control (IMC) (Garcia & Morari, 1982) which is gaining acceptance in the process control community. It has no inherent optimality characteristics but should provide a good engineering approximation to the optimal solution. The procedure consists of two steps

STEP 1: The controller is designed for good nominal performance disregarding model uncertainty.

STEP 2: The controller is detuned in the high frequency range to satisfy the robustness constraint.

The key idea is that a special controller parametrization is employed which automatically guarantees nominal closed loop stability during the detuning. We will first introduce the new controller parametrization and then discuss Steps 1 and 2. We will limit the subsequent discussion to open-loop stable systems emphasizing that a complete development for unstable systems is available.

IMC Structure

The block diagram of the IMC loop is shown in Fig. 9A. Note that the complete control system to be implemented through the computer software and/or analog hardware is contained in the dashed box. Because in addition to the controller q it includes the plant model $\tilde{p}$ explicitly we refer to this feedback configuration as Internal Model Control.

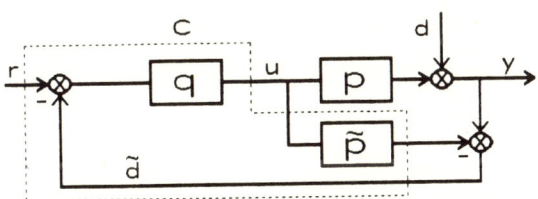

Fig. 9A. Alternate representation of the IMC structure.

The feedback signal is

$$\tilde{d} = (p-\tilde{p})u + d \qquad (27)$$

If the model is exact ($p=\tilde{p}$) and there are no disturbances ($d=0$), then the model output $\tilde{y}$ and the process output y are the same and the feedback signal $\tilde{d}$ is zero. The control system is open-loop when there is no uncertainty, i.e. no model uncertainty and no unknown inputs d. This demonstrates very instructively that for open loop stable processes feedback is only needed because of uncertainty. If a process and all its inputs are known perfectly there is no need for feedback control. The feedback signal $\tilde{d}$ expresses the uncertainty about the process. Because of the open-loop nature of the IMC structure the conditions for stability of the nominal system follow trivially.

Theorem 3: Assume that the model is perfect ($p=\tilde{p}$). Then the IMC system in Fig. 9A is internally stable if and only if both the plant p and the controller q are stable.

The manipulations necessary to transform the block diagram in Fig. 9A into the one in Fig. 9B leave the signals u and y unaffected. If we combine the two blocks q and $\tilde{p}$ in Fig. 9B which are both part of the control system, into one block c we obtain a classic feedback control system with

$$c = \frac{q}{1-\tilde{p}q} \qquad (28)$$

Thus in the way the outputs u and y react to inputs r and d the classic feedback structure and the IMC structure are entirely equivalent and the controllers c and q are related through (28).

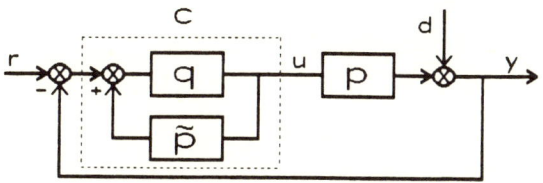

Fig. 9B. Alternate representation of the IMC structure.

Theorem 3 and (28) have profound implications for control system design. Usually a design procedure involves the search for a controller c such that the closed loop system has certain desired properties. This search is greatly complicated by the fact that only those controllers c are allowed for which the closed loop system is stable. (28) provides a simple parametrization of all stabilizing controllers c for the plant p in terms of the stable transfer function q. Thus instead of searching for c it is possible to search for q without any loss of generality. This search is much simpler because stability of q automatically guarantees the stability of the closed loop system with c determined from (28).

All the performance and robustness conditions derived earlier can now be expressed in terms of the controller q and the process model $\tilde{p}$. The sensitivity function and complementary sensitivity function of the nominal system become

$$\tilde{\varepsilon} = 1 - \tilde{p}q \qquad (29)$$

$$\tilde{\eta} = \tilde{p}q \qquad (30)$$

and the performance/robustness specifications are

Nominal Performance (cf. (16) and (18)):

$$||\tilde{\varepsilon}w||_\alpha = ||(1-\tilde{p}q)w||_\alpha < 1, \quad \alpha = 2, \infty \qquad (31)$$

Robust Stability (cf. 22):

$$||\tilde{\eta}\bar{\ell}_m||_\infty = ||\tilde{p}q\, \bar{\ell}_m||_\infty < 1 \qquad (32)$$

Robust Performance (cf. 25):

$$|\tilde{\varepsilon}w| + |\tilde{\eta}\vec{\ell}_m| = |(1-\tilde{p}q)w| + |\tilde{p}q\,\tilde{\ell}_m| < 1 \qquad (33)$$

Nominal Performance

For SISO systems we will generally choose $\tilde{q}$ such that it is H_2-optimal for a particular input d. Thus $\tilde{q}$ has to solve

$$\underset{\tilde{q}}{\text{Min}}\ ||e||_2 = \underset{\tilde{q}}{\text{Min}}\ ||(1-\tilde{p}\tilde{q})d||_2 \qquad (34)$$

subject to the constraint that $\tilde{q}$ is stable and causal. (34) reaches its absolute minimum (zero) for

$$\tilde{q} = \frac{1}{\tilde{p}} \qquad (35)$$

However, the model inverse is an acceptable solution only for minimum phase (MP) systems. For nonminimum phase (NMP) systems the exact inverse (35) is unstable and/or noncausal. The objective function cannot be made zero and an "approximate inverse" of $\tilde{p}$ has to be found such that the weighted 2-norm of the sensitivity function is minimized. Note that for NMP systems the optimal solution depends on the weight (the input d), while for MP systems (35) is optimal independent of the weight.

The following theorem based on the Wiener-Hopf factorization procedure provides the optimal $\tilde{q}$ in the general case.

Theorem 4: Assume that $\tilde{p}$ is stable. Factor $\tilde{p}$ into an allpass portion $\tilde{p}_A$ and a MP portion $\tilde{p}_M$.

$$\tilde{p} = \tilde{p}_A \cdot \tilde{p}_M \qquad (36)$$

so that $\tilde{p}_A$ includes all the RHP zeros and delays of $\tilde{p}$ and

$$|\tilde{p}_A| = 1 \qquad \forall \omega \qquad (37)$$

In general $\tilde{p}_A$ has the form

$$\tilde{p}_A = e^{-s\theta} \prod_i \frac{-s+\zeta_i}{s+\bar{\zeta}_i} \qquad Re(\zeta_i), \theta > 0 \qquad (38)$$

where ζ_i are the RHP zeros of $\tilde{p}$ and the super bar denotes complex conjugate. Factor the input d similarly

$$d = d_A \cdot d_M \qquad (39)$$

The controller $\tilde{q}$ which solves (34) is given by

$$\tilde{q} = (\tilde{p}_M d_M)^{-1} \{\tilde{p}_A^{-1} d_M\}_* \qquad (40)$$

where the operator $\{\cdot\}_*$ denotes that after a partial fraction expansion of the operand all terms involving the poles of $\tilde{p}_A^{-1}$ are omitted.

Robust Stability and Performance

The controller $\tilde{q}$ defined by (40) is often improper and always unsuitable for implementation because of its high high-frequency gain. To satisfy the robustness constraint (32) $\tilde{q}$ is augmented by a low-pass filter f

$$q \stackrel{\Delta}{=} \tilde{q}f \qquad (41)$$

where f typically has the form

$$f(s) = \frac{1}{(\lambda s+1)^n} \qquad (42)$$

The order n is chosen to make q proper. λ is the adjustable filter parameter which has to be made large enough so that $\tilde{\eta}$ satisfies (32)

It can be shown that a filter of the form (42) with $f(0) = 1$ guarantees error free tracking (no offset) of asymptotically constant inputs (e.g. steps).

The function of f is to detune the controller, to sacrifice performance (increase $|\tilde{\varepsilon}|$) for robustness (by decreasing $|\tilde{\eta}|$). Clearly, an open-loop stable system can be made arbitrarily robust, simply by turning off the controller. Thus (32) can always be satisfied. However, because of the associated loss of performance (33) will generally not be met. To design the filter for robust performance (33) is difficult. Indeed, there might not exist <u>any</u> filter to satisfy (33) for the particular choices of weights w and $\bar{\ell}_m$. In our design procedures we will tend to relax performance (decrease w) in exchange for simple search techniques for the appropriate filter. This will yield robust stability and reasonable, albeit not optimal, performance.

APPLICATION: IMC DESIGN FOR A FIRST ORDER SYSTEM WITH DEADTIME

We will first specify all the necessary information and then demonstrate the suggested design procedure.

Information:

1. Process Model: $\tilde{p} = \dfrac{ke^{-s\theta}}{\tau s+1}$

2. Type of input: Step

3. Performance specification: $w = \alpha \left| \dfrac{\theta_{max}s+2}{2\theta_{max}s} \right|$

As shown in Fig. 10, w is infinity for $\omega = 0$ and reaches an asymptotic value of 1/2 for $\omega \to \infty$. It can be shown that if (33) is satisfied for this choice of weight then the ISE for step changes is approximately bounded by θ/α^2 for <u>all</u> plants in the family Π defined by

(6). The larger the model uncertainty $\bar{\ell}_m$ the smaller α has to be selected for a controller to exist such that (33) holds.

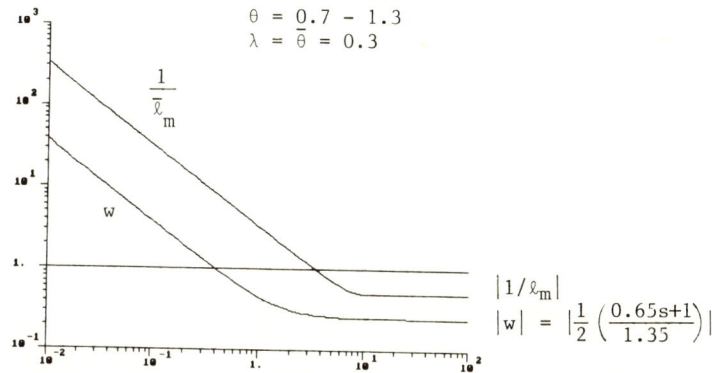

Fig. 10. Magnitude of performance weight w and multiplicative uncertainty $\bar{\ell}_m$ versus frequency.

4. Uncertainty information: We will assume only deadtime uncertainty

$$\theta = \tilde{\theta} + \delta \quad \text{where} \quad |\delta| < \bar{\delta}$$

$$\frac{p-\tilde{p}}{\tilde{p}} = e^{-s\delta} - 1$$

$$\ell_m(\omega) = |e^{-i\omega\delta}-1| \quad |\delta| < \bar{\delta}$$

It is easily seen that

$$\bar{\ell}_m(\omega) = \begin{cases} |e^{-i\omega\bar{\delta}}-1| & \omega\bar{\delta} \leq \pi \\ 2 & \omega\bar{\delta} \geq \pi \end{cases}$$

$\bar{\ell}_m(\omega)$ is shown in Fig. 10. Note the characteristic increase with frequency.

Design Procedure

STEP 1: Nominal Performance

For step inputs the H_2- optimal controller is provided by Thm. 4.

$$\tilde{p}_A = e^{-s\tilde{\theta}}$$

$$\tilde{q} = \tilde{p}_M^{-1} = k^{-1}(\tau s+1)$$

$$\tilde{p}\tilde{q} = e^{-s\tilde{\theta}}$$

STEP 2: Robust Stability and Robust Performance

Filter:
$$f = \frac{1}{\lambda s+1}$$

A first order filter is sufficient to make q proper.

Robust Stability:

$$\left|\tilde{p}\tilde{q}f\bar{\ell}_m\right| = \left|f\bar{\ell}_m\right| < 1$$

Fig. 10 immediately suggests a filter with a corner frequency $\omega = \pi/3\tilde{\delta} \approx 1/\tilde{\delta}$ or $\lambda = \tilde{\delta}$.

Robust Performance:

The constraint

$$\left|\tilde{p}\tilde{q}f\bar{\ell}_m\right| + \left|(1-\tilde{p}\tilde{q}f)w\right| < 1$$

or

$$\left|f\bar{\ell}_m\right| + \left|(1-e^{i\omega\tilde{\theta}}f)w\right| < 1$$

is shown in Fig. 11 for a 30% maximum deadtime error and the factor $\alpha = 0.5$ ($\tilde{\theta} = 1$, $\tilde{\delta} = 0.3$).

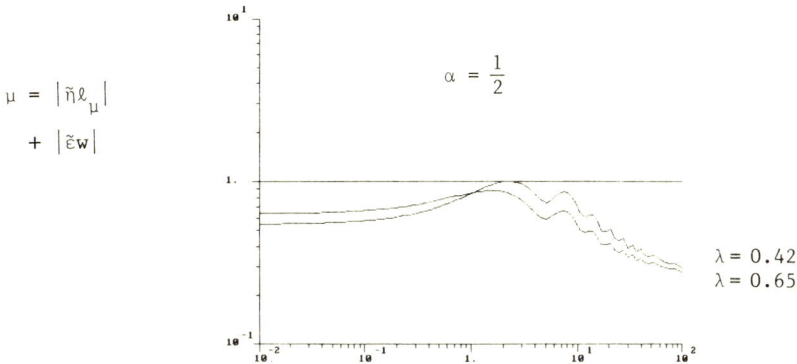

Fig. 11. Robust performance test for Smith Predictor controller for filter parameters.

Note that λ for robust stability depends only on the deadtime error δ but not on the absolute value of the deadtime $\tilde{\theta}$. For robust performance both δ and $\tilde{\theta}$ have to be taken into consideration. The parameters which are known exactly, k and τ, do not affect the filter design.

Conclusion

The proposed controller guarantees that for all deadtimes in the range $0.7 < \theta < 1.3$ the ISE for a step input will not exceed approximately $\theta_{max}/\alpha^2 = 5.2$. Note that $\lambda \cong \delta$, a rule of thumb suggested originally by Brosilow (1979).

Simple rearrangements of the block diagram in Fig. 9A for our specific example show that IMC is equivalent to a Smith Predictor (Fig. 1) with a PI controller.

with
$$c = k_c(1+\frac{1}{\tau_I s})$$

with

$$k_c = \frac{\tau}{\lambda k}$$

$$\tau_I = \tau$$

Thus we have developed a systematic design procedure for a Smith Predictor Controller in the presence of deadtime uncertainty. For more general uncertainty structures the multiplicative uncertainty bound $\bar{\ell}_m$ has to be constructed first. Then the identical design procedure can be followed.

FORMULATION OF CONTROL PROBLEM FOR MIMO SYSTEMS

The main concepts here carry over from the SISO case. Therefore this section is structured in the same manner as the SISO section and emphasis will be placed on the differences only.

For SISO systems the definition of "magnitude" for signals and transfer functions is straightforward: For signals we employed the 2-norm with a frequency dependent weight, for transfer functions the ∞-norm (the absolute value maximized over frequency) because it is induced by the 2-norm and is therefore a natural choice. In the MIMO case we have to deal with vector signals and transfer matrices. As measures of magnitude the norms generalize readily to MIMO systems. Indeed, if we just look at magnitudes, there is very little difference between SISO and MIMO systems. Unfortunately this simplistic approach does not allow to deal with many practical problems, where "directionality" is as important as "magnitude". For example, in addition to the magnitude of the input vector it is critical to know which components (input signals) are most significant. Alternatively, in addition to the magnitude of the uncertainty of a transfer matrix, it is important to specify in which transfer matrix elements the main uncertainties are located. In the problem formulation the directionality will be expressed in the form of weighting matrices. For the analysis of robust stability and performance special tools will be required which are fundamentally different from what is needed for SISO systems.

Model Uncertainty Description

In analogy to the SISO case the simplest possible uncertainty description is the unstructured multiplicative input (L_i) and output (L_o) uncertainty

$$P = \tilde{P}(I+L_i) \tag{43}$$

$$P = (I+L_o)\tilde{P} \tag{44}$$

where

$$L_i = w_{Li}\Delta \tag{45}$$

$$L_o = w_{Lo}\Delta \tag{46}$$

Here w_L is a scalar weight to shape the frequency content of the uncertainty and Δ is a matrix which is magnitude bounded. As a measure of magnitude the ∞-norm generalizes to the multivariable case in the following manner

$$||\Delta||_\infty = \sup_\omega \bar{\sigma}(\Delta(j\omega)) \leq 1 \tag{47}$$

where $\bar{\sigma}$ denotes the maximum singular value. (43) expresses, for example, uncertainty about the manner in which the manipulated variables act on the system. (44) describes uncertainty about the sensors.

The descriptions (43) and (44) suffer from a variety of limitations: The scalar weight w_L does not allow to emphasize the uncertainty of one actuator (sensor) over another, nor does it reflect the fact that the individual actuators (sensors) are independent. In this case the uncertainties are uncorrelated and Δ should be diagonal and not full as assumed in (47). Furthermore (43) and (44) could occur simultaneously and there could be other uncertainty which cannot be expressed in the form (43) and/or (44). Finally, for MIMO systems, an unstructured uncertainty at one point of the system becomes structured at another. Consider, for example, the unstructured input uncertainty (43), (45). Expressing L_O in terms of L_i we find easily

$$L_O = \tilde{P} L_i \tilde{P}^{-1} \tag{48}$$

Thus, if L_i is of the unstructured form (45), L_O is not of the form (46). We can <u>force</u> (48) into the form (46)

$$L_O = \kappa(\tilde{P}) \, w_{Li} \Delta \tag{49}$$

where the condition number $\kappa(\tilde{P})$ is defined as

$$\kappa(\tilde{P}) = \frac{\bar{\sigma}(\tilde{P})}{\underline{\sigma}(\tilde{P})} \tag{50}$$

The unstructured output uncertainty (49) includes all the uncertainty reflected in (45) but depending on the plant condition number $\kappa(\tilde{P})$, (49) can be arbitrarily conservative, i.e. it can include many more plants than are included in (43), (44).

(48) suggests that we should allow the more general structured uncertainty description

$$L = W_L^O \Delta W_L^I \tag{51}$$

where W_L^O and W_L^I shape not only the frequency content but also the spatial directions of the inputs and outputs of the unit norm perturbation Δ, respectively.

Nominal Performance

<u>Sensitivity and complementary sensitivity operator.</u> Interpreting the block diagram in Fig. 7 for MIMO systems the most important relationships between the input and output vectors are

$$e = (I+PC)^{-1}(r-d) = E(r-d) \qquad (52)$$

$$y = PC(I+PC)^{-1}r = Hr \qquad (53)$$

where E is the sensitivity and H the complementary sensitivity operator.

<u>Linear Quadratic (H_2-) Optimal Control.</u> The controller C is determined such that the weighted L_2-norm of the error

$$||e'||_2^2 = \int_0^\infty e'^T e' dt \qquad (54)$$

is minimized for a particular input d. Substituting for e' in (54) the optimization problem can be expressed in the frequency domain as

$$\underset{C}{\text{Min}} \ ||W_2^{-1}EW_1||_2^2 = \underset{C}{\text{Min}} \ \frac{1}{2\pi} \int_{-\infty}^\infty \text{tr} \ (W_2^{-1}EW_1)^*(W_2^{-1}EW_1) d\omega \qquad (55)$$

(55) can be interpreted as the minimization of the 2-norm of the sensitivity operator E weighted by the input weight W_1 and the output weight W_2^{-1}.

<u>H_∞- Optimal Control.</u> As a performance specification the designer requires maximum singular value of the sensitivity function (appropriately weighted) to be norm bounded.

$$||W_2^{-1}EW_1||_\infty = \sup_\omega \bar\sigma(W_2^{-1}EW_1) < 1 \qquad (56)$$

The controller design objective

$$\operatorname*{Min}_C ||W_2^{-1}EW_1||_\infty = \operatorname*{Min}_C \sup_\omega \bar\sigma(W_2^{-1}EW_1) \qquad (57)$$

is the minimization of the ∞-norm of the sensitivity function E weighted by the input weight W_1 and the output weight W_2^{-1}.

If W_1 and W_2 are scalars which implies that the inputs and outputs don't have any preferred directions, then (56) becomes

$$\bar\sigma(E) < \frac{1}{|w(i\omega)|}$$

or

$$\underline\sigma(I+\tilde PC) > |w(i\omega)| \qquad (58)$$

where we have set $w = w_1/w_2$. For low frequencies where the controller gains are large the specification (56) can be approximated by

$$\underline\sigma(\tilde PC) > |w(i\omega)| \qquad (59)$$

The loop gain expressed by the minimum singular value has to be shaped to fall above the performance weight $|w|$.

SISO TYPE ROBUSTNESS CONDITIONS FOR MIMO SYSTEMS

For conceptual simplicity it would be desirable to develop a framework within which the controllers for MIMO systems could be designed like those for SISO systems. Naturally this will only be possible if the expected inputs and the performance specifications do not show any directional dependence, i.e. if all input components have the same magnitude and if all output components are equally important. This implies that the input and output weights W_1 and W_2 defined in the last section should be scalars. Also the model uncertainty must be unstructured and restricted to act either at the plant input (43) or output (44). Then one can derive the following conditions for robust stability and performance (Stein, 1985)

Theorem 5: (1) (Input Uncertainty) Assume that all plants in the family of plants defined by (43), (45) and (47) have the same number of RHP poles and that a particular controller C stabilizes the nominal plant $\tilde{P}$. Then the system is robustly stable if and only if

$$||\tilde{P}^{-1}\tilde{H}\tilde{P}w_{Li}||_\infty < 1 \tag{60}$$

(2) (Output Uncertainty) Assume that all plants in the family of plants defined by (44), (46) and (47) have the same number of RHP poles and that a particular controller C stabilizes the nominal plant $\tilde{P}$. Then the system is robustly stable if and only if

$$||\tilde{H}w_{Lo}||_\infty < 1 \tag{61}$$

or

$$\bar{\sigma}(H(i\omega)) < w_{Lo}^{-1} \quad \forall \omega \tag{62}$$

We note immediately that output uncertainty imposes a bound on the magnitude of the complementary sensitivity operator in the same way as for SISO systems (22). Input uncertainty restricts the magnitude of another operator. However, by designing the controller as to make $\tilde{H} = \tilde{h} \cdot I$ where $\tilde{h}$ is a scalar, (60) and (61) become equal in form.

Theorem 6: (1) (Input Uncertainty) Assume that the system is robustly stable. Then the perturbed system meets the performance specification (56) if

$$\bar{\sigma}(\tilde{P}^{-1}\tilde{H}\tilde{P})w_{Li} + \kappa(\tilde{P})\bar{\sigma}(\tilde{E})w < 1 \tag{63}$$

(2) (Output Uncertainty) Assume that the system is robustly stable. Then the perturbed system meets the performance specification (66) if

$$\bar{\sigma}(\tilde{H})w_{LO} + \bar{\sigma}(\tilde{E})w < 1 \tag{64}$$

Again, for output uncertainty the condition (64) is completely analogous to the corresponding condition (25) for SISO systems with the exception that (64) is only sufficient while (25) is necessary and sufficient. In the first term of (63) the plant cancels if the controller is designed such that $\tilde{H} = \tilde{h}I$. The plant condition number which appears in the second term makes (63) fundamentally different from (25). It proves that MIMO systems are inherently different from SISO systems. In particular, it shows that for ill-conditioned plants ($\kappa(\tilde{P})$ large) design techniques which transform an MIMO system into a series of SISO systems have to fail. Let us assume that a decoupling controller is used and that $\tilde{H} = \tilde{h}I$. Then, if (22) is satisfied by the SISO system, (60) is satisfied for the MIMO system, i.e. robust stability of the SISO systems guarantees robust stability of the MIMO system. On the other hand, robust performance of the SISO systems (25) does not imply robust performance of the MIMO system (63) because the plant condition number does not appear in (25). In principle, it is possible to "correct" (25) by including the condition number but unfortunately (63) is only sufficient and arbitrarily conservative as we will show. Thus even for a MIMO system with the simplest possible performance specifications and uncertainty assumptions different techniques are called for to analyze robust stability and performance.

Example 2 (Distillation) revisited

We will make assumptions about the uncertainty present in the model and impose performance specifications. Then we will apply the robustness tests derived above to the system equipped with controllers (A) and (B).

We will assume the uncertainty to occur at the plant input and be described by

$$w_{Li} = \frac{0.4s+1}{5(0.04s+1)} \qquad (65)$$

This implies an actuator error of up to 20% in the low frequency range ($w_{Li}(0)=0.2$) and it indicates that uncertainty about the process dynamics starts to set in at $\omega \simeq \frac{1}{0.4}$ rad hr^{-1}.

The performance specifications are expressed through the weight

$$w = \frac{s+0.05}{2s} \qquad (66)$$

Thus we require high low frequency performance ($w(0)=\infty$), i.e. integral action and allow a maximum amplification by a factor of 2 of the disturbances at high frequency ($\lim_{s \to \infty} w(s)=1/2$).

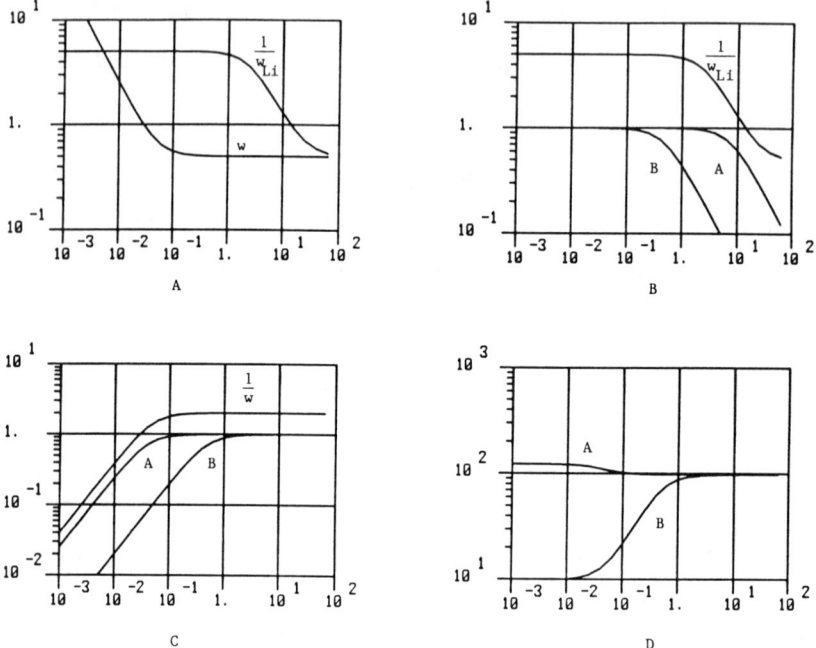

Figure 12(A). Performance (w) and robustness weight (w_{Li}),
(B) Complementary sensitivity functions for controllers A and B satisfying robust stability constraint $1/w_{Li}$.
(C) Sensitivity functions for controllers A and B satisfying nominal performance constraints $1/w$.
(D) Both controllers A and B fail robust performance constraint (63).

The weights are pictured in Fig. 12A. Fig. 12B, C and D show that both controllers (A) and (B) satisfy the robust stability condition (60) and the nominal performance condition (56) but <u>neither</u> meets the robust performance test (63). Indeed, it seems impossible to find <u>any</u> controller which satisfies (63). In the next section we will derive a necessary and sufficient test for robust performance and demonstrate that (63) is excessively conservative.

THE STRUCTURED SINGULAR VALUE FRAMEWORK FOR MIMO DESIGN

A key step in the development of the new analysis technique is the representation of the system in the form shown in Fig. 13. Here Δ is a block diagonal perturbation matrix

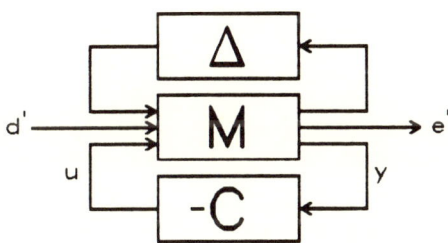

Fig. 13. General Interconnection Structure.

$$\Delta = \text{diag}(\Delta_i) \qquad (67)$$

which is norm bounded (47). M includes not only the information on the plant model $\tilde{P}$ but also on the weighting matrices associated with the uncertainty description and the performance specifications. This form of representation turns out to be remarkably general. A wide variety of practical problems can be represented in this way. Consider, for example, the multivariable design problem we have been studying in the last section with the unstructured input uncertainty (represented as a structured output uncertainty of the form (51)) and the performance weighting matrices w_1 and w_2. For this problem the 3x3 block transfer matrix M has the form

$$M = \begin{bmatrix} 0 & 0 & W_L^I P \\ W_2^{-1} W_L^O & W_2^{-1} W_1 & W_2^{-1} P \\ W_L^O & W_1 & P \end{bmatrix} \qquad (68)$$

and Δ is a full matrix.

In order to state the new robustness condition we first close the control loop in Fig. 13 to get the closed loop system in Fig. 14. The blocks of the 2x2 transfer matrix F(P,C) are defined in terms of the blocks of M as follows

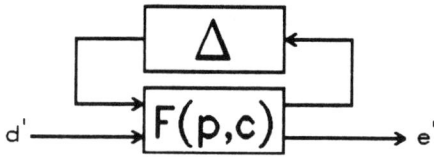

Fig. 14. Interconnection Structure with controller included in F(P,C).

$$F_{ij}(s) = M_{ij} - M_{i3}[I+CP_{33}]^{-1}CP_{3j} \quad i,j = 1,2 \qquad (69)$$

In particular

$$F_{11} = -W_L^I H W_L^O \qquad (70)$$

$$F_{11} = W_2^{-1} E W_1 \qquad (71)$$

Then we can state the following result

<u>Theorem 7</u>: Suppose that the system in Fig. 14 is stable without the perturbation Δ. Then

1. Nominal Performance (56) is satisfied if and only if

$$\sup_\omega \bar{\sigma}(F_{22}(i\omega)) < 1 \qquad (72)$$

2. Stability is robust if and only if

$$\sup_\omega \bar{\sigma}(F_{11}(i\omega)) < 1 \qquad (73)$$

3. Performance is robust if and only if

$$\sup_\omega \mu(F(i\omega)) < 1 \qquad (74)$$

where μ is the Structured Singular Value of F and will be defined in more detail shortly. (72) and (73) are simple restatements of (56) and (57) and do not require any further explanations. The significant result is (74). Though a rigorous proof is beyond the scope of this expository paper the following qualitative discussion should provide some insight. The performance specification requires the ∞-norm of the operator between d' and e' to be less than unity for all plant perturbations Δ. But this condition is also necessary and sufficient for stability if we choose to connect a second norm-bounded perturbation (Δ_p) between d' and e' as shown in Fig. 15. Thus robust performance is equivalent to robust stability in the presence of the two perturbations Δ and Δ_p. The two perturbations can be combined into one block diagonal perturbation $\bar{\Delta}$. The

SSV μ was introduced by Doyle (1982) to test the robust stability of systems with block diagonal perturbations. If the perturbation is unstructured (the matrix Δ is full) then the maximum singular value provides the stability condition (cf. (73)). For __structured__ perturbations the __structured__ singular value has to be used. μ depends on the particular structure of the perturbation. Thus, the block diagonal perturbation $\bar{\Delta}$ should also be specified in (74) to make the expression meaningful.

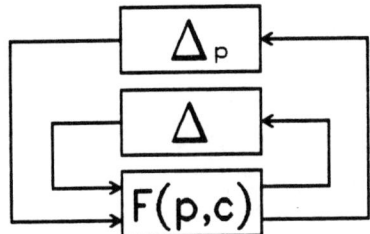

Fig. 15. Interconnection Structure with performance specification expressed as perturbation block Δp.

The techniques for the computation of μ are not as refined as those for σ, but very fast approximate techniques which provide excellent upper and lower bounds on μ are available. Not surprisingly, for SISO systems, the SSV which has to be tested for robust performance is equal to the LHS of (25).

Example 2 (Distillation) revisited:

Figure 16 shows the structured singular values for the high purity distillation column with the performance weight w, the uncertainty weight w_{Li} and the two controllers A and B specified previously. Controller (A) is seen to meet the robust performance criterion while (B) does not. This was confirmed by the simulation results shown at the beginning of this paper.

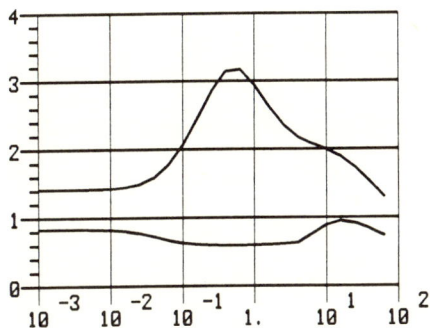

Fig. 16. Structured Singular Values for distillation columns with controllers A and B.

CONTROLLER DESIGN FOR MIMO SYSTEMS

A nonconservative test for robust performance is a very important tool for practical control system design. Even more desirable would be a technique which yields automatically the controller that optimizes robust performance, i.e. minimizes µ. The development of such a technique is a topic of active research (Doyle, 1983) and has not progressed to a stage where it can be applied in a routine manner.

An alternate approach is to extend the basic concepts of IMC outlined in this paper to the MIMO case. It is then necessary to search numerically for the filter parameters which minimize µ. In general, at least one adjustable parameter has to be included in each element of the filter transfer matrix to provide enough degrees of freedom. Otherwise the simplistic filter structure makes it impossible to reach small values of μ. An efficient gradient search algorithm for the minimization is under development (Zafiriou & Morari, 1986).

CONCLUDING REMARKS

This paper represents a first attempt to bring the rich and powerful mathematical theory for robust control system design into the reach of the practicing engineer. The high relevance of the new framework for chemical engineering is evident from the applications to the Smith Predictor and the distillation column problem. It was felt that in an expository review of this type, specific examples and cases should be emphasized rather than the generality of the theory. Thus, at places, the assumptions stated are much more restrictive than required. For example, the minimization of the ∞-norm of the sensitivity operator is appropriate for a number of other input types and performance specifications, which were not mentioned here. Also, the types of model uncertainties which can be treated by the SSV approach are most general and extend far beyond the unstructured multiplicative input uncertainty which was used here: For example, there could be independent uncertainties in each transfer matrix element or the uncertainties could be correlated in some complex manner.

The new theory has started to impact areas beyond robust control system design: It allowed the definition of a new interaction measure to guide the design of decentralized controllers (Grosdidier & Morari, 1986).

Though there are still gaps in the theory, especially as far as synthesis is concerned, what is needed now are applications, to provide feedback to the theoreticians and thus to guide the further developments in the most relevant direction.

ACKNOWLEDGEMENT

Financial support from the National Science Foundation and the Department of Energy is gratefully acknowledged. Dan Laughlin and Sigurd Skogestad provided Examples 1 and 2 respectively.

REFERENCES

Bristol, E. H. On a New Measure of Interaction for Multivariable Process Control. IEEE Trans. Autom. Control. $\underline{AC-11}$, 133-134 (1966).

Brosilow, C. B. The Structure and Design of Smith Predictors from the Viewpoint of Inferential Control. Proc. Joint Automatic Control Conf., Denver, CO (1979).

Cook, G. and M. G. Price. Comments on "A Comparison of the Smith Predictor and Optimal Design Approaches for Systems with Delay in the Control". IEEE Trans. Ind. Electronics and Control Instrumentation, $\underline{IECI-25}$, 180 (1978).

Doyle, J. C. Analysis of Feedback Systems with Structured Uncertainties, IEE Proc. $\underline{129}$, Pt. D, 242-250 (1982).

Doyle, J. C. Synthesis of Robust Controllers and Filters. Proc. Conf. on Decision and Control, San Antonio, TX, 109-114 (1983).

Garcia, C. E. and M. Morari. Internal Model Control. 1. A Unifying

Review and Some New Results. Ind. & Eng. Chem. Proc. Des. Dev., 21, 308-323 (1982).

Grosdidier, P. and M. Morari. Interaction Measures for Systems Under Decentralized Control, Automatica, May, 1986.

Horowitz, I. M. Synthesis of Feedback Systems. Academic Press (1963).

Ioannides, A. C., G. J. Rogers and V. Latham. Stability Limits of a Smith Controller in Simple Systems Containing a Time Delay. Int. J. Control, 29, 557-563 (1979).

Kestenbaum, A., R. Shinnar and F. E. Thau. Design Concepts for Process Control. Ind. Eng. Chem. Process Des. Dev., 15, 2 (1976).

Kwakernaak, H. and R. Sivan. Linear Optimal Control Systems, Wiley (1972).

MacFarlane, A. G. J., (ed.). Complex Variable Methods for Linear Multivariable Feedback Systems, Taylor & Francis (1980).

Owens, D. H. Robust Stability of Smith Predictor Controllers for Time-Delay Systems. IEE Proc., 129, Pt. D, 298-304 (1982).

Palmor, Z. Stability Properties of Smith Dead-Time Compensator Controllers. Int. J. Control. 32, 937-949 (1980).

Rosenbrock, H. H. Computer Aided Control System Design, Academic Press (1974).

Skogestad, S. Manuscript in preparation (1986).

Stein, G. Beyond Singular Values and Loop Shapes, unpublished manuscript (1985).

Zafiriou, E. and M. Morari. Design of the IMC Filter by Using the Structured Singular Value Approach. Proc. American Control Conf., Seattle, WA (1986).

Zames, G. Feedback and Optimal Sensitivity: Model Reference

Transformations, Multiplicative Seminorms and Approximate Inverses. IEEE Trans. Autom. Contr., AC-26, 301-320 (1981).

IMPACT OF MODEL UNCERTAINTIES AND NONLINEARITIES ON MODERN CONTROLLER DESIGN PRESENT STATUS AND FUTURE GOALS

Reuel Shinnar
City College of New York, New York, NY 10031

Abstract. The paper presents a review of robustness and model sensitivity in controller design. The review discusses the impact of this problem on the historical development of modern control algorithms. A rigorous method to experimentally identify and quantify model uncertainty is presented, which can also be applied to deal with non-linear effects in linear controller design. It is shown that robustness is not only important in tuning controllers but should be a primary consideration in the design of the system. Choice of manipulated and measured variables can strongly affect the robustness of the final design. The paper also discusses important problems that require further research.

Keywords. Controller design, robustness, model uncertainty, model sensitivity, nonlinear models, stochastic inputs, dead-time compensator, ARIMA noises, internal model control, model space, frequency response, pseudolinear frequency response, stability analysis.

I. INTRODUCTION

Dealing with model uncertainty in modern process control has in recent years received increasing attention. Handling uncertainties is more crucial in process control since many process models are complex and nonlinear; there are consistent significant differences between the model used for controller design and the actual behavior of the process. However, when modern control theory was first introduced into modern controller design this problem was often neglected. In the first conference which was held almost ten years ago in Asilomar, the author was almost a lone voice in emphasizing its crucial importance for successful design.

As dealing with model uncertainty was and is an important part of my work in developing advanced design methods for chemical reactor design and control, I welcome the opportunity to tie together my various attempts to deal with this subject. I will try in the following to review the problems faced in the past; I will then outline some of the challenges and problems that require solution, as well as some approaches that might be useful in meeting them. The reader should excuse the fact that this review is to a large part based on my own work and approach. There is no intention to minimize the contributions of others but rather to present a personalized view of the subject.

II. INVERSION OF TRANSFER FUNCTIONS AND ROBUSTNESS

Let me start with some history. My own work in control started with an attempt to look at the advantages of stochastic control approach to process control. Initially, we were fascinated by attempts to use a Wiener predictor (Newton, Gould, and Kaiser, 1957), and by the work of Box and Jenkins which was later summarized in their book (Box and Jenkins, 1970). Our interest in these problems arose from an interest in applying probability methods to chemical engineering problems which was the main thrust of our research program (for a review, see Shinnar, 1985).

All of the stochastic approaches for control have their origin in Wiener's original work on antiaircraft gun controllers (Wiener, 1949). If one wants to fire at an aircraft, one would not aim at the aircraft because by the time that the projectile reaches the position aimed at, several seconds have passed and the aircraft will have moved far away. Therefore, one must estimate where the aircraft will be when the projectile arrives. One does this by matching two separate estimates, an estimate of the future flight of the aircraft based on its previous trajectory and an estimate of the trajectory of the bullet as a function of time. A skilled hunter aiming at a running deer does a similar calculation intuitively.

We can look at control in the same way (Fig. 1b, Newton et al., 1957; Palmor, 1978, 1981; Shinnar, 1985). If we change a manipulated variable in a process input, the output does not respond immediately; there are often delays present. We can look at the process model as a predictor of what will occur to Y(t) as a result of any control action U. Moreover, process disturbances are not totally random (i.e., they are not "white noise"). They show a correlation with time, and we can guess, based on the observed history of Y(t), what its value will be in the next time interval. Just as in the antiaircraft problem this guess will be imperfect but it will provide a reasonable basis for a control action. In practice, what this means is that we must build a mathematical model for the disturbance n_t that is based on the past knowledge of the disturbance and predicts its future behavior. We try to match the impact of this disturbance at some future time, t+k, with a control action now that exactly compensates for this disturbance.

Fig. 1b seems to contrast strongly with the conventional control approach (Fig. 1a) in which one measures a controlled output overall and adjusts a manipulated input to maintain a process at a desired level in the face of disturbances. If we look a little closer we find that the difference between the two schemes is less than it seems. In order to predict the future impact of the disturbance, the scheme in Fig. 1b uses past values of the disturbance. But since these are not directly measurable, one must estimate them from past measurements of Y. If the controller in Fig. 1a has an integral control action, it also computes the necessary control action based on the past values of Y.

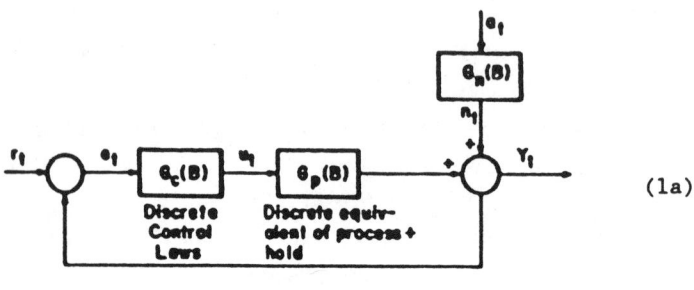
(1a)

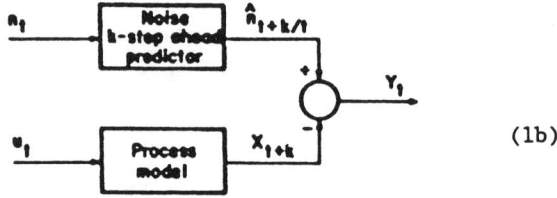
(1b)

Figure 1: Control schemes for process control
  1a: Conventional control scheme
  1b: Control schemes based on matching two predictions

The mathematical model used to estimate n_t is normally called a filter to which white noise is fed. Once we choose the mathematical form of this filter we can design a completely equivalent feedback controller and represent that feedback controller in purely classical terms without noting that it was conceived as a noise predictor.

Let me just give a simple example. A popular model for a process disturbance is the one formulated by Box and Jenkins (1976) and called an Arima (0.1.1) disturbance. It has the form:

$$n_t = \frac{1 - \lambda B}{1 - B} a_t \qquad (1)$$

where B is the <u>backward operation</u> ($By_t = Y_{t-1}$) and a_t is a white noise signal.

Consider a first order system with a delay that has the continuous transfer function

$$G_p(s) = \frac{K_c e^{-\theta s}}{1 + s} \qquad (2)$$

For a sampling time T the discrete transfer function becomes:

$$G_p(B) = \frac{\omega_0 - \omega_1 B}{1 - \delta B} B^{k+1} \qquad (3)$$

Where ω_0, ω_1 and δ are system parameters which are uniquely determinable from τ, θ, K_c and the sampling time T. Use of Fig. 1b will lead to the following controller:

$$G_c(B) = \frac{[(1-\lambda)/\omega_0](1-\delta B)}{(1 - \frac{\omega_0}{\omega_1} B)[1-\lambda B+(\lambda-1)B^{k+1}]} \qquad (4)$$

This is really a simple proportional integral controller with a dead time compensator and a filter (see Fig. 2). The gain of this controller is a function of λ, the only free parameter in eq. (4). If $\lambda=1$, the gain is zero. This is sensible because for $\lambda=1$ the disturbance is really white noise. If there is no correlation between successive values of n_t, there is no sense in using control.

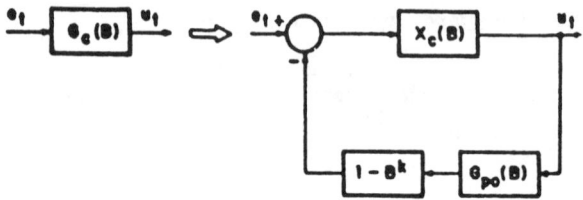

Figure 2: Alternative structure of controllers for simple process with dead time. (Sampled data control.) Process model without dead time G_{po}. Dead time k sampling intervals. B backward operator defined by $n_{t-1} = Bn_t$

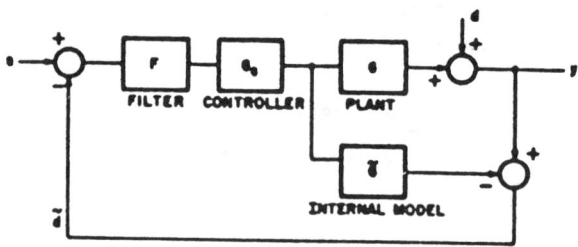

IMC STRUCTURE WITH FILTER.

Figure 3: Translation of control schemes in Fig. 1b into a conventional control scheme representation (internal model control or inferential control)

While the controller in Fig. 1b always has an equivalent formulation in terms of conventional feedback control, that formulation has some specific unique properties that made a very valuable contribution to conventional feedback controller's design. If we reformulate Fig. 1b in a form similar to Fig.1a, we will always end up with a controller with a structure similar to the one given in Fig. 3. It will contain the equivalent of the process predictor in Fig. 1b. If we observe a deviation in the output variable and compute a required compensating action, it makes sense to wait until this action is fully felt at the outlet. However, we cannot wait because other disturbances may appear while we are waiting. What we can do (and is done in Fig. 3) is to first correct the measured value of Y by deducting from it the estimated impact of all the control actions that we have made, but, as yet, have not affected the measured Y. Such a scheme is called <u>either inferential control</u> or <u>internal model control</u> (IMC) (Brosilow, 1979; Garcia and Morari, 1982). It is really a translation of stochastic control theory into a conventional control formulation which was suggested in Palmor (1978) and has since then been applied to multi-variable control (Garcia and Morari, 1982; Palmor, 1983).

The scheme in Figs. 2 and 3 has a problem that it shares with all stochastic optimal controllers. The controllers obtained contain the inverse of the process transfer function. Even if the transfer function is not invertible it contains something which is very similar to an inverse (Palmor, 1978; Kestenbaum, 1976). This leads to a particular type of stability problem. In terms of conventional stability analysis such controllers are always stable, but very sensitive to small changes in G_p. We show this in Fig. 4 where we give stability limits for the controller given by eq. (4). Here, the stability limits are given in terms of the parameters of G_p, and θ (eq. 8a). We concern ourselves here only with the basic ideas; the reader interested in the details should consult Palmor (1978).

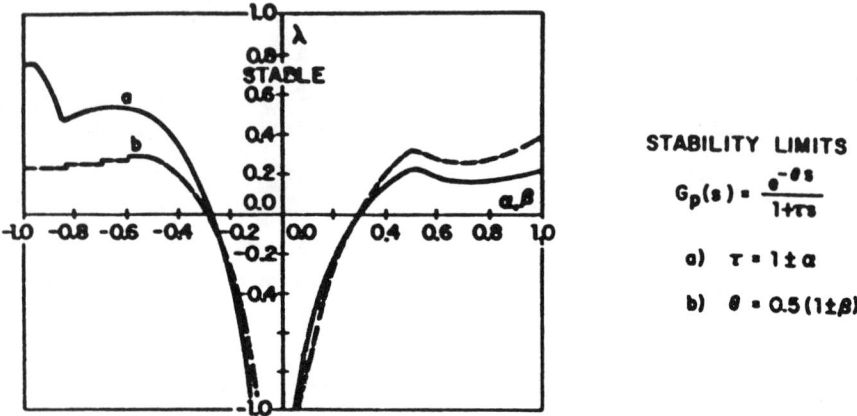

Figure 4: Stability limits for a Box and Jenkins sampled data (eq. 3) operating on a continuous process with the transfer function $G_p(s) = e^{-0.5s}/1+s$. Sampling time interval 0.25 time units. Stability limits are given in term of sensitivity to model parametrs as a function of λ. λ can be interpreted either as the noise constant in eq. 1 or underlying the design as the only free design parameter in the deterministic controller defined by eq. 3. Both the gain and the filter constant of this controller (See figs. 2, 3) are functions of λ (See Palmor, 1978).

In the noise formulation of Box and Jenkins, λ can vary between -1 and +1. We note however that for values of less than 0.5, the controller is very sensitive to variation in process parameters, or in other words, it is very sensitive to the accuracy of the process model.

The classical control engineer stayed away from any controller that contains an inverse of the transfer function because this cancels poles and zeros in the characteristic equation. If we do not know G_p exactly such cancellations are mathematical artifacts and consequently give us a false sense of security. However, such controllers are inherently sensible. Instead of dropping them, all we need to do is modify our stability analysis. We must take into account the fact that in Fig. 1b the prediction of the impact of a control action also involves uncertainties, just as much as the prediction of the future disturbance. We can take care of this if we desensitize the system by introducing a filter. Making the filter strong enough will always assure stability but will reduce the performance.

While this sounds simple in retrospect, it took me and the rest of the control community years to realize that if one wants to utilize the scheme in Fig. 1b (or Figs. 2 or 3), one has to modify the concepts of conventional stability analysis which becomes meaningless for this type of controller.

For an invertible process function this is obvious. If G_p is invertible then the controller obtained will have the form:

$$G(s) = \frac{K(s)}{G_p(s)} \tag{5}$$

and the characteristic equation will have the form:

$$1 + G_c G_p = 1 + K(s) \tag{6}$$

We can then choose $K(s)$ any way we want and obtain perfect control with complete stability. It is quite obvious that we cannot do so. If the process function chosen for design, G_{pd}, is not equal to G_p, then eq. (6) changes to:

$$1 + K(s) G_p(s)/G_{pd}(s) \tag{7}$$

stability will strongly depend on the difference between G_p and G_{pd}.

What is less obvious is that the same applies if the approach of Fig. 1b is applied to a G_p with noninvertible terms. This is discussed by Palmor and Shinnar (1978, 1981) and I will summarize it here. Consider the continuous form of G_p in eq. (2).

The controller in Fig. 3 will result in the controller:

$$G_c = \frac{K_c(1 + s)}{s + K_c(1-e^{-\theta s})} \qquad (8)$$

If $G_{pd} = G_p$ then this controller is stable for any K_c. It is interesting to see how it achieves it. Fig. 5 gives a Bode diagram of G_c for several values of K_c. The magnitude of G_c is small at all crossover frequencies, W_c, of G_{pd}. (W_c is defined as the frequencies for which ϕ is equal to $-\pi \pm 2n\pi$). The control effect is large wherever ϕ is $-2n\pi$. If $K_c \to \infty$ then G_c becomes infinite wherever $\phi = -2n\pi$. We note that this is a very sensible equivalent of inversion. The stability of a single loop controller depends totally on the magnitude of $G_p G_c$ at the crossover frequencies of $G_p G_c$. If G_p is known and invertible so that the controller in eq. (5) eliminates all crossover frequencies by introducing lead terms for any lag. If G_p is noninvertible we cannot prevent the existence of crossover frequencies. What we can do is put our control effort into the region where $\phi = -2n\pi$. We can use an arbitrary control effort and stay away from the dangerous region $\phi = -\pi \pm 2n\pi$. In fact, the controller in Fig. 5 has a gain margin of 2 and a phase margin of 60° at all frequencies.

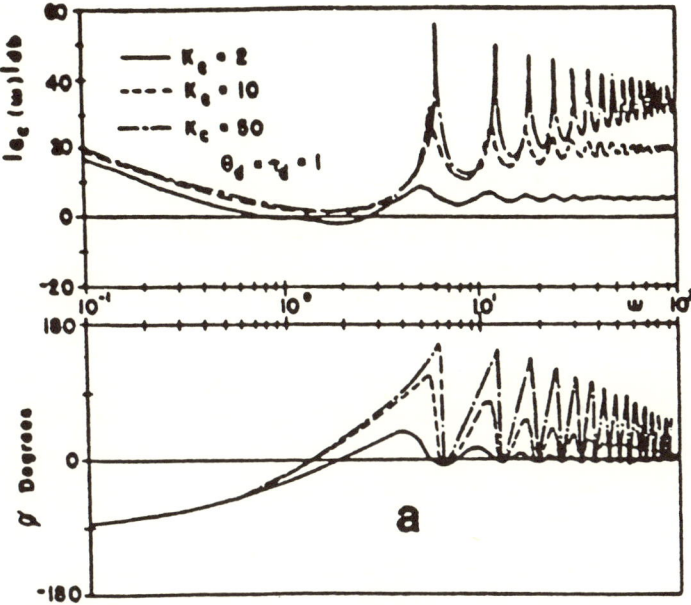

Figure 5: Bode plot for controller of eq. (8) with different gains

Gain and phase margins are very useful for controllers with a single crossover frequency. However, a 60° phase margin can be meaningless if $\phi = -10\pi$. Fig. 5 leads to the regrettable conclusion that for a controller such as eq. (8), gain and phase margins are meaningless.

There is a significant literature that deals with stability of optimal control designs for the case where $G_{pd} = G_p$, both single loop and multi-variable. These are useful in the sense that one must require any system to be stable for $G_p = G_{pd}$, but such an analysis gives only necessary instead of sufficient conditions for stability. Unless we introduce into the design procedure some information about potential deviations of G_{pd} from G_p the design might be totally non-robust. Good practitioners of optimal and stochastic control were aware of this and introduced some robustness by limiting the control effort. However, while this often works, it does not lead to a reliable design procedure. This will be discussed later.

III. **MATCHING DESIGN PROCEDURES TO AVAILABLE INFORMATION**

When I published my first work on model sensitivity in optimal control, I was told by several electrical engineers that I am using the wrong approach. Instead of desensitizing controllers I should improve the quality of the model information. Now while this sounds good and is often recommendable, it has several problems. a) In process control there are strong limits as to how much G_{pd} can approach the real G_p. First of all, the real process is nonlinear while we use linearized design technique. This not only affects the response, but even if the perturbation is small, the linearized approximation of G_p will change if we change the steady-state. Second, we approximate complex transfer function with simple models, which a priori limits the accuracy of a model.

b) Any successful design procedure depends on a match between identification procedures and design procedures. If I say I want a better model, I have to clarify and specify what that means in a quantitative way. One of the advantages of feedback control is that one can design a control system with minimum information. Information normally costs money and more accurate models may be expensive. One has to justify the cost of a more accurate model by the economic value of increased performance.

In complex systems like chemical reactors, there is one important design principle that is not taught sufficiently. That is the principle of optimal sloppiness proposed by Dwight Prater (1975). It is illustrated in Fig. 6. In any system we need information to obtain a better design. However, if the system is complex, a really

accurate description of all effects in an exact quantitative way may be far too expensive or totally unrealizable. Up to a certain point improving the process model and getting more information will pay. But at a certain point, the incremental value of additional information becomes small compared to the incremental cost increases. One wants to stay close to the region where the incremental ratio of value to cost is still large, which is the region of optimal sloppiness. However this optimum depends on two factors:
a) The complexity of the system
b) The specific requirements of the process

Now, while there are many problems in process control that are similar over all applications, there are very significant differences in the difficulty of identifying a proper process model. In some cases the correct process model is simple and only weakly nonlinear. Heat exchangers are a good example. In the other extreme, I would put crystallizers and polymerization reactors where models are complex and identification very difficult.

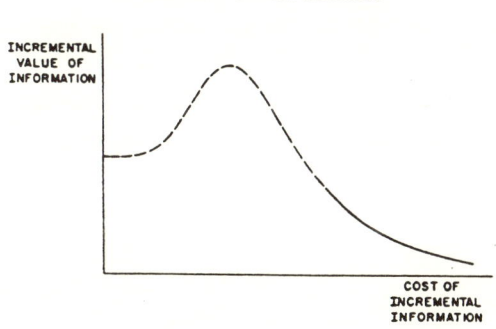

Figure 6:

There are also strong differences as to the needs of a control system and the value of improved control. I tend to classify industrial application into two main categories:

1) Specification oriented industries such as polymers, fibers, drugs, etc. All have the feature that specifications are exact, hard to meet and deviations cannot be corrected by mixing out the product.

2) Constraint oriented industries. Here, the main problem is to meet certain production goals facing process constraints. Optimization depends on the ability to work close to a limiting constraint, which may be shifting for different conditions. A refinery is a typical example. While it has to meet product specifications, in most cases one can correct for them. Thus, if the octane from the FCC is one day below spec one can compensate by increasing reformer severity.

Control needs and the gain of essential information are strongly different for these two cases, and there is a whole spectrum of problems in between. For example, in constraint driven applications steady-state control is far more important than dynamic control.

One therefore always has to be careful to generalize. Too often we judge problems only from the corner of our own experience, not realizing that a problem which we consider trivial in our own context may be central to another industry or application. Thus, the value of an improved model is strongly dependent on the application. Therefore, if we give an engineer a design method we have to be careful to present it to him in a transparent way such that he can judge what information is required and what its value might be. To translate this to our problem, what we need is a method that allows us to define the quality of model fit in a quantitative way and to judge the cost and value of additional information.

IV. MODEL SENSITIVITY IN CLASSICAL CONTROLLER DESIGN

While classical controls avoided inverting the process model, it was far from immune to model sensitivity. In fact there was a specific literature dealing with it (see, for example, Horowitz, 1963). Some of the more common design methods had it built in. Consider for example the quite common method of Ziegler Nichols (1942). The original tuning method was designed for a system described by eq. (2). We give G_c in Fig. 7a for a PI controller. We note that the integral controller is designed such that G_c is large for all frequencies much smaller than the first crossover frequency and becomes very small at the crossover frequency. For high frequencies the controller relies on the natural dampening of the system. The original Ziegler Nichols method is based on increasing the crossover frequency and the gain at the crossover frequency. This guarantees stability as long as the system has certain properties. The system is designed for the G_{pd} given by eq. (7) and the controller is stable as long as the process has the property:

$$G_p < G_{pd} \qquad (9)$$
$$\phi(G_p) > \phi(G_{pd})$$

Many units met in process control have this property. Some do not (see Fig. 7b), and there the method is not suitable. One can relax the conditions of the inequality in eq. (9) for a PI controller. It needs to hold only for frequencies below the first crossover frequency. For a PID controller (eq. 9) should hold for high frequencies. We do not know ϕ and $|G_p|$ for high frequencies, but we normally know it up to a frequency at which the process becomes strongly damped. We designate this frequency as the cutoff frequency and ask that (9) holds for $w < w(cutoff)$. In most cases $w(cutoff)$ is not very far from w_c and higher crossover frequencies are unimportant.

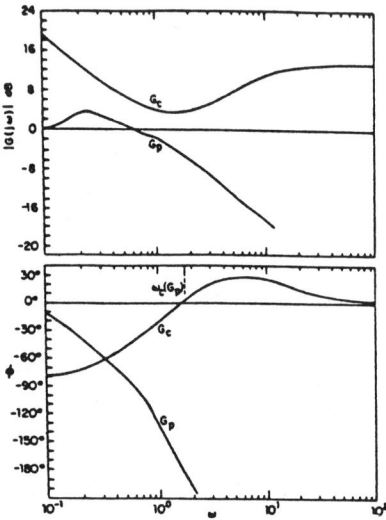

Figure 7a: Design of filtered PID controller

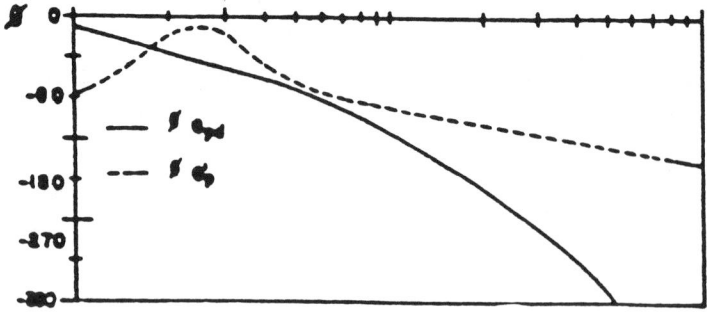

Figure 7b: An example of G_p not fulfilling eq. (9) at low frequencies.

Interestingly enough, Ziegler Nichols introduced another identification method mentioned in many textbooks, namely, the reaction curve method using a step response. This does not lead to robust controller design and can result in instability.

If we look at the controller in eq. (8) one notes that the problem is more complex as the stability criterion becomes two-sided. One has to replace inequality (9) by the criteria:

$$|G_p| < |G_{pd}|$$
$$\phi G_{pd}^2 < \phi(G_p) < \phi G_{pd}^1 \tag{10}$$

One way to ensure a reasonable stability is to test the controller for parameter sensitivity. The problem however arises that we do not know what sensitivity is satisfactory. We therefore need a method to define model accuracy.

In classical control some of these problems were overcome by better identification, and design based on the measured frequency response. As no inversion was involved using conventional safety margins, this normally led to stable designs. Furthermore, controllers were tuned on line, which is really an old form of adaptive control.

On the other hand, there is also considerable discussion in the classical control literature about how to deal with multiple crossover frequencies when $|G_p G_c|$ is not less than unity for all crossover frequencies. The Nyquist criterion which is mathematically correct suffers strongly from the model uncertainty and is seldom applicable to process control.

In cases of large model uncertainty, chemical controller design used the conservative criterion:

$$|G_p G_c| < 0.5$$

all frequencies for which the phase lag becomes significant. One can normally estimate a frequency low enough (or in sampled data control, a sampling time large enough) for which this leads to a safe stability criterion and which puts all the integral action below that frequency. For some cases this will always be a reasonable approach even if it leads to lower performance. One can use this for IMC controllers by using a proper filter.

However, to fully utilize modern controller design methods, we need a way to quantify the inaccuracy of the model in a way that can be handled by the design, and also be identified from the process.

V. DEFINING MODEL ACCURACY. THE CONCEPT OF MODEL SPACE.

Before we discuss the methods of quantifying model accuracy, let us first define the sources and nature of the difference between G_p and G_{pd}. These are caused by a combination of the following:
1) G_p is hard to identify due to complexity. This is often true for multiple input output systems.
2) G_p is mathematically too complex and is approximated by a simple transfer function.
3) G_p is nonlinear. This causes two problems:
   a) The linearized transfer function G_{pd} may change if the steady-state is changed, which means that a change in one or more of the set points changes G_p.
   b) The response is not adequately described by G_{pd}. For highly nonlinear systems, the deviations seldom stay inside the range where linearization holds.

There are therefore inherent limits as to how much G_{pd} can represent G_p and these limits depend on the system. There are similar problems in reactor design. For reactor design I proposed the following concept (Shinnar, 1978), which is also applicable to control:

Consider a complex system G_p^*. We are interested not in the exact form of G_p^* but rather in the way G_p^* acts on certain inputs. In the reactor we are interested in the product composition given a feed composition. Mathematically, we can write that we want to get a bounded estimate of the output Y as a function of the inputs X and G_p^*. We can consider G_p^* a nonlinear operation such that:

$$Y = G_p^* X$$

where Y and X are vectors. We cannot always get an exact estimate of Y but we are ready to accept bounds that mean we are looking for a set of operators G_{pd}^j such that for each element of Y there are two vectors in the set such that:

$$G_{pd}^k X < G_p^* X < G_{pd}^j X \qquad (11)$$

for each element of X. Y will then be within the bounds described by the set $G_{pd}^j X$. In general G_p^j will be nonlinear if G_p^* is nonlinear, but in certain special cases we can find a linear set $G_{pd}^j X$ that answers a specific problem.

For example, in reactor design the transport process in a reactor is highly nonlinear. However, if the chemical reactions are linear then the residence time distribution, which is a linear property of G_p^*, fully predicts the outlet composition. It even gives bounds for second order reactions.

Now, the fact that we have for decades been able to design controllers for nonlinear systems using a linear G_{pd} means that at least in the majority of cases we can find such a set for controller design. However, in complex cases it is advisable to use a nonlinear G_{pd} for stimulating the obtained controller design (Bristol, 1976).

If one chooses a set of G_{pd}^j such that the form stays constant and only the constants change, then this approach becomes either similar or sometimes identical to parameter sensitivity. However, there is no need to choose G_{pd}^j in such a way. Model uncertainty is often more significant at higher frequencies and one can incorporate this in the form of G_{pd}^j. For dynamic controller G_{pd}^j design is normally chosen as a set of linear functions. However, for evaluating the final design, one should choose nonlinear models. Thus, G_{pd}^j will be different for design as compared to simulation. Furthermore, for steady-state control G_{pd}^j should in most cases be nonlinear. We will return to this later.

As pointed out in the example in the previous section, the choice of G_{pd}^j depends strongly on the design method. There is a strong need for further research in this area, especially for multiple input multiple output systems.

Most design methods until now dealt with a specific choice of G_{pd} and used sensitivity for checking robustness. In Shinnar (1978) I suggested that one should develop methods which are a priori based on designing for the set of G_{pd}^j and I illustrated this later by several examples (Palmor and Shinnar, 1981; Silverstein, 1982). I am therefore very encouraged by the way Doyle and Morari in this volume have integrated this concept into a much more rigorous method. While this is a promising start, more work is needed to understand the proper choice of the set G_{pd}^j for different systems. Doyle and Morari also stress that proper choice of G_{pd}^j can take care of nonlinearities. Let me describe the impact of nonlinearities on the choice of G_{pd}^j in more detail.

VI. CHOICE OF LINEAR MODEL SET G_{pd}^j FOR HIGHLY NONLINEAR SYSTEMS

A large fraction of process control systems are nonlinear though the degree and type of nonlinearity varies. Valves are fairly linear over a wide range and then become saturated giving no further response. In other systems such as chemical reactors, the nonlinearities are very pronounced. Doyle and Morari in this volume make the assumption that by choosing a proper set G_{pd}^j one can take care of such nonlinearities. However, they do not show how this set is chosen, but their assumption strongly depends on the proper choice of G_{pd}^j.

Horowitz (1976) showed for trajectory control that if one can bound a nonlinear trajectory function by two linear functions a controller operating stably on the two functions will be stable for the nonlinear function. (Also see Zames, 1966.)

But Horowitz dealt with a defined nonlinear trajectory, whereas in process control we deal with a large class of disturbances, the form of which is unknown. From experience, we know that it works in most cases. Let me, however, present some heuristic arguments as to why this is so (see Silverstein, 1982).

Consider an adiabatic catalytic packed bed reactor with an exothermic reaction (see Fig. 8). If the reaction is strongly exothermic and the activation energy high, then the system is highly nonlinear. However, these nonlinearities have some special features that are illustrated in Figs. 9 and 10. In Fig. 9 we give the results of step inputs of different magnitude. We note that they are very similar only if the gain changes. And the gain can vary by a large factor. This suggests that here we can use a linearized set of G_{pd}^j with variable gain. Knowing the activation energy and the approximate kinetics of the system, we can bound this gain and use the highest gain as our criterion for stability. If flow rates vary, the time constants of G_{pd}^j will also change. A suitable set might be obtained by deriving the form of G_{pd}^j (which here is non-minimum phase with an initial response in the "wrong" direction) either experimentally or by modelling, estimate the maximum gain by computation, (we would not be allowed to measure it in the plant), and estimate limits for the time constants by estimating the variability of the flow that we would permit in safe operation. This would give us a set of G_{pd}^j. We could also use the equivalent of a frequency response. Now what does a frequency response mean in a nonlinear system. Fig. 10 gives a simulation of the response of the above reactor for sinusoidal inputs with reasonably large amplitude.

For a low frequency the nonlinearity of the response is clearly evident. For high frequencies the output is very similar to a sinusoid. However, the amplitude is non-symmetric around the expected average. However, even at low frequencies, the response does not contain any strong peaks. Higher harmonics while evident are strongly damped. This is true for a large class of systems encountered in process control. It is this dampening of higher harmonics that allows us to use linear methods in highly nonlinear systems. Now, not all nonlinear systems encountered in process control will necessarily have this property. But one could identify the possibility of exceptions by simple process modelling For such systems one can then plot a pseudo-linear frequency response which is given in Fig. 11. We note that its form is extremely similar to the linear case except that the gain changes by a constant multiplication factor (phase behavior is also similar).

ADIABATIC CATALYTIC PACKED BED REACTOR
EXOTHERMIC REACTION

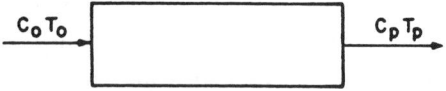

Figure 8:

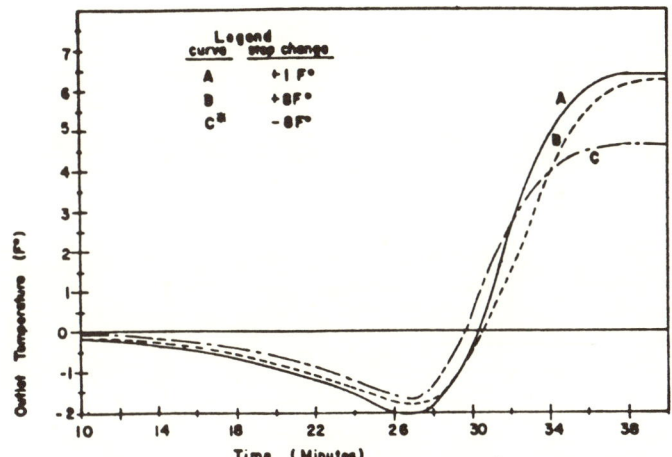

Figure 9: Response to a Step Change in Inlet Temperature of a Packed Reactor with an Exothermic Reaction (from Silverstein and Shinnar, 1977)

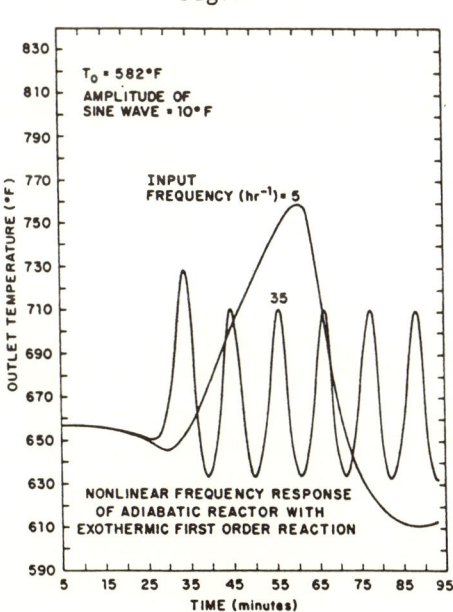

Figure 10:

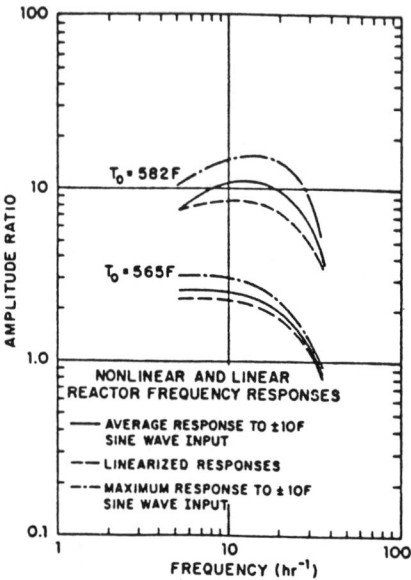

Figure 11:

The increase in the gain can be large. In some cases, however, the nonlinear gain is lower than the linearized one. In that case, the linearized response is sufficient to ensure stability. In the case where the gain increases with amplitude, we have to provide for the maximum possible gain for any disturbance or, if this is impossible, we need to decide on the maximum permissible disturbance and provide for it. This, however, means that we have to use other measures to protect the system such as limiting input disturbances. In many cases of highly nonlinear systems, this is not only feasible but also desirable.

There is another feature of nonlinearity that applies to a much wider class of systems. If we change the steady-state, then the linearized transfer function changes, even for systems which otherwise are nicely linear. The transfer function of all systems is nonlinear in the throughput which always appears in the time constant of transfer functions. And in almost all processes we have to be able to adjust throughput. The potential changes in the transfer function due to set point changes can normally be guessed by simulation and can be incorporated in G_{pd}^j.

If the changes are large we can often reduce the sensitivity of the system by some form of feed forward control which incorporates the expected impact of changes in set points on the transfer function. This latter is a kind of nonlinear internal model control and might be an interesting research area. However, one has to be careful to properly choose the applications as few applications justify such complexity. For changes in throughput, however, this is not difficult as one can guess the changes in the time constants and the crossover frequencies quite accurately.

VII. IDENTIFICATION OF G_{pd}^j MATCHING IDENTIFICATION TO DESIGN

For simple input output controllers, choosing a proper set of G_{pd}^j is not difficult. One can identify the form of the response by either step input or frequency response as outlined in the previous section, and fit it with a properly fitting function. One should, however, be careful not to use an oversimplified fitted function. The problem with Ziegler's reaction curve method was that the function was too simple. Personally, I find that a function of the type:

$$G_{pd}^j = \frac{A(s + a)e^{-\theta s}}{s^2 + 2\tau\xi s + \tau^2} \tag{12}$$

has the properties that are required to fit most cases as it is rich enough to give both underdamped and overdamped as well as wrong-way responses.

The limits on A and the time constants θ, τ and $1/a$ can be derived from looking at the potential changes in steady-state and at the nonlinearities involved. One has to be careful in the high frequencies. We normally have very little knowledge about the high frequency domain and this is not easy to introduce into G_{pd}^j. (Doyle's method in this volume is a promising start). One should therefore always filter out high frequencies. However, one must be careful. If one filters too strongly, very little is left of any modern controller. If one filters the dead time compensator in Fig. 5, it resembles a PID controller quite closely. If one has good information at low frequencies, one should use higher order filters with sharper cutoff. Design of these filters is an interesting research area. It depends both on G_{pd}^j and on the properties of the noise model.

One also has to look carefully at how sensitive the controller is to the form and constants of G_{pd}^j. This is a strong function of controller design and if we cannot get good identification, then a simple PI controller with a lead lag filter may be preferable.

For multi-variable systems proper identification of G_{pd}^j is still an open problem and while we made some progress, it is still an area of research. If we develop methods that allow robust design over a set of G_{pd}^j, we need to match this effort to the method of identifying G_{pd}^j and its possible inaccuracies.

One also has to ask how much can one gain from better identification. Doyle and Morari's method could be used here in a reverse way. By looking at the performance index of a robust design for a given set G_{pd}^j, one can compute how much the performance would improve if the bounds on G_{pd}^j are narrowed. Proper design should not only be matched to identification but interpreted to act in an interactive way.

One last comment. In measuring performances quadratic integral measures, while useful, are not very informative. Control is a minimax problem. We want to increase the bandwidth while reducing the maximum amplification. In the frequency domain one can plot $1/(1+G_p G_c)$ (see Fig. 12). Doyle integrates this into his criteria.

If one wants to use a stochastic noise such as in the Box and Jenkins method, one can achieve a similar goal by looking at performance over the whole range of noise parameters (Palmor and Shinnar, 1978).

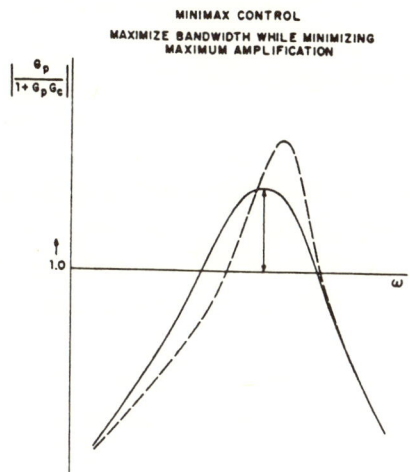

Figure 12:

In both cases one must make some trade-offs. The availability of interactive computers and graphic terminals allows one today to visualize what these trade-offs are, as one can construct these curves for the whole set of G_{pd}^j. One can then incorporate into the trade-off the knowledge one has about the nature of the disturbances. There are many process control problems in which the process model is well known, and the bounds on G_{pd}^j are narrow. In such cases in designing the filter in G_c, one can give emphasis on the structure and properties of the disturbance. (See Palmor and Shinnar, 1978). There is wide range of control problems in which identifying the nature of the disturbance leads to gains in performance.

In a similar vein, the success of feed forward schemes depends strongly on the quality of one's knowledge of G_p. However, a set G_{pd}^j that is sufficient for a specific type of design for feedback control may not have the proper properties for feedforward control.

VIII. EFFECT OF DESIGN ON ROBUSTNESS

In complex systems the design of the control and the proper choice is increased and manipulated variables have a strong effect on robustness, especially if identification is difficult. Let me illustrate this with an example which is probably one of the most nonlinear and hard to identify systems one is likely to encounter in process control, namely, a crystallizer. Crystallizers are hard to control and to scale up as they exhibit limiting cycles, and often operate close to a bifurcation point (Sherwin et al., 1967). For a mixed crystallizer one can write down a simplified process model that has the form given in Table 1. Its main parameters are the particle size distribution $f(r)$, the growth rate $G(r)$, the nucleation rate B, and the supersaturation C. The main feature of the instability is the fact that the nucleation rate B is affected by both the particle size distribution as well as the supersaturation. B has a strong impact on the particle size distribution $f(r)$ as well as on the supersaturation but there is a strong time lag between the instant a crystal is nucleated until the time that it reaches a size which impacts on the supersaturation or on B itself. In our first paper (Sherwin, et al., 1967) we used a simplified model which neglects the impact of $f(r)$ on B and G and assumes that these are functions of c only. One can demonstrate from this model the possibility of limiting cycles and also simulate cycles which are very similar to those observed (see Fig. 13). For this case that stability is a function solely of $d(\ln B)/d(\ln G)$, which is a measurable parameter. Regrettably, these measurements do not predict the behavior of real crystallizers (see Liss and Shinnar, 1976) in which B is a strong function of f. Regrettably, it is very difficult and time consuming to identify this dependence. One ,however, can overcome this by properly designing the control system (Shinnar, 1976; Lei et al., 1971). This design is schematically described in Fig. 14. In crystallizers one can control B by destroying part of the crystal in a, so called, nuclei trap. A stream of crystal magma is passed through an elutriator leg which removes all particles above a certain size, r_c. The ensuing dispersion of fine particles is passed either through a heater or a filter and the fine particles are either dissolved or removed. There are various ways in which one can measure the concentration of particles with size below r_c. If r_c is small this concentration is very strongly related to B. One can then use this measurement to control the flow through the nucleii trap and thereby directly control B. Here, we measure and control a parameter that is directly related to B with a very small time lag. This does not require any knowledge of the dependence of B on f and c, and thereby bypasses the complex problem of model identification. Such controllers have recently been introduced successfully. Their success depends on the introduction of instrumentation which allows on line measurement of the concentration of small particles.

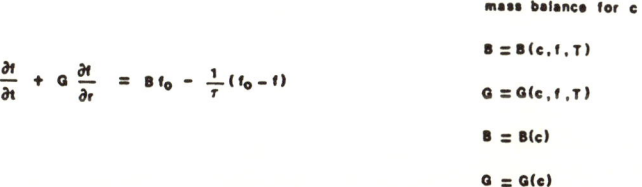

Table 1: Stirred crystalizer, equations describing the particle size distribution

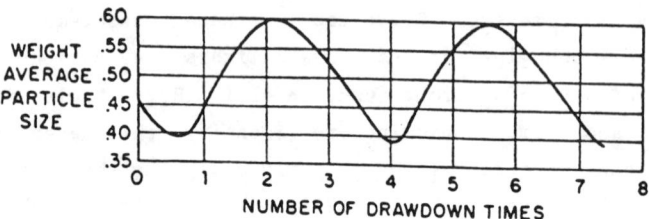

LIMITING CYCLE INSTABILITY
IN CRYSTALLIZER

Figure 13:

Controlling B in the above way requires that the nucleation at all times is larger than the set point of B. One can overcome this limitation by adding seeds to the crystallizer preferably of a size larger than the critical size that passes the elutriator. This allows stabilization and efficient control with a minimum of information.

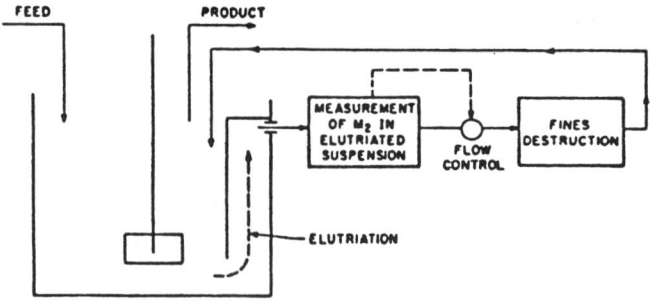

Figure 14: Crystalizer controll by fines destruction.

One can achieve this here because the nucleation rate is a dominant variable. In complex systems with many state variables, it is important to choose dominant variables for control. One can often achieve sufficient dynamic (or even steady-state control) in a complex system with many state variables using only one dominant variable in a single control loop. Obviously, one cannot keep more than one variable constant in such a case but one can often keep all of them in an acceptable range. And, if the loop is properly chosen one can do so with minimum model information. One should always remember that one of the great advantages of feedback control is the ability to run a process with minimum information, and while more information and more measurements can lead to improved control and higher profits, one should retain the features that allow one to run a process with minimum information.

This is especially important if one wants to stabilize a system at either an unstable steady-state or operate it close to a bifurcation point. Here, one wants to eliminate righthand poles without exactly knowing where they are. One should point out that in the case of the crystallizer the kinetic model uncertainty is not the only unknown nature of $B(c,f)$. We also assume here that we can represent the crystallizer as a stirred tank, or that we can represent a distributed parameter system by a simple lumped parameter system. While the assumption is quite good for many stirred tank reactors (Evangelista, 1969), it is not a very good assumption for a crystallizer. Small differences in supersaturation or particle concentration due to mixing effects around the feed nozzle or the cooler can have a large impact on B. Similarly, the assumption of ideal mixing is not very good for fast reactions (Evangelista et al., 1969). In practice we can still stabilize such cases provided we choose a control variable which is sufficiently dominant such that it compensates for these model errors. In fast exothermic reactions temperature is a suitable dominant control variable. In the crystallizer one suitable dominant variable is B.

I would like to expand this concept of the impact of design on robustness with two further examples. The first is a precipitation polymerization reactor such as used in the production of polyacrylonitrile. A standard control scheme is given in Fig. 15. Molecular weight as measured by the viscosity of the polymer in a standard solution, is controlled by controlling the oxidation rate of catalyst. Further steady-state control can be achieved by the reactor temperature and the feed rate of monomers.

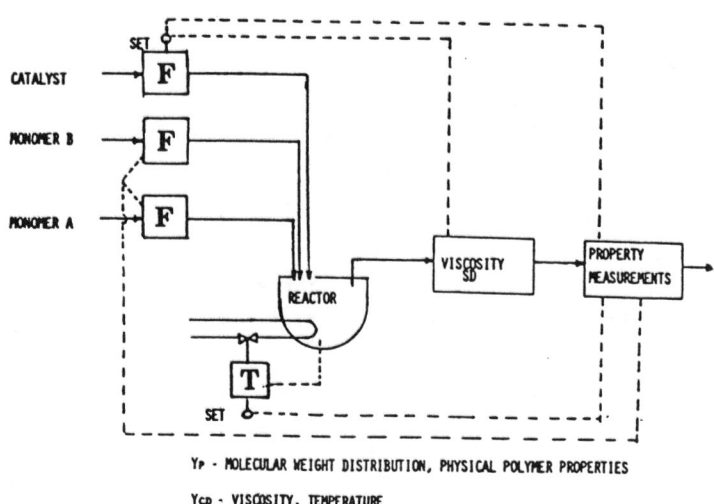

Y_P - MOLECULAR WEIGHT DISTRIBUTION, PHYSICAL POLYMER PROPERTIES
Y_{CD} - VISCOSITY, TEMPERATURE

Figure 15:

However, the important polymer properties are not just molecular weight, but also the nature of the molecular weight distribution. The form of the latter is strongly dependent on the particle size of the precipitated polymer, as the polymerization reaction mainly occurs in the monomer adsorbed in the solid phase. Regrettably, the particles here are too small to use the scheme in Fig. 14, as we cannot separate the nucleii from the dispersion. However, a modified scheme can be useful here (Shinnar and Katz, 1972). It is described in Fig. 16 in which a separate nucleation reactor is added to the system. Nucleation rate can be adjusted by controlling a stream from the main reactor. Here, addition of a measurement and manipulated variable strongly improves our ability to control the process without requiring very detailed model information. It is enough that we have qualitative information about the nature of the nucleation phenomenon which in this case is very similar to the case discussed by Liss and Shinnar (1976).

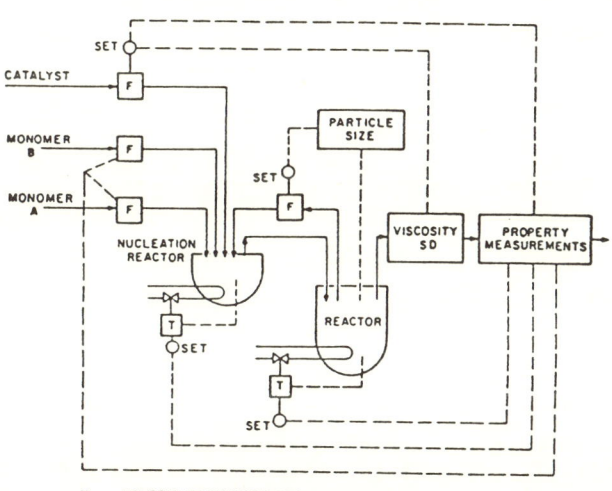

Figure 16:

As a last example, I want to discuss the case of a fluid catalytic cracker (FCC), which has been widely discussed in the control literature (Kurihara, 1970; Lee and Weekman, 1976; Shinnar, 1972). In an FCC reactor, a heavy oil fraction is brought into contact with hot catalyst, forming gasoline distillate, gases and coke. The spent catalyst together with the coke is recycled to a regenerator where the coke is combusted off the catalyst supplying the heat required to run the total reactor (see Fig. 17).

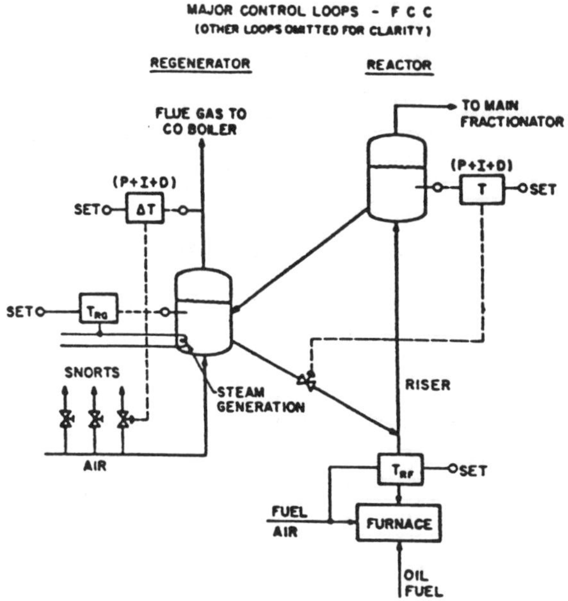

Figure 17: Modified FCC control, increased number of manipulatable variables.

The dominant factor of the control is the heat balance. The main measured variables are temperatures and the two manipulated variables are catalyst circulation rate and air feed rate to the regenerator. Reactor temperature and catalyst circulation rate affect not only the heat balance but also the product composition. But dynamic control is solely based on maintaining the temperatures. In some old units catalyst circulation rate is fixed and can be changed only over a small range. This strongly limits the ability to adjust the reactor temperature and conversion in the reactor. Modern units have added manipulated variables which allow the heat balance to be maintained over a wider range of conditions. These are either a preheater furnace for the feed or a steam generator in the regenerator. Fig. 17 contains both, though in practice one often suffices. This allows to safely maintain the heat balance over a much wider range of operating conditions. One can also look at it as allowing robust control with less information. The FCC example illustrates another aspect of this problem. Fifteen years ago most FCC operated with incomplete coke combustion in the regenerator, as the regenerator was operated with complete oxygen combustion but incomplete CO combustion. Any oxygen reaching the reactor outlet would react with CO and cause excess temperature. To completely react the oxygen the catalyst had to contain excess coke. Today most FCC units operate at conditions that allow almost complete combustion of the coke and result in a completely regenerated catalyst. They achieve it either by a catalyst modification (combustion promoters) or by modification of the regenerator design. This change has some important control implications. As pointed out by Kurihara (1967) the amount of coke remaining on the catalyst strongly affects coke formation in the reactor. This creates a coupling between the reactor and the regenerator, and a strongly interacting control scheme. I discussed some of the implications of this interaction at the first meeting (Shinnar, 1976). Burning the catalyst clean has completely eliminated this interaction and allows the two main control circuits (air rate and catalyst circulation) to operate independently, reducing the interaction.

We note here the strong impact of process design as well as of the choice of manipulated and measured variables. One should realize that this is not simply a control problem. Manipulated variables are chosen when the process is designed. Taking control into account in the initial phases of process design is therefore very important.

X. STEADY-STATE CONTROL OF CHEMICAL REACTORS AND TOTAL PLANTS

Most of the academic control effort is directed towards dynamic control problems, and I do not want to minimize their importance. However, in the chemical industry 80% of all control problems and 90% of the potential financial gain is in steady-state control. This applies both to constraint driven and specification driven industries. There are quite a number of academic studies which try to put economic optimization into dynamic control problems, but I have never encountered a control problem where this is realistic. (This does not mean that such cases don't exist). In that sense process control differs strongly from aerospace problems. Deviations have to stay within limits and there is little economic value to reducing deviations further. In many industries the main value of dynamic control in addition to taking care of disturbances is to allow fast changes of set points and to allow operating at steady-state conditions, which otherwise are not feasible (such as an open loop unstable system or a system close to a bifurcation point), which is a common problem not only in crystallization and polymerization, but also in operating adiabatic exothermic reactors.

The financial gain in steady-state control is due to two items:
a) The ability to maintain a production within specifications.
b) The ability to find a more profitable operating steady-state either for the reactor or for the plant.
If we can't get an accurate model, we want it to have the correct features as it deals with our dominant outputs and inputs.

While dynamic control deals with maintaining a control variable at a given set point, steady-state control uses the set points to achieve the desired catalyst goal. Here, one also uses variables such as feed composition catalyst properties or addition rates, etc., that do not appear at all in the dynamic control matrix. An outline of a systematic approach to such problems was given in Shinnar (1981).

Design of a control strategy must involve the overall steady-state control as well as the dynamic control. The main problem is generally not in tuning but in understanding the overall control problem and choosing the right measured and manipulatable variables. As the latter are chosen when the plant is designed, improved control strategies will have to be integrated into the process design.

For dynamic control we can get away with linearized models. For steady-state control this in many cases is not practical and steady-state control, therefore, often requires nonlinear models to find the proper set points. We should remember that a nonlinear model does not represent the real model G^* but is just a different degree of approximation. Here, we also have problems of model sensitivity. If we designate the steady-state approximation of G^* as G_s, then again we can use a set G_s^j to characterize our model uncertainty, G_s^j will have completely different properties than G_{pd}^j used in the previous section. Its properties and identification will also depend on the goals of the control, the specifications and constraints.

There are some challenging problems in steady-state control, and how to improve them on line by using process information. There is a place here for steady-state adaptive control. And the fact that such models are imperfect should be an underlying thought accompanying any such effort.

Another area of interest is model sensitivity of optimization schemes dealing with total plant optimization, either on line or as tools used in guiding the generation of the plant. While some progress has been made, it has yet to be translated into practice. Studies of relevant realistic examples could be helpful here.

X. CHALLENGING RESEARCH PROBLEMS

In all of the previous sections, I tried to discuss not only the past and present but also outline some of the challenges for the future. I will try to briefly summarize some of them. In doing so it occurred to me that in process control there is very little difference in trying to describe the research problems, which relate to model sensitivity, and research problems facing process control in general. I do not mean to imply that research in model sensitivity is the only important area. Rather, by proposing solutions that acknowledge the fact that our model are imperfect (and are going to stay that way), we can take advantage of the tremendous improvements in hardware to obtain better plant control and increase profits.

So, let me just close with a wish list of important problems relating to control system design.

1) Relation between G_{pd}^j, and G^* for different design strategies.
2) Identification of G_{pd}^j and G_s^j for multiple input output controllers using IMC. It is not enough to design methods of control given a G_{pd}^j. We also need to know how to identify G_{pd}^j and what part of G_{pd}^j has the strongest impact on performance and robustness.
3) How does robustness relate to the control structure. This again relates to another unsolved problem. Namely, how do we systematically develop better control strategies with minimum information for both steady-state and dynamic control of complex systems.
4) How does process design affect controllability, and how can one modify a design to get good robust control with minimum information.
5) Incorporating model sensitivity in steady-state optimization schemes both on and off line.

6) The concept of practical controllability or sufficient control. If we have a complex and imperfectly known system, such as a multi-variable system with many state variables, how can we find a minimum set of manipulated variables to maintain the system in a desirable operating region. We may not be able to maintain the desired state variable at an exactly prescribed point but we may keep it within acceptable bounds. Should we have more than one measurement for each manipulated variable. This is both a dynamic and steady-state control problem. It is crucial not only in chemical reactor control but also in economics.

7) Use of strong or dominant control variables to maintain a system in a desirable region in the face of large model uncertainty. This is strongly tied to item 6.

8) Stabilization of unstable systems in the presence of imperfect models.

9) Steady-state adaptive control and optimization systems using approximate models.

The profit margin of American Industry is shrinking due to international competition. Better and tighter control are essential to maintain it. The recent developments in hardware and software give us tools that we never had. This provides both a challenge and an opportunity and requires a much larger effort in process control research than we have now. But, for it to be successful, we have to understand the problems facing industrial control, integrate them with modelling of complex process and chemical reactor and process design.

REFERENCES

Box, G.E.P. and G.M. Jenkins (1976). Time Series Analysis, Forecasting and Control, Holden Day, San Francisco.

Bristol, E. (1976). Control Conference, Asilomar.

Brosilow, C.B. (1979). JACC Proceedings, Denver.

Evangelista, J.J., S. Katz and R. Shinnar (1969). AIChE Journal 15, 843.

Evangelista, J.J., R. Shinnar and S. Katz (1969). Twelfth International Combustion Symposium, The Combustion Institute, p. 901.

Garcia, C.E. and M. Morari (1982). I&EC Process Design 21, 308.

Horowitz, I.M. (1963). "Synthesis of Feedback Systems," Academic Press, New York.

Horowitz, I.M. (1976). Proc. IEEE 64(1), 123.

Kalman, R.E. (1960). Trans ASME Basic Eng. 82, 35.

Karihara, H., L.A. Gould and L.B. Evans (1970). Automatica 6, 695.

Kestenbaum, A., R. Shinnar and F.F. Thau (1976). "Design Concepts for Process Control," I&EC Process Design & Development, 15, p. 2.

Lei, S.J., R. Shinnar and S. Katz (1971). AIChE Symposium Series 67, p. 129.

Liss, B., R. Shinnar (1976). Symposium Series, vol. 72, No. 153.

Newton, G.C., Jr., L.A. Gould and J. Kaiser, Jr. (1957). "Analytical Design of Linear Feedback Controls," Wiley, New York.

Palmor, Z. and R. Shinnar (1978). I&EC Process Design and Development, 16, p. 1.

Palmor, Z. and R. Shinnar (1981). AIChE Journal 27, 793.

Palmor, Z. and R. Shinnar (1979). "Design of Sampled Data Controllers," Ind. Eng. Chem. Process Design and Development, 18, No. 1, pp. 8-30.

Palmor, Z., Y. Halevy (1983). Automatica 19, 255.

Prater, D. (1975). Private communication.

Rosenbrock, H.H. (1974). "Computer Aided Control System Design", Academic Press, New York.

Sherwin, M.B., R. Shinnar, and S. Katz (1967). AIChE 13, 1191.

Shinnar, R. (1978). Chemical Reactor Modelling, Chemical Reaction Reviews, ACS Symposium Series, No. 72, p. 1.

Shinnar, R. (1976). U.S. Patent No. 3970587.

Shinnar, R. and S. Katz (1972). Advances in Chemistry No. 109, 56.

Shinnar, R. (1978). "Chemical Reactor Modelling - the Desirable and the Achievable," ACS Symp. Ser., 72.

Shinnar, R. (1976). AIChE Symposium Series, vol. 72, No. 159, p. 155.

Shinnar, R. (1985). AIChE Symposium Series, vol. 80, No. 235.

Silverstein, J.L. and R. Shinnar (1977). "The Effect of Design on the Stability and Control of a Hydrocracker," AIChE Meeting, New York.

Silverstein, J.L. and R. Shinnar (1982). I&EC Process Design and Development 21, p. 241.

Wiener, N. (1949). The Extrapolation Interpretation and Smoothing of Stationary Time Series," Wiley, N.Y.

Ziegler, J.G. and N.B. Nichols (1942). "A Controller to Overcome Dead Time," Transactions of the ASME, p. 759.

AN INDUSTRIAL VIEW OF ADVANCED PROCESS CONTROL

Robert G. Wilhelm, Jr.
AccuRay Corporation, Columbus, OH 43230

Perhaps the clearest lesson taught by Session II, if it was not already obvious, was that there is no uniquely "industrial view" of advanced process control. Indeed, there emerges no consensus even on the criteria which distinguish "advanced" control. We may conclude, however, that the use of process models is the dominant common factor in control development. The following paper by Doss, Downs, Roat and Vogel, for example, focuses upon the application of rigorous, detailed simulation models of distillation processes as a means of evaluating the effectiveness of various control schemes, which themselves turn out to be fairly conventional. It is said, however, that the modeling efforts represent only the first stage in an ongoing research program, and that many more advanced control schemes may now be evaluated quantitatively without risking poor process performance with on-line experimentation. The authors' devotion to the use of highly detailed, rigorous models serves, therefore, to highlight one of the primary reasons that industry is so slow to adopt more advanced approaches to control: experimenting with production units is potentially very costly, and concern for short term profitability nearly always dominates the desire for a less certain long term payback accompanying advanced control schemes. Detailed dynamic models may be difficult and costly to build, but for complex chemical processes they may pay off handsomely if they make it possible to comparatively evaluate different control strategies reliably and quantitatively. Whether more advanced control strategies are an outgrowth remains to be seen.

Authors Hamer and Juba, in "Progress and Challenges in Batch Process Control" give a tutorial on some of the key differences between continuous and batch process control. They, like Doss, Downs et. al., illustrate the importance of modeling, but from quite a different perspective. Process heat transfer models, for example, are used to produce a reaction rate observer and to develop feedforward and nonlinear feedback control strategies which lead to tighter regulation, improved safety, and maximum throughput. The authors see the use of nonlinear time-varying process models and new sensor developments as key elements in further progress with batch process control. In the discussion following this paper a question was raised concerning the possible use of adaptive control to compensate for the nonlinear and time-varying process behavior. The authors responded that the process response often changes too rapidly for adaptive control, and one member of the audience commented on the folly of throwing away useful process knowledge in favor of adaptation. I would like to speculate these comments actually only reflect on the relatively narrow and immature view we currently have of adaptive control.

The paper by Treiber and Hoffman presents a novel frequency domain approach for designing multivariable controls to account for constraints which temporarily disable certain control variables. This technique, like those presented in the earlier papers, also relies heavily on modeling, although in this case relatively simple transfer function models are used, being derived from time series analysis. The design method, based on the multivariable frequency domain approaches of Rosenbrock and MacFarlane, naturally yields excellent insight into the stability of the candidate control approaches. (Recall that MacFarlane presented the basis for the present paper to the first CPC conference in 1976, also at Asilomar.) It is interesting to compare Treiber's approach to Brosilow's treatment of a similar problem as presented in a later session at CPC-III.

Session Summary

Concluding this session we heard "A Practitioner's View of the Control Literature" by Robert Otto. The author has stated that his intent was to foster discussion, and I suspect that his presentation served to enliven more break-time discussions than most others. Dr. Otto takes the academic community to task for failing to recognize or deal with such real world problems as steady-state offset, nonstationary disturbances, nonminimum phase dynamics, and the need for simple setup and tuning. He observes that the vast majority of the control literature is of little or no use in the daily practice of control engineering. Unfortunately I suspect that the same situation exists in most technical fields of endeavor, and yet, slowly and ever so painfully, useful technology percolates through the mire and finds its way into practice. Although many practitioners may empathize with the cynicism expressed by Dr. Otto, only the most foolish would be willing to dispense with ANY of the modern control techniques he mentions: state variable control, model-based control, stochastic control, and self-tuning control.

THE INTEGRATION OF RIGOROUS DYNAMIC MODELING AND CONTROL SYSTEM SYNTHESIS FOR DISTILLATION COLUMNS: AN INDUSTRIAL APPROACH

Suzanne D. Roat
James J. Downs
Ernest F. Vogel
James E. Doss
Tennessee Eastman Company, Kingsport, TN

Abstract. In this paper we illustrate a reliable methodology for designing distillation control schemes. Interaction measures, accurately calculated from a rigorous steady-state distillation simulation, are used to screen potential control schemes and eliminate undesirable strategies. Rigorous dynamic simulation is used to evaluate the remaining strategies. The use of interaction measures alone, or measures which have been inaccurately calculated, can lead to specification of unworkable control schemes. This control design methodology is illustrated with three examples which demonstrate proper use of the various interaction measures.

Keywords. Distillation Control; Dynamic Distillation Simulation; Singular Value Decomposition; Relative Gain Array; Interaction Measures; Distillation with Reaction.

INTRODUCTION

Within the last decade, there have been a number of advances in the areas of rigorous steady-state and dynamic simulation of distillation columns. Simultaneously, the Relative Gain Array (RGA) (Bristol, 1966) and Singular Value Decomposition (SVD) (Klema and Laub, 1980), have been developed to assist in the analysis and synthesis of complex, multivariable control systems. However, there has not been a focused effort to integrate these rigorous simulation tools and interaction measures into a cohesive distillation control design strategy. Distillation control researchers have frequently resorted to use of simplified distillation modeling techniques in the interest of mathematical tractability.

The objectives of this paper are two-fold:
- Through the use of several case studies, we illustrate that rigorous steady-state and dynamic simulations for distillation are a necessary part of the control design problem.
- Second, we illustrate that no single analysis tool is sufficient to guide the distillation control designer in a totally reliable fashion. Each of the tools has its individual merits, as well as its own set of weaknesses and modes of failure. We intend to illustrate that successful control system design is a composite result obtained from examining all of the measures available, and then testing the candidate control schemes with a rigorous dynamic simulation.

RIGOROUS STEADY-STATE AND DYNAMIC DISTILLATION SIMULATION

Current industrial practice for multicomponent distillation design generally involves two separate but related areas:
- Steady-State Staging Simulation
- Steady-State Hydraulics Simulation

The staging effort usually requires reasonable descriptive models of the vapor and liquid phase nonidealities which occur. Vapor phase nonidealities are handled by modified equations of state, such as Soave-Redlich-Kwong (1974). Multicomponent liquid nonidealities are described sufficiently well by use of models such as the Wilson equation (1964), the NRTL equation (1968), or the UNIQUAC equation (1975). The other major requirement of the steady-state staging effort is the use of a staging technique which is numerically stable, fast, and robust. Suitably constrained modifications of Newton's method (Naphtali-Sandholm, 1971), together with some modification to improve storage and computational efficiency (Downs, 1982), generally suffice.

Over the last ten years, the lack of good numerical techniques has been removed as a major hindrance to steady-state simulation for column design. Removal of this restriction is important, because the ultimate limitation to accurate steady-state descriptions has been shifted back to the fundamental VLE models, i.e., staging runs are now limited by the state of the art in vapor-liquid equilibrium thermodynamics.

Reliable prediction of hydraulic behavior is an area which has also undergone fundamental improvements. Due to extensive work by Fractionation Research, Inc. (FRI), and others, we can now predict most of the major hydraulic parameters. The one parameter most resistant to prediction is tray efficiency - although two recent papers (Chan and Fair, 1984a, 1984b) shed some new light in this area.

Dynamic simulation of distillation has also benefited from advancements in numerical integration methods. Holland and Liapis (1983) list a number of techniques for dynamically simulating distillation columns. An interactive dynamic simulation is ideally suited for the study of control systems. Downs and Vogel (1985) describe the interactive dynamic simulation used in this work.

The end result of these improvements is that most of the steady-state column designs today incorporate techniques which are fundamentally sound from both a thermodynamic and hydraulic viewpoint. In addition, these rigorous techniques can also be extended to dynamic distillation modeling. These developments eliminate any need to rely on shortcut techniques.

PROPER USE OF INTERACTION MEASURES

Distillation control designers and researchers of today do not appear to be making sufficient use of the fundamentally sound information available through rigorous distillation simulation. There are any number of distillation control papers which assume "heavy-key, light-key" staging, constant relative volatility, equimolal overflow, and no hydraulic contributions to the overall problem.

The examples in this paper demonstrate how misleading such assumptions can be in apparently innocuous cases. We suggest a control strategy design approach which is an extension of the steady-state staging simulation. Based on the steady-state results, the designer calculates a number of interaction and stability measures: the Relative Gain Array, Niederlinski's index (1971), and the Singular Value Decomposition. It is likely that a workable control strategy can be developed using only simple loop-pairing types of strategies. Furthermore, the SVD will indicate whether it is desirable to combine several inputs to influence a particular output. The dynamic simulation is then used to evaluate the candidate loop-pairing strategies or multivariable control strategies. Because the interaction measures and dynamic simulations can be computed quickly, the control designer can evaluate a large number of these strategies in a short period of time.

In perspective, each of the interaction measures is more suitable for rejection of unacceptable control strategy alternatives, as opposed to acceptance of or descrimination among several good alternative strategies. Final choice is the made by comparison of the candidate strategies' performance on a rigorous dynamic distillation simulation -- that job being made much easier and smaller in scope by the presence of a considerably pared-down list of alternatives.

THE DESIGN METHODOLOGY
1. Using a rigorous steady-state modeling program, a simulation of the desired operating point is obtained.
2. From the rigorous steady-state simulation, the steady-state gains relating the column temperatures (or other controlled variables) to the possible manipulated variables are calculated.
3. Using the steady-state gains, the SVD, RGA, and Niederlinski indices are calculated.
4. Undesirable schemes are then eliminated based on interaction measures.
5. Finally, the remaining schemes are evaluated using rigorous dynamic simulation.

EXAMPLE 1 - DISTILLATION WITH SIMULTANEOUS CHEMICAL REACTION

Example 1 - Column Description

To illustrate the limitations of using only steady-state techniques for control system design, a 25 tray reactive distillation column was studied. The reaction in the column involved two reactants, B and D, being fed into the column along with a catalyst, E, yielding two products, A and C, via a second order, reversible, equimolar reaction:

$$COMP\ B + COMP\ D \underset{COMP\ E}{\rightleftarrows} COMP\ A + COMP\ C$$

The order of volatility from most volatile to least volatile was A, B, C, D, E. The desired product was component A. Component A formed low boiling azeotropes with both component B and component C. The column had three feed streams: reactant B at 50.4 moles/hr on tray 7, reactant D at 49.0 moles/hr on tray 21, and the catalyst, E, at 0.5 moles/hr on tray 17 (see Figure 1-1).

This arrangement allowed a countercurrent contacting of the feed reactants in the reaction section between stages 7 and 17. Above stage 17 the catalyst concentration was too low for reaction to occur. The column control objectives were to obtain at least 97.5 mole % pure product A in the distillate, and to obtain 96% conversion of reactants to products.

The steady-state temperature profile is shown in Figure 1-2. The profile was not monotonically decreasing up the column. The increase in temperature from tray 7 to tray 12 occurred because high boiling reactant D was reacted away in this region at a high enough rate to cause the concentration of the lower boiling products, A and C, to be higher on successively lower trays.

Example 1 - Control Configurations and Dynamic Simulation Results

For this column, a singular value analysis indicated that trays 3 and 12 are the most sensitive independent temperature measurements in the column. Table 1-1 lists the gains relating these temperatures to various manipulated variables. A control scheme that holds reflux rate constant at the top of the column was selected. The RGA suggested that controlling the temperature on tray 3 by manipulating component B feed flow rate on tray 7 and controlling the temperature on tray 12 by manipulating component D feed flow rate on tray 21 should be acceptable. The relative gain for this 2 x 2 scheme was 1.022. The Niederlinski index for this scheme, 0.979, was positive indicating local stability for the control system. The control scheme, labeled scheme 1 is illustrated in Figure 1-3. Note that, in general, a negative Niederlinski index establishes only sufficiency conditions for instability. However, for SISO and 2 x 2 systems, a negative index is necessary and sufficient for instability. See Grosdidier et al (1985).

Dynamic simulation of control scheme 1, illustrated in Figure 1-4, indicated that the control system was satisfactory for small (less than 1-2%) changes in production rate, as set by steam rate to the reboiler. However, for changes greater than this, the control system did not prevent the column from moving to a different and undesirable steady-state operating point where the set points were satisfied but the reactant B conversion was only about 30%. This phenomenon is illustrated in Figure 1-5.

Figure 1-6 shows that control scheme 1 did not add reactant D in a timely enough manner to prevent reactant B from "breaking through" to the upper section of the column. When this breakthrough occurred, the column moved to a completely different and undesirable operating condition where the temperatures were at their set point, but the tower composition profile was not satisfactory. This dynamic simulation illustrates the point that, even though the steady-state measures indicated that control scheme 1 should behave reasonably, the nonlinearities of the problem caused the control system to fail. Furthermore, the failure was discovered only by using the full nonlinear dynamic simulation. Simply having a rigorous steady-state model was not sufficient.

To have reactant D respond in a more favorable manner, the flow of reactant D was ratioed to the flow of reactant B and the ratio was reset based on column temperatures. With this approach, relative gain information suggested manipulating the reboiler duty to control tray 3 temperature and manipulating the reactant ratio to control the tray 12 temperature. This pairing had a relative gain of 0.966 and a Niederlinski index of 1.034. Dynamic simulation, however, indicated that for increases in production rate (as set by reactant B feed rate), this control scheme had the same problems as scheme 1.

Three solutions were found to the robustness problem. First, the reflux ratio was fixed at the top of the column instead of the reflux rate for scheme 1 described above. This change gave rise to the same left singular vectors in the U matrix of the singular value decomposition as the constant reflux rate case. Hence the most sensitive independent temperature measurements remained tray 3 and tray 12. The resulting control scheme, labeled scheme 2 and shown in Figure 1-7, had a relative gain of 0.976 and a Niederlinski index of 1.025. Dynamic simulation of scheme 2, illustrated in Figure 1-8, showed that the scheme performed well for the 5% production rate increase used previously. Furthermore, the control scheme also performed well for catalyst rate and reflux ratio disturbances.

Two alternate solutions (not shown) were based on controlling a temperature difference, $T_{12} - T_7$. This temperature difference was controlled by manipulating the reactant D feed rate on tray 21. Manipulating either reboiler duty or reactant B feed rate on tray 7 to control tray 3 temperature gave good control system performance. The relative gains for these control schemes were 0.792 and 1.605, and the Niederlinski indices were 1.263 and 0.621, respectively. The idea of using a temperature difference came from observing the temperature profile during the breakthrough of reactant B. Each time reactant B broke through to the upper section of the column, the temperature on tray 7 increased to levels significantly higher than tray 12 which was counter to the desired temperature profile. Dynamic simulation indicated that the temperature difference scheme with either reboiler duty or reactant B feed rate being manipulated to control tray 3 temperature performed as well as scheme 2 with no appreciable difference in dynamic behavior.

Example 1 - Summary

The design of a robust control system for a process as complex as this example is not a simple task. Shortcut modeling techniques would obviously be ineffective in describing the process. Even having a rigorous steady-state model of the process was insufficient. The design tools based on steady-state information would lead the designer to believe that all of the control schemes studied would be acceptable. Locally, they were. Only when these control schemes were tested on the rigorous dynamic model did the ineffectiveness of several of these designs surface. The understanding of column behavior and the subsequent solution to the control problem were facilitated by the ability to view what was happening in the column dynamically. All of the control schemes that were studied had reasonable relative gains and condition numbers. However, it was not possible to discriminate among the candidate schemes based strictly on the RGA and SVD information.

EXAMPLE 2 - FOUR-COMPONENT COLUMN WITH SIDEDRAWS

Example 2 - Column Description

The control system design tools discussed above were applied to the third column of a three-column, four-component complex distillation configuration (Elaahi and Luyben, 1985).

The column investigated was used to separate four hydrocarbons: propane, isobutane, n-butane, and isopentane. The column had 79 trays, three feed streams, and two sidedraw streams (see Figure 2-1). The control objective was to maintain 95% pure propane in the distillate stream, 90% pure isobutane in the upper sidedraw stream, 90% pure n-butane in the lower sidedraw stream, and 95% pure isopentane in the bottom stream. Only stream product compositions were considered as controlled variables. Control schemes involving temperature were not considered because of the lack of significant steady-state tray-to-tray temperature change in the middle section of the column. The steady-state tray-to-tray temperature change for tray 10 to tray 70 was about 0.3 degrees C but below tray 10 and above tray 70, the steady-state tray-to-tray temperature change varied from 3.0 to 5.0 degrees C.

Elaahi and Luyben (1985) and McAvoy (1985) investigated control schemes for this column and did not achieve consistent results using steady-state and dynamic modeling tools different from those presented in this paper. Luyben's dynamic modeling procedure differs from ours in that we allow tray pressure drop to vary dynamically as a function of variable tray loadings, based on modified FRI tray hydraulic correlations.

Example 2 - Control Configurations

The investigation began by determining the steady-state operating profile for the column. For this mixture of components at the pressures considered, both the vapor and liquid phases behave ideally. The steady-state gains for six different sets of manipulated variables (see Table 2-1) were determined using the steady-state model. To begin the selection of an appropriate control scheme, the steady-state gains were used to perform SVD and RGA analyses for each of the six manipulated variable sets. Also calculated were the condition numbers from the SVD singular values. Table 2-1 lists the condition numbers for each manipulated variable set. The condition numbers were used to reduce the number of manipulated variable sets to be considered for dynamic simulation. Note that the SVD is not scale-independent and the tentative rejection of any strategy must occur after an examination of the units used for input and output variables. Manipulated variable sets 1, 2, and 3 were eliminated because the condition numbers for these sets were much higher than for sets 4, 5, and 6. Set 5 was eliminated from further consideration because the column base level to steam flow control loop was not considered dynamically attractive for this example.

The two remaining manipulated variable sets, sets 4 and 6, were considered for dynamic simulation. Control scheme 6 consisted of manipulated variable set 6 and the product stream compositions as the controlled variables. Control scheme 6 was the scheme selected by Elaahi and Luyben to control the column. (This scheme is labeled Scheme A in Elaahi and Luyben's paper). Control scheme 4 consisted of manipulated variable set 4 and the product stream compositions as the controlled variables. The difference between schemes 6 and 4 was that the level in the reflux drum was controlled by manipulating distillate flow in scheme 6 and by manipulating reflux flow in scheme 4. The loop pairings for the two control schemes were obtained by performing individual SVD and RGA analyses using the appropriate four by four steady-state gain matrices for composition to flow loops. Tables 2-2 and 2-3 list the steady-state gain array, RGA, SVD, Niederlinski index, and condition number for control schemes 6 and 4, respectively. Using the RGA, the loop pairings for each scheme were obtained. Figures 2-2 and 2-3 illustrate the loop pairings for control scheme 6 and 4, respectively. The various interaction measures for both control schemes did not indicate that either scheme would perform poorly when simulated dynamically. Specifically, both had positive Niederlinski indices (of no help in this 4 x 4 case), most of the RGA pairing values were near unity, and the condition numbers were relatively low.

Example 2 - Dynamic Simulation Results

Control schemes 6 and 4 were dynamically simulated with two forcing functions: simultaneous +10% feed rate changes (at constant feed composition) for all three feeds, and a -20% change in the isobutane feed composition. The results of the dynamic simulations are illustrated in Figures 2-4 to 2-7. Figures 2-4 and 2-5 illustrate the results for a 24 hour simulation of control schemes 6 and 4 respectively with +10% feed rate changes. The results indicated that both control schemes 6 and 4 were stable for the +10% feed rate change. Figures 2-6 and 2-7 illustrate the results of a 24 hour simulation of a -20% change in the isobutane feed composition for control schemes 6 and 4, respectively. The results indicated that both control schemes were also stable for a -20% change in the isobutane feed composition.

The dynamic simulations for both of the applied disturbances indicated that both of the control schemes were stable. The dynamic simulation results supported the control strategies suggested by the steady-state interaction measures given in Tables 2-2 and 2-3. Comparison of the individual SVD and RGA analyses for the two control schemes indicated that control scheme 4 was preferred over control scheme 6. Specifically, the RGA pairing values were closer to unity, indicating less loop interaction, for control scheme 4 than for scheme 6. Also, scheme 4 had a lower condition number than scheme 6 indicating better conditioning for scheme 4. Comparison of the dynamic simulations indicated that with control scheme 4, the column experienced less oscillation and the product compositions suffered less deviation from set point than with control scheme 6. The steady-state and dynamic results were consistent in indicating that control scheme 4 was preferable to control scheme 6.

The robustness of both control schemes to change in operating conditions was tested by moving the column to a new operating point. This change was accomplished by increasing the tray 64 sidedraw flow rate by 5% and holding the distillate rate and tray 15 sidedraw rate constant. This change decreased the bottoms flow rate by 2.9%. The resulting distillate composition was 97% propane and the upper sidedraw composition was 87% isobutane. A new steady-state operating profile was determined and the steady-state gains were recalculated for the column. The new steady-state gains were used to perform SVD and RGA analyses for control schemes 6 and 4. Tables 2-4 and 2-5 list the recalculated steady-state gains, SVD, RGA, Niederlinski indices, and condition numbers for control schemes 6 and 4 respectively.

For the new operating condition, the Niederlinski index for control scheme 6 was negative whereas the index was positive for control scheme 4. For each control scheme the RGA and condition numbers were similar to those calculated for the previous column condition. Simultaneous 5% feed rate changes were made to all feeds for each control scheme operating at the new conditions. Figures 2-8 and 2-9 illustrate the dynamic simulation results for control schemes 6 and 4 respectively. The results of a 45 minute simulation using control scheme 6, depicted in Figure 2-8, indicated that it was unstable when the feed rates were decreased 5% from the new operating condition. Figure 2-9 illustrates the results for a two hour simulation using control scheme 4 and indicated that this scheme was also unstable for -5% feed rate changes. The dynamic simulation results illustrated that the negative Niederlinski index calculated for scheme 6 correctly predicted instability for control scheme 6, but the positive Niederlinski index calculated for scheme 4 was of no help in predicting the observed instability for scheme 4.

Example 2 - Summary

This example showed that the steady-state interaction measures can be used to discriminate among various control schemes. Specifically, the steady-state control systems design tools indicated that control scheme 4 was preferred over control scheme 6 when operating at the original operating point. The dynamic simulations supported the steady-state results for this case.

This example also demonstrated that the steady-state interaction measures do not always provide enough information to guide the control system designer in the correct direction. This point was illustrated by changing the operating point of the column. The Niederlinski index was negative for control scheme 6 and positive for control scheme 4 at the new operating point. However, the dynamic simulations indicated that both control schemes were actually unstable at the new set point.

EXAMPLE 3 - BINARY DISTILLATION WITH TRAY PRESSURE DROP EFFECTS

Example 3 - Column Description

This example demonstrates some of the problems that can occur in control system design when dealing with columns with significant changes in tray pressure drop. In these columns, tray temperature changes frequently occur as a result of a change in tray pressure drop, as opposed to a change in composition. Vacuum columns frequently exhibit this behavior. Ignoring the pressure drop effect in a steady-state model will result in the calculation of incorrect steady-state gains. Incorrect steady-state gains can lead to inaccurate SVD and RGA calculations which in turn can lead to gross errors in developing column control schemes.

The column used to demonstrate the effect of pressure drop variations was a 28 tray light component purifier. The binary system consisted of one feed stream on tray 7 with a composition of 90% light and 10% heavy and a total feed flow rate of 100 moles/hr (see Figure 3-1). The control objective was to maintain the top product stream composition above 99.95% light component. Bottom composition was allowed to float within the range of 35-65% lights. The two components had both vapor phase and liquid phase nonidealities which were accounted for appropriately in the steady-state model by using the Redlich-Kwong equation to model the vapor phase and the Wilson equation to calculate liquid phase activity coefficients.

Example 3 - Control Configurations

To begin the analysis, a steady-state model of the column was determined. From this model, the steady-state gains were calculated neglecting the changes in tray pressures (and tray temperatures) with respect to changes in column vapor rates. These steady-state gains were used to calculate the SVD, the RGA, and the condition number for six different manipulated variable sets. Table 3-1 lists the manipulated variables along with the condition number for each of the six candidate column control schemes.

Selection of an appropriate 2 x 2 control scheme began by examination of the condition numbers. Column schemes 1, 2, and 3 were eliminated from further consideration because their condition numbers were much larger than the condition numbers for schemes 4, 5, and 6. The RGA's for the remaining control schemes were examined to determine the best candidates for dynamic modeling. The best RGA's for double-ended control for the remaining three control schemes are given in Table 3-2. Also included in Table 3-2 are the Niederlinski indices and condition numbers for each scheme. Both the RGA and condition number for control scheme 6 were poor and this scheme was eliminated from further consideration. Scheme 4 was eliminated because the pairings recommended by the RGA, i.e., distillate flow to temperature on tray 1 and steam flow to temperature on tray 11, were not considered to be dynamically desirable in spite of the positive Niederlinski index and RGA pairing values near unity.

The remaining scheme, scheme 5, had RGA pairing values near unity, a positive Niederlinski index, and an acceptable condition number. The recommended pairings were temperature on tray 11 to reflux flow rate and temperature on tray 1 to bottom flow rate. Because all of the interaction measures were acceptable, this scheme was chosen for dynamic simulation.

Example 3 - Dynamic Simulation Results

The control scheme is depicted in Figure 3-2. The results of the dynamic simulation of scheme 5 with double-ended control are illustrated in Figures 3-3, 3-4, 3-5, and 3-6. Figures 3-3 and 3-4 illustrate the results of a sixteen hour simulation with a +10% feed rate disturbance. Figures 3-5 and 3-6 illustrate the results of a sixteen hour simulation with a -15% change in the light component feed composition. Again, the results indicated an unstable response to the applied disturbance.

Because scheme 5 was not stable, the steady-state gains were recalculated with consideration for the changes in tray pressures with respect to changes in column vapor rates. These tray hydraulics were taken into account using FRI tray loading correlations. Using the new steady-state gains, the SVD and RGA were recalculated for control scheme 5. The new SVD, RGA, Niederlinski index, and condition number for scheme 5 are listed in Table 3-3. The Niederlinski index remained positive but the recalculated RGA suggested that double-ended control scheme 5 might now be less desirable.

With the new RGA in mind, a single-ended control scheme was chosen. Based on the desire to manipulate the smaller of two product streams for composition control, the loop selected for single-ended control was temperature on tray 1 to bottom flow rate. The reflux flow control loop was not selected because it was considered dynamically undesirable to adjust reflux flow to control a temperature more than halfway towards the bottom of the column at tray 11.

The new control scheme was simulated dynamically with the same disturbances used previously. The single-ended control scheme is illustrated in Figure 3-7. The results of the dynamic simulations are illustrated in Figures 3-8, 3-9, 3-10, and 3-11. Figures 3-8 and 3-9 illustrate the results of a sixteen hour simulation with a 10% feed rate change. The results of a -15% change in the light component feed composition are depicted in Figures 3-10 and 3-11. The dynamic simulations indicated that this single-ended control scheme was stable for both disturbances and the composition of the light component product remained above the minimum specification of 99.95% purity in the overheads.

Example 3 - Summary

This example demonstrated that an unworkable control scheme was chosen using SVD and RGA analyses calculated from inaccurate steady-state gains. Specifically, incorrect steady-state gains were obtained by ignoring the effect of tray pressure drop variations. Dynamic simulation of the control scheme recommended by SVD and RGA analyses calculated from the incorrect steady-state gains showed that the double-ended scheme was unstable. When FRI tray loading correlations were used to correct the steady-state gains, the RGA calculated from the corrected gains indicated that the original double-ended control scheme was less desirable than originally thought. Because none of the corrected SVD and RGA analyses indicated that double-ended control was particularly attractive, a single-ended control scheme was implemented for the column. Dynamic simulation of the new control scheme indicated that the single-ended control scheme was stable and resulted in adequate control of the light component purifier.

SUMMARY

In this paper, we have illustrated a control design methodology which combines rigorous dynamic distillation modeling together with the guidance provided by several interaction measures. Each of the three examples illustrates the point that any single interaction measure will occasionally provide misleading information. From the viewpoint of the industrial control designer, the various interaction measures are useful for elimination of unworkable strategies, as opposed to discrimination among workable strategies. Information gained from the various interaction measures and the dynamic model allows the designer to develop simple, workable distillation control system designs in a reasonably short period of time.

REFERENCES

Abrams, D. S., and J. M. Prausnitz (1975). Statistical Thermodynamics of Liquid Mixtures: A New Expression for the Excess Gibbs Energy of Partly or Completely Miscible Systems. AIChE J., 21, 116.

Bristol, E. H. (1966). On a New Measure of Interaction for Multivariable Process Control. *IEEE Trans. Auto Con.*, **AC-11**, 133.

Chan, H., and J. R. Fair (1984). Prediction of Point Efficiencies on Sieve Trays. 1. Binary Systems. *Ind. Eng. Chem. Proc. Des. Dev.*, **23**, 814.

Chan, H., and J. R. Fair (1984). Prediction of Point Efficiencies on Sieve Trays. 2. Multicomponent Systems. *Ind. Eng. Chem. Proc. Des. Dev.*, **23**, 820.

Downs, J. J. (1982). The Control of Azeotropic Distillation Columns, Ph.D. Dissertation, The University of Tennessee, Knoxville, Tennessee.

Downs, J. J., and E. F. Vogel (1985). An Interactive Dynamic Distillation Simulator for Analysis and Control Systems Development. *Proceedings of the 1985 Summer Computer Simulation Conference*, 337.

Elaahi, A., and W. L. Luyben (1985). Control of an Energy-Conservative Complex Configuration of Distillation Columns for Four-Component Separations. *Ind. Eng. Chem. Proc. Des. Dev.*, **24**, 368.

Grosdidier, P., M. Morari, and B. R. Holt (1985). Closed Loop Properties from Steady-State Gain Information. *Ind. Eng. Chem. Fundam.*, **24**, 221.

Holland, C. D., and A. F. Liapis (1983). Modeling of a Distillation Column and Its Control System. *Computer Methods for Solving Dynamic Separation Problems*, McGraw-Hill, New York, 269.

Klema, V. C., and A. J. Laub (1980). The Singular Value Decomposition: Its Computation and Some Applications. *IEEE Trans. Auto. Con.*, **AC-25**, 164.

McAvoy, T. J. (1985). Communications to *Ind. Eng. Chem. Proc. Des. Dev.*, in press.

Naphtali, L. M., and D. P. Sandholm (1971). Multicomponent Separation Calculations by Linearization. *AIChE J.*, **17**, 148.

Niederlinski, A. (1971). A Heuristic Approach to the Design of Linear Multivariable Interacting Control Systems. *Automatica*, **7**, 691.

Renon, H., and J. M. Prausnitz (1968). Local Compositions in Thermodynamic Excess Functions for Liquid Mixtures. AIChE J., 14, 135.

Soave, G. (1974). Equilibrium Constants from a Modified Redlich-Kwong Equation of State. Chem. Eng. Sci., 20, 847.

Wilson, G. M. (1964). Vapor-Liquid Equilibrium XI: A New Expression for the Excess Free Energy of Mixing. J. Am. Chem. Soc., 86, 127.

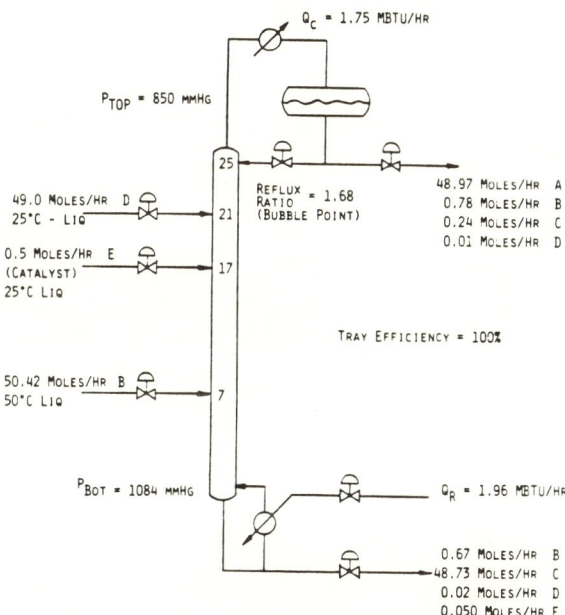

Figure 1-1. Example 1 Column Description

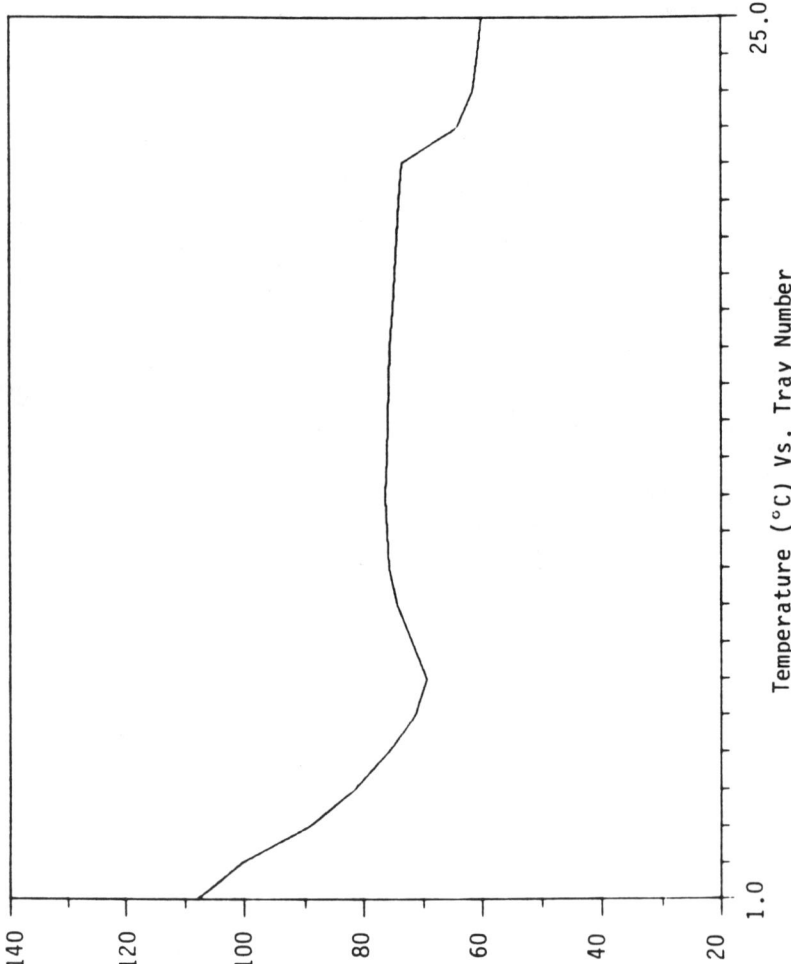

Figure 1-2. Steady-State Temperature Profile for Example 1

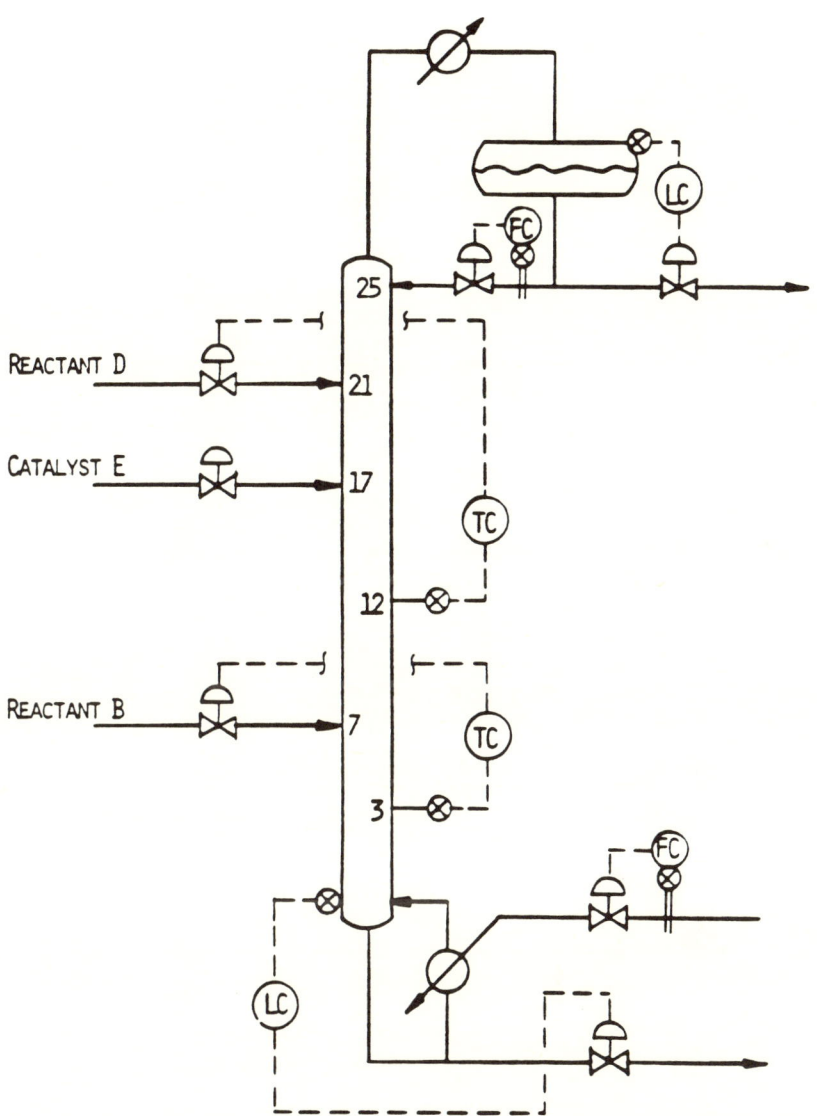

Figure 1-3. Control Scheme 1 for Example 1

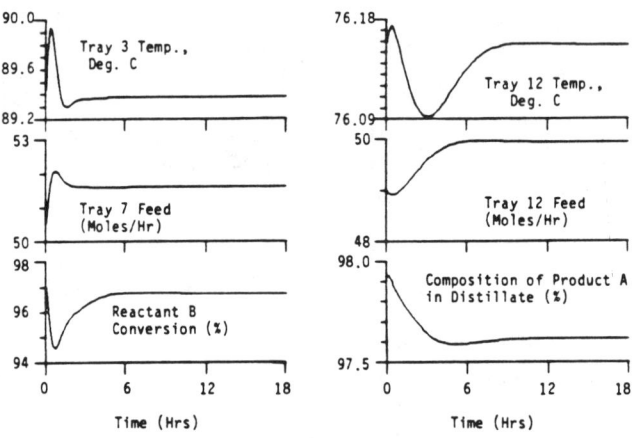

Figure 1-4. Example 1, Control Scheme 1 with a 1% Increase in Reboiler Duty

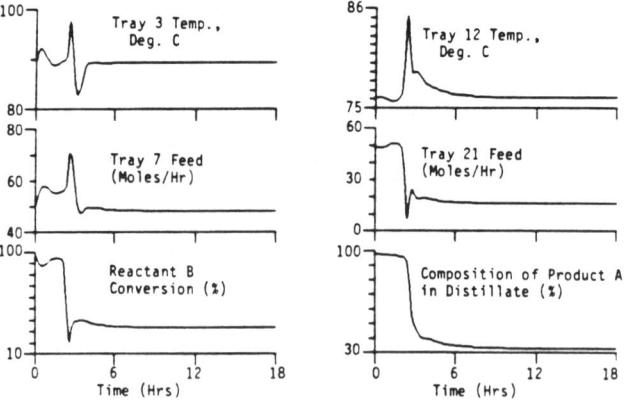

Figure 1-5. Example 1, Control Scheme 1 with a 5% Increase in Reboiler Duty

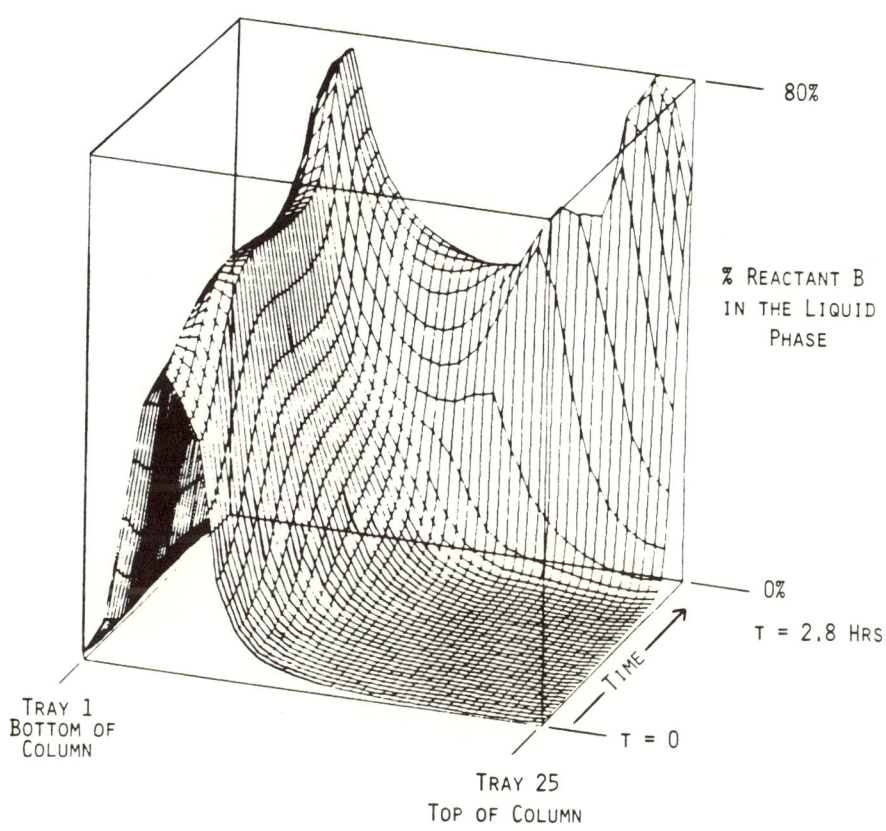

Figure 1-6. Reactant B Composition Profile Vs. Time For Example 1, Control Scheme 1 with a 5% Increase in Reboiler Duty

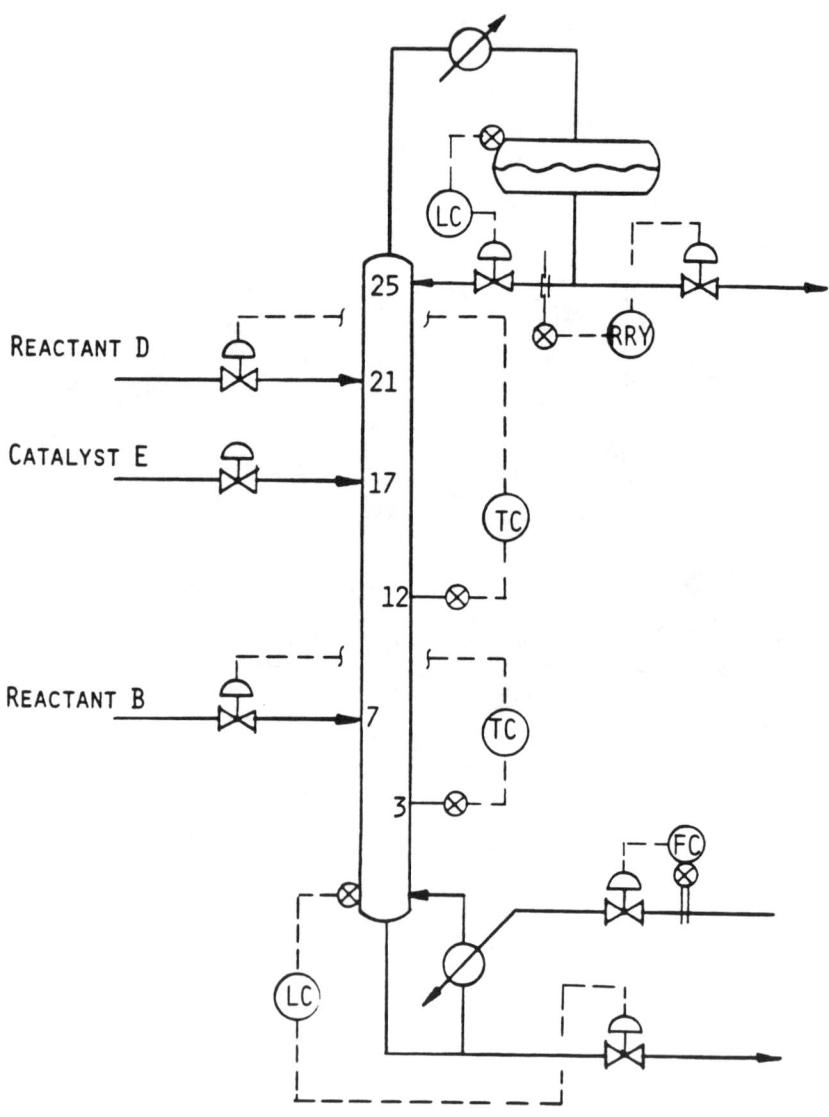

Figure 1-7. Control Scheme 2 for Example 1

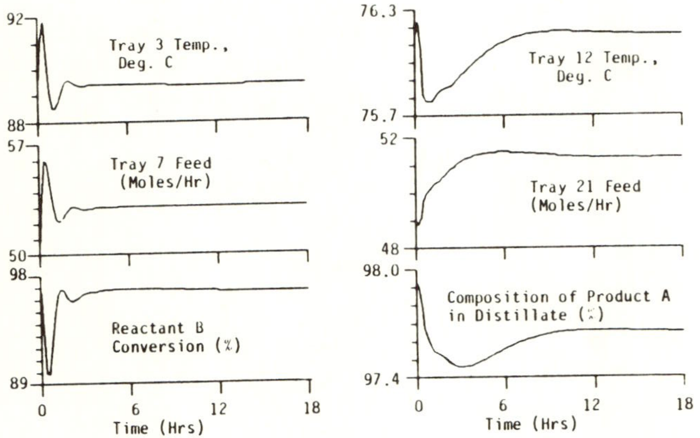

Control System Synthesis for Distillation Columns 123

	Reactant B Feed Rate	Reactant D Feed Rate	Reboiler* Duty
Reflux Rate Constant			
Temp Tray 3	-27.0202	-13.845	31.5868
Temp Tray 12	0.0094152	0.22814	0.018051
Reflux Ratio Constant			
Temp Tray 3	-34.390	-7.4833	13.3909
Temp Tray 12	-0.029673	0.26188	-0.078457
Temp $T_{12} - T_7$	-0.22799	0.18859	-0.12724

*Gains with respect to reboiler duty have been normalized by by the heat of vaporization of the liquid in the reboiler.

Table 1-1. Steady-State Gains (Deg. C/Mole/Hr) for Example 1

Figure 1-8. Example 1, Control Scheme 2 with a 5% Increase in Reboiler Duty

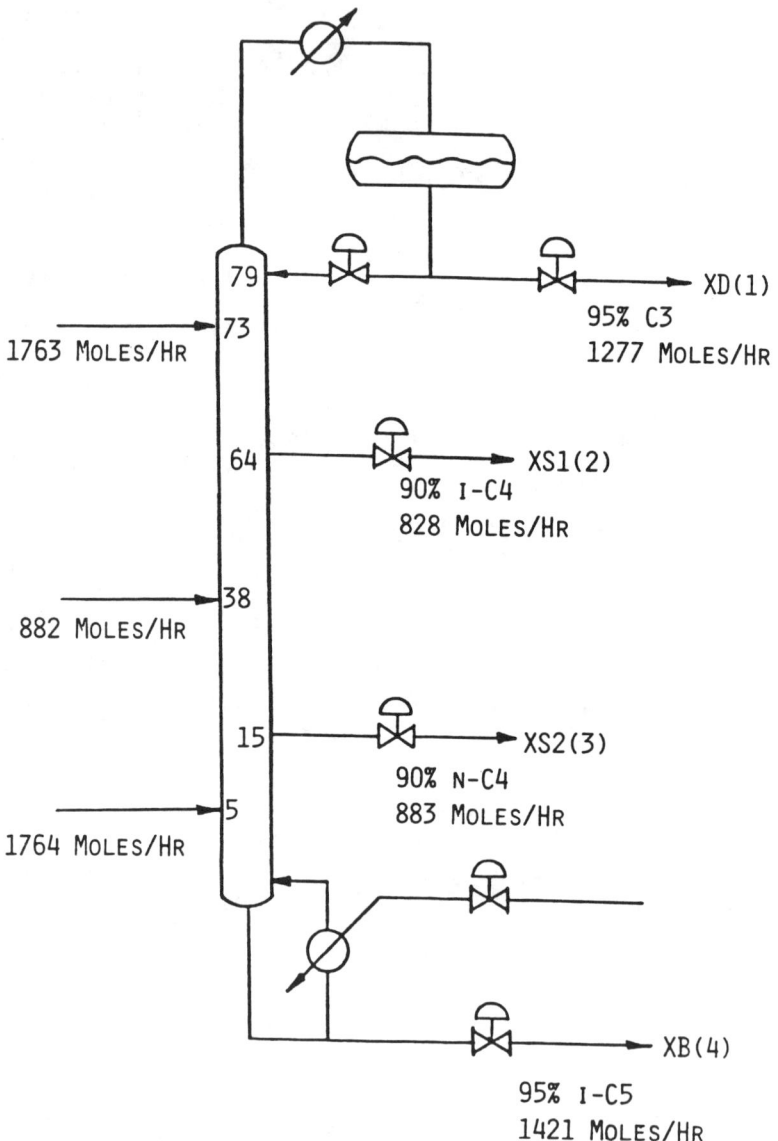

Figure 2-1. Diagram of Complex, Four-Component Distillation Column for Example 2.

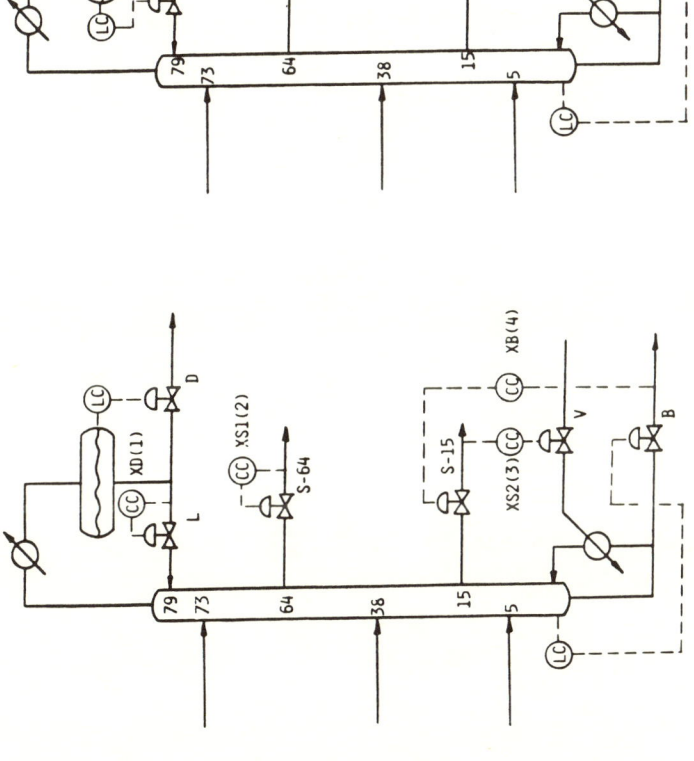

Figure 2-3. Diagram of Control Scheme 4 for Example 2.

Figure 2-2. Diagram of Control Scheme 6 for Example 2.

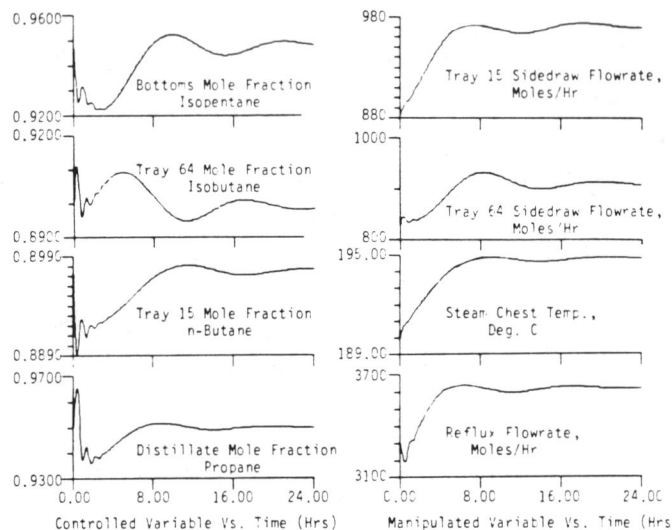

Figure 2-4. Controlled and Manipulated Variable Responses for a +10% Feed Rate Change for Control Scheme 6 (Figure 2-2), Example 2.

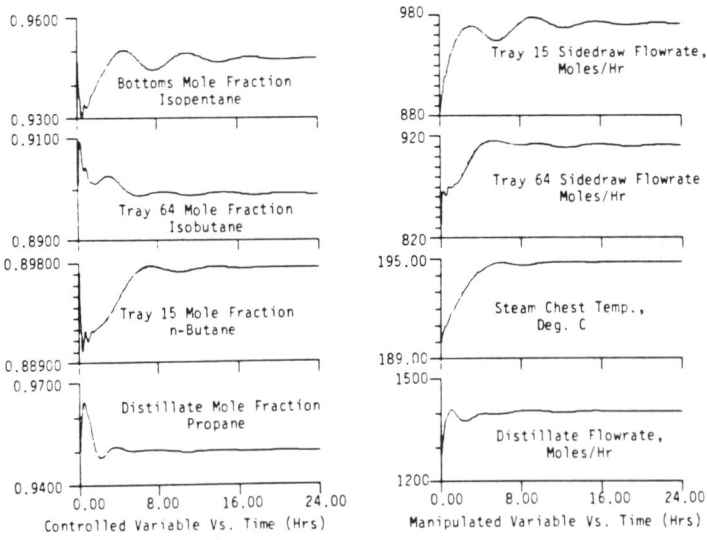

Figure 2-5. Controlled and Manipulated Variable Responses for a +10% Feed Rate Change for Control Scheme 4 (Figure 2-3), Example 2

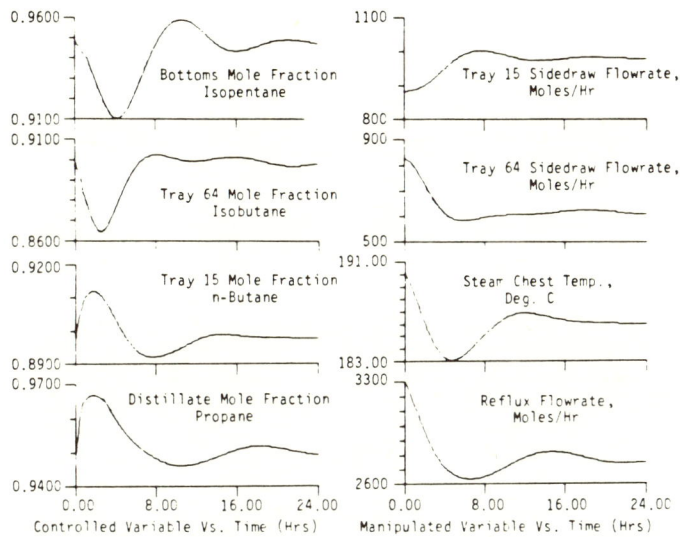

Figure 2-6. Controlled and Manipulated Variable Responses for a -20 Isobutane Feed Composition Change for Control Scheme 6 (Figure 2-2), Example 2.

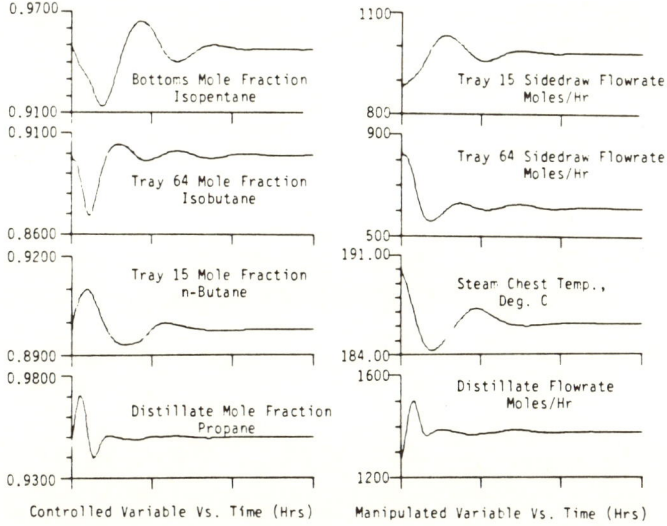

Figure 2-7. Controlled and Manipulated Variable Responses for a -20 Isobutane Feed Composition Change for Control Scheme 4 (Figure 2-3), Example 2.

128   AN INDUSTRIAL VIEW OF ADVANCED PROCESS CONTROL

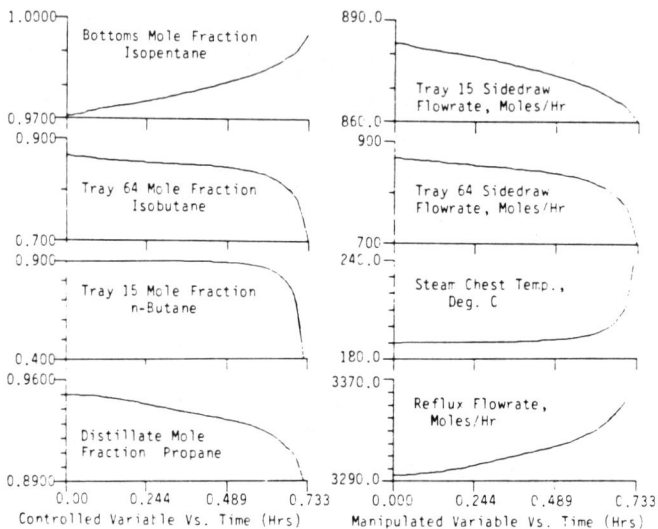

Figure 2-8. Controlled and Manipulated Variable Responses for a -5 Feed Rate Change for Control Scheme 6 (Figure 2-2) at the New Operating Condition, Example 2.

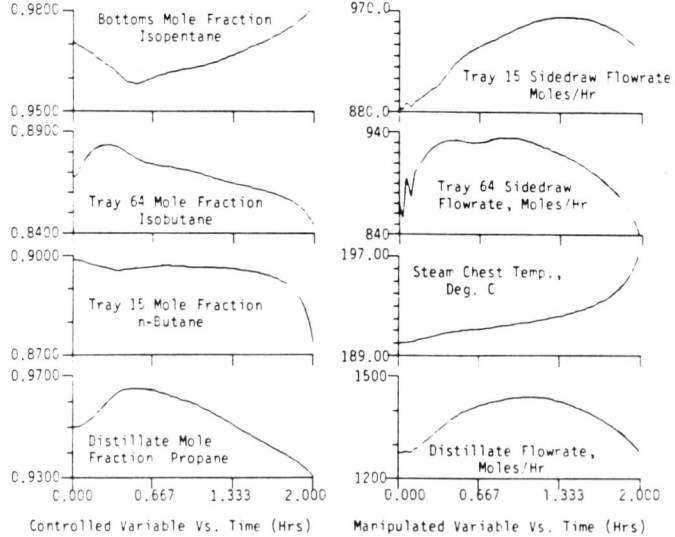

Figure 2-9. Controlled and Manipulated Variable Responses for a -5% Feed Rate Change for Control Scheme 4 (Figure 2-3) at the New Operating Condition, Example 2.

Controlled Variables

XD (1) - Mole fraction component 1 (propane) in distillate D
XS1 (2)- Mole fraction component 2 (isobutane) in upper (tray 64) sidedraw
XS2 (3)- Mole fraction component 3 (n-butane) in lower (tray 15) sidedraw
XB (4) - Mole fraction component 4 (isopentane) in bottoms B

Set Number	Manipulated Variables	Condition Number
1	L/D, S-64, V, S-15	9030
2	V/L, S-64, V, S-15	60100
3	D/V, S-64, V, S-15	116,000
4	D, S-64, V, S-15	51.5
5	L, S-64, B, S-15	57.4
6	L, S-64, V, S-15	53.8

*Notes:
1. Notation differs from that used by Elaahi and Luyben
2. S-64 represents upper sidedraw flow rate at tray 64
3. S-15 represents lower sidedraw flow rate at tray 15
4. Control schemes consist of corresponding items from lists of controlled and manipulated variables, i.e., for scheme 6, the first controlled variable XD (1) is paired with the first manipulated variable L, and so on.

Table 2-1. Condition Numbers for the Gain Matrix Relating Column Controlled Variables to Various Sets of Manipulated Variables for Example 2

AN INDUSTRIAL VIEW OF ADVANCED PROCESS CONTROL

Steady-State Gain Array

	L	S-64	V	S-15
XD(1)	0.274E-03	0.374E-04	-0.253E-03	0.454E-04
XS1(2)	0.248E-03	-0.685E-03	-0.255E-03	0.341E-04
XS2(3)	-0.658E-04	0.201E-03	0.983E-04	-0.412E-04
XB(4)	-0.186E-03	0.586E-03	0.211E-03	0.577E-03

Relative Gain Array

	L	S-64	V	S-15
XD(1)	0.672	0.056	0.279	-0.006
XS1(2)	2.890	0.714	-2.640	0.035
XS2(3)	-2.270	0.218	3.010	0.047
XB(4)	-0.288	0.013	0.351	0.924

SVD Singular Values

| 0.110E-02 | 0.470E-03 | 0.301E-03 | 0.204E-04 |

V^T Matrix

	L	S-64	V	S-15
	-0.2995	0.8225	0.3279	0.3555
	0.2769	-0.1722	-0.2905	0.8996
	0.5215	0.5419	-0.6084	-0.2533
	0.7494	0.0151	0.6618	-0.0141

U Matrix

YD(1)	-0.0935	0.3613	0.9272	0.0315
XS1(2)	-0.6434	0.6196	-0.3172	0.3186
XS2(3)	0.1841	-0.2521	0.0846	0.9463
XB(4)	0.7372	0.6496	-0.1804	0.0458

Niederlinski Index = 0.367
Condition Number = 53.83

Table 2-2. Steady-State Gain Array, RGA, SVD, Niederlinski Index, and Condition Number for Control Scheme 6 at the Original Operating Condition for Example 2

Steady-State Gain Array

	D	S-64	V	S-15
XD(1)	-0.742E-03	-0.265E-06	0.195E-06	-0.332E-07
XS1(2)	-0.248E-03	-0.727E-03	0.253E-04	-0.162E-04
XS2(3)	0.218E-03	0.212E-03	0.241E-04	-0.279E-04
XB(4)	0.618E-03	0.618E-03	0.105E-05	0.615E-03

Relative Gain Array

	D	S-64	V	S-15
XD(1)	1.001	-0.0004	-0.0002	0.000
XS1(2)	-0.002	0.757	0.262	-0.017
XS2(3)	0.002	0.230	0.737	0.032
XB(4)	0.0002	0.013	0.002	0.985

SVD Singular Values

| 0.159E-02 | 0.565E-03 | 0.382E-03 | 0.309E-04 |

V^T Matrix

	D	S-64	V	S-15
	0.7833	0.5716	-0.0076	0.2441
	-0.5379	0.4267	0.0075	0.7270
	0.3116	-0.7006	0.0267	0.6414
	0.0016	0.0199	0.9996	-0.0208

U Matrix

XD(1)	-0.3661	0.7064	-0.6048	-0.0331
XS1(2)	-0.6693	0.2131	0.6366	0.3184
XS2(3)	0.1795	-0.0825	-0.2569	0.9460
XB(4)	0.6211	0.6699	0.4037	0.0502

Niederlinski Index = 1.327
Condition Number = 51.48

Table 2-3. Steady-State Gain Array, RGA, SVD, Niederlinski Index, and Condition Number for Control Scheme 4 at the Original Operating Condition for Example 2

Control System Synthesis for Distillation Columns

Table 2-4. (left side)

Steady-State Gain Array

	L	S-64	V	S-15
XD(1)	0.234E-03	0.753E-05	-0.266E-03	0.164E-04
XS1(2)	0.283E-03	-0.760E-03	-0.300E-03	0.884E-05
XS2(3)	0.707E-04	-0.225E-03	-0.548E-04	-0.351E-03
XB(4)	-0.136E-03	0.429E-03	0.158E-03	0.420E-03

Relative Gain Array

	L	S-64	V	S-15
XD(1)	0.768	0.012	0.224	-0.003
XS1(2)	1.725	0.884	-1.619	0.010
XS2(3)	2.500	-0.202	-1.701	0.403
XB(4)	-3.993	0.307	4.096	0.590

SVD Singular Values

| 0.107E-02 | 0.469E-03 | 0.301E-03 | 0.162E-04 |

V^T Matrix

	L	S-64	V	S-15
	-0.3237	0.8253	0.3465	0.3066
	0.2458	-0.1274	-0.2803	0.9191
	0.5198	0.5499	-0.6052	-0.2473
	0.7514	0.0168	0.6596	0.0026

U Matrix

XD(1)	-0.1460	0.3109	0.9386	0.0314
XS1(2)	-0.7656	0.5504	-0.3058	0.1317
XS2(3)	-0.3129	-0.5564	0.1101	0.7619
XB(4)	0.5428	0.5393	-0.1154	0.6334

Niederlinski Index = -0.600
Condition Number = 66.12

Table 2-4. Steady-State Gain Array, RGA, SVD, Niederlinski Index, and Condition Number for Control Scheme 6 at the New Operating Condition for Example 2

Table 2-5. (right side)

Steady-State Gain Array

	D	S-64	V	S-15
XD(1)	-0.742E-03	-0.213E-06	0.202E-06	-0.109E-07
XS1(2)	-0.899E-03	-0.770E-03	0.221E-04	-0.111E-04
XS2(3)	-0.707E-04	-0.227E-04	0.257E-04	-0.356E-03
XB(4)	0.433E-03	0.433E-03	0.322E-05	0.429E-03

Relative Gain Array

	D	S-64	V	S-15
XD(1)	1.000	-0.0003	-0.0002	0.000
XS1(2)	-0.002	0.894	0.119	-0.013
XS2(3)	-0.002	-0.204	0.797	0.409
XB(4)	0.003	0.013	0.083	0.604

SVD Singular Values

| 0.151E-02 | 0.546E-03 | 0.384E-03 | 0.245E-04 |

V^T Matrix

	D	S-64	V	S-15
	0.8118	0.5521	-0.0145	0.1896
	-0.4787	0.4433	-0.0131	0.7577
	0.3344	-0.7058	0.0136	0.6244
	0.0009	0.0234	0.9997	0.0042

U Matrix

XD(1)	-0.3980	0.6509	-0.6462	-0.0198
XS1(2)	-0.7637	0.1472	0.6146	0.1319
XS2(3)	-0.2481	-0.4821	-0.3562	0.7610
XB(4)	0.4437	0.5676	0.2791	0.6349

Niederlinski Index = 1.236
Condition Number = 61.78

Table 2-5. Steady-State Gain Array, RGA, SVD, Niederlinski Index, and Condition Number for Control Scheme 4 at the New Operating Condition for Example 2

132   AN INDUSTRIAL VIEW OF ADVANCED PROCESS CONTROL

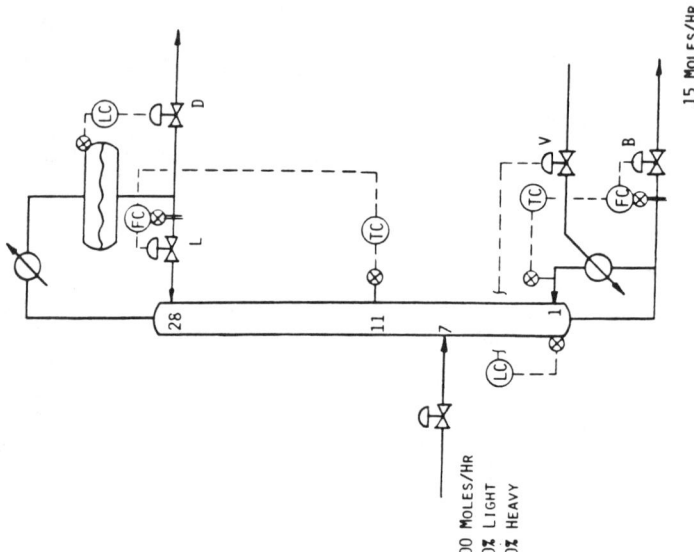

Figure 3-2. Diagram of Control Scheme 5 Recommended by RGA and SVD for Light Component Purifier in Example 3.

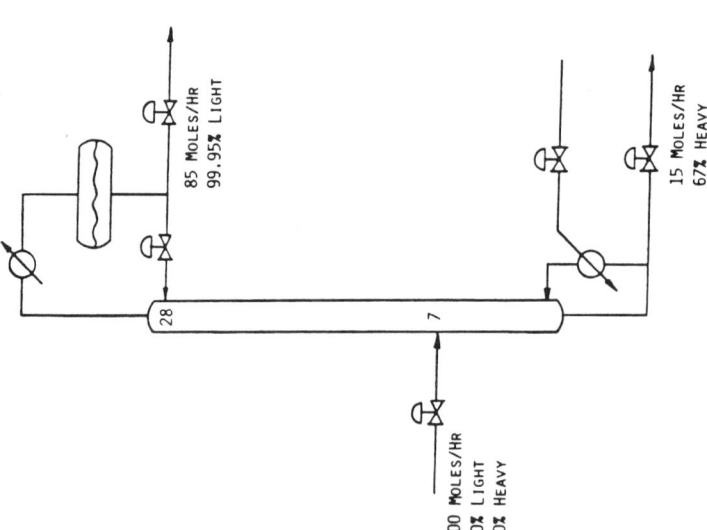

Figure 3-1. Diagram of Light Component Purifier for Example 3.

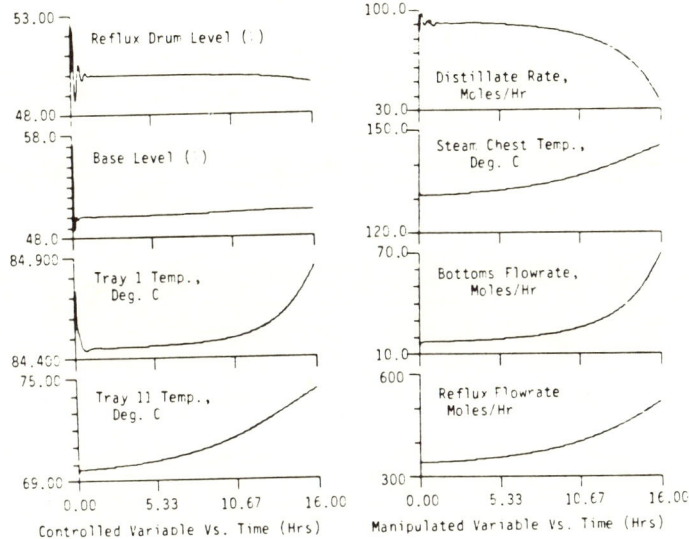

Figure 3-3. Controlled and Manipulated Variable Responses for a +10% Feed Rate Change for Control Scheme 5 (Figure 3-2), Example 3.

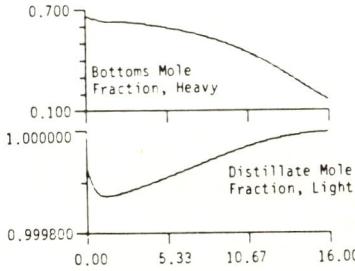

Figure 3-4. Bottom Heavy and Distillate Light Composition Responses for a +10% Feed Rate Change for Control Scheme 5 (Figure 3-2), Example 3.

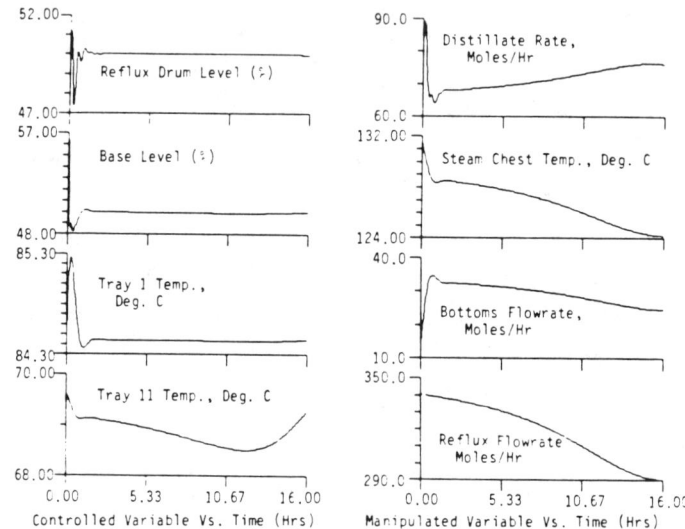

Figure 3-5. Controlled and Manipulated Variable Responses for a -15% Disturbance in the Light Composition in the Feed for Control Scheme 5 (Figure 3-2), Example 3.

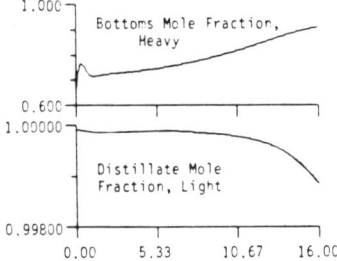

Figure 3-6. Bottom Heavy and Distillate Light Composition Responses for a -15% Disturbance in the Light Composition in the Feed for Control Scheme 5 (Figure 3-2), Example 3.

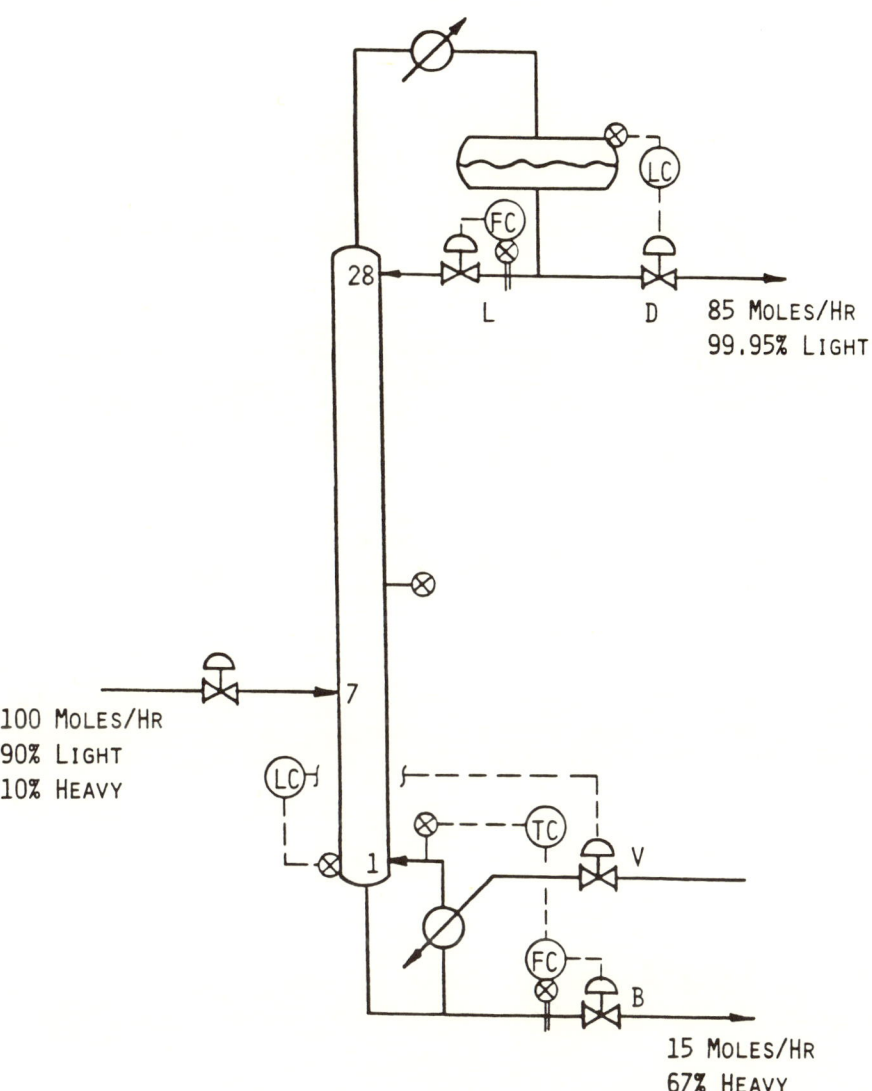

Figure 3-7. Diagram of Control Scheme with Single-Ended Control for Light Component Purifier in Example 3.

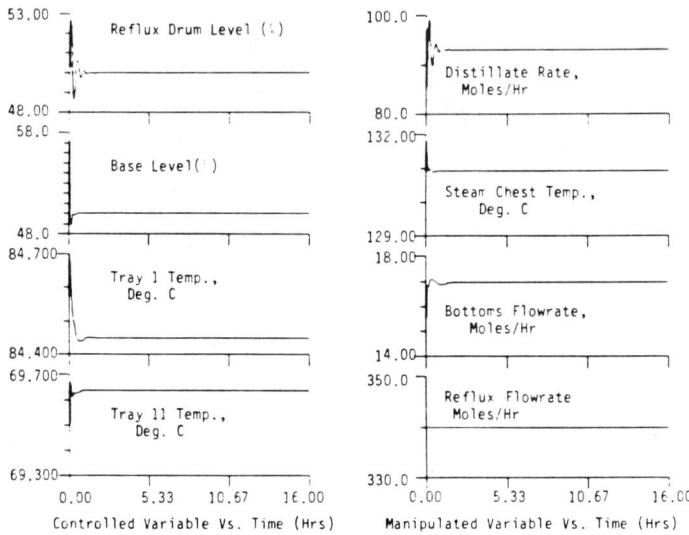

Figure 3-8. Controlled and Manipulated Variable Responses for a +10% Feed Rate Change for the Single Ended Control Scheme (Figure 3-7), Example 3.

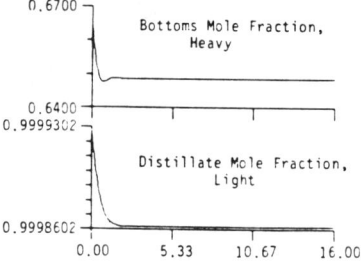

Figure 3-9. Bottom Heavy and Distillate Light Composition Responses for a +10% Feed Rate Change for the Single-Ended Control Scheme (Figure 3-7), Example 3

Control System Synthesis for Distillation Columns 137

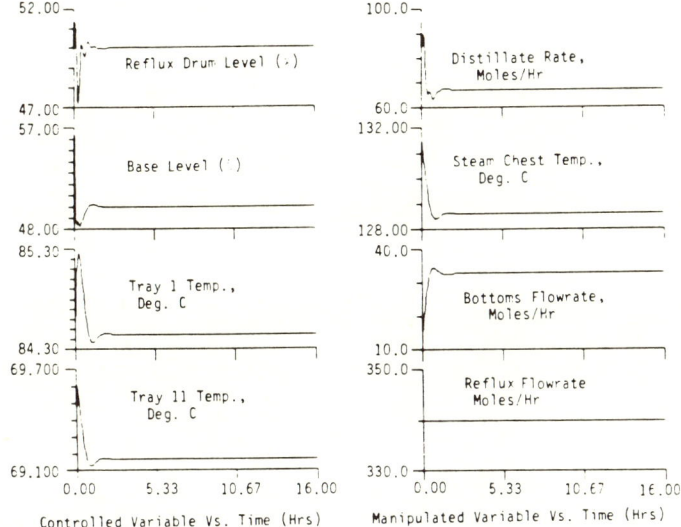

Figure 3-10. Controlled and Manipulated Variable Responses for a -15% Disturbance in the Light Composition in the Feed for Single-Ended Control Scheme (Figure 3-7), Example 3.

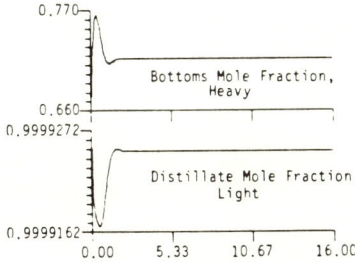

Figure 3-11. Bottom Heavy and Distillate Light Composition Responses for a -15% Disturbance in the Light Composition in the Feed for Single-Ended Control Scheme (Figure 3-7), Example 3.

Set Number	Manipulated Variables	Condition Number
1	L/D, V	3650
2	V/L, V	58400
3	D/V, V	91300
4	D, V	210
5	L, B	290
6	L, V	700

Table 3-1. Condition Numbers for the Gain Matrix Relating Column Temperatures to Various Sets of Manipulated Variables for Example 3

Set Number 4

	D	V
T(1)	0.9972	0.0028
T(11)	0.0028	0.9972

Niederlinski Index = 1.0028
Condition Number = 278.8

Set Number 5

	L	B
T(11)	0.9973	0.0027
T(1)	0.0027	0.9973

Niederlinski Index = 1.0027
Condition Number = 390.6

Set Number 6

	L	V
T(11)	43.13	-42.13
T(1)	-42.13	43.13

Niederlinski Index = 0.02318
Condition Number = 931.0

Table 3-2. Relative Gain Array, Niederlinski Index and Condition Number for Control Schemes 4, 5, and 6 for Example 3

	B	L
T(1)	1.5265	-0.5265
T(11)	-0.5265	1.5265

Niederlinski Index = 0.6551
Condition Number = 11.7

Table 3-3. Relative Gain Array, Niederlinski Index and Condition Number for Control Scheme 5 Calculated from New Steady-State Gains for Example 3

PROGRESS AND CHALLENGES IN BATCH PROCESS CONTROL

Mark R. Juba
John W. Hamer
Eastman Kodak Company, Rochester, NY 14650

ABSTRACT

   Batch processing is an important segment of the chemical processing industry and presents some unique process-control challenges. This paper discusses the features that distinguish batch process control from the control of continuous processes and reviews the major advances in batch process control. The major area of focus, control of batch reactors, is broken into two sections. The first deals with exothermic reactions with unknown kinetics, which are difficult to control and potentially dangerous. A simple reactor model is used to obtain a reaction-rate estimator, which is then used with the model to design feedforward and nonlinear feedback controllers that greatly improve isothermal operation. The model and rate estimator also permit the implementation of nonisothermal and fed-batch policies that increase reactor productivity. The second type of problem discussed is the optimal control of a batch reaction for which the kinetics are known, to maximize some function of composition. Although these techniques are well developed, the major challenge remains in developing simple methods for identifying the nonlinear and sometimes time-varying models for these processes so that the optimal control methods can be applied.

KEYWORDS

batch process control, batch reactor control, process control, optimal control, batch-reactor, fed-batch reactor, semibatch reactor, identification, nonlinear models, feedforward controllers, nonlinear controllers

I. INTRODUCTION

Batch operations have always been part of chemical engineering technology, and their major advantages, great flexibility and rapid response to changing market conditions, are well known. Over the next decades, batch chemical processing will increase in importance as industry pursues the manufacture of low-volume, high-value-added chemicals (Anderson, 1984). Our discussion of batch process control is divided into four sections:

Section II. How batch process control differs from the more familiar concepts of continuous process control.

Section III. Progress in the development of batch process control technology, including how computerized operation and control will play a major role in the renewal of this technology.

Section IV. Application of model-based analysis and control methods to batch reactors with exothermic reactions. This is the main focus of the paper and shows some solutions to the problem of exothermic reaction control using simple modeling and analysis.

Section V. Control of batch reactors to optimize composition functions. Here the greatest remaining challenge is to develop methods for cases where the reaction kinetics and mechanisms are unknown.

II. BATCH VS. CONTINUOUS PROCESS CONTROL

In contrast to the continuous manufacture of commodity chemicals, batch processes are characterized by both a very different manufacturing environment and different dynamic behavior. For example, the multiple product environment of batch processes requires that ingredients, control loop setpoints, and tuning parameters must be changed frequently (Mehta, 1983; Shaw, 1983). In addition to the changes required for preparing different chemicals, controller configurations and tuning parameters must be adjusted when different unit operations (reaction, distillation, extraction, crystallization, etc.) take place in the same piece of process equipment.

Batch processing also involves the movement of batches through multiple streams or paths in the process equipment (Armstrong and Coe, 1983). For example, at the same time, a batch plant could be involved in different stages of manufacture for several different products. Although controlling a single task in the manufacture of a particular chemical may not necessarily be difficult, coordinating the various operations and allocating process resources, such as energy and raw materials, could be complex.

The frequent product and process changes associated with a batch chemical manufacturing facility also present some special challenges for control system design (Smith, 1984; Thome et al., 1979). For example, the requirement for startup and shutdown regulatory control in batch processes demands good dynamic response over the entire operating range of the controlled variable. This contrasts with the precise control over a small range that is required in many continuous processes. In addition, because idle time between batches is normal, special control system configurations, such as antireset windup and accommodating nuisance alarms, are often required.

Accuracy requirements for physical and analytical measurements are usually much higher in batch than in continuous processes (Allen, 1977). The continuous process can be "lined out" or tuned to obtain the desired productivity and product quality, permitting considerable grace in measurement accuracy. Also, the required measurement range is usually far greater for batch vs. continuous process monitoring. For example, compare a continuous reactor train in which a reactive ingredient may undergo a 10% conversion per reactor with a batch reactor in which the conversion often exceeds 90%. If one wishes to monitor the conversion in each reactor to within 1%, the continuous reactor monitor would require a scale of only ±5% about the operating point, resulting in a required accuracy of 10% of full scale (10% conversion). The batch reactor monitor, on the other hand, would require a 0-100% scale with an accuracy of 1% of full scale.

The same wide operating ranges and nonstationary behavior that cause difficult sensor problems in batch reactors also influence control system design. For example, during process identification, the linearized approximate models that are so common for continuous processes are not applicable, because of the lack of a nominal steady state about which the process can be linearized. As a result, nonlinear models based on actual values rather than deviation variables must be used. Similar difficulties are encountered during controller design, where variable-gain controllers are often required to compensate for time-varying process dynamics. The controller design is further complicated by asymmetric penalties, such as in composition control where the formation of unwanted byproducts is irreversible. This contrasts to continuous processes, where upsets eventually wash out of the system and the process can return to the desired steady-state.

III. DEVELOPMENT OF BATCH PROCESS CONTROL SYSTEMS

Most of the batch process control systems that have been reported on use a combination of sequence control and direct digital servo and regulatory control (DDC) (e.g., Bacher and Kaufman, 1970; Balzer, 1982; Fraade, 1980; Johnson, 1977; Ketchum et al., 1979; Robson, 1981).

Sequence control coordinates operation of the equipment required to execute the various steps of a batch process recipe. Most of the reported applications of sequencing control to batch chemical plants compare the throughput and product quality improvements observed in new "grass roots" facilities with those of older plants that used a combination of manual and analog control (e.g., Baker and Weber, 1969; Law, 1977; Smith, 1977; Spellman and Quinn, 1975; Tsai and Lane, 1982; Uhlig, 1977). These comparisons are often misleading, because process design changes and improved process recipes are often introduced along with computerized operation. The simultaneous introduction of many improvements in new computerized plants makes it difficult to determine which of the reported benefits actually resulted solely from computerized operation and to estimate the potential benefits of retrofitting computers into an existing plant. The difficulty of quantifying the benefits of sequence control leads to two common misconceptions.

The first misconception is that computerized repetition of a recipe automatically results in significantly less process and product variability. If the major sources of variability are raw material quality, variations in process utilities, or lack of available downstream equipment, process and equipment design changes will be needed to obtain the desired variance reduction. For example, the common equipment modifications often mentioned as part of sequence control systems include additional process sensors (e.g., load cells, metering devices for reagent additions, extra temperature, pressure, and flow measurements) and elimination of process bottlenecks (e.g., variations in process utilities such as cooling water temperature or steam supply pressure, lack of available downstream equipment which often requires the installation of intermediate storage between process units) (Clifton, 1983; Ghosh, 1980; Gidwani, 1982).

In addition, reducing the variance for a particular parameter, such as process yield, does not necessarily increase the average yield. Our experience suggests that if a parameter such as yield is normally distributed, the computer-controlled process tends to produce the same average value of the process parameters but with a smaller variance. This implies that if the variance of the process yield is reduced, the probability of making either a high- or low-yield batch is reduced equally.

The second misconception is that an established process recipe for a manually controlled plant is easily converted to computer software (Sparrow and Rippin, 1974). The difficulty with this idea is that an experienced operator can quickly observe and react to process information that either may be difficult for a computer to sense or may require special software to respond properly. Typical examples are foaming and/or solid entrainment during reflux and distillation operations, sloppy interfaces during liquid/liquid extractions, and differences in filtration rates for solid products.

These examples highlight the strong interaction between process analysis/design and software development, which is needed to achieve all of the potential benefits of sequence control. In the rapidly changing, highly flexible environment of the multiproduct batch chemical plant, however, this interaction is not an easy matter.

For example, in the design of a continuous chemical manufacturing process, the process chemistry is often well defined, and the process design and analysis can consume many man-months or several man-years of effort. For the flexible, multiproduct batch plant, however, the process chemistry is constantly changing and may not have been envisioned during the original equipment design. As a result, process analysis and software development is an ongoing, often iterative effort. Since the volume and even the lifetime of each batch chemical are far smaller than those of most continuously produced chemicals, the economic impetus for determining the kinetics of every batch reaction is often not present. This means that in controlling batch reactors, often the greatest challenge is the development of a strategy that requires a minimum of kinetic knowledge and consequently can be applied to many batch processes.

In addition to sequencing control, most computerized batch chemical plants also have provisions for direct digital servo and regulatory control (DDC). These capabilities are particularly useful when different control strategies are needed to account for different unit operations being done in the same process equipment. The DDC capabilities are also essential for implementing model-based control schemes to account for the unique nonlinear and time-varying dynamics of batch processes.

IV. APPLICATION OF MODEL-BASED CONTROLLERS TO THE CONTROL OF EXOTHERMIC REACTIONS IN BATCH REACTIONS

Although many different unit operations (reaction, distillation, extraction, crystallization, etc.) are commonly used in batch chemical synthesis, we will limit our discussion to the control of reactors. In the control of these vessels, there are two related but distinct challenges.

(a) The first (discussed in this section) is the control of vessels with highly exothermic reactions. Here safety-related concerns such as adiabatic runaway are primary, followed by productivity concerns if the heat-release rate is far from constant. Composition control, if important, is usually handled by operation under isothermal conditions, since small-scale product development and scale-up work is almost always done isothermally. In these cases, there is typically a poor idea of the reaction kinetics and mechanism, and only the realization that conventional temperature controls are inadequate as the product is scaled up.

(b) The second control challenge (discussed in Section V) is that of controlling the batch reactor to optimize a composition function such as yield or selectivity. These optimal control problems have received considerable attention in the literature and are restricted to cases where the kinetics of the desired and undesired reactions are known. For these reactions, stability and safety are rarely important issues, since composition control includes the subproblem of reaction rate control.

Our discussion of the control of exothermic reactions is divided into three parts. The first discusses the stabilization and control of isothermal batch reactors. The second extends this to nonisothermal operation so that the productivity might be increased. The third discusses achieving productivity gains through isothermal semibatch operation.

1. Isothermal Batch Reactor Control for Exothermic Reactions

Because temperature has a highly nonlinear influence on the rate of many chemical reactions, temperature control in exothermic batch reactors has long been recognized as an excellent candidate for the application of model-based process controllers (Anonymous, 1984; Hopkins and Alford, 1973; Lake, 1977; Luyben, 1975b; Marroquin and Luyben, 1972). We will use the specific example of regulatory temperature control for an exothermic, autocatalytic polymerization to illustrate the four main parts of the problem: development of the necessary process models, synthesis of a good controller for a pilot plant reactor, robustness of this controller to perturbations in the reaction chemistry and model parameters, and scale-up of the control strategy to larger reactors.

1.1 Development of process models. To design good temperature controllers for exothermic batch reactors, three models are important (Guy, 1982; Luyben, 1973). First, an equipment model describes how the manipulated variables, typically steam and cooling water flow rates, affect the reactor temperature. Second, a kinetic model is used to describe the influence of the reactor temperature on the reaction chemistry. This model is also useful for determining temperature trajectories that optimize selectivity or productivity. Finally, a calorimetric model is used for the influence of reaction rate on the reactor temperature. This model predicts the major disturbance acting on the temperature control loop.

For most problems, these three models do not exist, yet it is through these that one is able to understand why the batch-reactor temperature-control problem is so challenging (Luyben, 1975a). The following simple analysis shows how these effects combine to give open-loop unstable systems.

The energy balance for the reactor contents in the absence of a chemical reaction is:

$$MC_p \frac{dT_c}{dt} = -UA_j(T_c - T_j) \tag{1}$$

(model 1 - equipment model)

For controller design purposes, it is convenient to use Laplace transforms and express Eq. (1) in transfer-function form:

$$\frac{T_c(s)}{T_j(s)} = \frac{1}{\tau s + 1} \tag{2}$$

where

$$\tau = \frac{MC_p}{UA_j}$$

Since the root of the transfer-function denominator, the process pole, is negative, the batch-reactor heat-transfer control loop is always stable in the absence of a chemical reaction (Coughanowr and Koppel, 1965).

If the reactor energy balance [Eq. (1)] is modified to include an exothermic reaction, the heat-transfer control loop is no longer guaranteed stable. For the example of a simple first-order reaction, the kinetic model (model 2) is given by Eqs. (3) and (4), and the calorimetric model (model 3) is given by Eq. (5).

$$A \xrightarrow{k_1} B \qquad (3)$$

$$r = -\frac{da}{dt} = k_1 a \qquad (4a)$$

$$k_1 = k_{10} \exp[-E_a/R_g T_c] \qquad (4b)$$

$$Q = (-\Delta H) v r \qquad (5)$$

The batch-reactor energy balance [Eq. (6)] is obtained by combining the three models in Eqs. (1), (3), (4), and (5)

$$MC_p \frac{dT_c}{dt} = -UA_j(T_c - T_j) + (-\Delta H) v r_0 \exp\{\frac{\gamma T_{c0}(T_c - T_{c0})}{T_c}\} \qquad (6)$$

where r_0 is the reaction rate at temperature T_{c0} and γ is $E_a/R_g T_{c0}^2$.

To perform a stability analysis, Eq. (6) is linearized about T_{c0}. Define T_c' to be the deviation of T_c from T_{c0} and T_j' to be the deviation of T_j from the jacket temperature required to maintain steady state. The transfer function becomes

$$\frac{T_c'(S)}{T_j'(S)} = \frac{1/\tau}{S + 1/\tau - \frac{(-\Delta H)vr_0\gamma}{MC_p}} \tag{7}$$

Since the root of the transfer-function denominator can be positive for certain combinations of the reaction parameters, a feedback controller is required to make the heat-transfer loop stable. For example, a proportional controller with a gain of K_p could be used to make the closed-loop pole more negative (Shinskey, 1979).

$$S = \frac{-1}{\tau}(1 + K_p) + \frac{(-\Delta H)vr_0\gamma}{MC_p} \tag{8}$$

This simple linear model shows that as r_0 increases, K_p must also increase to maintain good closed-loop performance. Hence, controllers that are tuned by using step tests in the absence of reaction may perform poorly with the reaction present. For autocatalytic reactions, where r_0 increases dramatically as the reaction progresses, the fixed-gain control system becomes unstable.

The simulation and analysis of industrial reactor/heat-transfer configurations requires the same three models, equipment, kinetic, and calorimetric, as our simple first-order example. The deterministic equipment models for these systems, however, are often complex and may require model parameters that are difficult to measure or estimate (Marroquin and Luyben, 1973).

Empirical models are often used to circumvent these limitations. These empirical models are low-order (typically first- or second-order) approximations to the complete deterministic models and are usually expressed as transfer functions in either the Laplace or the discrete z domain.

Once an equipment model has been determined, the two models for the temperature/chemistry interaction (kinetic and calorimetric models) must be determined. These models are important because the price of designing and tuning controllers based only on nonreacting equipment models can be poor temperature control.

Good kinetic models, however, are usually difficult to determine for most problems of industrial interest. Our experience suggests that development of these models can require as much as a man-year of laboratory investigation and that the models can still lack robustness to impurities and procedural variations found in the plant environment. Because of these difficulties, we have elected to use a heat-release-rate estimator for the kinetic model and empirically identify the calorimetric model by simulating reactions inside the plant equipment. This, along with a rough estimate of the activation energy, effectively gives us both $((-\Delta H) v r_0 \gamma)$ and its coefficient $(1/MC_p)$ in Eq. (8).

The equations for the heat-release estimator can be developed by using the deterministic or the empirical approach. Several authors have suggested that a deterministic on-line energy balance can be used to derive an online energy-generation estimator (e.g., Beckingham et al., 1979; Gordon and Weidner, 1978; Lackmeyer and Kempfer, 1977; Nilsson et al., 1982; Valtyni et al., 1979). The difficulty with this approach can be illustrated by considering the major energy sinks for a jacketed batch reactor (Fig. 1). The reactor wall, jacket wall, and

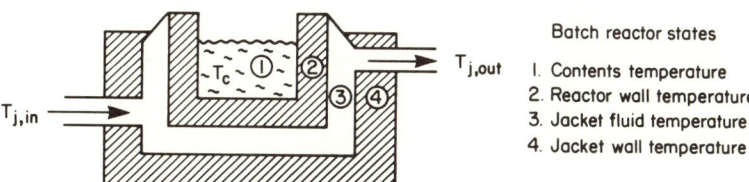

Fig. 1. Batch reactor schematic.

heat-transfer fluid are actually distributed states. If this is ignored, the following four-state lumped energy-balance model results.

Reactor-contents energy balance:

$$M_c C_p \frac{dT_c}{dt} = -U_1 A_1 (T_c - T_{w2}) + (-\Delta H) vr \qquad (9)$$

Reactor-wall energy balance:

$$M_{w2} C_{pw2} \frac{dT_{w2}}{dt} = U_1 A_1 (T_c - T_{w2}) - U_2 A_2 (T_{w2} - T_j) \qquad (10)$$

Circulating-fluid energy balance:

$$M_j C_{pj} \frac{dT_j}{dt} = U_2 A_2 (T_{w2} - T_j) - U_3 A_3 (T_j - T_{w4}) + F_j \rho_j C_{pj} (T_{jin} - T_j) \qquad (11)$$

Reactor-jacket-wall energy balance:

$$M_{w4} C_{pw4} \frac{dT_{w4}}{dt} = U_3 A_3 (T_j - T_{w4}) \qquad (12)$$

Four main difficulties are encountered in the solution of this system of coupled differential equations. The first is that the reactor-wall and reactor-jacket-wall temperatures (T_{w2} and T_{w4}) are unobserved and must be estimated from other available measurements. The second problem is the modeling uncertainty that results from the difficulty of accurately estimating the various overall heat-transfer coefficients. Third, good numerical algorithms are required for real-time computation of the temperature derivatives in Eqs. (9)-(12). Finally, any time delays between the fluid entering and leaving the reactor jacket ($T_{jin} - T_j$) must be included in the online energy balance.

To overcome these difficulties, the model for the heat-release estimator was developed empirically by using a discrete-time transfer-function model of the reactor. This method was based on the fact that the reactor-contents temperature was a function only of the jacket-inlet temperature and the heat generated by the exothermic reaction. With steam injection used to simulate the heat generation, the identification procedure for our 40-liter pilot plant reactor involved: filling the reactor with 30 liters of water, turning on the DDC temperature controllers, introducing square pulses of steam into the reactor contents with varying height and duration, using a dip tube to remove the steam condensate, monitoring the reactor-contents and jacket temperatures.

Figure 2 shows a typical example of the steam input sequence and the resulting reactor-contents and jacket-inlet temperature profiles.

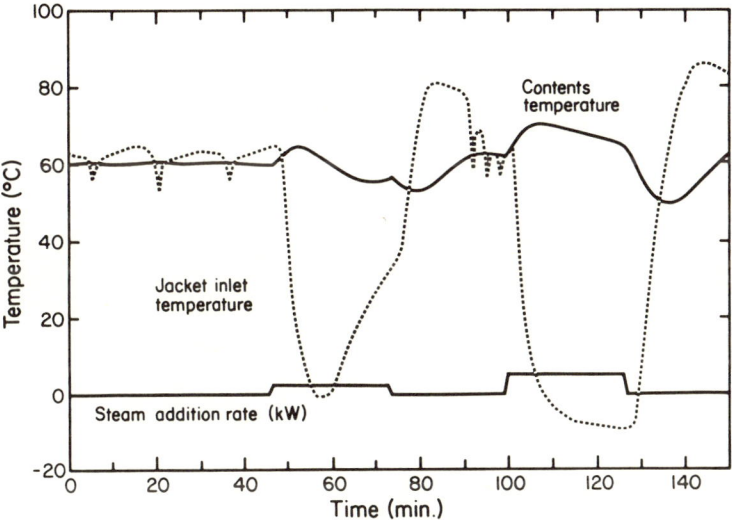

Fig. 2. Reactor identification experiment.

The experimental input/output data were sampled at 1-min intervals, and the discrete time transfer functions in Eq. (13) were identified using time series analysis methods (Box and Jenkins, 1976; Liu and Hudak, 1983).

$$T_c = \frac{(0.0125B + 0.0219B^2)}{(1 - 0.9671B)} T_{jin}$$
$$+ \frac{(0.2793B + 0.0840B^2)}{(1 - 0.9671B)} Q - 5.076 + \text{Noise} \quad (13)$$

where B is the backshift operator. These simple first-order transfer functions were adequate to simulate the experimental data.

The heat-generation estimator was obtained by rearranging Eq. (13) to obtain

$$\hat{Q}(k) = 2.637 T_c(k+1) - 2.550 T_c(k)$$
$$- 0.03296 T_{jin}(k) - 0.05775 T_{jin}(k-1) \quad (14)$$
$$+ 0.4403 - 0.2215 \hat{Q}(k-1) + \text{Noise}$$

Figure 3 illustrates the performance of the heat-generation estimator for tracking a series of steam injection steps. A careful inspection

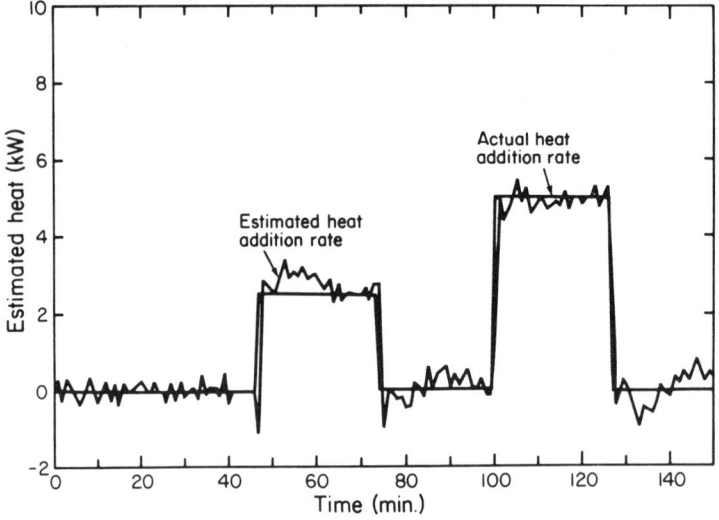

Fig. 3. Raw estimate of the heat-release rate inside the reactor.

of Fig. 3 reveals two important features of the heat-estimation algorithm. First, there is an implicit pure delay of one sampling period between the measurement of the contents temperature and the estimation of the heat generation. This is evident at the edges of the step changes. Second, the estimator is corrupted with noise from neglecting higher-order dynamics in the transfer-function modeling. Since the frequencies of the signal and the noise are so different, the high-frequency noise was easily eliminated simply by using a discretized second-order Butterworth filter (Oppenhiem and Schafer, 1975). Figure 4 compares the raw and filtered heat-release estimates.

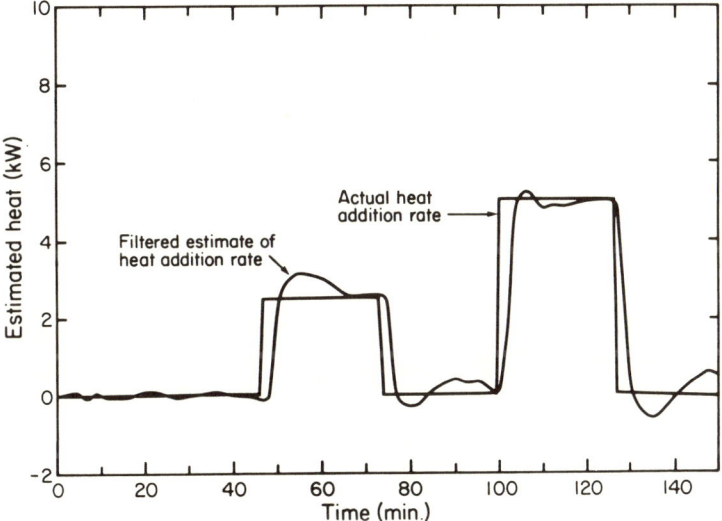

Fig. 4. Filtered estimate of the heat-release rate inside the reactor.

The period of delay can be removed by simple extrapolation, once the heat-release rate curve is smoothed. Alternatively, the reaction rate observer, noise filter, and predictor could have been formulated by using a Kalman filter (Hamilton et al., 1973). This would make use of the structure of the noise model identified with the process to improve the estimates. Future observers will probably be done this way.

At this point a first iteration of the modeling process is complete. We have models relating manipulated variables and the heat-release rate to the contents temperature [Eq. (13)]. The third model, the chemistry-dependent model relating temperature, concentration, and time to reaction rate, has been replaced by a reaction rate estimator [Eq. (14)].

1.2 <u>Controller design</u>. Once the appropriate process models have been determined, one can proceed with the controller design. For exothermic reactions, the heat release due to reaction is usually the major temperature disturbance affecting the system. Since we estimate this quantity [Eq. (14)], it is possible to design a feedforward controller to reject most of this disturbance (Jutan and Uppal, 1984; Wu, 1985) and a nonlinear feedback controller to correct for "plant-model mismatch".

(1) Feedforward Controller Design

To design a feedforward controller, the dynamics of the process must be inverted. To avoid noninvertability problems, a steady-state model of the process is used. This is derived from the transfer-function model in Eq. (13).

$$T_c = 1.046 T_{jin} + 14.08 \hat{Q} - 5.076 \quad (15)$$

This simple expression agrees well with operator experience for our 40-liter reactor. For example, to achieve a contents temperature of 60°C in the absence of an exothermic reaction requires a jacket temperature controller setpoint of 63°C. Equation (15) also provides a quick calculation of the maximum heat-removal capacity for this reactor, about 6 kW. The feedforward controller necessary to counter the heat release disturbance from the reaction is

$$\Delta T_{jin,FF} = -13.47 \hat{Q} \quad (16)$$

(2) Feedback Controller Design

The main challenge of the feedback controller design is to account for the change of the process transfer function with increasing reaction rate [Eq. (8)]. This effect can be illustrated by replacing the heat-generation rate in Eq. (13) with the linearized form of the Arrhenius rate expression

$$Q = Q_0(1 + \gamma T_c') \tag{17}$$

$$\frac{T_c'}{T_{jin}'} = \frac{B(0.0125 + 0.0219B)}{1 - (0.9671 + 0.2393Q_0\gamma)B - 0.0840Q_0\gamma B^2} \tag{18}$$

where Q_0 is the heat release rate at the reference conditions. For this particular reactor, the process becomes unstable for $Q_0\gamma > 0.07$. This corresponds to a heat release rate of only 0.80 kW at 60°C with an activation energy of 20 kcal/mole. If the reactor contents temperature were controlled with a proportional feedback controller with gain K_p, the characteristic equation for the closed-loop process would be given by

$$1 - (0.9671 + 0.2793Q_0\gamma - 0.0125K_p)B$$
$$- (0.0840Q_0\gamma - 0.0219K_p)B^2 = 0 \tag{19}$$

Figure 5 shows the magnitude of the process poles as a function of $Q_0\gamma$. These results agree with the observation that a proportional temperature controller with a gain between 2 and 4 gives good performance to setpoint changes in the absence of a chemical reaction. When the chemical reaction is included, however, a controller with a gain of 3 would be unstable for a heat-release rate of 2.5 kW (T_{c0} = 60°C, E_a = 20 kcal/mole). Hence, the controller gain must increase as the heat-release rate increases in order to maintain stability and good control. The rate of change of the controller gain would be a function of the activation energy for the reaction. The following example illustrates the application of the feedforward/feedback controller design.

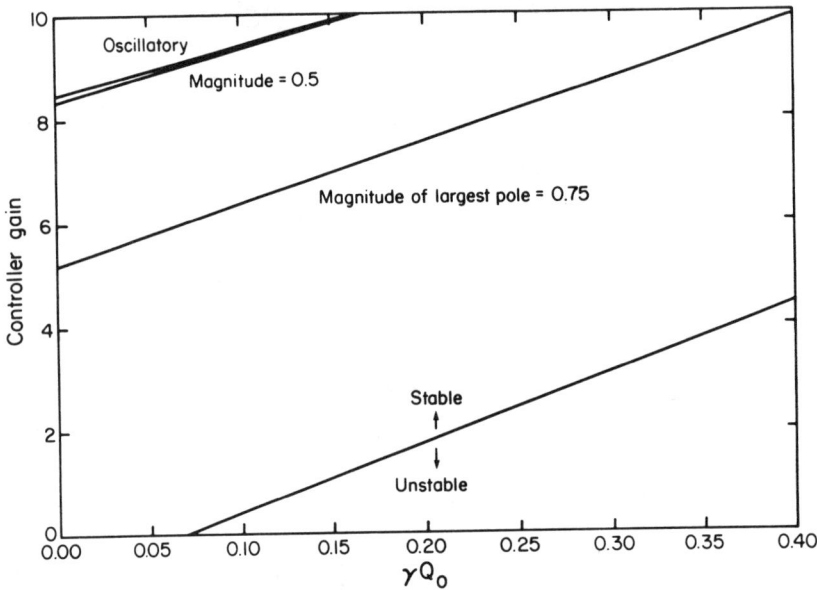

Fig. 5. Closed-loop pole location as a function of controller gain and the product of activation energy times reaction rate.

(3) Experimental Results

The feedforward/feedback controller designs were tested on the suspension polymerization of methyl methacrylate initiated by azobisisobutylronitrile. This reaction was selected because of its severe autoacceleration, high activation energy, and induction period of unpredictable duration. The experiments were performed in a 40-liter glass-lined jacketed reactor, using the experimental conditions listed in Table 1. The first experiments involved polymerization under classical feedback control. The controller configuration used a PI contents-temperature controller cascaded to a PID jacket temperature

TABLE 1 Recipe for Suspension Polymerization in a 40-Liter Reactor

8.30 kg monomer
165 g surfactant
41.0 g azobisisobutylronitrile (AIBN)
19.3 kg water

Procedure:

Charge cold monomer, surfactant, and initiator and dissolve at 10°C.
Charge water.
Evacuate four times.
Circulate through high shear pump for 30 min.
Raise setpoint to 60°c.

controller. Both controllers were tuned to get good response to setpoint changes. Figure 6 shows the reactor contents temperature profile along with the estimated heat generation rates observed during the polymerization experiments. The data in Fig. 6 have four interesting features:

(a) The reactor-contents temperature exceeded its setpoint by 18°C.

(b) The maximum estimated heat-release rate was 12 kW, which is greater than the cooling capacity of the vessel (6 kW). This maximum heat-release rate is greater than the value that would have been predicted based on isothermal operation.

(c) The jacket inlet temperature saturates at its lowest allowable value.

(d) The highest heat-generation rate occurs toward the end of the reaction. As a result, when the reaction is complete, the jacket inlet temperature is at its minimum value, and the contents temperature will undershoot the setpoint. This is of little consequence, because isothermal operation is important only while the reaction is proceeding.

In the second set of experiments, the heat-release estimate was used in a feedforward controller to reduce the reactor jacket-inlet temperature according to Eq. (16). A single-step extrapolation based on the two previous measurements was used to compensate for the delay between the actual and filtered heat-release rates. Figure 6 also shows the results of the feedforward controller experiments.

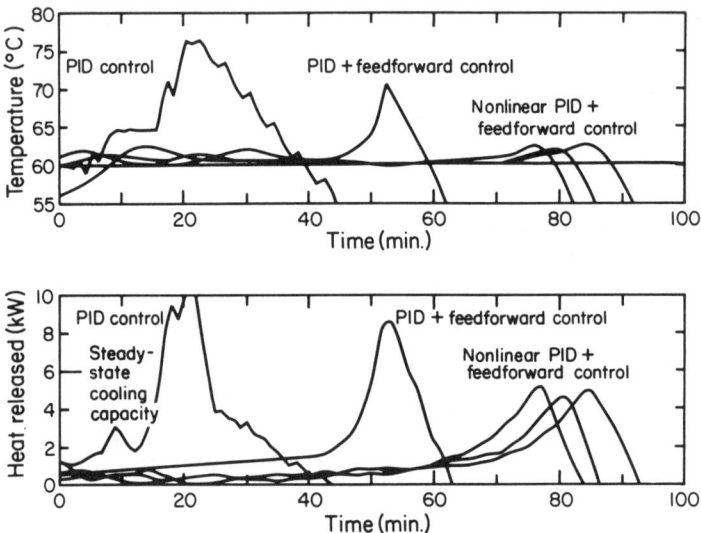

Fig. 6. Suspension polymerization of methyl methacrylate in 40 L batch reactor under various control schemes.

The final set of experiments combined the feedforward controller with an adaptive-gain feedback controller. The adaptive gain controller ensures stability of the feedback portion of the control loop with changes in the load disturbance, heat-generation rate. The integral gain was held constant at 0.01, and the proportional gain-adaptation rule was

$$K_p = 3 + 2\hat{Q} \tag{20}$$

where the value for the heat-release-rate coefficient comes from the slope of the stability boundary shown in Fig. 5. These results are also shown in Fig. 6. Note that only for the runs with both the adaptive feedback controller and the feedforward controller is the temperature overshoot within acceptable limits. These are also the only runs for which the heat-release rate is within the heat-removal capability of the reactor.

1.3 Evaluation of controller robustness. Because batch reactors do not have long periods of steady-state operation with the luxury of time for online controller tuning, and exothermic reactions have the potential for dangerous adiabatic runaways, the design engineer is often challenged with assessing control-system robustness before actual implementation (Hugo, 1980). For exothermic reactions, the most frequently asked robustness question is "what is the sensitivity of a given control system to variations in reaction chemistry?"

To test the robustness of our feedforward/feedback temperature controller, we conducted suspension polymerization experiments at the same reaction conditions, this time adding 2% ethylene glycol dimethacrylate as a crosslinking agent to the methyl methacrylate. The crosslinker causes a substantially larger gel effect, autoacceleration of the reaction rate, and consequently a more severe load disturbance. Figure 7 shows the reactor-contents temperature profile observed in these experiments, and Fig. 8 compares the heat-generation-rate profiles for the polymerizations with and without the crosslinking agent. These results indicated that the feedforward/feedback temperature controller was robust to variations in the load disturbance.

1.4 Scale-up to larger reactors. The question of control system scale-up from small pilot-plant reactors to larger production vessels is a special type of robustness issue. In this case, one is interested in tuning the process controllers and guaranteeing their safe operation before introducing a chemical reaction. These questions can be easily answered by using the following modifications to our steam-injection experiments:

Obtain a heat release profile for a reference temperature either from dynamic simulations or from experiments in smaller reactors.

Correct the reference profile for measured deviations of the contents temperature from the reference temperature by using an Arrhenius temperature dependence.

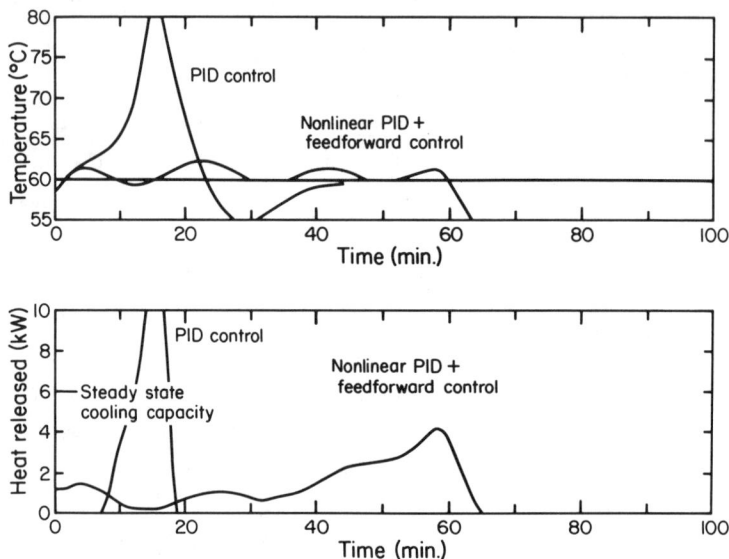

Fig. 7. Suspension polymerization of methyl methacrylate and ethylene glycol dimethacrylate in a 40 L batch reactor under various control schemes.

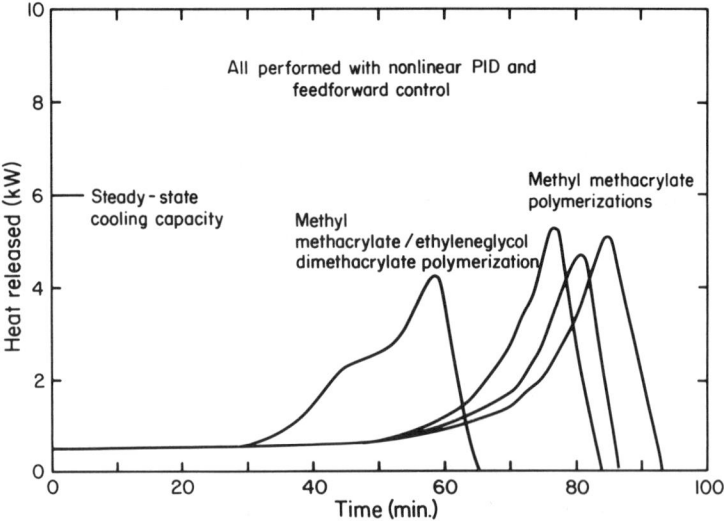

Fig. 8. Comparison of the heat-release-rate profiles with and without the crosslinking agent.

Use steam-injection experiments to identify the dynamic model for the larger tank, and determine the maximum heat removal capability of the jacket [cf. Eq. (15)].

Scale up the recipe from the smaller reactor, based on the ratio of the heat-removal capabilities of the large and small tanks. Test the proposed control agorithm on the large tank, using a real-time computer program to open and close the steam-injection valve according to the predicted heat-release profile, opening and closing the steam-injection valve additional amounts to simulate the effect of the observed deviations of the temperature from the reference temperature.

Finally, test the controller with the actual chemical reaction.

This procedure was used to scale up the suspension polymerization of methyl methacrylate to an 80-liter reactor based on reaction-rate data obtained from experiments in a 40-liter reactor. Figure 9 illustrates the experimental contents-temperature and heat-release rate profiles from the 80-liter reactor. The large tank had much better heat-removal capability than the small tank because the small tank was

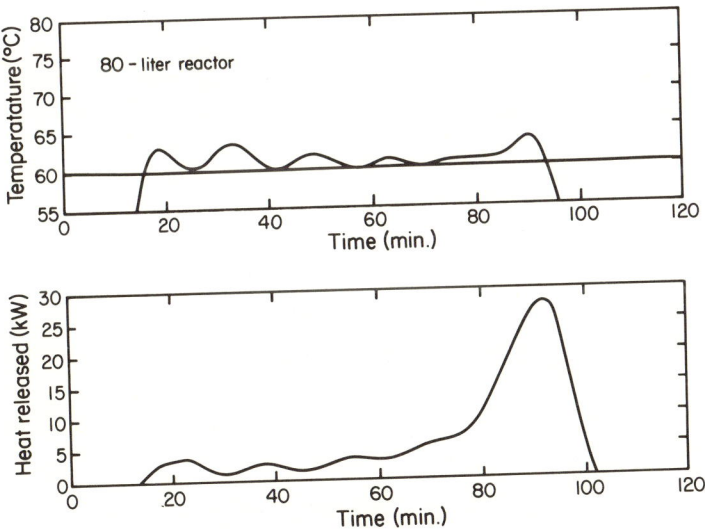

Fig. 9. Suspension polymerization of methyl methacrylate in a 80 L batch reactor under feedforward and nonlinear feedback control.

glass lined and the large tank was not. The heat-removal capability of the large tank was estimated at 35 kW. Hence it was possible to increase the monomer charge beyond the ratio of the tank capacities, complicating the task of altering the recipe for the larger tank. Note, however, that the feedforward/nonlinear-feedback control system successfully maintained almost isothermal operation in the larger reactor, with the maximum heat-release rate using 80% of the maximum cooling capacity.

2. Nonisothermal Batch Reactor Control for Exothermic Reactions

Although our discussion of regulatory temperature control has focused on isothermal reactor operation, nonisothermal temperature trajectories are a common means of optimizing the throughput of many batch reactors (Clarke and Gawthrop, 1981; Horak and Jiracek, 1983; Kiparissides and Shah, 1983). This is because the productivity is limited by the maximum heat-generation rate, and this typically occurs for only a short time. For exothermic reactions, the temperature trajectories are often calculated to maintain a constant reaction rate, which is within a specified safety limit of the maximum heat-removal capacity of the reactor. Figure 10 illustrates that, for simple first- and second-order reactions, significant throughput advantages result from constant-rate operation.

The constant-rate temperature trajectories are easily calculated for a simple rate expression, but more complex rate equations require experimental determination of the trajectories or the solution of a variational optimization problem (Millman, 1968). The temperature profiles are then implemented as a series of open-loop setpoint changes in the reactor-contents temperature controller. A major problem with this open-loop strategy is that induction effects in the chemical reactions introduce uncertainties in the timing of the setpoint changes. An alternative approach to maintaining constant-rate operation in the face of unknown or uncertain kinetics is:

(1) Obtain a real-time measurement of the reaction rate or use an estimator such as Eq. (14).

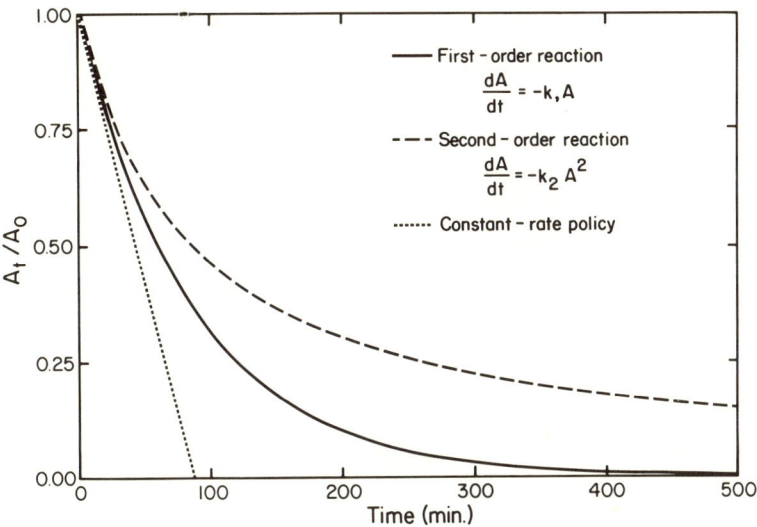

Fig. 10. Demonstration of the reduction of the batch-reaction time possible when a constant-reaction-rate policy is implemented.

(2) Compare this reaction rate with the desired value and compute an error.

(3) Use an appropriate control agorithm to adjust the reactor temperature setpoint, based on the process error.

The simplest constant-rate control agorithm is based on a Taylor-series expansion of the Arrhenius expression for the reaction-rate equation. For arbitrary kinetics, the reaction rate expands as [cf. Eq. (17)]

$$r(T_c) = r(T_{c0})(1 + \gamma(T_c - T_{c0})) \qquad (21)$$

where γ contains an estimate of the activation energy. This immediately suggests a controller of the form

$$T_{c,sp} = T_c + K_p(r_{sp} - \hat{r}) \qquad (22)$$

where

$$K_p = \frac{1}{\hat{\gamma r}} = \frac{R_g T_c^2}{E_a \hat{r}} \qquad (23)$$

Equation (21) is actually a nonlinear proportional controller for the reaction rate. The gain depends explicitly on the contents temperature and the observed reaction rate.

To apply the constant-reaction-rate controller to solution polymerization experiments, we used the gel-effect model of Balke and Hamielec (1973) to include the effect of conversion on the apparent activation energy

$$\gamma = \frac{2E_p + E_d - E_t}{2R_g T_c^2} - 0.1082 x_m f_m + 0.0785 (x_m f_m)^2 \qquad (24)$$

where E_p, E_d, and E_t are the activation energies of polymerization, initiator decomposition, and chain termination, respectively, x_m is the conversion of monomer, and f_m is the initial monomer fraction. The algorithm now not only uses the estimate of the instantaneous reaction rate, but also the integral of the rate to compensate for the autoacceleration or conversion dependence of the reaction rate.

3. Isothermal Semibatch Reactor Control for Exothermic Reactions

Nonisothermal operation as discussed in the previous section is a method of increasing productivity by reducing batch time. The disadvantage with the method, however, is that different products may be produced under a range of temperatures. Also, since most products are developed by using isothermal recipes, this nonisothermal operation may be unacceptable. An alternative method of increasing the productivity of the batch reactor is to increase the amount of material produced per batch. By metering in the limiting reagent, the heat-release rate can be kept within the heat-removal capability of the equipment, and product can be prepared isothermally. The large advantage with semibatch operation comes when the feed to the reactor is cold; hence the effective heat-removal capability is increased. This has the potential to dramatically increase the productivity of the reactor, since in the exothermic reaction considered here, the reactor is heat-transfer limited.

Several factors make this problem much more difficult than the design of a temperature controller for an exothermic batch reactor (Kladko, 1971):

The control policy must satisfy coupled mass- and energy-balance constraints.

The control objective function is more complex and includes both throughput and safety constraints (batch time and heat removal).

The reactor heat-removal capacity varies as a function of time.

The reactor can exhibit nonminimum phase behavior, i.e., stopping the cold feed can result in tremendous overshoot in the temperature before the reactor temperature decreases due to the decrease in reaction rate.

Once reactant is added, there is no way to remove it.

The objective is to develop an online control agorithm that results in optimum productivity. As before, we are interested in solutions that require a minimum of kinetic information, since this information rarely exists. If good kinetic information existed, one would formulate and solve the problem as a minimum-time variational-optimization problem for the open-loop feed-rate profile. This trajectory could then be used as setpoints to flow controllers.

This optimal control approach can also be used in the case when kinetics are unknown. The coupled mass and energy balances for the reactor are

$$\frac{dA}{dt} = -R + Fa_f \tag{25}$$

$$\frac{dv}{dt} = F \qquad (26)$$

$$v\rho C_p \frac{dT_c}{dt} = -F\rho C_p (T_c - T_f) - UA_j(T_c - T_j) + (-\Delta H)R \qquad (27)$$

where A is the quantity of the limiting reagent in the reactor, R is its rate of disappearance due to reaction, and F is the feed flow rate with concentration a_f. Along with the initial conditions for v and A, there are also final conditions for both, namely, that the tank is full and that a specified fraction of A be converted.

This problem is linear in the control F; hence the minimum-time isothermal operating policy (maximum throughput) involves operating at one of the problem constrains:

$$F \geq 0 \qquad (28)$$

$$F \geq \frac{S_1 R}{a_f} - S_0 \qquad (29)$$

where

$$S_0 = \frac{UA_j(T_c - T_{jmin})}{\rho C_p (T_c - T_f)} \qquad (30)$$

$$S_1 = \frac{(-\Delta H)a_f}{\rho C_p (T_c - T_f)} \qquad (31)$$

Here T_c is the desired isothermal contents temperature and T_{jmin} is the minimum reactor jacket temperature.

Physically, S_0 represents the feed flow rate so that the feed stream removes as much sensible heat as does the jacket, and S_1 represents the adiabatic temperature rise divided by the contents-to-feed temperature difference. Only when $S_1 < 1$ does semibatch operation increase the effective heat-removal capability of the reactor.

The minimum-time solution will be composed of three sections. The center section, which is usually the longest, is spent at the heat-removal constraint [Eq. (29)]. On the state plot (Fig. 11) this is the region for t_{0f} to t_1 along the optimal solution curve. Thus the heat-generation rate is just balanced by the maximum jacket cooling and the feed flow. This phase ends when the flow rate on the constraint goes negative. Thus begins the batch phase at constraint 28, in which the tank is full and the conversion increases to the desired level. This is the section from t_1 to t_f in Fig. 11. The first section involves filling the reactor as quickly as possible to reach the heat-removal constraint. This is the region from t_0 to t_{0+} and depends on the specific initial conditions. Note that for initial conditions above and to the right of the optimal heat-removal constraint, no isothermal path exists to the desired final conditions.

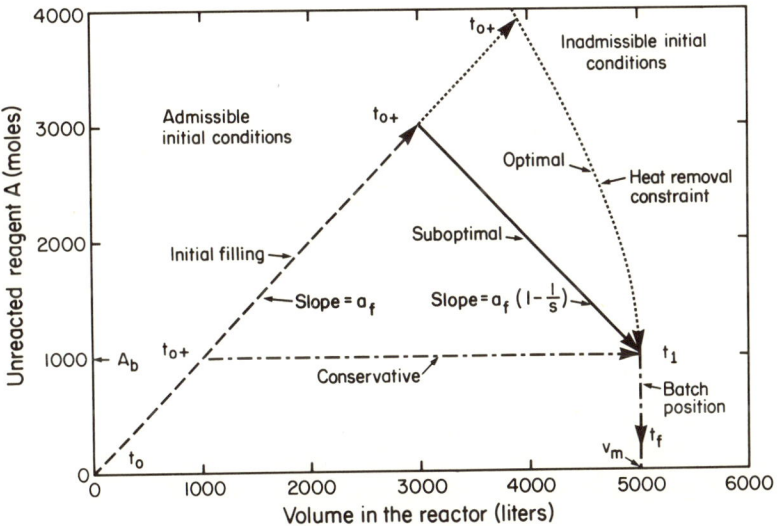

Fig. 11. State plot for semibatch reactor operation.

When the kinetics are known, the volume and composition during each of these sections can be solved for easily, starting from the final conditions. If the kinetics are unknown, but instead a reaction-rate estimate is available, a suboptimal policy can still be developed, which in most cases closely approaches the optimal answer. In addition to the reaction-rate estimator, it is necessary to know the maximum quantity of material that can be reacted isothermally using only the jacket for cooling. Usually this quantity A_b is available from batch experience or calorimetry. The difficulty with operating at the heat-removal constraint [Eq. (29)] is that it contains no information about the finite volume of the reactor. To solve this problem, integrate Eq. (29) over the remaining fed-batch time τ_{fb}, with final conditions $A = A_b$ when $v = v_m$ at $\tau_{fb} = 0$.

The optimal fed-batch policy is defined by

$$A \leq A_b + (v_m - v)a_f(\frac{1}{S_1} - 1) + \frac{S_0 a_f \tau_{fb}}{S_1} \quad (32)$$

The quantity of unreacted material in the reactor, A, can be estimated by integrating the reaction rate and feed flow rate according to Eq. (25). Although Eq. (32) respects the finite volume of the tank, the remaining fed-batch time τ_{fb} is not known with certainty. Setting $\tau_{fb} = 0$ results in a conservative suboptimal policy.

$$\hat{A} \leq A_b + (v_m - v)a_f(\frac{1}{S_1} - 1) \quad (33)$$

This suboptimal policy allows one to increase productivity by increasing the reaction rate beyond the heat removal capability of the jacket while still guaranteeing that isothermal operation can be maintained.

To implement this suboptimal policy, adjust the feed flow rate to try to maintain the equality in Eq. (33). If the desired conversion has not been reached by the time the reactor is full, the reaction continues batchwise until it is reached.

The following example demonstrates these results. The objective is to achieve 95% conversion of 5000 moles of A to product using a

5000-liter tank and 1 M starting and feed solutions. Suppose the reaction is first order with a time constant of 20 min. Say that S_0 is 25 liters/min and S_1 is 0.5. Under these conditions, the final batch phase begins with 1000 moles of unreacted A in the tank and continues for 27.7 min.

During the fed-batch portion of the reaction, three operating modes are considered. In the first, the reaction rate is held at a level so that the jacket is capable of removing all the released heat. This means that the feed could be discontinued at any point and isothermal conditions maintained. Under this conservative mode of operation, the reactor starts 20% full and the fed-batch portion lasts 80 min. When the optimal constraint as defined by Eq. (29) is observed, the reactor starts 78% full and the fed-batch portion lasts only 36 min. In the suboptimal policy described by Eq. (33), the reactor starts 60% full and the fed-batch portion lasts 44 min. These three cases are shown in Fig. 12.

As this example demonstrates, using the cooling capacity of the feed stream during semibatch operation can greatly increase the productivity of the reactor. It also demonstrates that there is little penalty in using the suboptimal constraint, which uses no kinetic information but needs a real-time reaction-rate estimator.

V. CONTROL OF COMPOSITION IN BATCH REACTORS

In this section, we consider the problem where several reactions are occurring simultaneously and kinetic expressions are known that accurately describe them. The objective is to manipulate the temperature or reactant addition to optimize some function of the composition (e.g., productivity, yield, or selectivity). (This contrasts with the previous section, where the kinetics were not known and the challenge was to control exothermic reactions so the heat-removal capability of the equipment was not exceeded.)

These problems are formulated as optimal control problems, with the solution being an open-loop temperature or flow-rate trajectory.

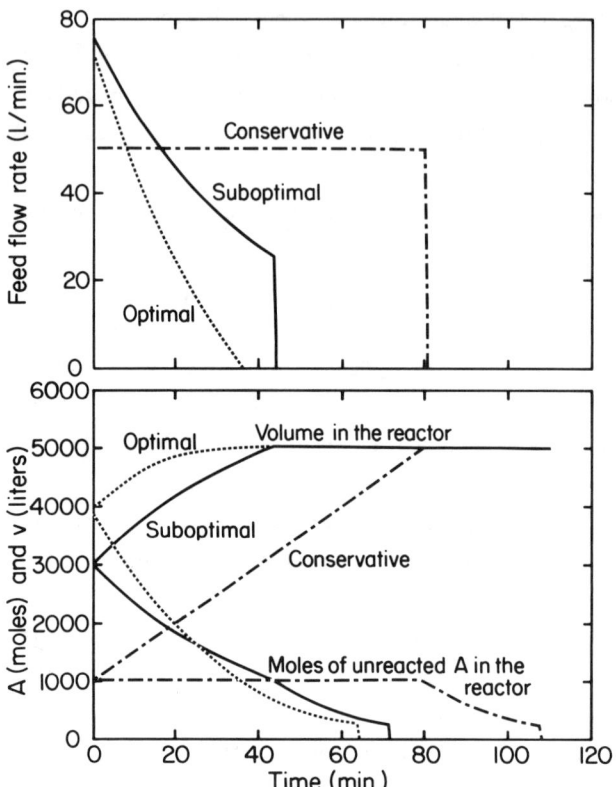

Fig. 12. Feed-flow rate, unreacted reagent concentration, and reactor volume for three methods of semibatch reactor operation.

The problem of calculating an optimal temperature profile differs significantly from that of finding optimal addition rates in that the former is a nonsingular problem, whereas the latter is almost always singular.

1. Optimal Composition Control in Batch Reactors Using Temperature Trajectories

The general form of the optimal temperature profile has been calculated for most simple series and parallel reactions. These are well summarized by Rippin (1983), who also summarized the works where the optimal temperature profiles have been calculated for more complex systems, e.g., polymerizations, batch distillations, and batch crystallizations.

2. Optimal Composition Control in Semibatch Reactors Using Feed-Rate Trajectories

This problem differs significantly from the calculation of optimal temperature profiles in that the Hamiltonian is almost always linear in the feed flow rate, making the problem singular. Simple gradient methods, which are used to solve most nonsingular problems, converge too slowly or not at all on singular problems, and hence improved methods are sought.

Three basic approaches have been proposed for the solution of singular optimal control problems: modified gradient methods, a series of nonsingular problems, and transformations. Of these three, the most promising is the gradient or modified gradient approach.

The approach of solving a series of nonsingular problems whose limit is the singular problem (e.g., Yeo, 1980) suffers because the rate of convergence to the solution of the associated nonsingular problems is strongly dependent on the magnitude of the added terms, with nonconvergence as the problem approaches singularity. These methods involve substantial calculation on a problem that is not the problem of interest.

The transformation methods involve problem-specific algebra before the problem can be solved numerically. All involve repeated differentiation of the gradient of the Hamiltonian. This gives a problem that can then be solved by using gradient-type algorithms (Soliman and Ray, 1972) or where the singular arc trajectories are solved for explicitly and a method such as quasilinearization (Aly, 1978) is used to solve the remaining problem. These methods seem appropriate for small-dimensional problems but are difficult to automate when there are several control and state variables.

The modified gradient methods, on the other hand, are relatively easy to use, even for rather large problems. These techniques include conjugate gradient and quasi-Newton methods such as Davidon-Fletcher-Powell and Broyden methods. The challenge with these methods is to handle the constraints, usually on the feed flow rates. Methods for doing this have been developed by Pagurek and Woodside (1968) and improved by Quintana and Davison (1974). The quasi-Newton algorithms seem to converge faster on singular arc problems with bounded controls (Edge and Powers, 1976); however, conjugate gradient methods can be simpler to implement and when coupled with coarse line searches are perhaps the most efficient in terms of computation time (Jones and Finch, 1984; Murty et al., 1980).

Included in the class of gradient approaches is the method where the optimal control problem is solved directly as a nonlinear programming problem by approximating the differential equations using collocation on relatively few points (Biegler, 1984). This reduces an infinite-dimensional optimization problem to a low-order problem easily solved by methods such as successive quadratic programming. For a simple nonsingular problem, Biegler showed a significant reduction in computational time compared to a conjugate method. However, no singular problems were solved.

To demonstrate the efficiency of conjugate gradient methods in solving fed-batch optimal control problems, consider the example of Young (1980). It has two parallel second-order reactions:

$$A + B \xrightarrow{k_2} C \tag{34}$$

$$2A \xrightarrow{k_3} D \tag{35}$$

The objective is to find the feed-rate profile for A that maximizes the concentration of C after 2 h. The mass balances are given by

$$\frac{da}{dt} = -k_2 ab - 2k_3 a - \frac{F}{v}(a - a_f) \tag{36}$$

$$\frac{db}{dt} = -k_2 ab - \frac{Fa}{v} \qquad (37)$$

$$\frac{dc}{dt} = k_2 ab - \frac{Fc}{v} \qquad (38)$$

$$\frac{dv}{dt} = F \qquad (39)$$

where a, b and c are the concentrations in the reactor. Initially, the reactor contains 0.5 L of 0.2 M B. The feed solution is 0.2 M A and has a maximum flow rate of 0.01 L/min. When the rate constants for both reactions are 0.5 L/(mole min), the solution has three sections (see Fig. 13). The first, which lasts for 10 min, has the feed rate at its maximum value. For the next 76 min, the feed rate follows

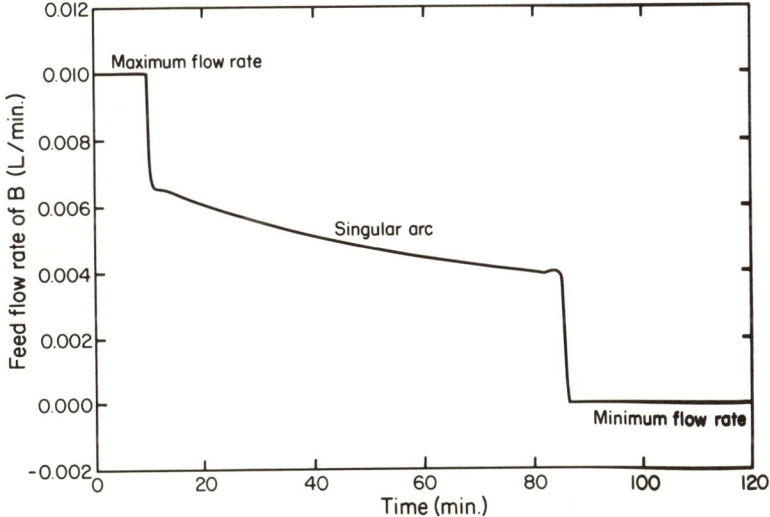

Fig. 13. Optimal feed-flow-rate trajectory.

a singular arc, i.e., the gradient of the Hamiltonian with respect to flow rate is zero, and the optimal feed-flow rate is not at either its upper or lower limit. During the final period, the flow rate is at the minimum value, zero. The associated composition profiles are shown in Fig. 14.

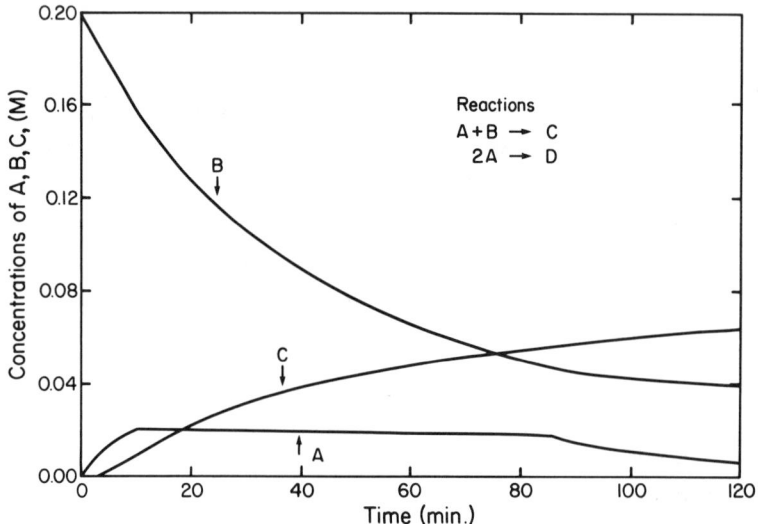

Fig. 14. Concentrations of A, B, and C that result when the optimal feed-flow-rate trajectory is followed.

To compare methods, Young solved the problem using a gradient procedure requiring 80 iterations and 160 integrations. The same solution can be achieved by using only 6 conjugate-gradient search directions and 24 integrations. This uses the Quintana and Davison (1974) method of handling the constraints on the inputs, which guarantees improvement at each step. A course line search followed by quadratic interpolation was used to get the optimal step size in each direction.

In summary, the conjugate-gradient method requires no problem-specific preparations and yet gives much enhanced convergence to the solution of the singular fed-batch problems. Having an efficient method of solution requiring a minimum of preparatory algebra is particularly important in industrial problems, where there can easily be ten mass balances and several feed streams.

3. Remaining Challenges in Composition Control

It is clear from the previous two sections that good techniques are in hand to solve for the optimal temperature and feed-rate trajectories once good process and kinetic models are known. The remaining challenge is to develop efficient identification procedures that will give the nonlinear time-varying models for these processes. Most of the identification techniques have been developed to obtain linear models for continuous processes and hence assume an underlying steady state and no time-varying features. More work must be done to develop better methods for identifying nonlinear batch reactor models and batch reaction chemistry models.

VI. SUMMARY

In summary, the increasing interest in the manufacture of low-volume specialty chemicals in batch reactors will continue to present many challenging opportunities to process control engineers. Sequence and recipe control will benefit from advances in computer hardware, such as distributed process control systems, and from the increased interaction between process design and analysis and process control (Fruit et al., 1974; Joglekar and Reklaitis, 1984; Knopf et al., 1982; Liou and Shei, 1984; Morris, 1983; Overturf et al., 1978a, 1978b; Squires and Reklaitis, 1980; Wells and Chester, 1978; Wright, 1984).

We have shown how to easily obtain a simple transfer-function model for process equipment that can be inverted and used as exothermic reaction rate observer. This can be used to track reaction progress or to solve scale-up problems. The observer also allows the design of feedforward and nonlinear feedback controllers that are easy to implement and greatly improve the quality of the control, the safety of the operation, and the maximum possible batch size. The observer has also been used to extend these ideas to nonisothermal and semibatch schemes for increasing productivity. In short, highly exothermic reactions with unknown kinetics are no longer so difficult to control and thus can be done safely on large scales with higher productivity than ever before.

Finally, composition control in semibatch reactors remains an area of major research challenges, particularly for nonlinear model identification and nonlinear controller design methods. The emerging interests in biochemical reaction engineering will undoubtedly continue to provide a stimulating list of potential composition control problems (San, 1984; Stutts, 1983). Looking ahead, we see new sensor technology (Haggin, 1984) and nonlinear controller design offering the potential for online closed-loop implementation of the optimal policies.

REFERENCES

Allen, P. (1977). Automation of batch reactors. Chem. Ind. 1977(8), 300-305.

Aly, G. M. (1978). The computation of optimal singular control. Int. J. Control, 28(5), 681-688.

Anderson, E. (1984). Specialty chemicals mixed bag for growth. Chem. Eng. News, June 4, 20.

Anonymous (1984). Temperature control for batch processes. Chem. Eng. (London), 402, April, 29-31.

Armstrong, W. S., and B. F. Coe (1983). Multi-product batch reactor control. Chem. Eng. Prog., 79, January, 56-61.

Bacher, S., and A. Kaufman (1970). Computer-controlled batch chemical reactors. Ind. Eng. Chem. 62(1), 53-61.

Baker, W. J., and J. C. Weber (1969). Direct digital control of batch processes pays off. Chem. Eng. (N.Y.), 76(27), 121.

Balke, S. T., and A. E. Hamielec (1973). Bulk polymerization of methyl methacrylate. J. Appl. Polym. Sci., 17, 905-949.

Balzer, D. (1982). Process automation in the chemical industry. Chem. Tech. (Leipzig), 34(1), 54-55.

Beckingham, B. F., B. N. Hendy, and J. V. Simons (1979). Br. Patent 1,549,841, August 8.

Biegler, L. T. (1984). Solution of dynamic optimization problems by successive quadratic programming and orthogonal collocation. Comp. Chem. Eng., 8(3/4), 243-248.

Box, G. E. P., and G. M. Jenkins (1976). Time Series Analysis, Forecasting and Control, rev. ed., Holden-Day, San Francisco, CA.

Clark, D. W., and P. J. Gawthrop (1981). Implementation and application of microprocessor-based self-tuners. Automatica, 17(1), 233-244.

Clifton, D. M. (1983). Batch computer control of a bulk pharmaceutical chemical process. Adv. Instrum., 38, part 2, 1131-1136.

Coughanowr, D. R., and L. B. Koppel (1965). Process Systems Analysis and Control, McGraw-Hill, New York, p. 323.

Edge, E. R., and W. F. Powers (1976). Function-space quasi-Newton algorithms for optical control problems with bounded controls and singular arcs. J. Optimiz. Theory Appl., 20(4), 455-479.

Fraade, D. J. (1980). Some user aspects of computer control of batch reactors. Instrumentation in the Chemical and Petroleum Industries, Proceedings of the 1980 Symposium, 16, 75-94.

Fruit, W. M., G. V. Reklaitis, and J. M. Woods (1974). Simulation of multiproduct batch chemical plants. Chem. Eng. J., 8(3), 199-211.

Ghosh, A. (1980). Checklist for batch process computer control. Chem. Eng., 87(4), 88-91.

Gidwani, K. K. (1982). Batch control methodology, Adv. Instrum., 37(2), 681-691.

Gordon, D. L., and K. R. Weidner (1978). Computer control of emulsion reactor. Organic Coatings, Plastics, and Chemicals, 43, 840-852.

Guy, J. L. (1982). Dynamic modeling of tank-type reactors. Chem. Eng. (N.Y.), 89(13), 97-100.

Haggin, J. (1984). Faster, smaller integrated sensors in offing for process control. Chem. Eng. News, June 4, 7; also April 2, (1984) 7, and May 21, (1984) 7.

Hamilton, J. C., D. E. Seborg, D. G. Fisher, and D. Grant (1973). Evaluation of Kalman filtering. AIChE J., 19(5), 901-909.

Hopkins, B., and G. H. Alford (1973). Temperature control of polymerization reactors. Instrum. Technol., 20(5), 39-43.

Horak, J., and F. Jiracek (1983). Adaptive nonisothermal computer control of a batch reactor. Collect. Czech. Chem. Commun., 48(2), 449-463.

Hugo, P. (1980). Anfahr and Betriebsverhalten von exothermen Batch-Prozessen. Chem. Ing. Tech., 52(9), 761.

Joglekar, G. S., and G. V. Reklaitis (1984). Simulator for batch and semi-continuous processes. Comp. Chem. Eng., 8(6), 315-327.

Johnson, P. (1977). Safety considerations in design and operation of batch reaction systems. Chem. Ind., 1977(8), 305-308.

Jones, D. I., and J. W. Finch (1984). Comparison of optimization algorithms. Int. J. Control, 40(4), 747-761.

Jutan, A., and A. Uppal (1984). Combined feedforward-feedback servo control scheme for an exothermic batch reactor. Ind. Eng. Chem., Process Des. Dev., 23, 597-602.

Ketchum, Jr., L. H., R. L. Irvine, and R. W. Dennis (1979). Sequencing batch reactors to meet compliance. AIChE Symp. Ser., 75(190), 186-191.

Kiparissides, C., and S. L. Shah (1983). Self-tuning and stable adaptive control of a batch polymerization reactor. Automatica, 19(3), 225-235.

Kladko, M. (1971). The case of a real engineering design problem. CHEMTECH, 141-147.

Knopf, F. C., M. R. Okos, and G. V. Reklaitis (1982). Optimal design of batch/semicontinuous processes. Ind. Eng. Chem., Process Des. Dev., 21(1), 79-86.

Lackmeyer, R. J., and D. G. Kempfer (1977). Computer simulation of a batch reaction which utilizes an inferential conversion monitor. 1977 Joint Automatic Control Conference Proceedings, IEEE, New York, 1523-1529.

Lake, R. J. (1977). Practical simulation and scale-up methods for assessing the cooling of exothermic batch reactions. Chem. Ind., 1977(7), 250-255.

Law, W. M. (1977). Computer control of batch processes in ICI. Chem. Ind., 1977(8), 298-300.

Liou, C. T., and T. H. Shei (1984). Mathematical modeling of second order with nonideal mixing. J. Chin. Inst. Chem. Eng., 15(4), 301-310.

Liu, L. M., and G. B. Hudak (1983). The SCA System for Univariate-Multivariate Time Series and General Statistical Analysis, Scientific Computing Associates, Dekalb, IL.

Luyben, W. L. (1973). *Process Modeling, Simulation, and Control for Chemical Engineers*, McGraw-Hill, New York, p. 62.

Luyben, W. L. (1975a). Batch reactor control. *Instrum. Technol.*, 22(8), 27-34.

Luyben, W. L. (1975b). Introduction to batch reactors. *ISA Joint Spring Conference Proceedings*, ISA, Pittsburgh, PA, pp. 1-8.

Marroquin, G., and W. L. Luyben (1972). Experimental evaluation of nonlinear cascade controllers for batch reactors. *Ind. Eng. Chem. Fundam.*, 11, 552-556.

Marroquin, G., and W. L. Luyben (1973). Practical control studies of batch reactors using realistic mathematical models. *Chem. Eng. Sci.*, 28, 993-1003.

Mehta, G. A. (1983). The benefits of batch process control. *Chem. Eng. Prog.*, 79(10), October, 47-52.

Millman, M. C. (1968). Optimal temperature control in batch reactors. Ph.D. Thesis, City University of New York.

Morris, R. C. (1983). Simulating batch processes. *Chem. Eng.*, 90(10), 77-81.

Murty, B. S. N., K. Gangiah, and A. Husain (1980). Performance of various methods in computing optimal control policies. *Chem. Eng. J.*, 19, 201-208.

Nilsson, H., C. Silvegren, and B. Toernell (1982). A calorimetric reactor system for kinetic studies of heterogeneous polymerizations. A calorimetric reactor for kinetic studies. *Chem. Scr.*, 19(4), 164-171.

Oppenheim, A. V., and R. W. Schafer (1975). *Digital Signal Processing*, Prentice-Hall, Englewood Cliffs, NJ.

Overturf, B. W., G. V. Reklaitis, and J. M. Woods (1978a). GASP IV and the simulation of batch/semicontinuous operations: Single train process. *Ind. Eng. Chem., Process Des. Dev.*, 17(2), 166-175.

Overturf, B. W., G. V. Reklaitis, and J. M. Woods (1978b). GASP IV and the simulation of batch/semicontinuous operations: Parallel train process. *Ind. Eng. Chem., Process Des. Dev.*, 17(2), 166-175.

Pagurek, B., and C. M. Woodside (1968). The conjugate gradient method for optimal control problems with bounded control variables. *Automatica*, 4, 337-349.

Quintana, V. H., and E. J. Davison (1974). Clipping-off gradient algorithms to compute optimal controls with constrained magnitude. Int. J. Control, 20(2), 243-255.

Rippin, D. W. T. (1983). Simulation of single and multiproduct batch chemical plants for optimal operation and design. Comp. Chem. Eng., 7(3), 137-156.

Robson, M. (1981). Controlling batch production. Engineering, 221(12), 933-935.

San, K. Y. (1984). Studies on the on-line identification and optimal control of bioreactors. Ph.D. Thesis, California Institute of Technology.

Shaw, W. T. (1983). Structured design produces good batch control programs. Control Eng., 30(12), 72-76.

Shinskey, F. G. (1979). Process Control Systems, 2nd ed., McGraw-Hill, New York.

Smith, F. A. (1977). Design and control by computer of batch reaction systems. Chem. Ind., 1977(8), 293-297.

Smith, F. A. (1984). Development of remote auto-manual control system for batch organic processes. Chem. Ind., 1984, 396.

Soliman, M. A., and W. H. Ray (1972). A computational technique for optimal control problems having singular arcs. Int. J. Control, 16(2), 261-271.

Sparrow, R. E., and D. W. T. Rippin (1974). Computer-aided design of multiproduct batch chemical plants, an example of the use of the multibatch package. Proceedings of the European Federation of Chemical Engineers 7th Symposium, Frankfurt AM, Germany, pp. 155-165.

Spellman, R. A., and J. B. Quinn (1975). Computer control of batch reactors. ISA Trans., 14(4), 312-317.

Squires, R. G., and G. V. Reklaitis (1980). Computer applications to chemical engineering process design and simulation (symposium), 1979. ACS Symp. Ser., 124, ACS, Washington, DC.

Stutts, B. E. (1983). An improved algorithm for singular control problems with applications to the optimization of the fed-batch penicillin fermentation, Ph.D. Thesis, Purdue Univ.

Thome, R. J., M. W. Cline, and J. A. Grillo (1979). Batch process automation. Chem. Eng. Prog., 75(5), May, 54-60.

Tsai, T. H., and J. W. Lane (1982). Computer control of an existing chemical plant. Paper 29a, AIChE 1982 Winter Meeting, Orlando, FL.

Uhlig, R. J. (1977). Process computer applications in resin plants. Digital computer applications to process control. Proceedings of the 5th IFAC/IFIP International Conference. The Hague, Netherlands, June 14-17, 1977. North-Holland, New York, 1977, pp. 165-171.

Valtyni, J., J. Ilavsky, and M. Vanko (1979). Optimization of polymerization reactor heating. Chem. Prum., 29(3), 119-123.

Wells, G. L., and M. Chester (1978). Simulation of batch chemical plants. Proceedings of the UKSC Conference on Computer Simulation, April 4-8, 1978, IPC Science and Technology Press, Guildford, Surrey, England, pp. 150-157.

Wright, A. R. (1984). Application of modeling and computer simulation to pharmaceutical processes: The Williamson synthesis. Chem. Eng. Res. Des., 62(6), 391-397.

Wu, R. S. H. (1985). Dynamic thermal analyzer for monitoring batch processes. Chem. Eng. Prog., September, 57-61.

Yeo, B. P. (1980). A modified quasilinearization algorithm for the computation of optimal singular control. Int. J. Control, 32(4), 723-730.

Young, M. J. (1980). Semi-batch reactions offer optimal yields. Processing, March, 27-33.

NOMENCLATURE

Roman Letters

A	limiting reactant and its amount in reactor (mole)
A_b	amount of A that can be reacted isothermally batchwise (mole)
A_j	effective heat-transfer area of jacket (m^2)
A_x	effective heat-transfer area between two states, x = 1-4 (m^2)
a	concentration of A (mole/L)
a_f	concentration of a in the feed stream (mole/L)
B	backshift operator
B	chemical species
b	concentration of B (mole/L)
C	chemical species

C_p	heat capacity of reactor and contents (kcal/kg K)
C_{px}	heat capacity of state x, x = c, j, w2, w4 (kcal/kg K)
c	concentration of C (mole/L)
D	chemical species
E_x	activation energy, x = a, p, t, d (kcal/mole)
F	feed flow rate (L/min)
F_j	flow rate of coolant in jacket
f_m	fraction of monomer initially in the reactor
$-\Delta H$	enthalpy of reaction (kcal/mole)
K_p	gain of proportional feedback temperature controller
k_1	first-order reaction rate constant (min^{-1})
k_{10}	pre-exponential factor (min^{-1})
k_2, k_3	second-order reaction rate constants (L/mole min)
M	mass of the reactor and contents (kg)
M_x	mass of the state x, x = c, j, w2, w4 (kg)
Q	rate of heat release inside the reactor (kW)
R	rate of reaction in reactor (mole/min)
R_g	gas content (kcal/mole K)
r	reaction rate per unit volume (mole/L min)
S_0	feed flow rate needed to equal maximum jacket heat-removal rate (L/min)
S_1	adiabatic temperature rise of feed divided by feed-to-contents temperature difference
s	Laplace transform variable (min^{-1})
T_x	temperature of state x, x = c, j, jin, w2, w4, f (K)
t	time (min)
U	overall heat-transfer coefficient (kcal/m^2 K min)

U_x	overall heat-transfer coefficient, $x = 1\text{-}3$ (kcal/m² K min)
v	volume in the reactor (L)
v_m	maximum volume in the reactor (L)
x_m	conversion of the monomer

Greek Letters

γ	scaled activation energy
ρ	density of reactor contents and feed stream (kg/L)
ρ_j	density of coolant fluid (kg/L)
τ	time constant for open-loop system without reaction (min)
τ_{fb}	time remaining in the fed-batch portion of the run (min)

Subscripts

c	contents
FF	feedforward
f	feed
j	jacket
jin	inlet to the jacket
w2	reactor wall
w4	outside jacket wall
sp	setpoint
0	at reference temperature

MULTIVARIABLE CONSTRAINT CONTROL
USING A FREQUENCY DOMAIN DESIGN APPROACH

Steven Treiber
Treiber Controls Inc., Toronto, Ontario
Canada M5H 2M5

David W. Hoffman
Shell Canada Products Company, Calgary, Alberta
Canada T2P 4A6

Abstract. The problem of controlling a process with more inputs than outputs arises fairly often in industry. The operating strategies applied to many of these processes involve forcing some of the extra actuators to extreme limits. The control problem is further aggravated by the need to force different actuators to limits, depending on the operating mode, and under some circumstances to have all actuators free, but with the ability to constrain movement of selected actuators. Treiber (1984) has shown how the Direct Nyquist Array method of Rosenbrock (1974) can be extended to design of control schemes for systems with an unequal number of inputs and outputs. The technique involves design of a non-square compensator which reduces plant interaction and directly addresses the issue of constraint of actuator movement. This paper describes the application of this technique to a multi-draw distillation column. The resulting control structure is incapable of recognizing hard bounds on the inputs or outputs so the "alarm action" capability available on most commercial process control software packages has been used to address the bounding problem in an elegant way.

   Keywords. Multivariable Control; Constraint Control; Nyquist Array.

INTRODUCTION

   Processes with an unequal number of inputs and outputs are a small but very important class of industrial process control problems. Computer control of these processes usually yields profits that are one to two orders of magnitude greater than the profit available from controlling the more common single input, single output processes. Within this class of multivariable control problem the more common processes are those which have been designed to have more actuators available for manipulation than there are measurements to control. Examples of these processes are crude pipe stills, and fluid catalytic crackers, in the petroleum industry, and continuous digestors, lime kilns, and paper machines, in the pulp and paper industry.

The usual control strategy employed by human operators of these units is to square the control problem by treating some of the manipulated variables as "extra" actuators which are manipulated only occasionally. In the case of the crude pipe still the operator will manipulate each of the draw streams in order to control the composition of each of those streams. Ray (1981) gives a simplified example of this problem. The recirculating refluxes which provide staged heat removal in the distillation column are occasionally set through some recipe determined from off-line simulation in order to provide the yields of product required from each of the draws. The industrial approach to computer control of this process has evolved as a formalization of the human operators' control strategy. The draw flows are used to control the product stream compositions with some attention being paid to the interactions involved. The control scheme attempts to manipulate the process material balance. It recognizes that manipulation of a particular draw stream will affect composition in that stream and in other draw streams. The recirculating refluxes are set by recipes based on an on-line calculation of the unit heat balance. While this approach can be very profitable it leads to a very awkward, ad hoc control structure. While the basic rules of the human control strategy are simple they are governed by a set of complex heuristic principles which are not easily incorporated in a computer program. Among the many deficiencies of this approach to computer control are that the heat and material balances are treated as separate, non-interacting problems, and that the interaction in the material balance control is addressed through what might be called the "Doctor Dolittle" or "push me-pull you" control scheme where, if the flowrate in a draw stream is <u>increased</u> by a unit volume flow the flowrate in an adjacent stream is simultaneoudly <u>decreased</u> by an equal amount, without regard to a process model. The problem of basis weight and moisture control on a paper machine provides an exact analogy in the paper industry.

The arbitrary squaring of the control problem as described above is one of the more objectionable aspects of the solution. The process control problem described later will show that use of the full set of actuators can be very advantageous and in some circumstances is mandatory if control of the process is to be maintained at all times.

Among the mathematical approaches to solving the non-square control problem is to apply linear programming when the problem can be posed as a steady-state control problem. This can be done when high frequency disturbance rejection is not an important part of the control problem. Linear programming can also be applied to a control scheme based on a discrete dynamic model of the process however an objective function consisting of a linear sum of errors tends to lead to radical actuation which can only be suppressed by arbitrarily clamping controller moves and thus arbitrarily influencing control performance. In addition both methods require considerable computing power.

A better approach has been Dynamic Matrix Control (DMC) due to Cutler and Ramaker (1979). The method is somewhat more efficient than the linear programming approach in that it usually involves no on-line inversions of the process model and incorporates the ability to constrain movement of specific actuators. IDCOM by Richalet et al. (1978) is a method similar to DMC except that no controller penalty weights are incorporated in controller design. The controller is formed by an explicit, off-line, inversion of the process model and in the case of open-loop unstable and non-minimum phase systems will lead to unstable controllers. An error filter is used to constrain actuator movement and an intelligent selection of the control interval coupled with a very long filter time constant can be used to overcome the problem of instability at the cost of performance. In any event both methods involve multiplication of large matrices and have no direct way of addressing limits on actuator position. Ad hoc programs are super-imposed on these algorithms which test for the possibility of limits being violated in the future. If such violation will occur the actuator penalties in DMC or the error filters in IDCOM are modified and a new set of control moves is calculated. Both methods will iterate in this way until no constraint violations are detected. The controller will then implement the moves calculated. The penalty and filter adjustments are arbitrarily selected by experiment and in the case of DMC each such iteration will lead to an explicit on-line matrix inversion.

Recently Garcia and Morshedi (1984) and Morshedi et al.

(1985) have formulated the DMC problems as a quadratic programming problem (QDMC) and a linear programming problem (LDMC), respectively. Both methods involve minimization of a quadratic objective function, allow constraint of actuator moves, and directly address the issue of limits on the full value of inputs and outputs. Unfortunately, both methods exact heavy computational costs. In the case of high order multivariable control problems the 16 bit addressing limitation of most industrial process control computers will force severe compromises on the controller formulation. This could be somewhat alleviated by reformulating their controller in terms of impulse weights rather than step weights however the control structure will remain quite complex.

MacFarlane et al. (1978) and Kouvaritakis and Edumunds (1978) have developed a feedback control design method for non-square control problems. Their approach is an extension of the method of characteristic loci due to MacFarlane and Belletrutti (1973). It is based on frequency domain and state space design, and focuses on placement of the multivariable zeroes generated in the squaring down process. The resulting control structure is not much simpler than that for the numerical methods such as DMC, and the design method is more complex.

Treiber (1974) has developed a method of compensator design which allows extension Rosenbrock's (1974) direct Nyquist array method (DNA) to systems with an unequal number of inputs and outputs. The resulting control structure retains the relative simplicity of the Nyquist array method while permitting constraint of actuator movement. The method was demonstrated on a simulated system with fewer actuators than measurements to be controlled. This paper documents an application of the method on a real multi-draw distillation column, with more actuators available than measurements to control. Unlike LDMC and QDMC the method is incapable of recognizing limits on full value of inputs and outputs. This problem has been overcome by using the "alarm action" capability available in most commercial process control software packages. The resulting control scheme has all the feedback control properties of a standard Nyquist array design and permits constraint of actuator movement relative to other actuators, and constraint of actuator position to avoid violating limits on; actuator position, value of outputs, value of uncontrolled but associated variables. In addition, control points may be knocked out of the control scheme without reformulating the controller. These features will be discussed in relation to the application described later.

THE DIRECT NYQUIST ARRAY METHOD

The direct Nyquist array method can be applied to any compensated system, Q, where

$$Q(s) = G(s)D \qquad (1)$$

G is the process transfer function matrix, D is a pre-compensator, and (s) indicates the Laplace formulation. While Q must be square there is no such restriction on G. The DNA dominance criterion is applied to the return difference matrix, F, where

$$F(s) = I + Q(s) \tag{2}$$

and, I, is the identity matrix. F(s) is said to be diagonally dominant if, on its' Nyquist plot, the magnitude of the diagonal elements is greater than the sum of the magnitudes of the off-diagonal elements at all frequencies. Testing for dominance by columns

$$\left|f_{ii}(s)\right| > \sum_{j} \left|f_{ij}(s)\right| \quad , \; j \neq i \tag{3}$$

or by rows

$$\left|f_{ii}(s)\right| > \sum_{i} \left|f_{ji}(s)\right| \quad , \; i \neq j \tag{4}$$

Achieving dominance in F establishes a stability criterion which is similar to that for single-loop systems. The closed-loop system will be stable if and only if the number of counter-clockwise encirclements of the critical point (-1,0) by the Nyquist diagrams of the diagonal elements of Q is equal to the number of right-half plane poles of the open-loop system.

The Nyquist plot of Q can be used to perform a graphical test for dominance in F. Recognizing that

$$\left|f_{ij}(s)\right| = \left|g_{ij}(s)\right| \tag{5}$$

The Gershgorin disks of radius, $\sum \left|g_{ij}\right|$, are drawn on each of the Nyquist plots of the diagonals, g_{ii}, to form a Gershgorin band about each diagonal plot. The return difference matrix, F, is dominant if none of the Gershgorin bands encompass or touch the critical point (-1,0) on any of the diagonals. The necessary and sufficient condition for stability of open-loop stable systems is that neither the Gershgorin bands nor the Nyquist plots themselves shall encompass the critical point on any of the plots of the diagonal elements. The Gershgorin bands are a measure of the interaction in the closed loop system and represent a region of uncertainty in the frequency response of the diagonal elements. That uncertainty is introduced by loop interaction and thus the narrower the bands the less the interaction. The focus of the method is to design a compensator which shrinks the Gershgorin bands and allows for implementation of single loop controllers whose tuning constants are selected with regard only to the Nyquist plots of the diagonals and their attendant Gershgorin bands. Once the degree of interaction has been reduced satisfactorily any single-loop design method can be applied to the compensated system, Q, though P+I controllers are usually the type selected. Details of an application of the direct Nyquist array method to an industrial process have previously been provided by Treiber (1982).

COMPENSATOR DESIGN FOR NON-SQUARE SYSTEMS

The usual procedure in compensator design for square transfer function models is to first examine the Nyquist array of G, to determine whether interaction is strong enough to require compensation. This cannot be done for non-square G's as the concept of dominance does not apply. In this method one must begin with a compensator, D, which produces a square compensated system, Q. It is usually wise to examine some square subsets of G to determine whether or not the added complexity of dealing with a non-square problem produces a worthwhile improvement in control performance. As will be shown later by the application described, the decision to address the non-square control problem may be influenced by issues other than feedback control performance.

When designing compensators for square transfer function matrices the inverse gain matrix is often a good starting point. There is no inverse for non-square systems; however by analogy the psuedo-inverse gain matrix may prove to be attractive. The psuedo-inverse of a non-square matrix, K, is defined as

$$(K'K)^{-1}K' \qquad (6)$$

where prime, ', indicates transposition. A more flexible form of the pseudo-inverse compensator may be taken from the work of Boyle (1978), who developed a quadratic optimization technique to solve steady-state problems. A minor modification to Boyle's problem formulation allows his idea to be applied to designing a non-square compensator. For a system which has a process model of the form

$$e(k+1) = e(k) - Ku \qquad (7)$$

where $e(k+1)$ is next error, $e(k)$ is present error, u is change in control input, and K is the process gain matrix. It can be shown that a performance index

$$e'(k+1)Ce(k+1) + u'Bu \qquad (8)$$

is minimized by the following control law

$$u = \left[(K'CK + B)^{-1}K'C\right]e(k) \qquad (9)$$

where C is a diagonal matrix of error penalty weights, B is a diagonal matrix of penalty weights on control moves, u is the vector of control changes, and e is the error vector. The contents of the square brackets in equation (9) can be considered to be a compensator in a unity feedback gain control scheme, thus

$$D = (K'CK + B)^{-1} K'C \qquad (10)$$

is the pre-compensator applied in equation (1) to arrive at a square compensated system, Q, which can be examined using the direct Nyquist array.

For a process model, G, with m rows and n columns (m X n), D will have dimensions n X m. The compensator in equation (10) will be recognized as a modified formulation of the DMC control law. In the DMC formulation each element of the gain matrix, K, is replaced by a truncated, lower triangular matrix of step weights. The formulation suggested here is attractive in that the matrices involved are small. The design procedure is then to select some starting values for the penalty matrices C and B, usually the identity matrix, solve for the compensator, D, and then to apply the compensator to the array G. The Gershgorin bands on the compensated system, Q, can be manipulated by adjusting the values of the penalty weights in C and B. In this way the Nyquist array of Q can be manipulated to achieve a satisfactory degree of shrinking in the Gershgorin bands. Treiber (1984) used a simulated distillation column to demonstrate the technique. This paper describes a process control application for a vacuum distillation column in a refinery Luboil plant.

PROCESS DESCRIPTION

The vacuum distillation unit (VDU) is in the Luboil plant at Shell Canada's Montreal East refinery (see Fig. 1) is a three side-draw column with partial draw only. The column is 3 meters in diameter with 43 valve trays and 3 chimney trays. Auxiliary equipment includes top and middle pumparounds, a runback loop, and three steam side-strippers. The VDU feed is long residue from the bottom of an upstream crude oil distillation column and it is operated at a pressure of approximately 0.1 atmospheres. This single column processes all of Shell Canada's lubricating oil product as well as supplying product to other parties for resale. The product streams are called waxy distillates. After removal of the waxes in downstream units the three side-stream products are blended to make lubricating oils. The bottom product is sold as grease after wax extraction.

The basic operating goals of the unit are to maintain specified viscosity targets for each of the side streams while maintaining the flash point of the top side stream, labelled WD-100 in Fig. 1, above a specified value. To achieve these objectives the operators use measurements from

FIG. 1. VACUUM DISTILLATION UNIT

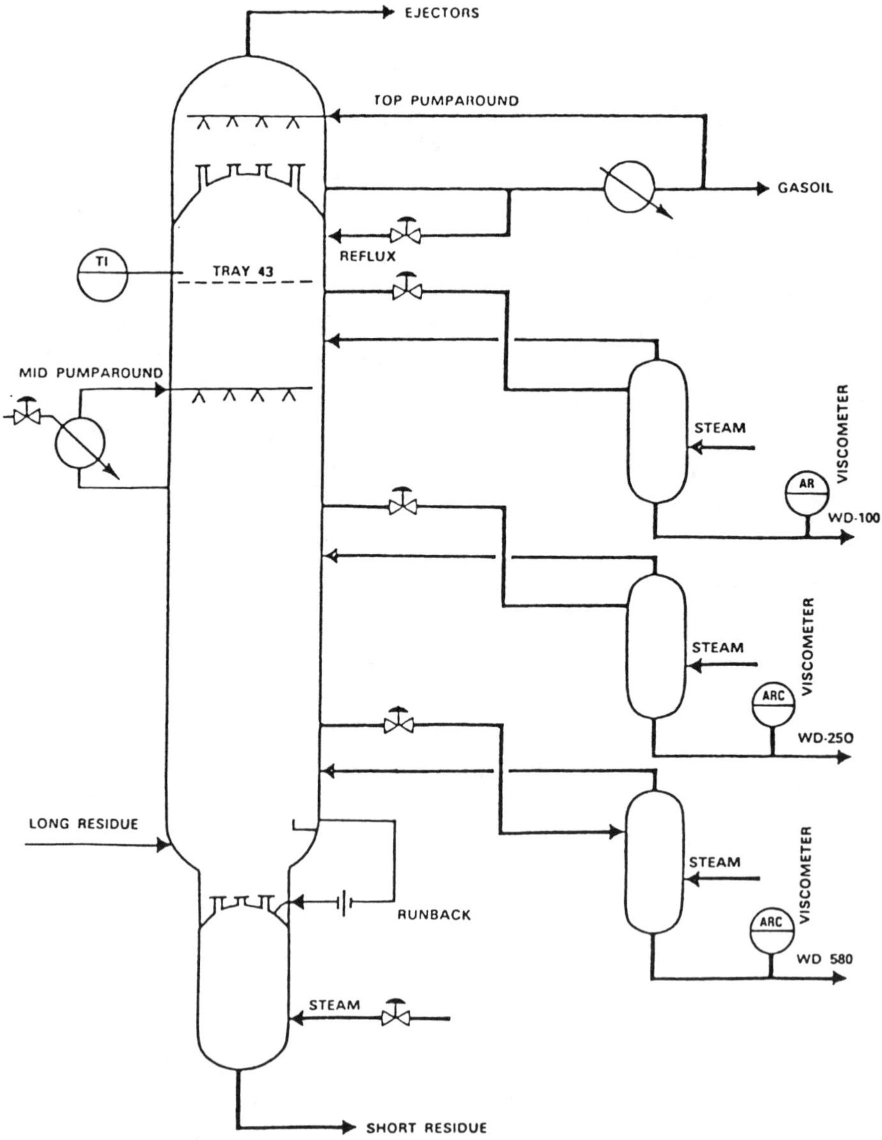

in-line viscometers in each of the side-draw streams, once daily laboratory viscosity measurements, and a laboratory measurement of the flashpoint of the WD-100 draw. The in-line viscometers are positioned downstream of the side-strippers introducing a significant delay between column conditions and measurement. When the unit is upset some of the operators use the draw tray temperatures as a rough indication of viscosity. Also, as the WD-100 flashpoint measurement is only available once per shift at best the operators use the temperature on tray 43 as an indication of WD-100 flashpoint.

The operators control the process by manipulating the sidestream flows in order to maintain the stream viscosities at their targets. The reflux is occasionally adjusted when the WD-100 flashpoint approaches the lower bound and the reflux is also sometimes adjusted to influence the sidestream viscosities. Recognizing the strongly interactive behaviour of the process, experienced operators have developed heuristic strategies for multiple manipulation of actuators, though none seem to use the same strategy. All the strategies seemed to work reasonably well under stable operating conditions allowing the operators to gradually steer the process to its targets. When the unit is upset by disturbances these schemes are totally ineffective and a completely different set of rules is used to manage the column. Unfortunately, the column is frequently disturbed by changes in feed composition, feed temperature, and column pressure. The slow response of the in-line viscometers, their limited measurement range, and the strong interaction in the process make the operators task extremely difficult. The task is further complicated by the operating schedule of the plant. There is a minimum of storage available for intermediate and product streams and so the plant operation is very closely tied to market demand for the various blended products. To accommodate the market demand and the availability of downstream storage the unit is operated in two different modes. When the demand for light oils is high the operators will attempt to maximize the flow of the WD-100 draw, and when the demand is for heavier oils they will attempt to maximize the flow of the middle sidestream, WD-250. This mode switch is accomplished by adjusting the middle pumparound and the column reflux. All this while maintaining the viscosities at their targets and the flashpoint within bounds. A mode switch is accompanied by numerous adjustments of the draw flows. The mode switch can be accomplished in as little as 6 hours when there are no disturbances and very much longer when there are upsets. The mode switches can be required as frequently as every few days though the unit will generally run in a single mode for several weeks at a time. The changing price and availability of crude oil feedstocks has caused relatively frequent feedstock changes which force radical changes in the operating conditions of the tower so as to provide products of consistent quality.

The operators control strategy attempts to accomodate a number of apparently conflicting demands. Mode changes are accomplished through recipe adjustments of reflux and middle pumparound. While there is some fine tuning to help meet the quality specifications these actuators are essentially fixed between mode changes and upsets. Given only the draw flows to control viscosities neither the WD-100 nor the WD-250 can be set to their maximum flows. Indeed the maximum flows are not known with any degree of certainty as that is determined by the feed flow, column pressure, and feed composition, which is unknown. The maximum allowable flow is sometimes the point beyond which the draw tray is being drawn dry and sometimes the point beyond which any further increase in draw flow will cause an excursion in the stream viscosity.

The main profit from computer control of this process comes from maximizing product yield in the valuable WD-100 and WD-250 streams as opposed to the less valuable WD-580 and short residue. In addition there is significant profit from eliminating the viscosity excursions caused by distrubances and mode changes as these tend to propagate downstream and disturb the extraction and dewaxing processes. The final lubricating oil product qualities are not affected by these disturbances as specifications are always met by blending of final product and reprocessing intermediate products.

PROCESS RESPONSE TESTS

Due to the severe and frequent distrubances to which the process is subjected, step response tests were not a viable method for determining a process model. The open-loop response tests were carried out by manipulating the inputs in a random binary pattern, the so-called pseudo-random binary sequence (PRBS). The transfer function models were then derived by using the Time Series Analysis method of Box and Jenkins (1969).

The manipulated variables were reflux flow, middle pumparound temperature difference (ΔT), and the draw stream valve positions. There are no flow controllers nor flow measurements in the draw streams. The side product flows are measured downstream of the steam side-strippers. The measured variables were tray 43 temperature, the temperatures of liquid on the three draw trays, and the viscosities indicated by the three in-line viscometers. The experiments were carried out by manipulating only one of the inputs at a time and measuring the response of all the outputs. Fig. 2 is a graphical plot of one test in which the pumparound ΔT is the manipulated variable shown as the lower plot and the tray 43 temperature is the output variable shown in the upper part of the plot. The plots are scaled so the relative magnitude of the changes shown are not meaningful however, notice that the output series shifts level during the experiment. This level shift was caused by plant disturbances. Also, the input series shows a level shift which was introduced by the authors to prevent the viscosities from going out of the measuring range of the in-line viscometers. All the experiments carried out were subjected to these kinds of disturbances forcing the experimenters to make occasional adjustments in the level of the PRBS input and adjustments in other variables. At times it was not possible to prevent the viscosities from going out of range of the viscometers and portions of the data series had to be eliminated. Despite these difficulties there were no repetitions of tests due to corruption of data by disturbances.

The first step in the Time Series Analysis method is to estimate the impulse response weights for the input/output series'. Fig. 3 shows the impulse response weights estimated for the series shown in Fig. 2 after first differencing both series' to account for the disturbance noted earlier. The impulse weights are estimated by a recursive least squares solution of the equation

$$\gamma_{xy} = \Gamma_{xx} v \qquad (11)$$

where γ_{xy} is the vector of cross-covariance of the series' between lag 0 and lag K, Γ_{xx} is a symmetric matrix of the auto-covariance of the input series between lag 0 and lag K, and v is the vector of impulse weights between lag 0 and lag K. Substituting estimates of the autocorrelation function, r_{xx}, and cross-correlation function, r_{xy}, for the auto- and cross-covariances the impulse weights are a crude estimate of the model which improves as the signal to noise ratio in the output series increases. The impulse weights are used only to provide a preliminary guess at the model structure, an initial estimate of the number of lags of delay, and the model parameters. This information is then used to initialize a non-linear least-square fit of the model to the data series. At this stage of the modelling procedure, noise models can be

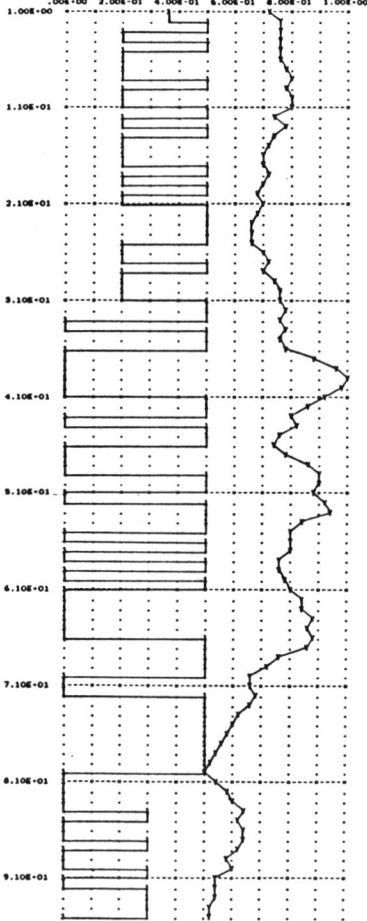

Fig. 2. PRBS Test

incorporated in the model structure if necessary and models for multiple input series can be estimated if desired.

After satisfying all the statistical tests of sufficiency the models were converted to the Laplace domain for the sake of convenience. The transfer function model derived from the data plotted in Fig. 2 was

$$\frac{-5.23e^{-s}}{22.7s + 1}$$

Fig. 4 shows a plot comparing the actual output series with that predicted by the model. The fit was excellent and typical of all the experimental data gathered.

Data from a single set of five experiments, one per input variable, were used to estimate transfer function models relating the tray temperature responses to the five inputs, the viscosity responses to the five inputs, and transfer function models treating each of the draw tray temperatures as inputs and their respective viscosities as outputs. These models were then used to design a control scheme for the process.

CONTROL DESIGN

The first step in the design was to examine the

FIG. 3. IMPULSE WEIGHTS AFTER DIFFERENCING DATA

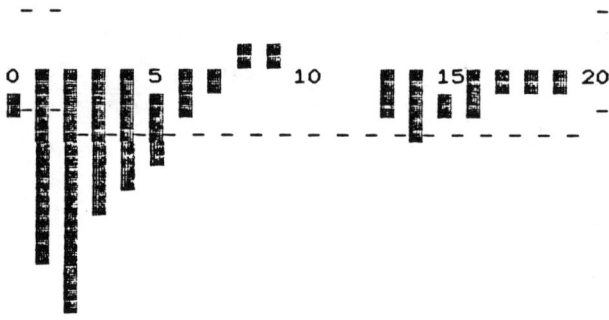

possibility of a control scheme that replicated the operators' control strategy. This might be justified on the grounds of familiarity and therefore ease of use. This control scheme would have the reflux and pumparound ΔT set by some recipe and only the draw flows manipulated in a feedback control scheme. The controlled variables are the sidestream viscosities. This leads to a transfer function matrix with 3 inputs and 3 outputs shown in Table 1. The time constants and delays are expressed in minutes and the gains are scaled in terms of ratios of fractions of full scale. Table 1 shows that this process is characterized by long time constants and long time delays. An examination of the DNA showed that while the system was essentially dominant no simple compensator could sufficiently shrink the Gershgorin bands. The bands must be extremely narrow here because the phase shifting introduced by the time delays not only causes the Nyquist plots to wrap themselves, but pulls the Gershgorin bands around, thus severely reducing the stable gain space. In addition, the nature of the process disturbances, the observed speed of response to those disturbances, and the limited measurement range of the in-line viscometers forced abandonment of direct control of the viscosities. Instead it was decided to apply multivariable control to the tray temperatures and then to add single loop viscosity controllers in a cascade, as shown in Fig. 5.

Table 2 shows the transfer function matrix for the system including all 5 inputs and 4 outputs, expressed in the same units as the model in Table 1. The square subset marked off by a dashed line in the lower right hand side of Table 2

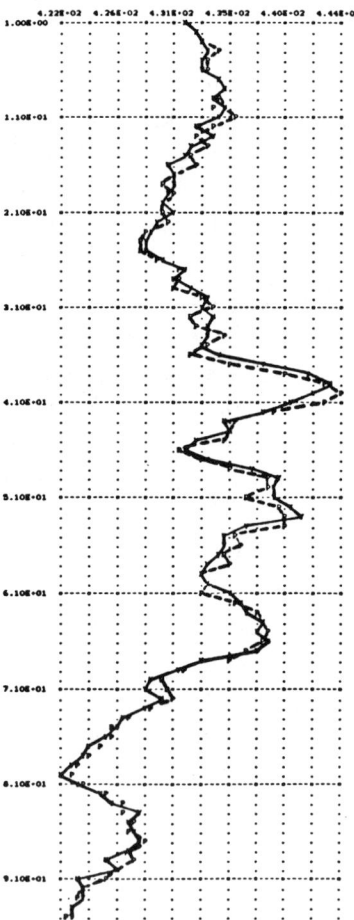

Fig. 4. Comparison of predicted and actual output

TABLE 1

VISCOSITY RESPONSE MODEL

	1A5105v	1A5167v	1A5168v
1A5105	$\dfrac{7.1E^{-18s}}{43.5s + 1}$	0	$\dfrac{0.93E^{-26s}}{10.2s + 1}$
1A5167	$\dfrac{6.76E^{-14.8s}}{23.8s + 1}$	$\dfrac{5.5E^{-21s}}{30.3s + 1}$	$\dfrac{1.55E^{-38s}}{20.4s + 1}$
1A5168	$\dfrac{8.57E^{-15.2s}}{21.3s + 1}$	$\dfrac{4.6E^{-11.3s}}{5.75s + 1}$	$\dfrac{21.4E^{-15.2s}}{29.4s + 1}$

MEASURED VARIABLES	MANIPULATED VARIABLES
1A5105 - WD-100 DRAW VISCOSITY	1A5105v - WD-100 DRAW VALVE POS.
1A5167 - WD-250 " "	1A5167v - WD-250 " " "
1A5168 - WD-580 " "	1A5168v - WD-580 " " "

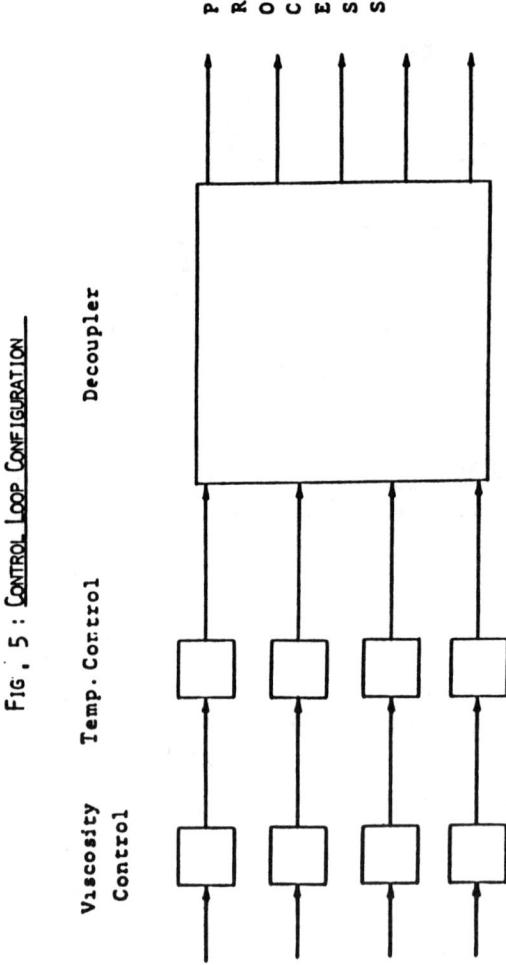

FIG. 5 : CONTROL LOOP CONFIGURATION

Table 2
Temperature Response Model

	1F5129	1T5136	1A5105v	1A5167v	1A5168v
1T5129	$\dfrac{-2.84e^{-s}}{16.6s+1}$	$\dfrac{-5.23e^{-s}}{22.7s+1}$	0	$\dfrac{0.41e^{-18s}}{7.9s+1}$	0
1T1209	$\dfrac{-2.82e^{-4s}}{15s+1}$	$\dfrac{-5.06e^{-2s}}{14.6s+1}$	$\dfrac{0.756e^{-3s}}{11.8s+1}$	0	$\dfrac{e^{-12s}}{3.2s+1}$
1T1208	$\dfrac{-1.5e^{-4s}}{13.7s+1}$	$\dfrac{-4.15e^{-3.3s}}{40s+1}$	$\dfrac{2.91e^{-s}}{14.4s+1}$	$\dfrac{1.35e^{-6s}}{19s+1}$	0
1T1207	$\dfrac{-1.79e^{-2.5s}}{4.2s+1}$	$\dfrac{-2.8e^{-3.2s}}{17.4s+1}$	$\dfrac{4.17e^{-5.2s}}{21s+1}$	$\dfrac{2.4e^{-2s}}{10.5s+1}$	$\dfrac{5.2e^{-s}}{20.9s+1}$

Measured Variables
- 1T5129 – Tray 43 Temp.
- 1T1209 – WD-100 draw tray Temp.
- 1T1208 – WD-250 draw tray Temp.
- 1T1207 – WD-580 draw tray Temp.

Manipulated Variables
- 1F5129 – Reflux flow
- 1T5136 – Pumparound ST
- 1A5105v – WD-100 draw valve position
- 1A5167v – WD-250 "   "   "
- 1A5168v – WD-580 "   "   "

is analogous to the model shown in Table 1. In Table 2 the outputs of the 3 X 3 sub-system are the draw tray temperatures rather then the draw stream viscosities as in Table 1. Comparing the two models one can see that the time constants and delays in the temperature response model are much smaller than those in the viscosity response model. An examination of the Nyquist array of this system and design of a few simple compensators confirmed that the choice of temperature control over direct viscosity control was a good one. The stable gain space expanded, however the Gershgorin bands about the second and third diagonal elements were extremely broad indicating that control of the WD-250 and WD-580 draw temperatures would be quite poor. A variety of compensators was applied but all attempts indicated that WD-100 draw control would need to be severely compromised in order to improve performance in the other loops.

It was then decided to examine the full non-square control problem represented by Table 2. The starting compensator was designed by taking the gain array

$$K = \begin{bmatrix} -2.84 & -5.23 & 0 & 0.41 & 0 \\ -2.82 & -5.06 & 0.756 & 0 & 1 \\ -1.5 & -4.15 & 2.91 & 1.35 & 0 \\ -1.79 & -2.77 & 4.17 & 2.4 & 5.2 \end{bmatrix}$$

and the error and effort penalty matrices, C and B, and the identity matrix. Solving equation (10) gives

$$D = \begin{bmatrix} -.079 & -.052 & 0.085 & -.020 \\ -.059 & -.065 & -.056 & 0.030 \\ -.201 & 0.028 & 0.233 & 0.010 \\ 0.218 & -.313 & 0.063 & 0.075 \\ -.011 & 0.082 & -.208 & 0.150 \end{bmatrix}$$

The diagonal elements of the Nyquist array for this system are shown in Fig. 6. The Gershgorin bands are reasonably narrow for the first two loops but tend to widen at higher frequencies for the bottom two loops, WD-250 and WD-580 draw trays. This is a distinct improvement over any square scheme however the interpretation of this result is that the reflux flow and pumparound ΔT influence the bottom draws too strongly. There was also some concern that radical actuation of these two variables might cause tray flooding. The next step was then to constrain movement in the reflux and pumparound ΔT by increasing their penalty weights. The diagonal elements of the matrix, B, become [10 10 1 1 1]. The pre-compensator solved for is

$$D = \begin{bmatrix} -.041 & -.035 & 0.006 & 0.004 \\ -.067 & -.062 & -.011 & 0.014 \\ -.196 & 0.036 & 0.25 & 0.004 \\ 0.222 & -.31 & 0.067 & 0.072 \\ -.006 & -.082 & -.226 & 0.156 \end{bmatrix} \qquad (12)$$

FIG. 6. LIGHT CONSTRAINT OF ACTUATORS

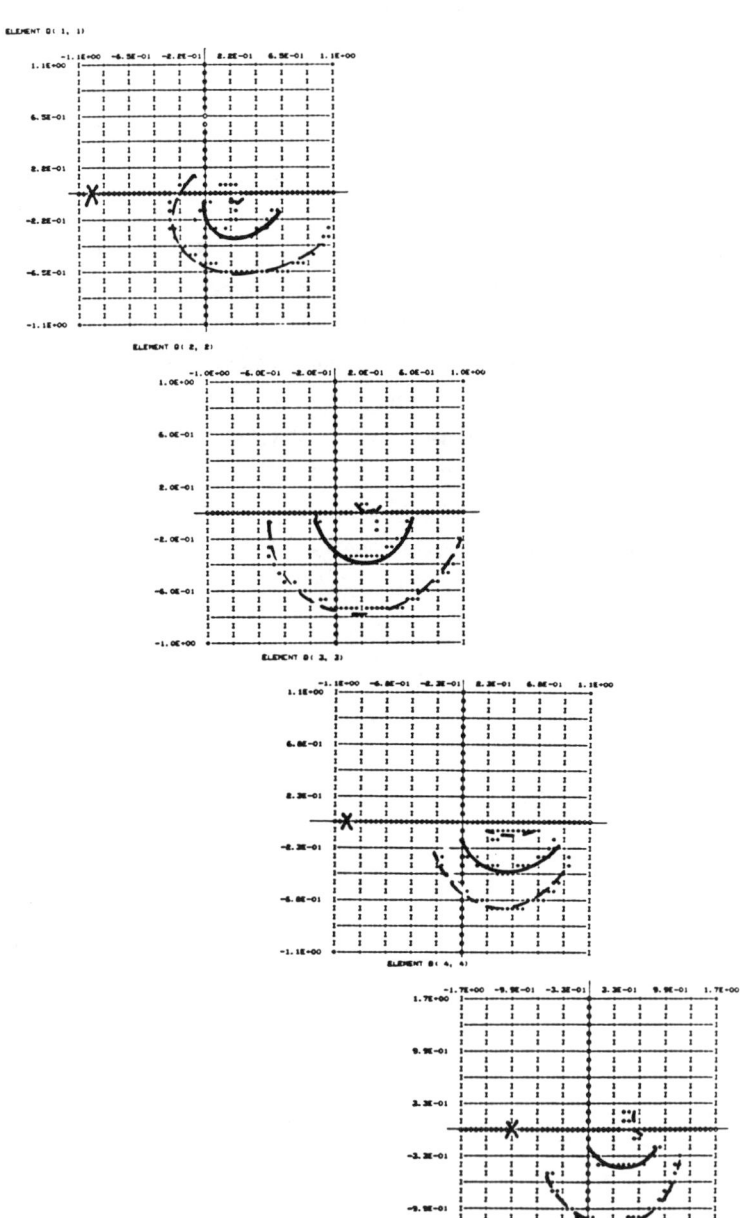

Fig. 7 shows the diagonals of the resultant Nyquist array. The high frequency band width has now been reduced in loops three and four without any measurable influence on the first two loops.

The compensator in equation (12) was then taken as the base case. Recognizing that models are rarely perfect it was expected that some on-line tuning of the implemented control scheme would be required, and that some of the tuning might include adjustment of the compensator it was decided to further simplify the compensator. The parameters in the upper right hand and lower left hand corners are relatively small. Starting from those corners the parameters were successively zeroed along the diagonals and each of these new compensators was applied to the transfer function matrix. This procedure was repeated until a significant deterioration in the Gershgorin bands was noted. The simplest compensator, which produced no significant loss in performance from that of equation (12) was

$$D = \begin{bmatrix} -.041 & 0 & 0 & 0 \\ -.067 & -.062 & 0 & 0 \\ -.196 & 0.036 & 0.25 & 0 \\ 0 & -.31 & 0.067 & 0.072 \\ 0 & 0 & -.226 & 0.156 \end{bmatrix} \quad (13)$$

Based on the process time constants the control inverval

216   AN INDUSTRIAL VIEW OF ADVANCED PROCESS CONTROL

FIG. 7. HEAVY CONSTRAINT OF ACTUATORS

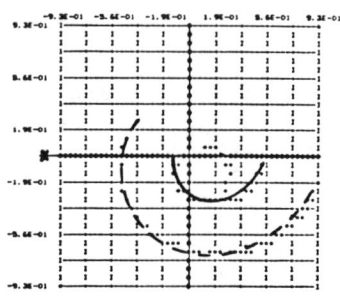
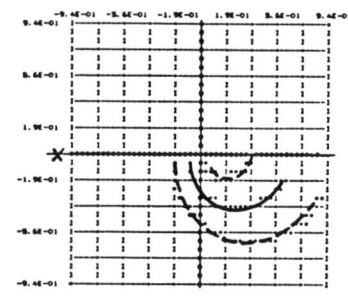
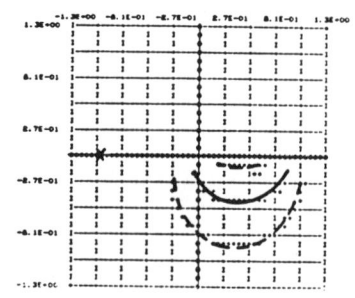

FIG. 8. P & I CONTROLLER TUNING CONSTANT ESTIMATES

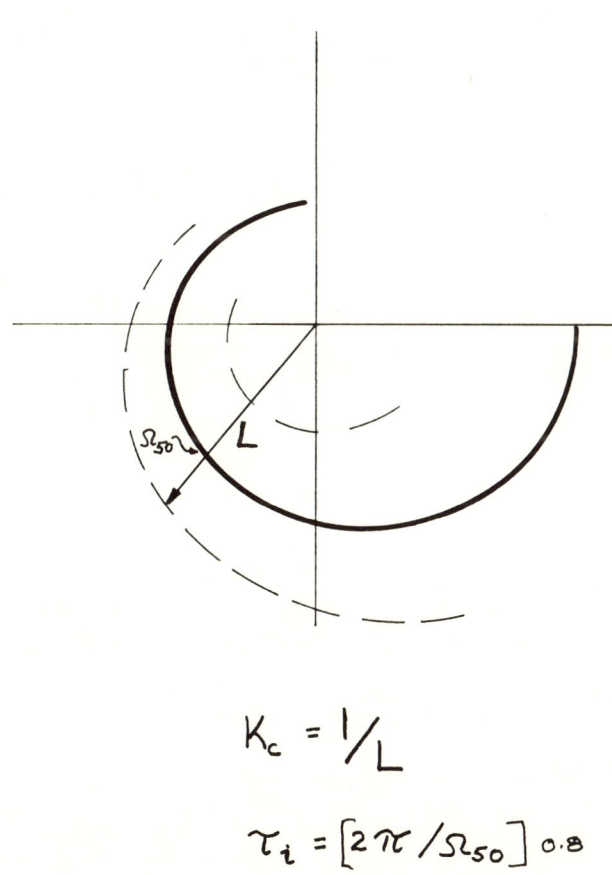

$$K_c = 1/L$$

$$\tau_i = [2\pi/\Omega_{50}]\, 0.8$$

was selected as 4 minutes. Before proceeding to design single-loop P+I controllers for the tray temperatures, and draw stream viscosities a zero-order hold was incorporated in the transfer function models. Both the explicit form of the zero-order hold and adding half of a control interval as time delay to each transfer function elements were tried. As the two approaches showed no significant difference in this case the latter was used to design the controllers.

The P+I controllers were designed to achieve a phase margin of $50°$. As shown in Fig. 8 the outer limit of the Gershgorin band is used to estimate controller gain

$$K_c = 1/L$$

and integral time   (13)

$$\tau_i = 0.8 \, (2\pi/w_{50})$$

where L is the magnitude of the Nyquist plot plus Gershgorin disk at a phase angle of $-130°$ and w_{50} is the frequency at that point, as shown in Fig. 8.

The initial tuning parameters of the temperature controllers are shown in Table 3. The viscosity versus tray temperature models, flashpoint versus tray 43 temperature model, and the viscosity controller tuning parameters are shown in Table 4. The purpose of the design is to achieve fast disturbance rejection from the temperature controllers. As the temperatures are not an absolute determination of draw

Table 3
Temperature controller tuning

Tray 43	$\left(1 + \dfrac{1}{5s}\right)$
WD-100 Tray	$\left(1 + \dfrac{1}{8s}\right)$
WD-250 Tray	$1.2\left(1 + \dfrac{1}{21s}\right)$
WD-580 Tray	$0.5\left(1 + \dfrac{1}{5s}\right)$

Table 4
Viscosity vs. Temperature Models

$$\frac{\text{Flash Pt.}}{1\text{T}5129} = 0.9$$

$$\frac{1\text{A}5105}{1\text{T}1209} = \frac{1.9e^{-4.4s}}{19s + 1}$$

$$\frac{1\text{A}5167}{1\text{T}1208} = \frac{2e^{-12s}}{19s^2 + 2s + 1}$$

$$\frac{1\text{A}5168}{1\text{T}1207} = \frac{2.8e^{-5s}}{11s^2 + 5s + 1}$$

Viscosity Controller Tuning Constants

Flash Pt.	$\dfrac{1}{20s}$
1A5105	$0.5\left(1 + \dfrac{1}{50s}\right)$
1A5167	$0.1\left(1 + \dfrac{1}{50s}\right)$
1A5168	$0.1\left(1 + \dfrac{1}{41s}\right)$

stream viscosity the viscosity controllers perform slow correction of the temperature targets. With this objective it might have been wiser to make the inner temperature loops proportional only controllers, however the viscosity analysers are more prone to failure than thermocouples. The configuration chosen provides a more convenient backup to full analyser control as good temperature control can be achieved on the loops where the analysers are unavailable.

CONSTRAINT OF CONTROL ACTION

The compensator design allows constraint of actuator movement and relaxation of performance requirements for any particular loop through adjustment of the penalty weights in the matrices, B and C. There is however no direct way of consistently recognizing that hard limits on inputs or outputs have been reached in a multivariable control scheme of this sort. A constraint action scheme has been imposed on the algorithms which accomplished the necessary clamping. The control strategy shown in Fig. 5 was implemented in a commercial process control software package which has two important capabilities that allow implementation of a clamping scheme. The algorithm blocks available in that software package were used to implement the compensation matrix and P+I controllers. These blocks allow users to set output clamp limits on any of the blocks. The limits may be fixed analog values such as 75% of scale as the maximum output and 25% of scale as the minimum output. The output clamp limits may also be digital bits in the computer database. In this case if the high limit bit is set the output will not be permitted to increase but will be permitted to decrease. If both bits are reset the output may change in any direction. The state of these bits may be changed through the user keyboard or through a user written program. The second important capability is related to alarming of measured values. Alarms may be invoked for any measured point in the computer database. For example a high limit and a low limit may be declared for a measurement. The points designated for alarm are constantly monitored and if a limit is exceeded a unique alarm condition is declared for that point. In addition the event of an alarm may be used to trigger a user written program called an "alarm action" program.

An alarm action program was written which accomplishes the appropriate clamping required in the control strategy. Alarm points were declared for variables in the control strategy. For example there are alarm limits on the reflux flow, draw stream valve position, calculated tray loading, pumparound ΔT, pumparound valve position. All these alarm points trigger the alarm action program. The function of the alarm action program is to set a digital bit that is one of the clamp limits of the control algorithm blocks. The program is completely generic. With no programming the user may direct the program to prevent increase of WD-100 draw flow if the valve position has reached its upper limit, the draw tray is being drawn dry, or the reflux flow has reached a lower limit. The user may select any combination of alarm points that may be grouped together and any number of points in that group may cause the state of a selected digital bit to be changed. The program is file driven and each record of that file represents a unique combination of alarm points and state change of bits. The effect of this program is to accomplish row and column knock-out in the compensator and prevent wind-up in the control scheme. Rows or columns of the compensator are effectively zeroed out when clamps are hit. Note that numerical values in the compensator are not changed. Clamping is merely equivalent to making a row or column zero. For example when the WD-100 flow hits a high limit the compensator is

$$D = \begin{matrix} -.041 & 0 & 0 & 0 \\ -.067 & -.062 & 0 & 0 \\ 0 & 0 & 0 & 0 \\ 0 & -.31 & 0.067 & 0.072 \\ 0 & 0 & -.226 & 0.156 \end{matrix}$$

This reduces the problem to a 4 X 4 control scheme. The actuators which are most likely to hit limits are the WD-100 and WD-250 valves, and the pumparound ΔT. Compensators representing these conditions were applied to the transfer function model to assure that control of the process would be satisfactory under these circumstances without modification of the controller tuning constants. This proved to be the case. In addition to cases of actuator knock-out it is also possible that under certain circumstances control of the WD-100 flashpoint or control of the WD-580 viscosity is not necessary. These cases result in knocking out compensator columns, reducing the control problem to controlling three variables with five actuators. The Nyquist arrays for these cases were also examined with satisfactory results.

CONTROL PERFORMANCE

The control scheme was implemented and tested without any alarm action clamps. The control loop performance was excellent and required no tuning. Constraint action was then implemented and the control scheme was monitored to determine its' performance in response to disturbances. Figs. 9 to 12 show the performance of the temperature control scheme in response to a 15% change in column pressure. Fig. 9 shows a trend plot of the pressure disturbance. The pressure trend is circled. Time progresses from the bottom of the trend and the time period shown is two hours. Fig. 10 shows the movement of four of the actuators. Reading the trend from left to right the plots are of the WD-100, WD-250, WD-580 valve positions, and the reflux flow. Fig. 11 shows the same set of draw stream valve positions and the pumparound ΔT. Fig. 12 shows the tray temperature responses. Reading from left to right the trend plots are for tray 43, WD-100, WD-250, and WD-580 draw tray temperatures. The arrows at the bottom of the trend plots indicate the controller targets. The control performance is excellent. Without computer control the pressure disturbance shown would have caused serious excursions in product specifications.

Comparable performance has been achieved for cases where actuators have saturated. By a judicious selection of the WD-100 flashpoint targets for the two modes of operation and slightly varying the viscosity targets the product yields have been significantly increased. There has been a substantial reduction in product variability. Viscosity responses have not been shown here to protect proprietary information. Mode changes are accomplished in less than two hours with no appreciable variation in product quality and the control scheme appears to have no difficulty handling feed composition changes including those resulting from feedstock switches.

CONCLUSIONS

A method of extending the direct Nyquist array method to the design of control schemes for systems with more inputs than outputs has been demonstrated. The compensator design technique is borrowed from optimization theory and permits constraint of actuator movement. The alarm action capability available in many commercial process control software packages has been used to accomplish constraint of actuator position by row and column knock-out in the multivariable compensator. The technique has been successfully applied to control of a multi-draw, vacuum distillation column in a lubricating oil plant.

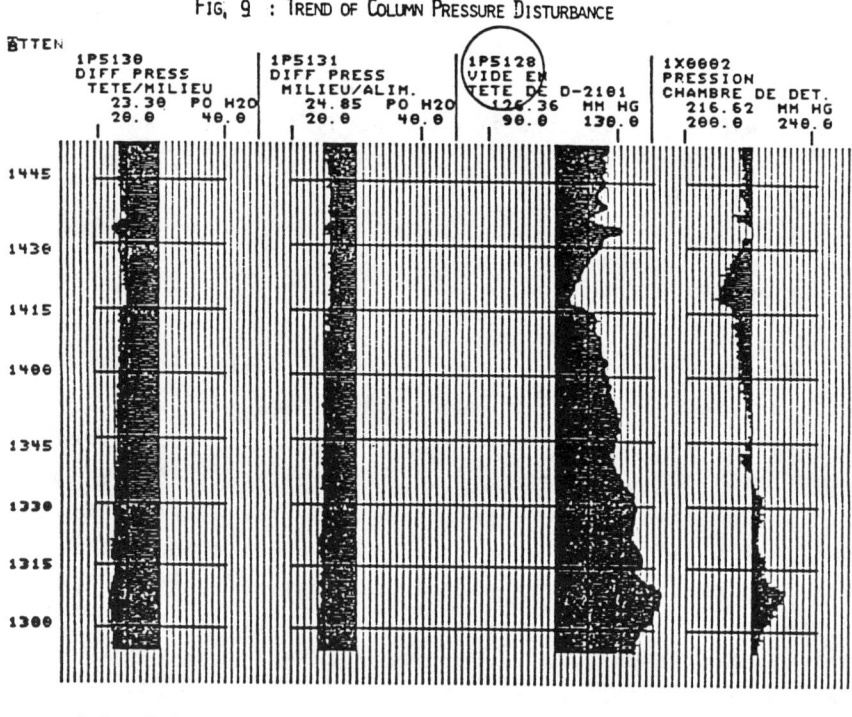

Fig. 9 : Trend of Column Pressure Disturbance

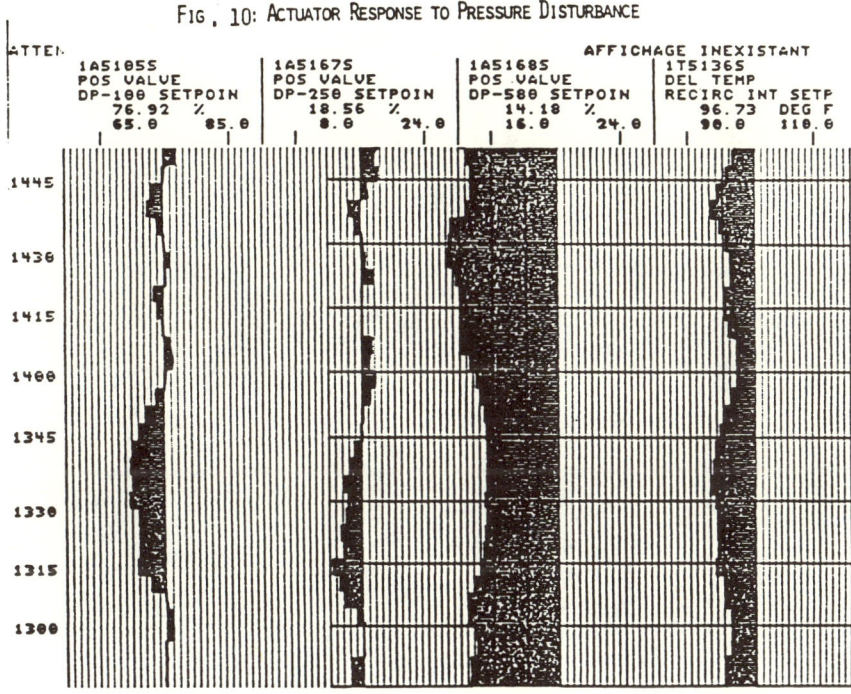

Fig. 10: Actuator Response to Pressure Disturbance

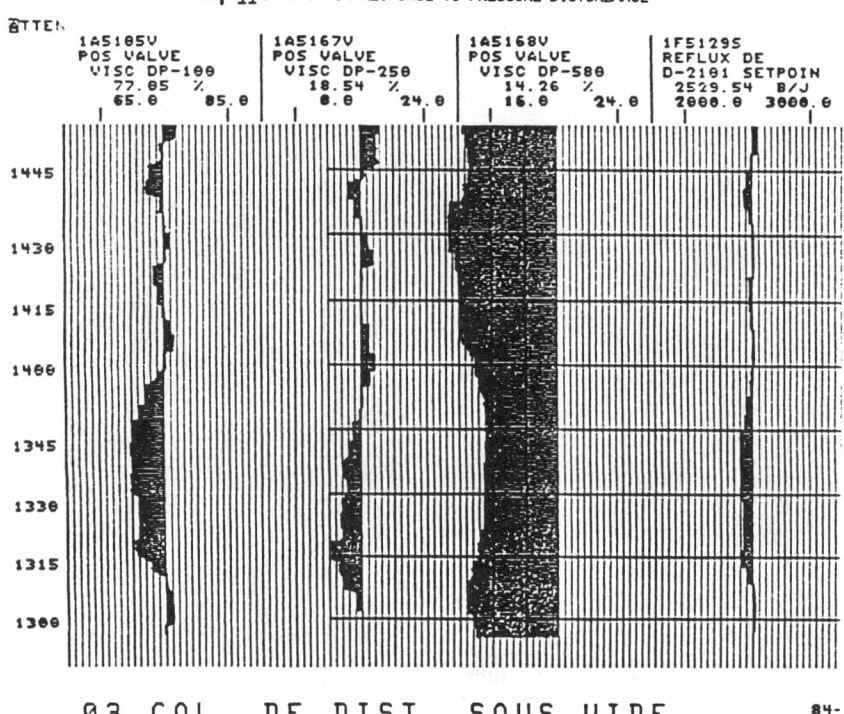

FIG. 11: ACTUATOR RESPONSE TO PRESSURE DISTURBANCE

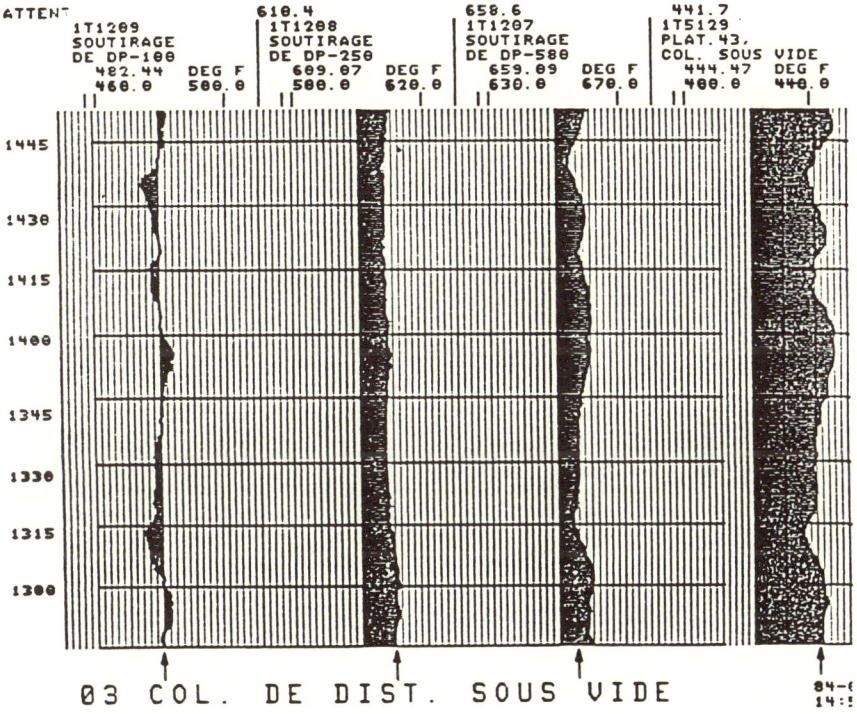

Fig. 12: Tray Temperature Behaviour Under Computer Control

REFERENCES

Boyle, T. J. (1978). Practical algorithms for cross-direction control. TAPPI, 61, 1, pp. 77-80.

Box, G. E. P., and G. M. Jenkins (1969). Time Series Analysis - forcasting and control. Holden-Day, San Francisco.

Cutler, C. R., and B. L. Ramaker (1979). Dynamic matrix control - a computer control algorithm. 86th AIChE National Meeting, Houston.

Morshedi, A. M., et al. (1985). Optimal solution of dynamic matrix control with linear programming(LDMC). Proc. JACC, Boston. Paper WA7.

Garcia, C. E., and Morshedi, A. M. (1984). A solution of the dynamic matrix control problem using quadratic programming (QDMC). Proc. 1984 Can. Conf. Ind. Comp. Systems, University of Ottawa, Ottawa, Ontario. Paper 13.

Kouvaritakis, B., and J. M. Edmunds (1978). In M. K. Sain et al. (Ed.), Alternatives for Linear Multivariable Control. National Engineering Consortium Inc., Chicago. Paper 4.2, pp. 229-246.

MacFarlane, A. G. J., et al. (1978). In M. K. Sain et al. (Ed.), Alternatives for Linear Multivariable Control. National Engineering Consortium Inc., Chicago. Paper 4.1, pp. 189-228.

MacFarlane, A. G. J., and J. J. Belletrutti (1973). The characteristic locus design method. Automatica, 9, pp. 575-588.

Ray, W. H. (1981). Advanced Process Control. McGraw-Hill, New York. Chap. 6, p. 320.

Richalet, J. et al. (1978). Model predictive heuristic control: applications to industrial processes. Automatica, 14, pp. 413-428.

Rosenbrock, H. H. (1974). Computer-aided Control System Design. Academic Press, London.

Treiber, S. S. (1982). Multivariable control of a propane-propylene splitter. Proc. 1982 Can. Conf. Ind. Comp. Systems., McMaster University, Hamilton, Ontario. Paper 5.

Treiber, S. S. (1984). Multivariable control of non-square systems. Ind. Eng. Chem. Proc. Des. Dev., 23, pp. 854-857.

MODEL PREDICTIVE CONTROL

Dale E. Seborg
University of California, Santa Barbara, CA 93106

Session Summary:

During the past two decades there has been a growing interest in model-based control strategies, that is, strategies which use a process model in the on-line control calculations. Early methods such as the Smith Predictor and feedforward control have been successfully applied for over 25 years. Multivariable control techniques such as decoupling and optimal control have also been available but apparently have not been generally adopted.

In recent years there has been considerable interest in a special class of model-based control techniques referred to as "predictive control" or "model predictive control". The distinctive feature of these techniques is that a process model is used to predict the future behavior of the process outputs. The control calculations are then based on these predictions as well as the current measurements. The two most popular techniques, Dynamic Matrix Control (DMC) and IDCOM, are based on discrete convolution models, but the same conceptual approach can be used in conjunction with other types of process models. Both techniques accommodate inequality constraints on inputs and outputs by solving a quadratic programming problem at each sampling instant.

In the first paper in this session, Garcia and Prett summarize the current state of the art of Shell's Dynamic Matrix Control. They also include a stimulating discussion of the need for a Unified Theory for process control and for an Integrated Technology to apply both process control and on-line optimization. The second paper, by Froisy and Richalet, provides an overview of the IDCOM approach and summarizes industrial applications. The final paper, by Popiel et al, proposes a model-based control strategy for control problems with a single controlled variable and multiple, constrained manipulated variables.

INDUSTRIAL APPLICATIONS OF IDCOM

J. Brian Froisy
Setpoint, Inc., Houston, TX

Jacques Richalet
Adersa Gerbios, Paris, France

Abstract. Predictive control techniques are being analyzed and applied in both academic and industrial environments today. These techniques are well suited to process control applications. One such technique, IDCOM, has been applied to the multivariable control of several refinery and petrochemical processes during the past decade.

An introductory discussion describes the basic IDCOM control structure. Process and control system descriptions for recent applications involving reactor controls are presented to show the general structure and strategies possible with IDCOM. These applications demonstrate the importance of dealing with constraints in complex control systems. Moreover, it is evident that economic incentives also need to be imbedded in the structures to fully realize potential benefits. These experiences indicate the need for predictive control algorithms that can solve constrained optimization problems with reasonable amounts of engineering and computing resources.

Keywords. Internal Model Control; Predictive Control; Computer Control: Digital Control; Industrial Processes; IDCOM.

INTRODUCTION

IDCOM (IDentification and COMmand) is a multivariable predictive control technique that has been applied to several industrial processes since the early 1970's. The technology was developed in France and initial applications were primarily in Europe with recent applications in the United States, Canada, and Japan. The basic philosophy of this control technique is to use an internal process model to predict the effects of past process inputs, both manipulated variables and measured disturbances, and to subsequently calculate values for future manipulated variables required to achieve a specified process response goal. IDCOM is usually implemented with linear process models obtained by process identification. The identified process is represented internally by discrete convolution models and therefore the methodology is applicable to arbitrarily complex (linear) dynamic systems. Extensions where process nonlinearities warrant nonlinear models are also possible.

Practical control applications often require strategies that must cope with constraints on manipulated and controlled variables and also on ancillary dependent variables that can override the normal control sequence. Similarly, the degrees of control freedom may vary dynamically for some multivariable control problems due to constraints or other factors. IDCOM provides a control structure whereby these types of practical problems can be accommodated in a flexible way which is usually dictated by operational or economic considerations.

IDCOM - A BRIEF DESCRIPTION

A brief description of IDCOM including comments on some similarities between IDCOM and other internal model-based controls is presented in this section. More detailed descriptions and comparisons are available in the literature (Martin, 1981; Mehra et al, 1980; Richalet et al, 1978.)

The internal model is structured as a discrete convolution model represented by the process impulse response. Model predictions are of the form:

$$s(n) = \Sigma a_i e(n-i) = A^T E(n) \qquad (1)$$

The elements a_i of the vector A are the coefficients of the discrete impulse response and the time dependent vector E represents the process input. The output at time n is s(n). The extension to the general multivariable case is reasonably straightforward but not necessary for understanding the basic control strategy.

The summation in (1) theoretically must be done over all non-zero terms. Since impulse response coefficients tend toward zero for asymptotically stable processes, a finite summation provides an adequate approximation in practice. With an unstable process, e.g. some level control applications, the problem can be restructured to avoid an infinite length impulse response model.

The convolution model is a non-minimal representation since there are more impulse response coefficients than parameters in an equivalent transfer function representation. For example, a first order process model is fully described by a gain and a time constant whereas a convolution model for the same process would theoretically require an infinite number of coefficients. In practice, the model is often approximated with fewer than 50 coefficients. The model structure is also order-independent since an arbitrary order process, including dead time or other non-minimum phase behavior, can be represented without additional effort. This characteristic is helpful when process identification tests must be run to establish a model without detailed knowledge of the process. In the final analysis, the impulse model structure provides a flexible way to handle general problems at the expense of extra computations and storage.

The idea behind the IDCOM control calculations can be explained by reference to the following equation.

$$s(n+k) = \Sigma_f a_i e(n+k-i) + \Sigma_p a_i e(n+k-i) \qquad (2)$$

The process output at k time steps into the future is s(n+k). This value depends on input contributions that have already occurred represented by the past sum, Σ_p, and on future inputs whose contribution is represented by Σ_f. The control algorithm must then determine the future values for e that will result in the output, s(n+k), satisfying the control goal. The goal used by IDCOM is that the output should be at some reference point at time (n+k). There will be at least one solution if k is greater than the dead time. A unique solution is found if the future inputs at (n,n+1,...n+k-1) are assumed to be equal. In this case, the control output is:

$$e_f = [r(n+k) - \Sigma_p a_i e(n+k-i)]/\Sigma_f a_i \qquad (3)$$

The desired reference point is r(n+k) and the term $\Sigma_f a_i$ is a constant. The control input, e_f, is applied at time (n) and the entire procedure is repeated at time (n+1). Thus, different values are calculated and applied at each time increment even though the calculation is done as if only a single step will be applied. Notice that the numerator in (3) is a predicted error at some time in the future.

The division by $\Sigma_f a_i$ in (3) shows that the algorithm must effectively invert part of the process model to determine the control action that will satisfy the goal. This inversion is possible when k is chosen greater than the dead time or non-minimum phase response time. (The criteria in the multivariable case is more complex, but the idea of inverting part of the model still applies.) Other model-based control strategies have similar requirements. For example, the design approach for internal model control (IMC) requires an inversion of the process model but this is limited to the "invertible part" of the model. Likewise, Dynamic Matrix Control (DMC) requires a prediction horizon that is greater than the dead time. All of these requirements are results of physical realizability considerations. The reference point used to calculate the control action is normally computed such that this point is on a <u>reference trajectory</u> which defines the path the process would take in returning to the setpoint if no further disturbances were

encountered. Figure 1 is a diagram that indicates how the response to
a setpoint change might look for a specified reference trajectory.
The reference trajectory is often specified to be a first order
response with a dead time equal to the process dead time. For the
multivariable case, each controlled variable may have a unique
reference trajectory. The reference trajectory used by IDCOM is
basically the same as the filter that is used in the IMC design
procedure. The reference trajectory time constant plays a vital role
in practical IDCOM applications. It is a tuning parameter that can
be adjusted in an intuitive way to establish the tradeoff between
control system performance and robustness. Also, the activity of the
manipulated variable which often enters into subjective tuning
criteria is directly affected by the reference trajectory time
constant. Simulation of the process and the control system has
proven to be an effective tool for establish initial tuning and for
gaining insight into response characteristics when there is mismatch
between the internal model and the process.

IDCOM APPLICATIONS

The following list summarizes the process control applications
of IDCOM.
- PVC Plant - Reactors and Distillation (1971)
- Fluid Catalytic Cracker Main Fractionator (1973)
- Power Plant Steam Generator (1976)
- Crude Oil Fractionators (1976-1978)
- Oil Hydrotreater/Hydrocrackers (1981-1985)
- Fluid Catalytic Cracker Regenerator (1985)
- Catalytic Reformer (1985)

Applications in other areas such as military weapons and a ship
control system have also been reported.

Some early industrial applications of IDCOM have been documented
by Richalet et al (1978). The first application involved control of
a PVC plant in France. The system involved cracking oven temperature
control with feedforward compensation based on measured and

calculated disturbances. Two distillation columns were also controlled. For one column, impurities of two components in the distillate were controlled by manipulating reflux. Composition feedback was provided by a GC and control was based on the impurity which was closest to its specification limit. For the other column, the IDCOM strategy was configured to supervise two analog temperature controllers that in turn manipulated boilup and reflux. Cloud point (a diesel oil property) control on a crude oil atmospheric distillation column was another early application of IDCOM. There were three levels of cascade control. The lowest level was simply an analog flow controller. Next, was an IDCOM controller, including feedforward and feedback, for a tray temperature. The highest level controller was an IDCOM feedback controller based on an on-line cloud point analysis. This cascade/feedforward structure demonstrates how IDCOM can be used in familiar control system arrangements.

A fluid catalytic cracking fractionator column is similar to a crude oil atmospheric distillation column. In this application, IDCOM was applied to the multivariable control of two key column temperatures that relate closely with product quality. An internal pan level in the column is affected by various disturbances and also by the manipulated variables (the side draw flow rates). An important objective of the control strategy was to ensure that this level did not fall below a safe minimum. This is a case where the predictive nature of the the IDCOM was used to allow the level to vary within bounds but also override the product quality control objective if predictions indicated that the level limit would be violated.

A power plant steam generator control system was built using IDCOM technology. The system provided control of steam pressure and two superheat temperatures. This was an interesting application in that there were non-minimum phase responses in the process.

There have been several recent applications of IDCOM. Among these is a hydrocracker control system described by Martin and Van Horn (1984). There are three applications of the basic process and control system shown in Fig. 2. The overall control objective is to maintain a weighted average bed temperature which relates to the reaction severity and is a key parameter that determines product quality and yield. Feed is heated in a fired heater before entering the first catalyst bed where an exothermic reaction occurs. Hydrogen is injected at the inlet to subsequent beds both to quench the reaction and add more reactant. Reactor products are processed in a downstream fractionation section. There are three levels of control cascade in this system. First, bed inlet temperatures are controlled using standard PID algorithms in a distributed control system. At the next level are three IDCOM controllers implemented in a minicomputer that control the outlet temperature of beds two through four. The deadtime between bed inlet and exit was one reason for choosing an advanced control technique at this level. The weighted average temperature control block supervises one first level controller and the other IDCOM controllers. The ability to handle valve position constraints with a dynamic prediction are an important part of this highest level control function. The strategy also provides feedforward compensation for changes in feed rate or hydrogen partial pressure.

An extension to this basic control system has been applied to a case where it was desirable to adjust the reaction severity to maintain a ratio between fresh feed and a recycle stream from the bottom of a downstream fractionator. Ratio control was implemented in a distributed control system and one task of the IDCOM controller was to maintain the level in a recycle surge drum within an acceptable operating zone. This was done by manipulating the weight average temperature setpoint of the previously described control system. The transfer function between the weight average temperature and the level included an integrator. In this case, it was necessary to effectively remove the integrator from the process by using an estimated derivative of the level as the controlled variable. This approach results in an internal model that is asymptotically stable.

IDCOM has recently been applied to the supervisory control of the regenerator in a fluidized bed catalytic cracking (FCC) process (McPherson et al 1985). A simplified process diagram including the basic instruments used by the IDCOM control algorithm is shown in Fig. 3. The process is designed to upgrade low value heavy oil into lighter products such as gasoline. Fresh feed and recycle oil are fed to the "risers" which are essentially plug flow reactors. Here, oil and solid fluidized catalyst are mixed and the endothermic cracking reactions quickly occur. A reaction by-product is "coke" which deactivates the catalyst. Reaction products are separated in the reactor (a misnomer since the reaction occurs in the risers) and routed to a fractionation column while the deactivated catalyst is returned to the regenerator. Here, the coke is burned off of the catalyst at a high temperature, typically 1,300 degrees F. Combustion air is supplied by a large air blower. An important operating objective is to supply the right amount of combustion air to satisfy the coke burning demand. Alternately, it is sometimes desirable to adjust the coke generation to match available air blower capacity.

There are many physical and economic interactions in the typical FCC process that can result in different control system objectives. In this application, there was incentive to control excess oxygen in the flue gas by adjusting either air rate or reactor severity. Riser temperatures, recycle rate, and feed preheat temperature are all operating parameters which affect severity and hence the amount of coke deposited on the catalyst. Therefore, the air/coke balance as measured by flue gas oxygen could be controlled by adjusting any of these quantities as well as direct control of air rate.

The IDCOM structure applied to this control problem is shown in Fig. 4. There are five potential manipulated variables and two potential controlled variables. Control logic determines which manipulated variables to use based on operator input, the availability of the individual manipulated variables, and on an economic priority. Flue gas oxygen is the primary controlled variable while the flue gas temperature becomes a controlled variable in case of high temperature excursions. The temperature is constantly being predicted so that preventative control action can be taken to avoid an actual temperature limit violation.

SOME OBSERVATIONS

There are important observations about these applications related to degrees of freedom and constraints. Industrial multivariable control problems involve more than simply regulating an N X N process. It is not unusual to have extra degrees of freedom for the manipulated variables. Operational and economic considerations determine how these degrees of freedom should be used. However, encountering manipulated variable constraints can result in degrees of freedom being changed in real time and the control system must accommodate this possibility. It is also possible for constraints to result in an uncontrollable system. In this case, the controller performance must be allowed to degrade by relinquishing control on one or more variables. Constraints also apply to dependent variables that are normally not controlled. These dependent variables effectively become controlled variables when they reach a limit and sometimes override the normal control sequence. The internal model must be maintained for these dependent variables both to predict potential constraint violations and to assure a smooth transition when an override is necessary. Since the optimum operating point often turns out to be on one or more constraint boundaries, it is common for the control system to have to be continuously dealing with constraints and varying degrees of freedom.

CONCLUSIONS

Observations in the previous section point out areas for further developments in multivariable predictive control. The IDCOM structure is very flexible and provides a framework for building control systems that deal with constraints and varying degrees of control freedom. While some applications involving these factors can be accommodated with little or no additional effort, other more complex ones require special attention. Additional control tools that can handle increasingly difficult problems with a general structure are a definite need. Furthermore, the tools should provide a means to translate physical and economic objectives into control system objectives in a straightforward manner. Control engineers need proven tools that can be applied without undue effort. End users need controllers that are robust and can be maintained and tuned by local personnel.

Industrial multivariable control applications pose difficult problems that demand both sound engineering and reasonable maintenance practices to ensure long term success.

REFERENCES

Martin, G.D. (1981). Long-Range Predictive Control. A.I.Ch.E. Journal, Vol. 27, No. 4.

Martin, G.D. and L.D. Van Horn (1984). Experience with hydrotreater computer control. Hydrocarbon Processing, March 1984.

McPherson, M., Nairn, I., Lee, P., Van Horn, L.D., Froisy, J.B. and T.A. Badgwell (1985). Experience with FCCU computer control. NPRA Computer Conference, paper No. CC-85-108.

Mehra, R.K. and R.Rouhani (1980). Theoretical considerations on model algorithmic control for non-minimum phase systems. Joint Automatic Control Conference Proceedings.

Richalet, J., Rault, A., Testud, J.L. and J. Papon (1978). Model predictive heuristic control: applications to industrial processes. Automatica, Vol. 14, 413-428.

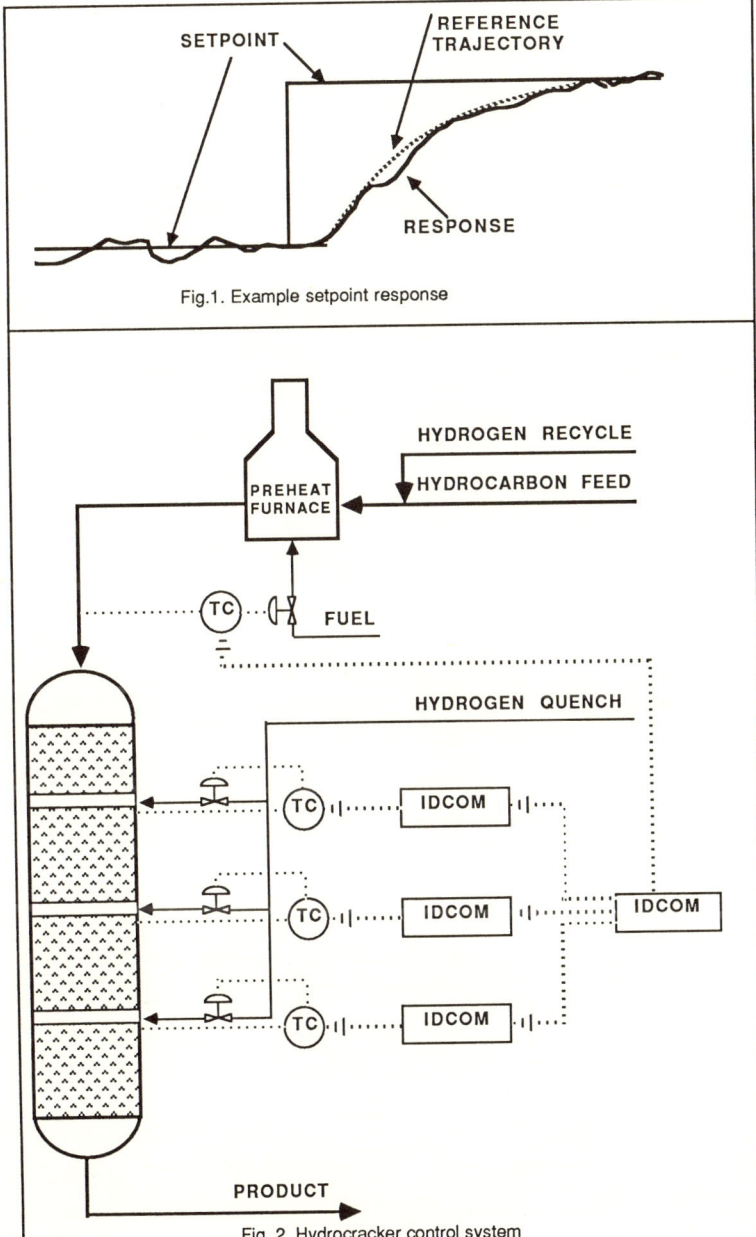

Fig.1. Example setpoint response

Fig. 2 Hydrocracker control system

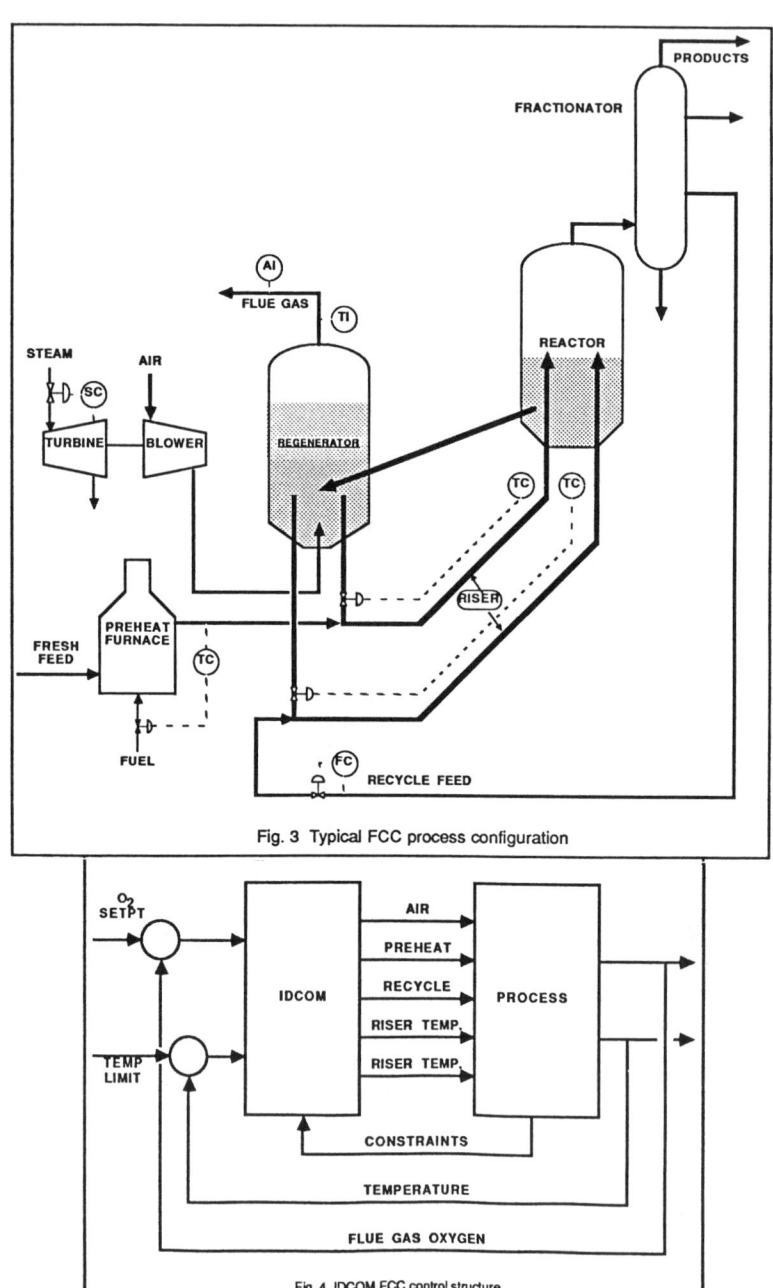

Fig. 3 Typical FCC process configuration

Fig. 4 IDCOM FCC control structure

ADVANCES IN INDUSTRIAL MODEL-PREDICTIVE CONTROL

Carlos E. Garcia
David M. Prett
Shell Development Company, Houston, TX 77001

Abstract. Due to the economically sensitive condition of the chemical and petroleum industries, we can no longer afford to operate with the inefficiencies of the past. Over the last 15 years we have found that on-line optimization implemented with model-predictive control recognizing process and operating policy constraints, provides the best means for achieving the profit potential of our plants. Only model-predictive controllers permit the flexibility required to handle the constantly changing performance criteria, in particular the enforcement of operating constraints. However, the performance criteria of today's problems are becoming harder to quantify, while optimization systems are driving the processes over a wider range of operating conditions than ever before. Therefore, there is a need to improve the model-predictive control techniques so that practical performance criteria based on engineering judgement can be transparently specified, and that model inaccuracies are considered explicitly in the problem formulation. In this paper the state-of-the-art in industrial model-predictive control is presented. An attempt is made to propose a path of evolutionary development in process control that will converge to a **Unified Theory** replacing many of the ad-hoc solutions developed over the last thirty years. New techniques for multi-objective optimization and robust control are described that have the potential to allow us to improve the current technology in order to solve the control problem at hand. It is concluded that the complex control problems of today can only be solved through the development of a **Unified Theory** along the concepts of model-predictive control. This **Unified Theory** of process control will then allow for the application of the **Integrated Technologies** of process optimization and control.

<u>Keywords.</u> Model Predictive Control; Dynamic Matrix Control; optimization; constraints; performance; robustness.

PREFACE

It is the purpose of this paper to determine a future direction for process control research within the context of model-predictive control methodologies that will allow for the evolution of a consistent, unified approach to problem solving in our field.

We recognize that history includes many ad-hoc solutions that did not pass the test of time. However, we have learned as much from historical failures in approach as we have from successes. Over the last thirty or more years there have been many significant contributions to our technology. Some were not appreciated because of the truly innovative nature of the contribution and some were overly praised but did not withstand the scrutiny of time. We wish that our paper be accepted in the sense of a contribution that is very dependent on the work of those that came before us. In presenting this paper we wish to acknowledge this work and credit our contribution to the fine work done by our predecessors, the large number of whom preclude individual recognition without the probability of omitting some. With the benefit of observation of the work of all our predecessors we therefore propose this forward plan for the development of a **Unified Theory** of process control as our contribution to the field.

INTRODUCTION

In order to ensure a competitive edge in today's economic climate the chemical and refining process industries need to extract the maximum profit from their processes in the face of constantly changing market conditions. An expensive way of realizing the achievable profits is to modify the process design and install modified equipment to take account of these changes. One then hopes that the changed market condition is maintained long enough to recoup the capital cost and also generate a resultant net profit. Alternatively, or in conjunction with modified equipment, plant profitability can be achieved with minimum capital costs by applying broad-based technology to on-line optimize our chemical plants and refineries.

On-line optimization systems relentlessly realize the achievable performance of the existing equipment by driving the process to its operating constraints while constantly updating the operating conditions based on the process economics. Its successful implementation generally involves two phases (Garcia, 1982) (Fig. 1):

- Optimization Phase:

Steady-state process representations are used to optimize the process based on an economic objective function subject to equipment and operating policy constraints; the process representations include the effect of disturbances of economic impact as are feedstock quality, equipment degradation, etc. which are assumed to occur with larger period than the inherent residence time of the process; the operating point is updated with a frequency comparable to the disturbance frequency.

- Control Phase:

Dynamic process representations are used to optimize a performance objective function which penalizes deviations of process variables from their optimum values given by the optimization phase; since the optimal operating point generally lies at the intersection of constraints, this phase has the responsibility of dynamically enforcing those constraints while rejecting fast disturbances; the control system updates manipulated variables at a faster frequency than the optimization phase executions.

One can argue about the justification of the division between the phases. It is obvious that in cases where the disturbances of economic impact change faster than the inherent plant dynamics both phases coalesce therefore requiring frequent dynamic on-line optimizations. Even in those cases where separation is not advisable, it has been done because it is not possible with today's hardware to solve on-line optimization problems with detailed dynamic process representations. Therefore, the two-phase on-line optimization approach is widely used.

Detailed process representations of large collections of algebraic and differential equations based on first principles avoid the issue of pre-biasing the solution of the optimization problem caused by the use of so-called simple models. Since there exists no technique today for choosing, a-priori, the correct degree of detail for optimization models, large computation requirements are inevitable. The field of expert systems is beginning to contribute significantly in providing a framework that allows us to collect the modelling skill of practitioners as new systems are implemented. Also, the great improvements in computing power provided by parallel processing might make this issue of model rigor disappear in the near future. More will be said about these two areas of research effort later.

Since the optimization problem is more tractable than the control phase and conceptually easier to understand, there exists a temptation to "implement" only the optimization phase without regard for the control phase. However, one must recognize that optimization problems are relatively easy to solve but the results are extremely difficult to implement. The control phase makes this implementation possible. Therefore, the achievable profits of optimization cannot be realized without good control and one cannot divorce one technology from the other. We have chosen to define this synergism of technologies as the **Integrated Technology** approach to process control. It is only in this realm that process control methodologies can even be considered for industrial application, and therefore, we have a clear way of assessing current technological advances and of directing future research efforts. Let us review the current state of the process control area in the light of this approach.

Over the last 25 years we have experienced a tremendous evolution in the techniques used to control our chemical and refining processes. If there is a single explanation for the changes observed it is that the applications have always been influenced by the control hardware available for concept implementation. Academicians as well as practitioners have been discussing for years the so called "gap" between theory and practice, giving what we consider to be an imprecise picture of the real situation. As we see it, from the researcher's side there exist two main reasons for the lack of solutions to the real problems. On the one hand, some theoreticians have either totally ignored the real problem and gone into theoretical developments that cannot be applied to chemical and refining processes, or when applicable the solutions have resulted in being practically impossible to implement with the available hardware . On the other hand, many researchers (both in industry and academia) have focused their attention on ad-hoc solutions to problems so that they fit the capabilities of the most commonly available hardware implementation modes, therefore necessarily limiting the scope of their research. While this may have been justified in the past it is hard to understand why this is still practiced to the present when on-line computing power has become relatively inexpensive and provides a framework for generalized solution conception replacing the ad-hoc approach.

However, we as practitioners should not be so quick as to completely blame theoreticians for such failure. Our greatest sin is to have led them to believe that ad-hoc solutions were all that were acceptable to us. Even when these techniques have failed, their failure was usually excused resulting in the development of "more advanced" ad-hoc solutions. As a consequence, this has created a situation where these pseudo-successful ad-hoc solutions have become the accepted theory. Although no significant economic penalty has been paid so far for these misdirected efforts, the current economically sensitive condition of our industry dictate the need to search for a **Unified Theory** of process control in order to ensure a competitive edge. It is our estimate that the profitability of our industry could be significantly improved through application of the **Integrated Technology** approach of optimization and control.

In recent years we have observed the surge of a new generation of researchers who have courageously made significant advances and proposed solutions to control problems which tend to provide some unification. It is interesting to realize that the catalyst for these advances has been provided by developments in industry. As early as the mid-seventies some visionary corporations realized the potential of a **Unified Theory** and embarked in the development of their own control algorithms in order to address the real problem of achieving improved profitability. By then process control practitioners had already abandoned the "modern control theory" approach to multivariable control. Except in few notably successful implementations (Ray, 1981) the Linear Quadratic Optimal Controller (LQC) so successful in aerospace control applications (Kwakernaak and

Sivan, 1972) was generally found to be difficult to tune and exhibited poor robustness to plant changes. Therefore the interaction problem was handled by the use of decoupling techniques for multi-loop systems (Ray, 1981). Independent developments in the USA by Shell Oil Co. (Cutler and Ramaker, 1979; Prett and Gillette, 1979) and in France by ADERSA/GERBIOS (Richalet et.al., 1978) brought forth new techniques based on a linear model prediction: Dynamic Matrix Control (DMC) and IDCOM, respectively. One can show that in the absence of process constraints these techniques are mathematically identical to LQC except that a simpler formulation allows easier on-line tuning and transparent specification of control objectives. However, it is the explicit inclusion of process constraints in the algorithm what has made these techniques so successful in the industrial environment. The development of DMC marked the first time a control technique was devised which could successfully handle process constraints, thus allowing the realization of achievable profits via implementation of results of on-line constrained optimizations. In fact, in the particular case of fluid catalytic cracking control no technique previously published had provided the performance required in the face of constantly changing control objectives and active constraints (Prett and Gillette, 1979; Morari, 1981).

One significant contribution associated with the introduction of DMC to the control literature was in setting a new tone by communicating industry's needs in a language that both industry and academia could understand, i.e. the language of mathematics and control engineering. It was hoped that this would lead to a renaissance of control research along more unified lines. Such renaissance has happened and we have witnessed numerous theoretical developments in the chemical engineering process control area over the last seven years which appear to be on the right track for finally achieving the desired unification.

Our objective in this paper is to define where we presently are technologically, and then point out what is necessary to promote a continuance of the current efforts toward solving the control problem. We strongly believe that the model-predictive control methodology provides the only framework for achieving this. In addition to its constraint handling capabilities, model-predictive control has performed reliably in applications because of its:

- multivariable features
- stability properties
- inherent deadtime compensation

These properties have been analyzed in the control literature (Garcia and Morari, 1982).

But the most important feature of model-predictive controllers, and in particular of DMC, is that they solve a new problem and thus produce a new controller **at each execution.** This makes the algorithm capable of taking measures on-line which could not be foreseen at the design stage therefore having unique constraint handling capabilities. In this sense, **model-predictive** controllers must be uniquely distinguished from other controller designs. For lack of a better terminology, in this paper we will classify as **recursive** controllers those techniques that use a fixed relationship between error and manipulated variable which is solved for only once, during the conceptual design phase.

The algorithm as proposed did not explicitly deal with the issue of model accuracy. The best available model was determined and the control engineer designed a control system to meet implementation performance standards. Robustness was achieved by an intuitive process of relaxing performance standards to recognize inherent model inaccuracies. Of course, as time went on, the model might become increasingly inaccurate due to changing process conditions. This was handled by having an expert work force in the field to perform on-line detuning until such time as a new model determination was demanded. However, in today's business environment it is not possible to justify the field assignment of such expert manpower. For this reason, as well as for control efficiency, we must begin to deal with the performance/robustness issues more methodically.

As a result of an increased number of installations of on-line optimization systems, recent experiences have indicated that even though the model-predictive control approach is still the best tool for multivariable constrained control, the traditional approach that assumes an exact, fixed, linear model does not allow us to achieve the desired control performance objectives. We are finding that the performance objective and constraints of modern plants are changing more frequently than before and are more difficult to translate to solvable mathematical forms. In addition, processes are being operated over a wider range of conditions than ever before. Since we can no longer afford the costs associated with maintaining these complex loops, we have to improve our control methods so that the desired performance is achieved with minimum demands on manpower.

In this paper we discuss some of the many areas of research which we are currently exploring in order to solve these problems while at the same time pointing out where the future developments in this area should be directed.

The paper is organized as follows. First we give a brief description of the current state-of-the-art in model predictive control as practiced in Shell pointing out the special properties that make it suitable for solving practical control problems. The issue of specification of control performance is discussed next in the light of the **Integrated Technology** approach. Finally, in the last part we deal with the issue of the effects of model error on performance loss and list the several methods currently under study to address the problem at hand. The conclusion will contain an assessment of the process control area as we see it, indicating where the technological breakthroughs need to happen in the light of the current and future condition of our petrochemical industry.

ELEMENTS OF MODEL-PREDICTIVE CONTROL: THE LINEAR CASE

The model-predictive control methodology consists of the following elements:

- Linear model relating the manipulated variables and measurable disturbances to the outputs of interest

- Prediction of the outputs of interest over the future time horizon, corrected via feedback

- Computation of future manipulated variable moves to make the prediction of the outputs and manipulated variables satisfy some performance criterion.

It is assumed that the model is open-loop stable, and that the algorithm is implemented in a sampled-data system. Therefore, a discrete-time model of the process is considered where values of variables are known only at discrete intervals of time k:

$$kT \leq time \leq (k+1)T$$

where T is the sampling time of the system.

A "moving horizon" approach is used where at every sampling interval the prediction is updated and the future manipulated variable moves are computed. However, only the manipulated variable move corresponding to the current sampling interval is implemented. One step of this approach is illustrated in Fig. 2 where the prediction of a single manipulated variable $m(k)$ and output variable $y(k)$ is obtained at the end of the computation phase in order to, for example, bring the prediction to a target value.

In this section all three elements are discussed based on current practice. This will allow us to point out where the advantages in the method may lie as well as where improvements are necessary.

The Model Formulation

Model-predictive control algorithms employ a linear model of the process to predict the effect of past changes of manipulated variables and measurable disturbances on the output variables of interest. In practice an input-output model description is used as follows:

$$y(k) = \sum_{\ell=1}^{\infty} a_\ell \, \Delta u(k-\ell) + d(k) \tag{1}$$

where a_ℓ is a matrix of unit step response coefficients, Δu are the changes or moves of the system inputs (manipulated variables and measurable disturbances) and $d(k)$ is a term containing all the contributions to the outputs that the model cannot describe as are unmeasurable disturbances. An equivalent model form is the impulse response model :

$$y(k) = \sum_{\ell=1}^{\infty} h_\ell \, u(k-\ell) + d(k) \qquad (2)$$

where h_ℓ is a matrix of unit impulse-response coefficients, and

$$a_\ell = \sum_{j=1}^{\ell} h_j \; .$$

The only difference between models (1) and (2) is that the step-response model contains an implicit integrator.

Of course, the properties of the method should not depend on the particular model formulation. One could even use a state-space model as follows:

$$\begin{aligned} x(k+1) &= A\, x(k) + B\, u(k) \\ y(k) &= C\, x(k) + d(k) \end{aligned} \qquad (3)$$

Since the system is stable, successive substitutions of the state $x(k)$ should yield exactly model (2). In order to obtain (1), we add an additional state which acts as our integrator as follows:

$$\begin{bmatrix} \Delta x(k+1) \\ q(k+1) \end{bmatrix} = \begin{bmatrix} A & 0 \\ CA & I \end{bmatrix} \begin{bmatrix} \Delta x(k) \\ q(k) \end{bmatrix} + \begin{bmatrix} B \\ CB \end{bmatrix} \Delta u(k) \qquad (4)$$

$$y(k) = q(k) + d(k)$$

where Δx denotes a change in the state. The resulting input-output model of (4) is the step-response model (1).

The Prediction Problem

Any of the model descriptions can be used to predict future values of the outputs $y(k)$ over a time horizon. For this discussion only model (1) will be considered. For any future interval of time ℓ the output prediction is given by:

$$\hat{y}(k+\ell) = \hat{y}^*(k+\ell) + \sum_{i=1}^{\ell} a_{mi} \Delta m(k+\ell-i) + \hat{d}(k+\ell), \qquad \ell = 1,2\ldots P \qquad (5)$$

where the contribution of past inputs to the projection is given by the term

$$\hat{y}^*(k+\ell) = \sum_{i=\ell+1}^{\infty} a_i \Delta u(k+\ell-i) \qquad (6)$$

and $\hat{d}(k+\ell)$ is the prediction of unmodelled effects. Note that equation (6) contains the effect of all past inputs: both manipulated variables and measurable disturbances while the summation in (5) only goes over future manipulated variable moves. The step-response coefficients a_{mi} relate manipulated variable moves to the outputs.

An important term in (5) is the prediction of the unmodelled effects $\hat{d}(k+\ell)$. This term can be obtained by using the most recent output measurement $y_m(k)$. Using equation (1) we can calculate $d(k)$ as follows:

$$d(k) = y_m(k) - \sum_{\ell=1}^{\infty} a_\ell \Delta u(k-\ell)$$

The predictions $\hat{d}(k+\ell)$ can then be estimated by filtering $d(k)$ appropriately.

In the particular case of DMC the prediction most commonly used is

$$\hat{d}(k+\ell) = d(k) \quad \text{for } \ell = 1,2..P. \tag{7}$$

This estimator is satisfactory only when the measurement noise level is small and the deterministic part of the unmeasured disturbances has fast dynamics and is infrequent. Otherwise, significant errors in the prediction could ensue.

One way to resolve this problem is to use optimal filtering theory (Kwakernaak and Sivan, 1972) to design an observer/estimator for $\hat{d}(k+\ell)$ if some knowledge of the noise dynamics of the process is at hand or if the deterministic part of the unmeasured disturbance model is available through some identification method. Such approach would not only provide the filtering of noise but also good estimation of slowly varying disturbances. However, one should be aware of the fact that these techniques are valid only under very stringent assumptions about noise dynamics (namely, zero mean Gaussian noise) and therefore, one is forced to resort to ad-hoc selection of the noise covariances for tuning. In the absence of noise the use of a model for the disturbances should improve the prediction.

The Structured Singular Value theory of Doyle (1982) gives a comprehensive controller design method under uncertainties in both model parameters and disturbance description. One advantage is that the analysis is not restricted to a Gaussian description of the noise. Consequently, it seems to be the best approach to date to resolve this issue of disturbance prediction. This is a topic of current research in our group.

The Control Problem

Once a model prediction of the outputs is obtained the following generalized control problem is solved:

> Find the sequence of M future manipulated variable moves $\Delta m(k)$, $\Delta m(k+1)$, ... $\Delta m(k+M-1)$ so that the prediction of the manipulated variables and outputs satisfy a set of performance criteria.

This problem is solved at every sampling time when a new prediction is updated based on the most recently obtained feedback measurements. The first move $\Delta m(k)$ is the only one that is implemented although the others could be used in case of loss of measured variables. However, the main reason for computing future moves is to allow for the imposition of desirable performance objectives and constraint handling on manipulated variables as well as on output variables.

Except for the linear model assumption, this control problem is general enough to allow the solution of most foreseeable control problems of industrial relevance. The challenge consists in being able to translate the desired performance objectives into an equivalent mathematical form that allows computation of the moves. One convenient form that we have used with success is the Dynamic Matrix Control algorithm (Cutler and Ramaker, 1979; Prett and Gillette, 1979) and more recently the Quadratic Dynamic Matrix Control (QDMC) algorithm (Garcia 1984). The QDMC formulation solves for performance objectives expressed as a sum of squared deviations of controlled outputs from target values and inequality constraints of outputs and manipulated variables.

In the particular case of QDMC the following quadratic program is solved on-line:

$$\min_{\substack{\Delta m(k) \ldots \\ \Delta m(k+M-1)}} \tfrac{1}{2} \sum_{\ell=1}^{P} \{ \, || \, \Gamma_\ell [\hat{y}(k+\ell) - y_S(k+\ell)] \, ||^2 + || \, \Lambda_\ell \, \Delta m(k+\ell-1) \, ||^2 \, \}$$

$$\text{s.t} \quad \hat{y}(k+\ell) = \hat{y}^*(k+\ell)$$

$$+ \sum_{i=1}^{\ell} a_{mi} \, \Delta m(k+\ell-i) \quad (8)$$

$$+ \hat{d}(k+\ell)$$

$$y_L(k+\ell) \leq \hat{y}(k+\ell) \leq y_H(k+\ell) \, , \, \ell = 1,2..P$$

$$m_L(k+\ell-1) \leq m(k+\ell-1) \leq m_H(k+\ell-1) \, , \, \ell = 1,2..M$$

$$\Lambda_\ell = 0 \, , \, \ell > M$$

where the prediction in equation (5) is used and $M \leq P$. Since a linear prediction is used, the resulting inequality constraints are linear in the solution variables $\Delta m(k) \ldots \Delta m(k+M-1)$ and therefore (8) is indeed a quadratic program.

Comparison with Classical Approaches

It is important at this point to reflect on what the predictive control problem methodology allows us to do and compare it with standard approaches which might appear similar to the uninformed observer. Here we will compare the QDMC algorithm with the LQC methodology on the basis of industrial control applicability. The observations on LQC should equally apply to most other classical algorithms.

Model-Predictive type controllers (QDMC). QDMC assumes one objective function to be minimized which reflects some of the desired performance criteria of the control system. Different objectives for each of the projections of variables are lumped into one objective function. The relative importance of each objective is specified by appropriate selection of weights. These weights are closely tied up with the variable scaling.

Additional performance criteria are specified by the use of inequality constraints in the predictions. These constraints impose hard bounds on the predictions of the variables in order to keep them within acceptable operating limits. For these hard constrained variables the corresponding weights in the objective function are zero.

Since the problem is solved at each sampling time, the controller is able to account for different scenario sets of active constraints which were not foreseen at the design stage. The transfer from one set of active constraints to the other is done smoothly and optimally by the quadratic programming algorithm. Therefore, it is in practice an on-line intelligent system albeit still with some limited capabilities.

In terms of numerical solution, a quadratic program can be customized to efficiently solve problem (8), which in the absence of constraints reduces to a simple least-squares problem (Prett and Gillette, 1979). Although complex by historical standards, a simple interface allows easy implementation of the algorithm.

<u>Recursive type controllers (LQC).</u> The discrete optimal control algorithm as described in Kwakernaak and Sivan (1972) employs a state-space model of the process as in (3) in order to solve the following optimization problem:

$$\min_{\substack{m(k) \ldots \\ m(k+P-1)}} \tfrac{1}{2} \sum_{\ell=1}^{P} \{ \, ||\, \Gamma_\ell \, y(k+\ell) \,||^2 + ||\, \Lambda_\ell \, m(k+\ell-1) \,||^2 \, \}$$

$$\text{s.t.} \qquad x(k+1) = A\, x(k) + B_m\, m(k)$$

$$y(k) = C\, x(k)$$

(9)

LQC also minimizes a weighted quadratic objective function of the model predictions of the outputs and manipulated variable values and assumes M=P. As formulated, the resulting LQC controller is not suited for handling non zero targets, measurable disturbances and non zero mean unmeasurable disturbances since it does not provide integral action (Garcia and Morari, 1982). However, several modifications can be made to resolve these issues. For example, one can formulate the objective function as in QDMC (equation 8) and perform the prediction using the state-space model (4) and the estimate of $d(k+\ell)$ in (7). As a result, the two problems become identical in the absence of process constraints.

The solution of problem (9) (or equivalent) produces a set of manipulated variable values. By using the LQC method, a set of Riccati equations can be solved to obtain a simple expression where the manipulated variable to be implemented is a linear combination of the states. Such relationship lends itself to a simple implementation which can be represented in the form of block diagrams and easily realized with standard hardware. The computational burden consists of calculating the controller gain matrix, which is done off-line. After the computation is performed, the controller is fixed.

Even though one can minimize the same objectives as in QDMC and influence them by using weights, the LQC formulation does not allow for the specification of constraints. In fact, the only way constraints can be handled by LQC or any other recursive type of controller is to add an intelligent decision maker on top that will decide which constraints become active and therefore restructure the control problem accordingly. Such approach has been suggested before by Arkun and Stephanopoulos (1980).

In this respect most controllers are similar to LQC and consequently different than model-predictive type controllers. In QDMC the computed set of moves in the manipulated variables depend not only on past history of plant dynamics but also on predicted violations of constraints. This is not possible to represent in any recursive / transfer function / block diagram type of formulation. At the same time one must also realize that if constraints are removed and future target trajectories are linear transfer functions, then QDMC is essentially no different than any other control method. In these cases, the manipulated variable moves depend exclusively on past plant history and therefore it is possible to analyze with classical techniques familiar to most control engineers.

It is in this constraint-free environment that QDMC was found to have an Internal Model Control structure (Garcia and Morari, 1982) and therefore enjoy many important properties, in particular, that it does not require a stability analysis. These important properties by themselves provide enormous incentive for the use of model-predictive controllers over other designs. However, as explained above, in the constrained on-line optimization framework the added feature of on-line updating of the moves based on predicted process constraint violations is what allows model-predictive controllers to realize the achievable profits.

After comparing the two main methodologies of process control as we see them, one can now start to see where the important theoretical developments are necessary in order to solve the real problem at hand. We have identified two important areas where significant advances in model-predictive control are needed before we can attempt to solve the real control problems:

1. Performance Specification:

In the cases above we have assumed that the mathematical objectives solved represent the true objectives we want the control system to satisfy. Even then, we have arbitrarily lumped several unrelated objectives into one single function by assigning appropriate weights to the individual objectives. In addition, we have imposed hard constraints even on variables for which no significant penalty is paid for violations; although there is provision in QDMC to weight these "soft" constraints into the objective function in the event of a predicted violation. However, this brings up the more practical issue of how to distinguish between objectives to be optimized and constraints to be satisfied and, in turn, which constraints are hard or soft. These decisions need to be based on engineering judgement and can change during the operation of a unit.

Therefore, there is a need to develop on-line systems which translate the qualitative decision making into quantitative problems to be solved by the control system.

2. Robustness to Modeling Errors:

Most methodologies for process control do not handle the issue of model inaccuracy in a satisfactory way. In the event of mismatches in the assumed process model and disturbance description a "tuning" procedure is employed which consists in modifying the true performance objectives so that the controller obtained assuming a perfect model will perform as desired on the true plant. There is a definite flaw in the design when we know that model errors exist but insist on design methodologies that use a perfect model description. The new Structured Singular Value (SSV) approach of Doyle (1982) was developed to address this particular issue of design in the face of uncertainties. However, it assumes a transfer function type of implementation. Therefore, there is a need to tie these new results with model-predictive type controllers.

In the rest of this paper we will discuss in some depth these two issues and show some research efforts currently in place to address them.

SPECIFICATION OF PERFORMANCE CRITERIA FOR PROCESS CONTROL

The most important step in process control design is to determine what the control system is meant to do and what performance is expected from it. Even though we can attach some monetary or economic value to every criteria, there is usually not enough data to allow us to assign costs for every situation. Therefore, it is seldom the case that the performance is given by a single criterion (i.e. cost) nor that it remains invariant with time.

As explained above, we will assume a two-phase approach to process control as illustrated in Fig. 1. Therefore, the function of the control phase must be not only to move the process to the new operating point dictated by the optimizer but also to keep it operating there despite disturbance changes. In order to achieve this, there are multiple criteria to be satisfied by the control system. In this framework we will classify all **practical** process control criteria as belonging to one of the following kinds:

- Economic:

These can be associated with either maintaining process variables at the targets dictated by the optimization phase or dynamically minimizing an operating cost function.

- Safety and Environmental:

Some process variables must not violate specified bounds for reasons of personnel or equipment safety, or because of environmental regulations.

- Equipment:

Physical limitations of equipment must not be exceeded by the control system even when not directly related to safety.

- Product Quality:

Consumer specifications on products must be satisfied.

- Human Preferences:

There exist excessive levels of variable oscillations or jaggedness that the operator will not tolerate. There can also be preferred modes of operation.

It will be assumed that every variable needed to evaluate the practical criteria stated above is measurable or can be inferred by secondary measurements or operating policy. Then it follows that any practical criteria can be stated as being one of two types of **mathematical** criteria:

- Objectives:
    Functions of variables to be optimized dynamically where optimal means best satisfaction of the criterion.

- Constraints:
    Functions of variables to be kept within bounds which in turn can be of two kinds:
    - Hard constraints: no dynamic violations of the bounds are allowed at any time.
    - Soft constraints: violations of bounds can be allowed for satisfaction of other criteria.

The challenge for the designer consists in the selection of the actual functions and in the evaluation of trade-offs between criteria for solving the problem. Also, confusion between constraints and objectives must be avoided at the design stage. In the following we explore each of these issues and give examples of proper selection of performance criteria for practical problems.

Selection of Objective and Constraint Functions

Different methodologies in process control are based on assumptions about the functionalities of objectives and constraints in order to simplify the solution of the problem. Besides making the problem mathematically manageable, the main reason for simplification is to allow implementation in the existing hardware.

For example, in QDMC a quadratic objective function is used for mathematical convenience since it yields a simple least-squares control problem in the absence of constraints. When constraints are added, a quadratic program results which can be solved very efficiently and conveniently with standard software. However, for some applications a quadratic objective might not be the appropriate criterion to use. Also, there might be some advantage in using frequency response objectives instead of time domain objectives.

Although these issues have enough relevance to merit further discussion and analysis, in our opinion there are other assumptions which have a more significant impact on performance. For instance, the most typical assumption made in process control is to "convert" constraint criteria into objective criteria. The reason is that problems with constraints are difficult to solve with the available hardware. On the other hand, conveniently formulated objectives can yield concise control designs which can be implemented easily with simple blocks. However, the performance of the control system will generally degrade when a true criterion is compromised for mathematical convenience.

For example, quality criteria in the form of composition specifications is a constraint criteria. However, in many situations it is changed into an objective that minimizes deviations from a target, having the predictable consequences. On the one hand, composition is kept at a target even when deviations away from the real bound and into the feasible region are not critical and might even be necessary in order to satisfy other conflicting criteria. On the other hand, since the constraint limit is not enforced, violations can occur and therefore the target must be kept within some tolerance of the bound. Therefore, there is a performance loss directly related to the mathematical simplification of control criteria.

There seems to be a need to develop methods for not only specifying the correct function to be used as a criterion but also for making the choice between expressing a practical criterion as either an objective or as a constraint.

Evaluation of Trade-offs and the Solution of the Control Problem

Once a satisfactory set of functions is chosen as objectives and constraints, there remains the problem of evaluating the trade-offs between criteria. It is obvious that by making a constraint hard it is understood that it takes precedence over any other objective or soft constraint. In the case of soft constraints, they can be handled as objectives when they become active during the operation of the controller. Since there exist many quantitative as well as qualitative reasons for the designer to prefer one criterion over another, the control technique must provide a way to allow the designer to influence the solution. This can be done in several ways.

Single objective function. One can lump all objectives and all active soft constraints into a single objective function by using weights. The importance of each objective is then influenced by the relative size of the weights. This objective is then optimized subject to the hard constraints. Such approach is done by QDMC as described in the previous section. In addition, in QDMC we have been able to add other objectives that can be expressed as quadratic functions to solve economic optimization and enforcing targets for manipulated variables (Prett and Gillette, 1979).

Although computationally convenient, there are some difficulties with this approach. First, the selection of weights is dependent on the particular scaling of the variables, therefore making the enforcement of an objective difficult. For example, selection of weights for the satisfaction of a target for a temperature vs. that of a composition will depend on each variable engineering units. Second, even when there are no scaling problems, the relative importance of different kinds of objectives cannot be exactly quantified for most problems (i.e. economic cost vs. variable deviations). Third, since the objectives are lumped together there is no guarantee that by increasing the corresponding weight of one objective the other objectives will not be affected, due to an inherent interaction between objectives through the lumping. And finally, those weights chosen today may not be appropriate later on when operating conditions have changed.

Despite these difficulties, the single objective function approach has worked satisfactorily in many practical cases. The reason is that it is computationally so simple that one can afford to be imprecise with the initial weight selection. Then the weights can be updated on-line according to the observed performance. Such approach has been followed in Shell since the initial applications of DMC. However, its success has been achieved at a high maintenance cost. Although this was cost effective in the past, it is becoming increasingly difficult and costly to maintain the current large number of QDMC loops. Recent technological advances indicate that these difficulties can be overcome through better formulation of the problem. This then allows for less control expert personnel being assigned to the loop maintenance role. Gradually, our technological advances are allowing for the entry of more advanced loops into the world of general control loop maintenance. This is of course the objective of the pursuit of the **Unified Theory**.

<u>Multiple objective functions.</u> From the above discussion it follows that the exact mathematical formulation of the control problem to be solved is a multi-objective optimization algorithm subject to multiple constraints, with the ability to distinguish and to handle hard and soft constraints. By performing a suitable multi-objective

optimization (also known as vector-valued function optimization) the issue of assigning weights for each criterion can be removed.

One such algorithm solves the following generalized optimization problem at each sampling time (Nye and Tits, 1985):

$$\min_x \{ \max_j \; f_j(x) \; | \; g_i(x) \leq 0 \} \qquad (10)$$

where j=1,2... are the number of objectives and active soft constraints, i=1,2... are the number of hard constraints and the solution vector consists of the manipulated variable moves. This problem minimizes the maximum of the objectives subject to the hard constraints. Since each objective has its own level of importance and satisfaction to the designer, one has to somehow influence the solution of (10) based on the designer's desires. One way of achieving this is to normalize the set of objectives and soft constraints conveniently as follows:

$$f_j' = (f_j - \text{fgood}_j) / (\text{fbad}_j - \text{fgood}_j) \qquad (11)$$

where the "good" and "bad" values of the objectives and soft constraints are specified by the designer. Note that values of 0 and 1 correspond respectively to the specified good and bad values. This implies that these normalized objectives are to be always minimized.

By normalizing individual objectives with respect to their good and bad values all the criteria can be compared on the same basis. For example, two objectives having the same normalized value equally satisfy the designer; both would be equally "good" or equally "bad". Therefore, the problem of selecting weights when the objectives are lumped in one function is replaced by the problem of selecting the good and bad values for each objective. Even though still requiring some effort to select, compared to weights these values will make more engineering sense and will be directly related to the actual problem.

In the original paper where this optimization technique is presented (Nye and Tits, 1985) the authors make the point that such an algorithm must be supported with a good man-machine interface so that the designer can evaluate the trade-offs and influence the solution accordingly based on engineering judgement. In an on-line environment (as we propose to use such a technique) these decisions will necessarily need to be automated so as to achieve a maintenance-free operation. The best technology available at present to achieve such automation is the expert system technology. As we see it there will be an intelligent decision maker on top of the optimizer updating the criteria (Fig. 3). These criteria will be updated based on quantitative as well as qualitative information about the operation.

Some Examples of Selection of Performance Criteria

In this section two examples of problems where correct specification of performance criteria is critical are presented.

The surge level control problem. Distillation column trains where the product flow(s) of one column feed a downstream unit pose an interesting situation in control objective specification. It is common practice to balance the material in these columns by controlling an accumulator level through manipulation of the outlet flow. Since the manipulated variable for controlling the level in one column is in turn a disturbance to the following unit, any upset can easily propagate downstream. In olefin plants, for example, there is a fractionation train as shown in Fig. 4.

If only one column in this train is considered, the practical criteria for this problem are:

1) Minimize variations in the outlet draw.
2) Maintain the level within the upper and lower bounds of the tank over a future time horizon.
3) Maintain the outlet flow within its upper and lower bounds over a future time horizon.
4) Bring the level back to the middle of the tank at the end of the time horizon.

Satisfaction of these criteria will minimize the propagation of upsets, will keep the level within the equipment constraints while manipulating the outlet flow within its limits, and bring the level back to the middle of the tank in order to guarantee surge capacity for the next upset.

Several issues are of importance here. First, if the control problem is designed individually for each column in the train, there might be a loss of performance. Note that the downstream columns will be perturbed with known inlet flow changes and therefore one could feedforward the predicted moves. More elegantly, however, one could solve for the set of criteria for the entire train yielding a multivariable control problem.

Second, let us consider the traditional control setup for this problem. Since the level in the tank must not exceed its bounds the common practice is to design a controller for keeping the level at the middle of the tank usually very tightly. As a consequence, any disturbance change to a column will propagate downstream with equal intensity or possibly amplified if the loops are not adequately tuned. In the particular case of an olefins plant, a furnace trip will cause a severe step decrease in the feed to the first column (Fig. 4) which can upset the process for several hours. Thus, a high price is paid for substituting the true criteria for an alternate criteria of minor relevance, namely, tight control of the level. In this problem, it is more important to minimize outlet flow changes regardless of the level, except when it violates the upper and lower bounds.

Finally, from criterion 4) it is obvious that we are assuming every inlet flow disturbance change to settle faster than the horizon chosen. Should a dynamic representation of the inlet disturbance be available, this information would allow the controller to make moves to conserve surge capacity for future upsets without having to design the controller to satisfy criterion 4). This illustrates the effect of model error in forcing the modification of performance criteria. This issue will be covered in the next section.

A semi-batch reactor control problem. The process discussed here is the block polymerization of synthetic rubbers (Garcia, 1984). The final product consists of a polymer composed of several blocks of different monomers. A sequence of batch operations is performed to produce the blocks. Since the polymerization proceeds via an anionic mechanism, chain termination follows an Arrhenius type decay whose rate increases with temperature. Successful polymerization of all blocks demands that a only a small fraction of the molecules terminate at every reaction step.

These reactions are carried out in the vessel depicted in Fig. 5. The heat of reaction is removed by circulating the reactive solution through exchangers which use refrigerated water. In addition, there is the capability of manipulating the rate at which monomer is added for control of the heat release. The initial batch charge and recipe for the reaction are assumed given. In general, there is a desired temperature target at which the reaction is to take place.

Following this description, the criteria to be satisfied are:

1) The cooling water flow must be within its upper and lower bounds.
2) The monomer addition rate must be within its upper and lower bounds.
3) The reaction temperature must be close to the target over the whole batch.
4) The fraction of terminated chains must be smaller than a given bound at the end of the batch.
5) The batch reaction time must be minimized.

It must be noted that satisfaction of these criteria guarantees that other important properties of the polymer are satisfied, and therefore these criteria cover all possible aspects of the problem. The last criterion is purely economic.

In order to solve this problem, a multi-objective constrained optimization of the type described in (10) is required. In the absence of constraints and transforming criterion 1) to an objective criterion, then an optimal control approach can be used. Alternatively, due to computing limitations, the method reported by Garcia (1984) consists in defining alternate criteria which indirectly satisfy the true criteria. These alternate criteria are then solved for by using the standard QDMC algorithm (although with some modifications due to the nonlinearities in the model). These are:

1) The cooling water flow must be within its upper and lower bounds.
2) The monomer addition rate must be within its upper and lower bounds.
3) The reaction temperature must be close to the target over the whole batch.

4a) The monomer addition rate must be at its maximum possible value during the batch.

In order to satisfy these criteria, a control problem can be set up with the reactor temperature as the controlled variable. Also, a target can be imposed on the monomer addition rate equal to its maximum bound. This forces the controller to add monomer at the maximum possible rate. The manipulated variables are the cooling water flow and the monomer addition rate.

Note that satisfaction of alternate criterion 4a) indirectly satisfies criteria 4) and 5), since adding monomer at the maximum rate possible minimizes the reaction time. This in turn minimizes the chain termination rate due to the properties of the anionic mechanism. Note that by doing this we have substituted a constraint criterion by an objective criterion while having to lump both target criteria 3) and 4a) into one single objective. As a result, there is a need to tune the weights in the objective function in order to satisfy the true criteria.

In the following section the important issue of robustness to model uncertainties is discussed as it affects the satisfaction of performance specifications.

MODEL INACCURACY CONSIDERATIONS IN PROCESS CONTROL

In previous discussions, we have not explicitly dealt with the issue of accuracy of the model of the process and of the disturbances in designing model-predictive controllers. An important property of model-predictive controllers is that no stability problems exist under perfect model conditions, even in the face of constraints on the manipulated variables (Garcia and Morari, 1985). However, there exist inherent limitations in satisfying arbitrary criteria due to specific process characteristics. As discussed by Garcia and Morari (1982) system zeroes outside the unit circle and deadtimes impose limitations on the satisfaction of targets which cannot be removed by any control system. Since satisfaction of active output constraints can be thought of as a target satisfaction criteria, output constraint criteria satisfaction will also be limited by the same process characteristics. As a result, these limitations must be taken into account when formulating the objectives and constraints. The factorization method (Garcia and Morari, 1982) takes care of this problem for target criteria and therefore, for most practical cases these limitations do not impose additional difficulties in the design.

However, the most important limitation to satisfying control criteria is imposed by inaccuracies in the model assumed for design. Models used for design will not be accurate for several reasons:

- The model is assumed linear when the process is nonlinear and therefore, the model will not describe the process when the operating point changes beyond a certain amount. This problem is exacerbated by the presence of optimization systems which generally increase the frequency of operating point changes.
- The equipment degrades or is changed (e.g. catalyst change, heat exchanger fouling, tray collapses, etc.).
- Disturbance characteristics are unknown and therefore assumptions are made at the design stage on their dynamic behavior.
- Techniques used for identification are not accurate enough or the measurements are not of enough quality to produce the model detail wanted.

In the face of significant model inaccuracies the control system is generally unable to satisfy all of the true performance criteria specified for the process. In this event, the designer (before the implementation) or the tuner (after implementation) is faced with the decision to trade one performance criteria for another.

For example, two commonly specified criteria for single variable control systems are to maximize the setpoint tracking speed while exhibiting smooth manipulated variable response. The tuning parameter selection (e.g. controller gain) influences the trade-off between these criteria. In the absence of model error it is possible to achieve any desired speed in setpoint tracking behavior (within some inherent system limitations as are deadtimes and system zeroes outside the unit circle (see Garcia and Morari, 1982)). However, the faster the response, the more "power" will the manipulated variable need to have. In order to satisfy the smoothness criterion on the manipulated variable the designer can detune the controller until an acceptable response is obtained. This tuning procedure yields the fastest tracking speed possible for the desired manipulated variable smoothness in the absence of model error.

Let us now assume that the controller designed as explained above is installed and a significant change in the process occurs. Invariably, the fastest speed of tracking achieved by the designer when the model is perfect will not be achievable in the face of model error without an increase in manipulated variable jaggedness. Therefore, the tracking speed must be sacrificed to satisfy the smoothness criterion. If this slower closed-loop response is acceptable, the model error has not imposed a restriction on satisfying the criteria. On the other hand, if the closed-loop response is not fast enough for the particular application, then this inaccuracy forces the tuner to accept a higher level of manipulated variable jaggedness, or most likely, to take the loop off control.

Controller robustness can then be defined as the ability of the control system to satisfy the desired performance criteria of the true process in the face of inaccuracies in the model used for design. In the example above, an expert tuner is assumed to be present to ensure robustness, and therefore this is a drawback in the way controllers are designed and tuned today. In the particular case of model-predictive controllers, this issue is amplified by the fact that it is not possible to foresee at the design stage all possible scenarios of active constraints and tune appropriately for all of them. Our goal, that is equally shared by all practitioners, is to design a controller that stays on-line longer while requiring minimal maintenance. Let us examine the traditional methods used to achieve robustness and also discuss a new technique aimed at solving these difficulties.

Traditional Methods of Handling Model Inaccuracies

The single variable tuning example discussed above illustrates the most common procedure used to handle the uncertainty issue. Out of lack of better information about the process, most control methodologies assume the model description of the process to be accurate at the design stage. The control performance is evaluated based on this model and a set of tuning parameters is obtained. However, model inaccuracies will cause a degradation in the expected design performance when the loop is implemented. The traditional way of handling this loss of performance is by modifying the tuning on-line.

Another interpretation of the traditional tuning process is to think of it as a procedure that modifies controller criteria to guarantee robustness to modelling errors. In the QDMC method, weighted penalties on the manipulated variable moves are added to the objective function as given in equation 8. Even though in many practical cases there is a true need to minimize manipulated variable variations, the main reason for penalizing them in QDMC is to improve the robustness properties of the algorithm. However, to guarantee robustness on-line tuning is required.

It is reasonable to think that if some knowledge of the uncertainties is available at the design stage, a more robust design could be obtained by, for example, performing studies on the tuning parameters. The problem with such a study is that the true process performance would need to be evaluated by performing simulations over all possible process representations generated by the uncertainty description. In addition, in the model-predictive environment, innumerable studies would be required to consider all possible scenarios of active constraints.

The issue is then to be able to easily measure how model inaccuracies limit the satisfaction of true process performance criteria (both objective and constraint criteria) given an uncertainty description. This measure can then be used in two ways. First, to evaluate the trade-offs at the design stage for the most typical scenarios of active constraints. This would indicate whether or not the controller will work for expected uncertainties without having to perform extensive simulation studies. Second, during the implementation, changing criteria can be evaluated on-line in the face of inaccuracies. As a result, a design can be produced which will stay in operation longer with minimum maintenance. These issues are central in a newly developed robust control theory discussed in the following.

Methods to Solve the Model Uncertainty Problem

We are currently investigating several methods to handle the issue of model inaccuracy in the design of model-predictive controllers. In this section we give a short description of each and propose how they can be used to solve the problem at hand.

Robust control techniques. We have found recent developments in the area of control to have promising possibilities in solving the model-predictive control problem in the face of uncertainties. Although its implementation assumes a recursive type of controller with no constraint handling, the Structured Singular Value (SSV) approach of Doyle (1982) employs the correct philosophy of approaching the robustness issue: use an uncertainty description of the process at the design stage to obtain a controller that will not perform worse than desired on the true process, therefore, minimizing on-line tuning. In this discussion we briefly describe this technique as applied to recursive type controllers and point out what research topics need to be addressed in order to extend these ideas to model-predictive controllers.

The SSV provides a measure of true process performance satisfaction in the face of model inaccuracies. The result is a refinement of the Singular Value result (Doyle and Stein, 1981) where now structured uncertainties can be considered. This means that it allows the designer to specify uncertainties in the gains, time constants, etc. in given elements of the system transfer function matrix. As a result, a less conservative stability criterion is obtained.

In Doyle's formulation, performance is described by weighted frequency response functions that reflect the designer's desired closed-loop characteristics. The trade-offs are influenced by the selection of these weights. Model inaccuracy is described in any form desired as long as it can be represented in block diagram form suitable for linear processes. In addition, an uncertainty description of the disturbance dynamics in frequency response is used making the method more appealing than LQC where disturbances have to be described as Gaussian noise.

Under these assumptions, the method consists of the solution of two important control problems:

The Analysis Problem: given a controller, one can evaluate a criterion (the SSV) whose value indicates whether or not the desired performance is achieved for the true process in the face of the described model and disturbance uncertainties.

The Synthesis Problem: given the desired performance of the true process and a description of model and disturbance uncertainties, a controller transfer function is found so that the SSV criterion is met.

The first problem involves an optimization procedure to evaluate the SSV where the second involves another optimization to compute the controller.

Even though the methodology has been developed for a recursive type of control implementation, we see much promise in it, particularly in the way model and disturbance uncertainties are described. Of course, this description is not always available. However, it is our experience that the designer has enough knowledge about the process to be able to specify ranges of values for gains and time constants. Even such crude description would produce a more intelligent and consequently more robust design than assuming the parameters to be accurate.

At present, this technique is useful to solve the analysis problem of QDMC at the design stage, but only if constraints are removed. Using the nominal model one can design QDMC for a set of tuning parameters (as is done presently). Then the SSV is computed for such controller to evaluate whether or not the desired closed-loop properties of the control system (e.g. disturbance rejection, manipulated variable smoothness, etc.) are satisfied for the given model and disturbance uncertainty description. Such analysis would give the designer a tool for tuning his controller, without having to perform simulation studies. We are currently incorporating such analysis tool in our design techniques.

Since it is not possible at the design stage to consider all possible scenarios of active constraints and other criteria, and therefore, all possible controllers generated by QDMC, the best we can hope for with the current SSV theory is to perform analysis studies on the most common operating conditions. This would involve substituting a constraint criteria for an equivalent objective criteria. For example, if we expect some variable constraint to be active, it could be considered as a controlled variable with the target as its bound and the analysis performed on the resulting controller. This would hopefully reduce the situations where on-line tuning is required to a minimum.

However, in an environment of changing criteria that do not translate easily to frequency response functions (as are inequality constrained responses in the time domain) this methodology would likely need to be modified. As discussed above only the on-line solution of the problem provides such flexibility. We must realize that at the design stage there seems to be no comprehensive way of evaluating all the possible operating scenarios and consequently, case studies must necessarily be performed.

Among the possible modifications, one could remove the weight selection process by using a multi-objective optimization method that allows direct evaluation of the performance criteria based on engineering judgement. The technique described previously seems to be a convenient tool for this (Nye and Tits, 1985).

But the best extension of this technique to the model-predictive framework is to solve the synthesis problem on-line for the manipulated variables moves directly rather than off-line to obtain a transfer function. This has the potential of producing a controller which can accept changing objective and constraint criteria. These criteria are then traded on-line in the face of model inaccuracies to produce the best moves of the manipulated variables. Although possibly computationally prohibitive with today's computers, this appears to be a sound approach to solve the real problems at hand.

Adaptive Control Methods. Much attention has been given in the literature to adaptive control schemes. These methods handle the model inaccuracy problem by on-line identifying the process model based on measurements of inputs and outputs of the plant (Astrom and Wittenmark, 1973). Besides solving for only an objective criteria, namely the minimum variance objective function, this is done under the assumption that a perfect model is obtained at every execution time.

Besides being limited in the criteria that are solved, the assumption of perfect model can create problems. Even though the model is updated continuously, on-line experiments have inherent limitations which do not allow complete identification of all process modes or of all disturbance characteristics. Therefore, the model updates might not be good enough to allow the control system to satisfy the specified objective and therefore a tuning procedure is inevitable.

In the framework of model-predictive control, we have already discussed how to best handle the criteria formulation problem, and therefore, out of adaptive control theory we will only be concerned with the on-line model identification aspects. In light of the robust control theory described above, two research topics are of interest:

- Use robust control analysis to find out which model parameters lack enough accuracy to allow the satisfaction of the desired criteria. Then an on-line identification procedure can be used to refine the estimates of such parameters.

- Develop on-line identification methods that not only find the nominal model but also identify the uncertainty bounds on plant and disturbance models.

Current work consists in finding suitable on-line identification schemes for dealing with these issues.

Nonlinear model description. In case a dynamic nonlinear model of the plant is available, the model-predictive control problem can be solved assuming perfect model description. Numerical problems aside, this seems to be the best approach to take. However, model inaccuracies will inevitably be present even using nonlinear models and therefore, a nonlinear robustness analysis is needed. An advantage of using nonlinear models is that now the parameters can have definite physical significance as for example, fouling rates, kinetic rate constants, etc.

We have had good success using nonlinear models for solving complex control problems. The reactor control problem described in the previous section was solved by using nonlinear ordinary differential equations to generate the predictions and step response coefficients (Garcia, 1984). The solution method assumed linear superposition of future predictions to compute the moves and therefore was not optimal. However, one can easily formulate a dynamic optimization problem which solves the problem exactly. In this particular application, computing power limitations prevented us from using a more rigorous solution method on-line.

CONCLUSION

In this paper an attempt was made to formalize a rigorous approach to research in process control. The pursuit of a **Unified Theory** was introduced within the context of the need for our industry to apply an **Integrated Technology** philosophy to maximize the synergism of separate component technologies such as control and optimization. The field of process control has been pursued in an ad-hoc fashion for many years. Many so called solutions have not stood the test of time nor have they been translatable nor generalizable to cover more than the immediate application. In technical developments too much attention has been paid to implementation constraints, such as hardware capability, leading to the situation we have today wherein much of our industry is not capable of exploiting the rapid growth of computing power available at the field level. Hence we have the sad state of choosing between ad hoc solutions that really do not meet the requirements of the application, or so called advanced technology that really does not address the real problem.

The concept of an **Integrated Technology** refers to the recognition that a control system, besides having to perform in the load rejection mode, will also most probably experience some manipulation of setpoints which is generalized here in the term **Optimization**. Whether this is performed using a rigorous first principles approach or in a more ad hoc fashion is really beside the point. Our control design methodologies must incorporate this fact and hence be able to recognize and deal with constraints. Ad-hoc approaches to constraint handling superimposed upon so called advanced techniques are no better than overall ad hoc problem solutions themselves. It is time to recognize this fact and get on with the pursuit of research activities that will lead to a continually evolving series of problem solutions growing in technical sophistication. Over time it is hoped that this will lead to a **Unified Theory** of process control that will be generally applicable to most all our problems. The authors feel that the Dynamic Matrix Control technique is an excellent first step down this path that incorporates many of the requirements discussed. It is however just a beginning. It is a beginning that at least provides the framework for evolutionary growth. We must deal with the following issues:

- Nonlinear model requirements
- Constraint handling
- Detailed performance analysis and multi-objective solution techniques
- Dynamic problem structuring techniques
- Requirement for low maintenance

and furthermore deal with them in a generalized framework so that long term we can look to a **Unified Theory** incorporating all aspects.

While involved in this endeavor it is also important for us to suitably exploit the growing areas of Artificial Intelligence and Super Computing. The field of Artificial Intelligence will facilitate the synergism of qualitative and quantitative knowledge bases. Much of our historic problem with specification of performance requirements has been due to the qualitative nature of the input information. Difficulty has been experienced in effecting the appropriate level of interaction between this and the more quantitative solution techniques available. This is being rapidly addressed by the new developments in Artificial Intelligence and process control researchers are well advised to stay up to date with this work. One caution in this area is that there has been a tendency to shroud the old traditional ad-hoc solutions discussed earlier in the mantle of respectability by framing them within the context of Artificial Intelligence Systems. This is considered unacceptable. The real contribution of Artificial Intelligence will be as a complementary aid in solving the real problem and not in facilitating the implementation of ad-hoc solutions.

The field of super computing is also very exciting as a prospective aid in implementing our proposed approach. It is reasonable to expect that within the next five years low cost high performing computer power will allow us to consider removing artificial boundaries between the areas of control and optimization and so allow us to truly achieve an **Integrated Technology** approach.

REFERENCES

Arkun Y., and G. Stephanopoulos (1980). Studies in the synthesis of control structures for chemical processes: part iv. design of steady-state optimizing control structures for chemical process units. AICHE J., 26, 975.

Astrom, K.J., and B. Wittenmark (1973). On self tuning regulators. Automatica, 9, 135-199.

Cutler, C. R., and B. L. Ramaker (1979). Dynamic Matrix Control - a computer control algorithm. AIChE National Mtg., Houston, TX.

Doyle, J. (1982). Analysis of feedback systems with structure uncertainties. IEE Proc., 129, 242-250.

Doyle, J., and G. Stein (1981). Multivariable feedback design: concepts for a classical/modern synthesis. IEEE Trans. Autom. Control., AC-26, 4-16.

Garcia, C. E. (1982). Studies in Optimizing and Regulatory Control of Chemical Processing Systems. PhD Thesis, University of Wisconsin, Madison.

Garcia, C. E. (1984). Quadratic Dynamic Matrix control of nonlinear processes: an application to a batch reaction process. <u>AIChE Annual Mtg.</u>, San Francisco, CA.

Garcia, C. E., and M. Morari (1982). Internal Model Control 1. a unifying review and some new results. <u>Ind. Eng. Chem. Process Des. Dev.</u>, <u>21</u>, 308-323.

Garcia, C. E., and M. Morari (1985). Internal Model Control 2. design procedure for multivariable systems. <u>Ind. Eng. Chem. Process Des. Dev.</u>, <u>24</u>, 472-484.

Kwakernaak, H., and R. Sivan (1972). <u>Linear Optimal Control Systems</u>. Wiley-Interscience, New York.

Morari, M. (1981). Integrated plant control: a solution at hand or a research topic for the next decade? <u>Chemical Process Control II</u>, Sea Island, GA.

Nye, W. T., and A. L. Tits (1985). An application-oriented, optimization-based methodology for interactive design of engineering systems. <u>Int. J. Control</u> (submitted).

Prett, D. M., and R. D. Gillette (1979). Optimization and constrained multivariable control of a catalytic cracking unit. <u>AIChE National Mtg.</u>, Houston, TX.

Ray, W. H. (1981). <u>Advanced Process Control</u>. Mc-Graw Hill, New York.

Richalet, J. A., A. Rault, J. L. Testud, J. Papon (1978). Model predictive heuristic control: applications to an industrial process. <u>Automatica</u>, <u>14</u>, 413-428.

NOMENCLATURE

a_ℓ step-response coefficient matrix with respect to process inputs for ℓ^{th} time interval.

a_{mi} step-response coefficient matrix with respect to manipulated variables for i^{th} time interval.

A   state-space model state matrix.

B   state-space model input matrix.

B_m state-space model manipulated variable matrix.

C   state-space model output matrix.

d   unmodelled contributions to the output

f^j j^{th} objective in a multi-objective optimization problem.

g^i i^{th} hard constraint.

h_ℓ impulse-response coefficient matrix with respect to process inputs for ℓ^{th} time interval.

k   discrete interval of time.

m   process manipulated variable vector.

M   number of discrete time intervals when manipulated variables are allowed to move in the controller computation.

P   number of discrete time intervals in the controller horizon.

q   integrator state vector.

T   sampling time.

u   process input vector (manipulated variables and measurable disturbances).

x   vector of system states.

y   process output vector.

y_m process output measurement.

y_S output setpoint vector.

Greek

Δ denotes a change in a variable over a time interval: $\Delta x(k) = x(k) - x(k-1)$.

Γ_ℓ matrix of penalty weights on outputs.

Λ_ℓ matrix of penalty weights on manipulated variables.

Subscripts

L   variable vector of low limits.

H   variable vector of high limits.

Superscripts

$\hat{\ }$ predicted variable.

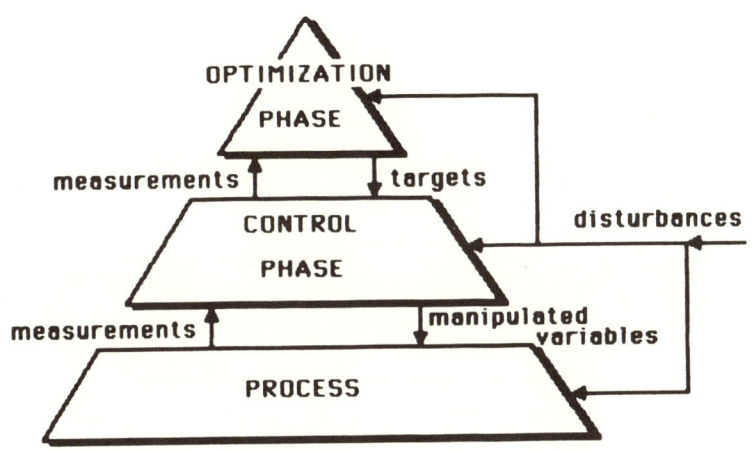

Fig. 1. The Integrated Technology Approach to process control.

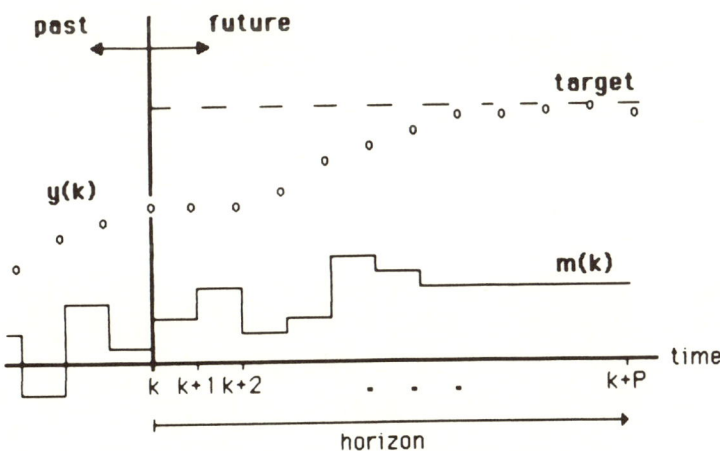

Fig. 2. The "moving horizon" approach to model-predictive control.

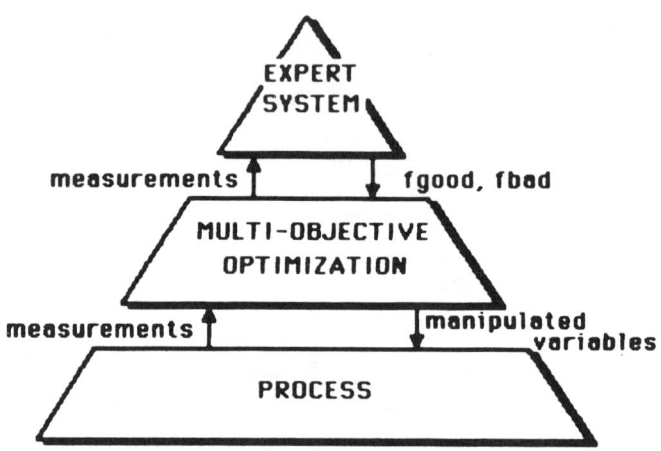

Fig. 3. On-line updating of the multi-objective optimization criteria by an expert system.

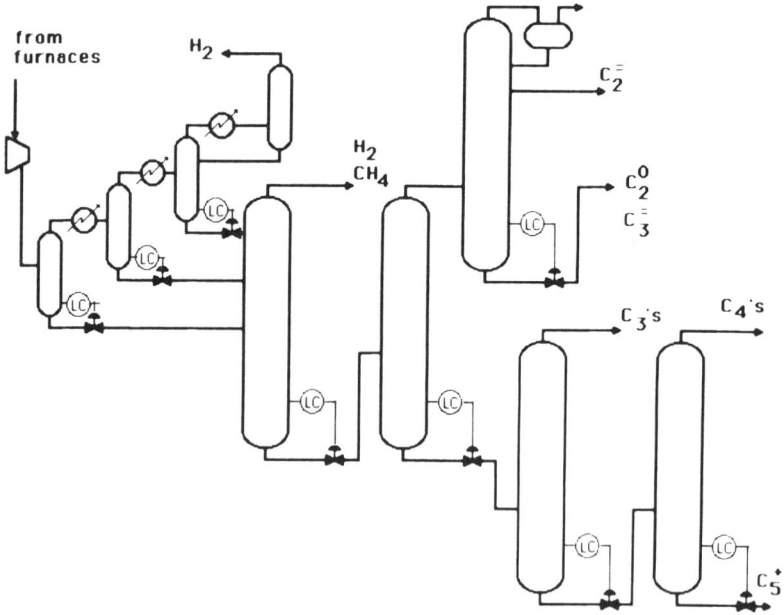

Fig. 4. Olefins plant fractionation train

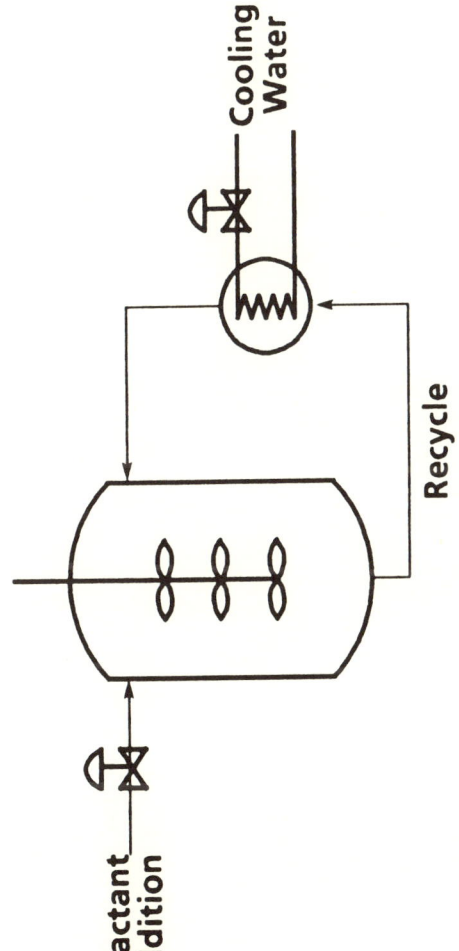

Fig. 5. Semi-batch polymerization reactor.

COORDINATED CONTROL

Larry Popiel*
Ted Matsko**
Coleman Brosilow
Case Western Reserve University, Cleveland, OH 44106

Abstract. Coordinated control is the technique of using multiple constrained control efforts to control a single measurable process output. This technique is designed to maintain the process variable at its setpoint while making the most efficient use of the control resources available. The real time implementation of the control algorithm in microprocessor based distributed controllers is presented. Implementation issues include bumpless transfer of the controller from manual override to automatic, accommodation of feedforward signals, and use as a setpoint supervisor to regulatory loops. The paper concludes with a discussion of the results of applying coordinated control to an industrial multiple effect black liquor evaporator.

Keywords. Coordinated Control; Model Based Control; Inferential Control; Internal Model Control; Multivariable Control; Multi-Input Single-Output Control.

* Presently at Exxon Chemicals, Baton Rouge, Louisiana
**Presently at Bailey Controls, Cleveland, Ohio

INTRODUCTION

Coordinated control is a technique for using multiple constrained control efforts to control a single measurable process output variable. The multiple control efforts are manipulated in concert to satisfy objectives on both the process variable and the control efforts. The approach provides a universal solution to constrained multiple-input, single-output control problems, heretofore solved by a variety of techniques such as valve position control and split range control (Shinskey 1981).

In a coordinated control system the objective is to maintain the process output at its setpoint while fulfilling a secondary objective on the control efforts. This secondary objective on the control efforts attempts to make the most efficient use of the control resources which are available. For example, if the resources differ in cost, the objective is to minimize the use of the more expensive resource while maintaining the process output at its setpoint. Another example is when resources of equal value effect the process variable through different dynamics. In this case the secondary objective is to maintain the largest possible degree of freedom on the control resource which effects the process variable most quickly.

A coordinated control system is an ordered combination of single-input, single-output model based controllers (e.g. Inferential, or Internal Model Controllers) configured to accommodate constraints on the control efforts. Such controllers have been discussed elsewhere (Parrish 1985; Brosilow 1984). The following section contains a brief review of these previous results along with a discussion of how to deal with constraints in adapting feedforward control to model based control systems. This latter subject is particularly relevant to coordinated controllers because control effort targets are treated in the same way as feedforward signals in the coordinated controller.

SINGLE-INPUT, SINGLE-OUTPUT, TRAJECTORY TRACKING, MODEL BASED CONTROLLERS

Figure 1 shows a model based controller structured so as to account for constraints on the control effort. The figure shows only the flow of information and suppresses data sampling and output projection functions of the various blocks. As usual, the function of the model is to remove the effect of the control effort from the measured output so as to obtain an estimate, $\tilde{d}_e$, of the influence of disturbances on the controlled process variable.

The role of the controller, C, is to force the output of model lag, g, to track either the filter output or its state, at a specified future time. We call the interval between the current and future time the 'time horizon'. The time horizon is always equal to or greater than the sampling time.

A single step controller computes the <u>constant</u> control effort required to make the model output (i.e. output of g of fig. 1) equal to the filter output at the end of the time horizon. If the time horizon is longer than the sampling interval, the computed control effort is held constant only for the sampling interval and the entire calculation is repeated at each sampling interval. Time horizons which are longer than the sampling interval are useful in stabilizing the computational feedback loop of fig. 1 when the model, g, is a second or higher order lag.

The control effort required to make the model output equal the filter output, f_m, after a time horizon of H is

$$m_k^*(H) = m_{k-1} + \frac{f(t_k+H, t_k) - y_m(t_k+H, t_k, m_{k-1})}{S_m(H)} \quad (1)$$

where $m_k^*(H)$ = the control effort to be applied over the interval $t_k < t \leq t_{k+1}$

m_{k-1} = the control effort applied over the previous interval $t_{k-1} < t \leq t_k$

$f(t_k+H, t_k)$ = the filter output at t_k+H given its states and inputs at t_k

$y_m(t_k+H, t_k, m_{k-1})$ = the model output at t_k+H, given its states at t_k and an input m_{k-1}

$S_m(H)$ = the step response of g evaluated at H

H = time horizon

$t_{k+1} - t_k = \Delta$ = sampling interval

An alternate form of (1) is

$$m_k^*(H) = \frac{f_m(t_k+H, t_k) - y_m(t_k+H, t_k, 0)}{S_m(H)} \qquad (2)$$

where

$y_m(t_k+H, t_k, 0)$ = the model output at t_k+H given its states at t_k and zero control input

The actual control effort, m_k, applied over the interval $t_k < t \leq t_k + \Delta$, is

$$m_k = m_k^*(H) \text{ if } m_L < m_k^*(H) < m_U \qquad (3)$$
$$= m_L \text{ or } m_U \text{ otherwise}$$

A necessary condition for the stability of constrained controllers around an <u>achievable</u> operating point is that the controller be stable for small perturbations around that operating point. This is equivalent to requiring that the unconstrained controller be stable. If in addition to the above, none of the future controls violate constraints, then the constrained controller is guaranteed stable. The foregoing leads to the following theorem.

<u>Theorem</u>: The single step algorithm is stable for any finite order linear constant coefficient model, g, if the time horizon, H, is long enough.

A proof for the above theorem is given by Brosilow (1986).

While single step controllers are always stable for a long enough time horizon, they are not optimal unless g is a first order lag. As will be discussed by Brosilow and Quigley (1986) multistep controllers improve performance for higher order process models.

The role of the filter in Fig. 1 is to stabilize the control system against modeling errors. It is generally a lag of the form $1/(\varepsilon s+1)^n$ for an nth order model lag. The filter time constant ε should be selected large enough so that an acceptable damping is obtained for the unconstrained control system for all possible operating conditions. Doyle (1982); Chen (1984) suggest procedures for carrying out this calculation when process uncertainty can be expressed as parametric bounds on the model, or gain and phase bounds in the frequency domain. Parrish (1985) suggests an on-line tuning procedure similar to that suggested by Ziegler and Nichols for PID controllers.

The filter must provide the controller with its projected output one time horizon into the future. To do this it is necessary to predict how $\tilde{d}_e$ and the setpoint will evolve over the time horizon. For this paper we make the simplest possible prediction that both $\tilde{d}_e$ and the setpoint will remain constant over the next horizon at their current values.

The diagram in Fig. 1 provides for bumpless transfer by having the filter state track the model state (i.e. the filter state is set equal to the model state) in the manual or the tracking mode of operation. In this way the value of the control effort computed by the controller will equal that supplied by the operator or the tracking signal and no 'bump' will occur when the controller is placed in automatic mode.

The bumpless transfer mode of operation can also be used to avoid controller windup. Windup can occur when the setpoint cannot be achieved because of constraints. In this case, the desired filter output can exceed that which is achievable and it may take some time for a change in disturbance or setpoint to bring the filter output to the point where the control effort leaves saturation. Thus, the filter 'windup' may introduce an undesirable delay in the controller response. This problem is eliminated by resetting the filter state to the model state at the start of each calculation if the control effort is saturated over the previous time interval. In this manner a change in disturbance or setpoint will bring the control away from saturation without a delay.

The foregoing method of windup control has the defect that in some applications it will needlessly slow down the system response. Windup in controllers like that of Fig. 1 is not as severe as windup in unmodified PI controllers and often is so mild as to not present a problem. In such situations the filter state should not be adjusted during saturation. When in doubt, however, the safer course is to use the anti-windup feature described above.

Figure 2 shows how we incorporate feedforward signals into a model based controller. Were it not for the constraints, the feedforward signal would enter the process but not the model. In this way the feedback controller would compensate for any disturbances not eliminated by the feedforward signal. Thus in Fig. 2, when the control m is not at constraints the signal to the model, m_m, is equal to the signal from the controller m_c. The presence of constraints complicates matters. Our approach to make the disturbance estimate $\tilde{d}_e$ respond in the same way to the feedforward signal ff in both the unconstrained and constrained cases. As shown in the appendix, the structure in Fig. 2 results in $\tilde{d}_e$ being independent of the measured disturbance when the feedforward compensation is perfect, regardless of whether or not the constraints are active. The effect of the constraints on the process variable is to distort its response, but only by the amount of the unrealized control effort.

The configuration of Fig. 2 can exhibit windup just as that in Fig. 1, and the same solution can be applied. That is, the filter state is reset to the model state whenever the control effort saturates. Note, that in this case, however, the control into the model, m_m, need not be constant since the feedforward signal, ff may be changing.

It is frequently desirable to use a model based controller as the outer loop of a cascade controller, where a PID controller is used in the inner loop. The recommended configuration is shown in Fig. 3. The lag in the feedforward path to the summing junction should reflect the dynamics of the actual inner loop when the constraints are not active. The lag will have a unity gain whenever integral action is present in the inner loop and is the closed loop response of the inner loop to a setpoint change. Often, the dynamics of this inner loop are sufficiently fast that the lag can be neglected.

The lag introduced by the inner loop of the cascade controller can necessitate a longer time horizon in a controller using a single-step algorithm. For example, if g were a first order lag and if there were not an inner loop, then a single-step algorithm would take the time horizon as the sampling time. The presence of the inner loop makes the effective model at least second order, thereby usually requiring a time horizon longer than the sampling interval to maintain the desired degree of stability.

COORDINATED CONTROL

Figure 4 shows a two control effort, single process variable Coordinated Control system. The primary loop, designated by the subscript 1, provides the long term control effort. The secondary loop (subscript 2) assists the primary control effort. The bias signal represents the desired steady state level for the secondary control effort. Both loops act simultaneously, but the computations for the secondary loop follow those of the primary, which computes its outputs first. If the model dead time in the primary loop exceeds that of the secondary then the output of g is delayed by the difference between the primary and secondary delays before being subtracted from the secondary filter output. This is done so as to coordinate the faster secondary loop with the primary loop. The net effect is to make the process variable respond as the filter f_2 with a delay T_2.

The secondary control effort bias shown in Fig. 4 should actually be implemented as the feedforward signal, ff, in Fig. 2. For simplicity of representation, Fig. 4 shows the bias directly influencing the process. When the control is not saturated, the two diagrams are equivalent. As shown, a change in bias will influence the process variable. To avoid having changes in bias effect the process variable, the bias should be treated as a measured disturbance which is compensated by a feedforward controller to the primary control effort. Such compensation is included in the evaporator example which follows this section.

To illustrate the operation of the coordinated controller shown in Fig. 4, consider the two exchanger heater of Fig. 5. The objective of control is to maintain the outlet temperature in the water stream. By its physical location the valve closest to the outlet will have the most immediate effect on the water outlet temperature. A reasonable control objective is to have the valve closest to the outlet compensate quickly for disturbances and to then back away to a nominal position, perhaps mid-range, as the more distant valve picks up the load. Thus, for this example the closer valve to the outlet is the secondary controller, the more distant valve is the primary controller and the secondary bias would be set at mid-range (9 psi for an air operated valve). Figure 6 shows the ideal heat exchanger response to a step change in setpoint. Notice that both controls m_1 and m_2 respond instantly to the change. If the setpoint change is too great to be accomplished by m_1 alone at steady state, then m_2 does not decrease to its bias value (which for the simulation was set at zero) but rather reaches whatever steady state is necessary to achieve the desired setpoint as shown in Fig. 7.

The assignment of controls as primary or secondary is problem dependent. When possible the primary control should be the least expensive control and/or have the slowest effect on the process variable. Conversely, the secondary control should be the more expensive and have the fastest effect on the process variable. There are, however, many important problems where the above rules must be violated as will be seen in the evaporator application.

To illustrate the selection of primary and secondary controls, consider the problem of furnace control with waste gas and natural gas as shown in Fig. 8. In this case there is no issue of response times since the furnace temperature responds equally rapidly to each control. Since the waste gas is the cheaper, it should be made the primary control while the natural gas is the secondary control. The bias on the secondary would be set as near zero as possible. Any disturbance entering the furnace which could not be suppressed as rapidly as demanded by the secondary filter using waste gas would call the natural gas into action. If one does not want to use extra natural gas to speed the system response, then the secondary filter time constant should be set equal to the primary filter time constant which in turn would be set long enough to allow most disturbances to be suppressed using only waste gas.

The above furnace control problem might also have been solved using a split range controller such as that shown for the reactor pressure control in Fig. 9. The objective of control is to maintain reactor pressure by manipulating the effluent and feed valves. In the split range strategy the feed valve begins to shut only after the effluent valve is wide open. A coordinated controller would assign the effluent valve as the primary controller and the feed valve as the secondary. The secondary bias would be either the desired steady state feed valve position, or more likely the desired feed flow rate, which in turn would set the valve position through an inner loop. In either case both valves would operate simultaneously if necessary to maintain reactor pressure. Thus, a coordinated controller can potentially yield safer operation than a split range controller.

In either of the above examples, the primary or secondary loop can be set on manual control while the other loop continues to function. However, when the secondary is set on manual, the signal from the secondary model must no longer be added to the output of the primary model so as to avoid steady state offset. When the primary controller is on manual the steady state value of the signal added to the output of the secondary model cancels with that subtracting from the secondary filter output and no changes are required other than those normally associated with bumpless auto-manual transfer.

Either or both primary and secondary controllers can supply the setpoints to the inner loop of a cascade configuration as shown in Fig. 10. In addition to the measured control efforts, each loop can accommodate a feedforward signal and a track or manual control effort signal. While not shown explicitly in either Figs. 4 or 10, the bias is treated like a feedforward signal.

Tuning the loop filters depends on expected modeling errors and desired system performance. If the secondary is the faster loop then a reasonable procedure is to tune it with the primary on manual. The primary can then be tuned with the secondary in automatic mode.

Generally, the filter time constant for the primary should not be reduced below that of the secondary (assuming that both filters are the same order lags). If the primary filter time constant is shorter than the secondary, then the secondary control attempts to undo some of the response caused by the primary and the two controls can fight each other.

One of the more interesting and potentially profitable uses of coordinated control is to simultaneously adjust the mass and energy input rates to a process (e.g. feed flow rate and reboiler duty in a distillation column). The next section describes such an application to an operating industrial process.

APPLICATION OF COORDINATED CONTROL TO AN INDUSTRIAL BLACK LIQUOR EVAPORATOR

The multiple effect evaporator is an essential unit operation in the recovery section of the pulp and paper plant. The evaporator functions to increase the weight percent of organic solids in a weak black liquor stream by evaporating water from that stream. The strong black liquor product is sent to a concentrator where the solids weight percent is increased to the point where the liquor can support combustion in the recovery boiler.

The multiple effect evaporator considered here is a unit at Boise Cascade and consists of five stages or effects in which the evaporation occurs, Fig. 11. As shown in Fig. 11 steam enters the evaporator through the first effect while the weak black liquor enters the evaporator through the fourth effect. In each effect the liquor is pumped upwards through vertical tubes. The steam flows into the steam chest and transfers heat to the liquor by condensing on the outside of the vertical tubes. Water is evaporated from the liquor and is sent to the next effect downstream as the heating medium. The remaining liquor is sent upstream to be further processed. The entire process operates at moderate pressures, with the first effect having a steam pressure of 30 psi. Each effect downstream operates at a lower pressure with the sixth effect operating at a vacuum of 22 in Hg. Both the steam and black liquor inputs to the process can be manipulated by control valves.

The evaporator has two different modes of operation. The first mode of operation is called maximum throughput. This mode of operation requires that the evaporator processes the maximum possible amount of feed. This philosophy is used at times when the level of weak black liquor in the feed storage tank becomes too high. In the maximum throughput operation the steam valve is maintained fully open while the feed valve is modulated to maintain the product at its setpoint. Tight control around the setpoint is crucial because the concentrator units which process the stream leaving the evaporator operates most efficiently when the stream is at or near it setpoint value. Also any black liquor too high in solids content causes clogging due to high viscosity, while liquor too low in solids content will not support combustion in the recovery boiler. The evaporator is run according to maximum throughput the majority of the time.

The second mode of evaporator operation is a steam modulation operation. This mode allows a nominal liquor feed rate (less than the maximum) to be set for the evaporator while the steam input is modulated to control product quality.

The evaporator is subject to a wide variety of upsets due to measurable and unmeasurable disturbances. The major measurable disturbances are 1) steam temperature and flow, 2) liquor feed density, 3) liquor feed temperature, and 4) degree of vacuum in the last effect.

In applying coordinated control to the black liquor evaporator, the steam flow rate to the evaporator is chosen as the primary (long term) control while the feed flow rate of weak liquor is the secondary control. In this application, PI controllers regulate the flow of steam and weak liquor around nominal setpoints. The coordinated controller supplies the setpoints to these regulatory control loops. The weak liquor feed has a desired steady state value which is the processing rate under steam modulated operation. The feed processing rate can be set to cause the steam valve to saturate, thus providing maximum throughput operation.

The models for the evaporator were obtained from step tests on the process. The results of several step tests on the steam flow indicated a wide range of values for the primary model parameters. The estimated deadtime (T1) varied between 130 and 428 seconds. The time constant varied between 240 and 430 seconds and the gain (K1) varied between 0.33 and 0.6. The primary model parameters actually used in the control algorithm were averages of all model parameters fitted from the individual tests. The primary model is

$$g_1(s) = \frac{.432\ e^{-340s}}{255s + 1}$$

The transfer function relating the product density to a change in liquor feed flow rate was obtained from step tests on the liquor feed rate with the steam flow held steady. The model parameters derived from these step tests also varied greatly. The deadtime (T2) varied between 330 and 840 seconds, the time constant varied between 262 and 560 seconds and the gain varied between -0.055 and -0.1. The secondary model parameters actually used in the control algorithm were also average values. The secondary model is

$$g_2(s) = \frac{-.077\ e^{-517s}}{438s + 1}$$

Special issues in the implementation of the coordinated controller to the evaporator were 1) process noise, 2) on-line filter tuning, and 3) response to liquor flow setpoint changes.

The coordinated controller is implemented using the steam flow rate as the input for the process model for the primary controller. This flow signal fluctuates at a high frequency. Direct use of this flow signal causes oscillatory behavior of the secondary control output (liquor feed setpoint). This situation occurs because the fluctuations in steam flow cause fluctuations in the primary model output which in turn causes the signal to the secondary controller to fluctuate. This fluctuating objective causes undesirable fluctuations in the feed flow rate. To reduce this problem the measured steam flow signal is "filtered" by restricting its rate of change to 3 Klb/minute.

The initial filter time constants selected for this application followed the previously suggested rule that the primary filter time constant should be equal to or greater than the secondary filter time constant. In this case they were chosen to be equal. For this application however, control system performance was improved by adding an additional 80 seconds to the secondary filter time constant. This adjustment of the secondary filter time constant was done on-line as a tuning measure. This tuning was done to try to account for the inner loop dynamics. The value of 80 seconds is used because it is the estimated time constant of the response of the steam flow to a step change in its setpoint.

The addition of the 80 seconds to the secondary filter time constant improved the control performance by reducing unneccessary secondary control effort movement. The addition improves performance because it enables the secondary filter output to better match the primary trajectory. Because the steam flow loop has significant dynamics, the primary trajectory lags the output of the primary filter. The time constant of the secondary filter is therefore increased to balance the control system. The dynamics of the weak black liquor flow loop was found to be insignificant and therefore required no additional tuning of the system.

Since changes in the weak liquor feed setpoints are made quite often, it is desirable to respond quickly to these changes. For this reason a feedforward signal based on the liquor setpoint (secondary bias) is sent to the primary controller as shown in Fig. 12. To reduce process upsets due to liquor flow setpoint changes, the setpoint for liquor feed flow is rate limited before being used as the bias for the secondary controller.

Results from implementing the coordinated controller on the black liquor evaporator for a four day test showed that the control system was able to maintain the product density to within 1 percent (plus or minus) of its setpoint, which is considered very good by the operator. Transition of the control system into and out of steam valve saturation was quite smooth.

The test period was terminated after four days, due to the shut down of a parallel production line. The evaporator set continued operation under the coordinated control algorithm during this critical period. Loss of availability during this time would force the mill to shut down. Following the start-up of the companion evaporator and recovery boiler, the mill experienced difficulty with the coordinated controller.

The density data available from the plant exhibited oscillatory behavior. The peaks were two to three percent away from setpoint, with a period of approximately one hour. This response provoked questions concerning system stability. To investigate this possibility, a simulation study was conducted varying the time constants and process order. The simulations did not replicate the plant behavior. Other roots to the problem were sought.

The data supplied indicated that the sensor has a tendency to spike infrequently. The measured density displays changes of 10% in the short time period of two to four minutes. The process cannot respond to any physical disturbance in this time frame. Persistent transmitter noise of this nature could cause the observed problem. A second plant trip disproved noise as a source of exitation for the density oscillations, but the coordinated controller was reinstated successfully at that time.

The coordinated controller was restarted with some minor changes, described below, but the original problems persisted. The plant data suggested that the oscillations weredue to a disturbance which eminates from the batch pulp digesters. The three digesters operate in parallel, sharing a receiving vessel. The gas that flashes into the shared receiving vessel is condensed with cooling water from the same source that supplies the evaporator condenser. The condenser creates a vacuum in the last stage of the evaporator to improve efficiency. When the cooling water is drawn away from the evaporator the vacuum drops, decreasing product density. Due to the process deadtimes, the coordinated controller could not respond in time to counteract this disturbance. By the time the control action affected the process variable, the disturbance had disappeared, causing the observed overshoot. Cooling water disturbances are dependent on process conditions, and they did not occur during the initial testing of the algorithm.

To counteract the cooling water disturbance, feedforward action on the condenser pressure was added to the coordinated controller. The input to the feedforward transfer function is the condenser pressure and the output is the steam flow. If the steam valve is saturated, the liquor flow automatically responds due to the coordinated controller. Figure 13 illustrates the improved operation obtained with the feedforward control. (The boxes next to the trend charts in Fig. 13 give the scale of the dependent variable. For example the product solids density scale ranges from 50 to 60 weight percent solids. The steam valve position (#513F801) ranges from full closed (0%) to full open (100%). The scale shows ranges from -5% to 105% to force full closure or full opening.) The density now increases with a disruption in the cooling water because of imperfect modelling in the feedforward control loop. The cooling water disturbances at 14:40 and 15:55 result in increased steam flow setpoints, but the steam valve is saturated. To maintain the density, the feed liquor flow is decreased. At 16:05 the operator ramped the secondary bias down to obtain a new production rate. The control system made a smooth transition to primary control. The overall performance of the control system was now acceptable.

The minor start-up changes referred to previously were to increase the sampling time from 1/4 second to 10 seconds and to introduce a one sample time delay into the secondary filter inputs. The increase in sampling time permits greater filtering of the process measurements and reduces the effect of noise on the controller. The insertion of a delay into the secondary filter is necessitated by the one sample delay caused by sampling the output of the cascade control loop in the primary. This delay serves to coordinate the signals from the secondary filter and the primary model lag. (A change in the output of the primary filter requires one sampling time to influence the output of the cascade loop and then another sampling time is required before the primary model lag responds to the observed change. There are alternate ways to resolve this computational problem.)

Some interesting features of the control system were noticed
following start-up. During a boiler trip the steam flow fell and the
coordinated controller droped liquor flow drastically. The actions of
the controller maintained liquor density at 51%. Had the density
fallen too low, the operators would have had to run the evaporators
off line, and then back down the load on the recovery boiler. This
would have created further problems in the steam supply. Indeed, a
slow manual response could have caused significant portions of the
plant to shut down. The coordinated controller approach maintained
plant production at normal levels. With restoration of the steam
flow, the evaporator returned to normal production rate and product
density. This smooth return to normal is enabled by matching the states in the filter and model of the primary controller.

At this time, (12/1/85) the coordinated controller has been in
operation over three months. This time period included a one week
plant shut down for routine maintenance. After start-up, the coordinated controller was still in use. The first few months for any
advanced control technique are the critical trial period. Plant
operators frequently prefer manual control and revert to that mode
during the introductory time period. Given that environment, the fact
that the coordinated controller has operated for three months indicates that it is now a permanent part of the plant.

OPTIMIZATION VERSUS COORDINATED CONTROL

This section compares the 'Coordinated Control' approach of the
authors to that of optimal model based predictive control [Garcia '86,
Richalet '86]. Superficially, in the latter method one simply defines
an objective function and then solves a nonlinear, or linear, program
to minimize (or maximize) the objective function. Let us assume that
the nonlinear program exists, always works perfectly, and requires
negligible computer resources. The problem of applying the optimization method then reduces to that of choosing the objective function to
get the desired control system behavior. Even this reduced problem is
not easy to solve.

First, the objective function must be able to yield stable control in spite of the inevitable modeling errors. This is usually accomplished in DMC by appropriately selecting the time horizon over which the optimization is performed and the number of control moves allowed during the time horizon. For the uninitiated it isn't obvious how to accomplish the foregoing. There is always a trade-off between desired performance and stability and some experience as necessary to adjust the time horizon and number of control moves to obtain a reasonable trade-off. In IDCOM or Model Algorithmic Control, the appropriate performance-stability trade-off is accomplished by specifying a desired process variable trajectory, which is similar to the method used in coordinated control. However, as is pointed out below, the objective function must also include at least one control effort. It is not at all clear what trajectory, if any, should be imposed on a control effort. Never mind, assume we overcome the above problems with a little experience.

The next problem is to choose the integrand of the integral objective function. In order for there to be a unique solution to the optimization problem, the integrand should be positive definite (i.e. never zero for nonzero trajectories). This means that the integrand can not be only a function of the process variable since the process variable can be maintained at set point for multiple nonzero values of the controls (If $y = g_1 m_1 + g_2 m_2$ then $y = 0$ simply specifies a relationship between controls m_1 and m_2.) One method of making the integrand positive definite is to include a penalty on one of the controls deviating from some desired level. This is the same as requiring a specification of a primary and secondary control effort for the two control effort Coordinated Controller. A reasonable objective might be

$$\underset{m_1(t),\, m_2(t)}{\text{minimize}} \int_0^T \left[(\text{setpoint} - \text{process variable})^2 + \beta(\text{Bias} - \text{Secondary Control})^2 \right] dt$$

The choice of the weight β presents a problem. Selecting β too small will mean that the secondary control will not return to its bias value and thus will not be able to supress disturbances as well as a Coordinated Controller. If β is selected too large then the secondary control will tend to remain near the bias and the process variable will drift further from its setpoint than in a Coordinated Controller. Finally, if the primary control saturates, then the process variable will never reach its setpoint unless β is set to zero, which is not allowed because then the primary control will not recover from saturation. Of course, life gets even more difficult as the number of controls increases beyond two.

In summary, the results of applying optimal model based predictive control to multi-input, single-output processes are likely to be inferior to those obtained from Coordinated Control for the reasons given above. Also, our assumption that the nonlinear program requires negligible resources is not generally correct. Finally, the functional minimization will not yield the same insight as that provided by Coordinated Control. Thus, it is our contention that Coordinated Control provides a valuable additional tool to solve industrial control problems.

REFERENCES

Brosilow, C., Q. Zhao and K. Rao (1984). A linear programming approach to constrained multivariable control. Proceedings of 1984 ACC, San Diego.

Brosilow, C. and Q. Zhao (1986). A linear programming approach to constrained multivariable control. Advances in Control and Dynamic Systems, 24.

Brosilow, C. and V. Quigley (1986). Adaptive horizon adjustment for model predictive control. Proceedings of the 1986 ACC, Seattle.

Chen, S. and C. Brosilow (1984). Control system design for multivariable uncertain processes. Proceedings of 1984 ACC, San Diego.

Doyle, J. (1982). Analysis of feedback systems with structured uncertainties, IEE Proc. 129, D(6), 242.

Garcia and Prett (1986). Advances in Industrial Model-Predictive Control. Proceedings of 3rd International Conference on Chemical Process Control, Asilomar, California, January 1986.

Parrish, J. and C. Brosilow (1985). Industrial Control Applications. Automatica, 21, 5, 527-538.

Richalet and Froisy (1986). Industrial Applications of IDCOM. Proceedings of the 3rd International Conference on Chemical Process Control, Asilomar, California, January 1986.

APPENDIX

Figure 14 shows the feedforward configuration of Fig. 2 redrawn with the limiter L replaced by an equivalent time function, $w(t)$, which is zero whenever $m_c + ff$ is within constraints and which makes m equal the constraint when $m_c + ff$ violates the constraint. The disturbance estimate $\tilde{d}_e$ is then given by

A1) $\tilde{d}_e = d_e + G(m_c + ff - w) - ge^{-sT}(m_c + ff - w - ff)$

$= d_e + Gff + (G - ge^{-sT})(m_c - w)$

For the case of a perfect model $ge^{-sT} = G$ and

A2) $\tilde{d}_e = d_e + Gff$

Let $d_e = d_e^* + G_d d_m$

where d_m = measured disturbance

G_d = transfer function between d_m and the process variable y

d_e^* = the effect of all other disturbances

For perfect feedforward compensation

A3) $ff = -(G/G_d)d_m$

Assuming that G/G_d is realizable and stable then from A2 and A3)

A4) $\tilde{d}_e = d_e^*$

and

A5) $y = d_e^* + G_d d_m + G(m_c w + ff)$

$= d_e^* + Gm_c - Gw$

Choosing the control m_c to make the model g follow the filter gives

A6) $f(v - \tilde{d}_e) = gm$

and

A7) $y = fe^{-sT}v + (1 - fe^{-sT})d_e^* - Gw$

Thus, for the case of a perfect model, saturation degrades the performance of the control system by the amount Gw, which generally is not severe.

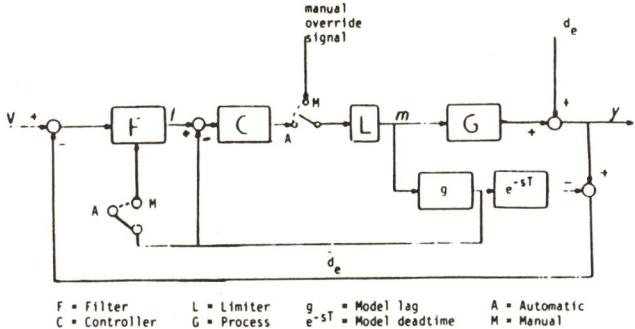

```
F = Filter        L = Limiter      g     = Model lag        A = Automatic
C = Controller    G = Process      e^-sT = Model deadtime   M = Manual
```

Fig. 1. Bumpless auto-manual transfer and anti-windup configuration.

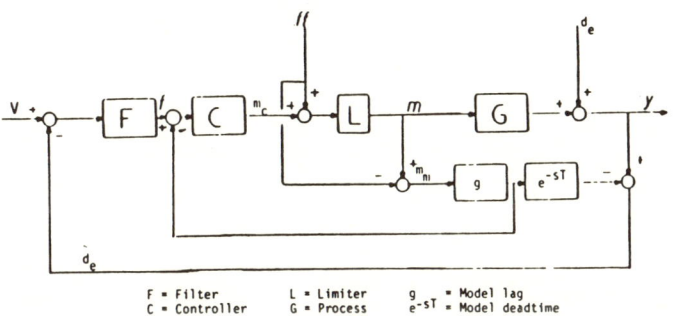

```
F = Filter        L = Limiter      g     = Model lag
C = Controller    G = Process      e^-sT = Model deadtime
```

Fig. 2. Accommodating feedforward signals.

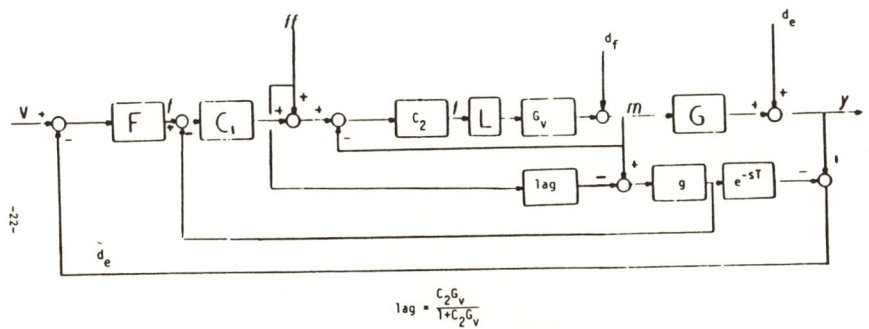

$$\text{lag} = \frac{C_2 G_v}{1 + C_2 G_v}$$

Fig. 3. Cascade control.

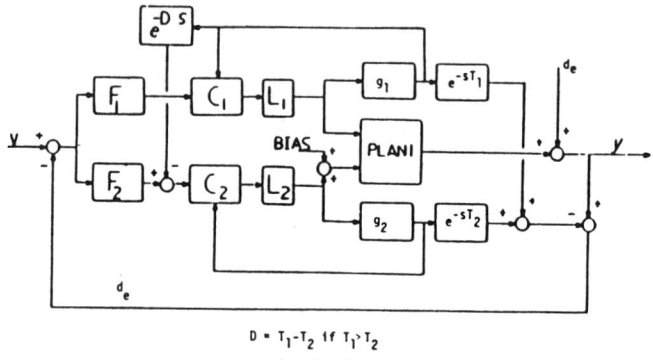

$D = T_1-T_2$ if $T_1 > T_2$
   $= 0$ otherwise

Fig. 4. Coordinated control system.

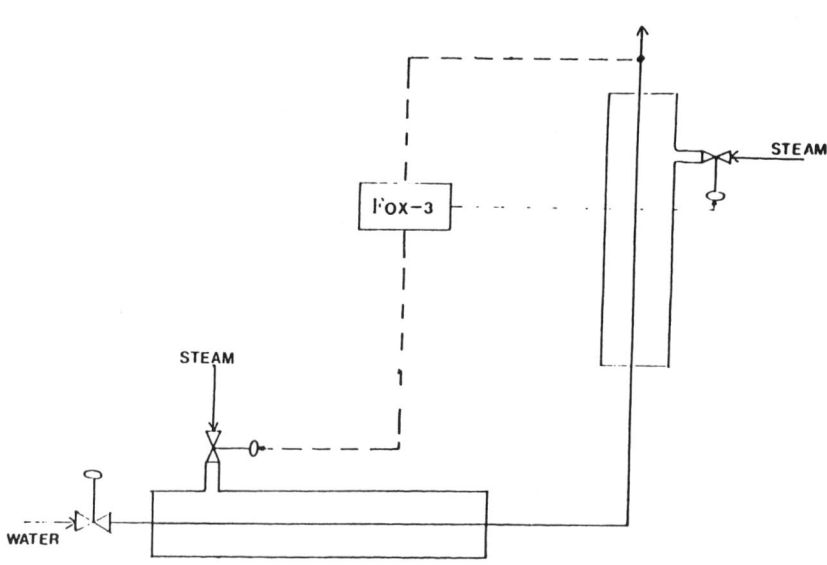

Fig. 5. Dual heat exchanger system.

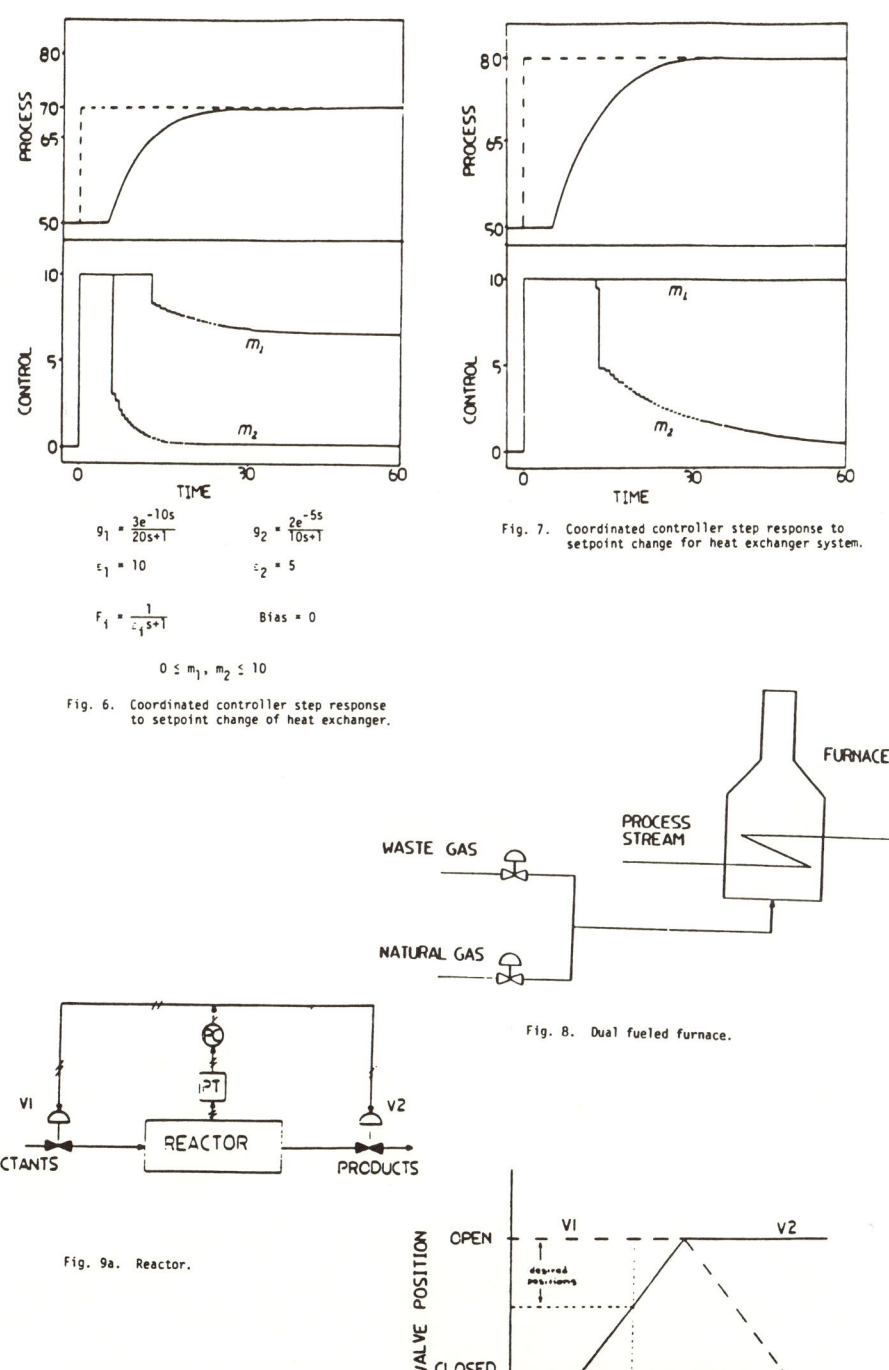

$g_1 = \frac{3e^{-10s}}{20s+1}$ $g_2 = \frac{2e^{-5s}}{10s+1}$

$\varepsilon_1 = 10$ $\varepsilon_2 = 5$

$F_i = \frac{1}{\varepsilon_i s + 1}$ Bias = 0

$0 \leq m_1, m_2 \leq 10$

Fig. 6. Coordinated controller step response to setpoint change of heat exchanger.

Fig. 7. Coordinated controller step response to setpoint change for heat exchanger system.

Fig. 8. Dual fueled furnace.

Fig. 9a. Reactor.

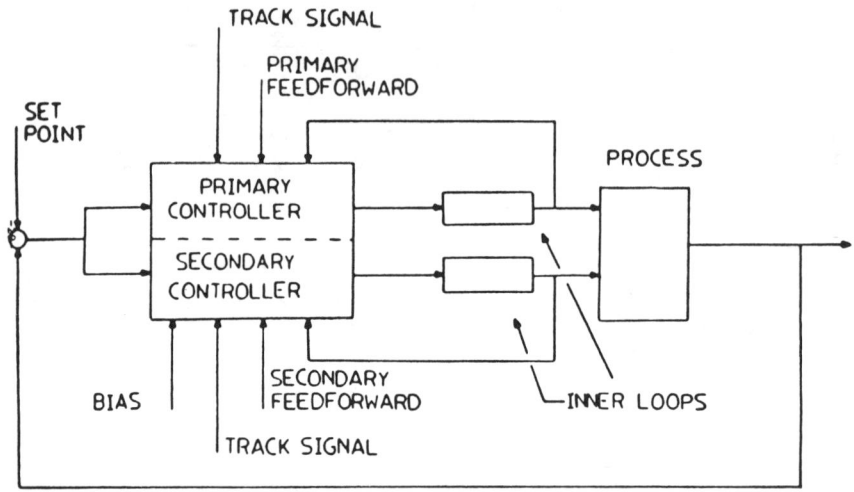

Fig. 10. Implementation of a two controls coordinated controller using feedforward and cascade configurations.

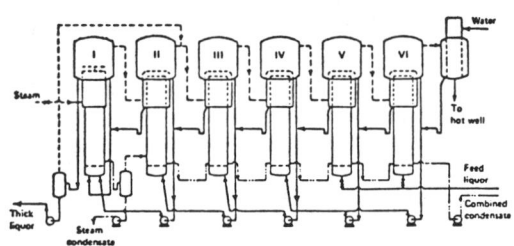

Fig. 11. Industrial black liquor evaporator.

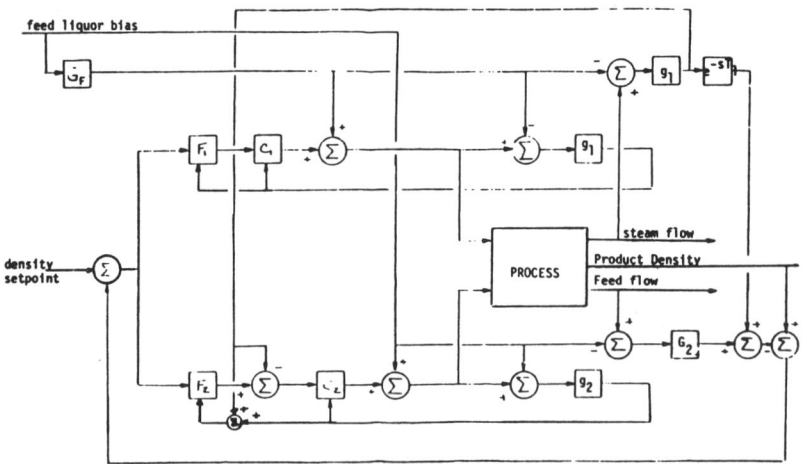

Fig. 12. Coordinated controller diagram for black liquor evaporator control.

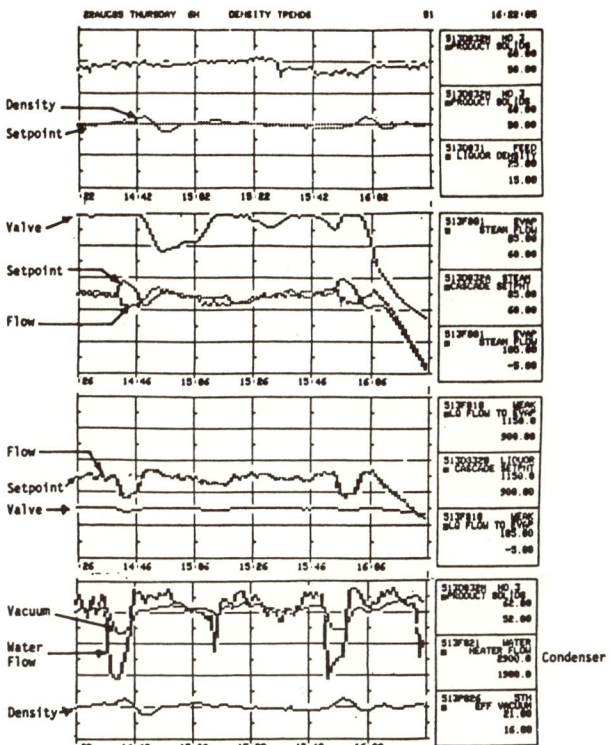

Fig. 13. Coordinated control with feedforward on vacuum.

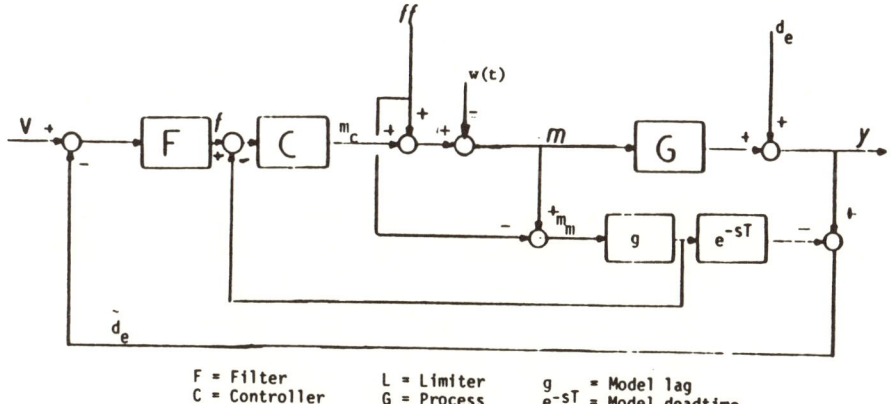

F = Filter   L = Limiter   g = Model lag
C = Controller   G = Process e^{-sT} = Model deadtime

Fig. 14. Accommodating feedforward signals.
Limiter replaced by function w(t)

PROCESS OPERABILITY

Irven H. Rinard
The Halcon SD Group, Inc., Montvale, NJ

Process Operability is a broad topic encompassing all aspects of the operation of a process plant, not just its moment-to-moment control. In addition to close adherence to the optimal operating point (or trajectory, in the case of a batch process), operability includes such considerations as safety, reliability, environmental acceptability, and ease of startup and shutdown. A more detailed discussion of operability, particularly its timely consideration with regard to the various stages of a process project, was given at the Second International Conference on Foundations of Computer-Aided Process Design (Rinard 1984).

Two important points were made, namely, that operability must be considered early enough in the project cycle to be incorporated into the process design and that, given the trend toward increased mass and energy recycling in processes, the entire plant must be considered as a whole rather than as a collection of individual unit operations. It was also stressed that time is critical due the pressure of the project schedule and that the methodology for operability analysis should be as simple as possible to accommodate both the time constraint and the need for the diverse members of the project team to understand the operability requirements.

The focus of this conference has been primarily upon control system methodology, particularly algorithms for model-based techniques. The two papers for this session have been chosen with this in mind. The two points just addressed, namely, timely consideration of operability requirements and evaluation of entire plant controllability, are also covered.

The first, by Yaman Arkun, is entitled "Dynamic Process Operability. Important Problems, Recent Results and New Challenges." First he looks at the relationship between the process design and controllability through the concept of robust dynamic operability. The key to the analysis is a suitable characterization of the uncertainty in the process model. Then he examines the problem of control system design, particularly the choice of control system structure, again using the concept of dynamic operability. The choice of decentralized control structures is attacked through the concept of the Block Relative Gain (BRG). The techniques presented in the paper, being multivariable, are well suited to studying the control of entire plants.

The second paper, by Tom Marlin and Tom McAvoy, is entitled "Short-Cut Methods for Process Control and Operability." As one may guess from the title, the intent is to provide the engineer with the means for making timely estimates of operability during the early design stages. Here the approach is to use simplified models and an integral error criterion to quickly examine operability. Entire plant control can be examined by using the simplified models to evaluate the Relative Gain Array (RGA) for instance.

As can be seen, the two approaches, while similar in goals, are quite different in style. Arkun's approach involves some fairly sophisticated mathematics and thus will require some further effort to make it accessible both intellectually and computationally to the practicing process design and control engineers. Marlin and McAvoy's approach, while readily useable, may be simplified beyond what is necessary given the amount of computing power that is being put on almost every engineer's desk these days.

Returning to the other aspects of process operability, one topic of considerable importance, namely process monitoring, should be mentioned. What should be measured and with what frequency, how the status of the plant can be inferred from the measurements, and what should be done when the status is abnormal comprise a topic about which much has been written. However, the occurrence of several catastrophic or near-catastrophic events in recent years can be attributed at least in part to inadequate process monitoring. Given the ramifications of these events, it is a topic which deserves much more attention by the process control community than it currently receives. There is one paper that deals with the subject, namely that by Moore and Kramer entitled "Expert Systems in On-Line Process Control." Process monitoring will also receive more attention at the Conference on the Foundations of Computer-Aided Process Operations scheduled for the summer of 1987.

DYNAMIC PROCESS OPERABILITY
IMPORTANT PROBLEMS, RECENT RESULTS
AND NEW CHALLENGES

Yaman Arkun
Georgia Institute of Technology, Atlanta, GA 30332

Abstract. In this paper dynamic process operability reflects the ability of the plant to maintain a satisfactory dynamic performance (i.e., good set-point tracking and disturbance rejection). The paper first describes the fundamental constraints on dynamic operability as imposed by process design and control system design, followed by a review of the analytical tools which are recently developed to quantify the concept of dynamic operability. Operability problems related to process design and control system design are discussed separately and, for each class of problems, new techniques are given to tackle the pertinent issues in a unified framework. Finally, questions and challenges are posed in the hope of stimulating innovative research in this important field.

Keywords. Dynamic process operability, process design, control system design, fundamental limitations to dynamic operability, internal model control, uncertainty, robustness measures, decentralized control.

INTRODUCTION

Today, chemical engineers face the challenge of designing chemical plants which can operate safely, smoothly and profitably within a <u>dynamic process environment</u>. For a typical plant, major contributions to such an environment originate from external disturbances such as changes in raw materials, product specifications, different market demands and energy sources. For higher productivity and better utilization of resources, chemical plants are expected to adapt easily to different operating conditions resulting from such environmental upsets.

<u>Dynamic process operability</u> refers to the ability of the plant to maintain satisfactory dynamic performance despite uncertainties in the process environment. From satisfactory dynamic performance, one usually understands effective recovery from disturbances, and fast and smooth transition from one set of operating conditions to another. In the light of recent concerns for feedstock and energy costs, increased uncertainty and process integration, and growing pressures to improve product quality, dynamic process operability has now become more paramount than ever before.

One can envision the overall design and operation of a chemical plant as depicted in Fig. 1. It comprises both the <u>design of the process</u> and the <u>design of the control system</u> within which different objectives are carefully evaluated and incorporated into the final plant design. It is clear that dynamic operability should be carefully addressed during all the development stages of a chemical plant, namely, during a) the synthesis of the chemical plant, b) the design of the individual processing units, and c) the control of the overall system. However, this has hardly been the practice in the past. Most of the time process design and control system design have been carried out as two isolated tasks without systematically exploring any beneficial interaction between the two activities which might lead to significant improvements in dynamic plant operation. Usually detailed dynamic simulations, empirical overdesign and trial and error methods have been used to detect and alleviate the potential dynamic operability problems. Although such techniques have proven to be quite acceptable in the past, they turn out to be very inefficient and costly for today's complex integrated plants.

Thus, there is a growing need for more rigorous techniques to assess and improve the dynamic operability characteristics of chemical plants. The practicing design engineer should be able to:
- (a) correctly quantify the dynamic operability of a process,
- (b) easily pinpoint the bottlenecks in the plant, and
- (c) systematically improve the dynamic operability of the process design and the control system design.

The purpose of this paper is to attempt a conceptual framework within which the above issues can be addressed by using the analysis and synthesis capabilities of some recently developed techniques. The paper is not intended to be an in-depth study of all the facets of dynamic operability but rather aims at providing perspective on some of the important problems while reviewing the recent results and pointing out new challenges for future research.

SOURCES OF DYNAMIC OPERABILITY PROBLEMS

Synthesis of chemical plants can be characterized as an interplay between the creation and assessment of alternatives. Several important decisions have to be made and evaluated before finalizing the design. These decisions can be classified as:
1. For Process Design
    - (a) Structure of the plant: the flowsheet.
    - (b) Design parameters: equipment sizes and operating conditions.
2. For Control System Design
    - (a) Control structure: manipulated inputs and measured outputs.
    - (b) Controller structure: interconnections among inputs and outputs.

(c) Control laws and the controller parameters.

Each of these decisions places different constraints on the system which will ultimately limit the dynamic operability of the final design. The goal of any dynamic operability study is then to be able to identify these constraints, quantify their effects, and propose design changes to remove them. In the sequel we will discuss how these fundamental limitations to dynamic operability are imposed by process design and control system design.

PROCESS DESIGN RELATED DYNAMIC OPERABILITY PROBLEMS

The major obstacle in the assessment of dynamic operability is the fact that dynamic operability, being a closed-loop property, is a function of both the process and its control system. Consequently, it becomes very difficult for the designer to distinguish between the effects of the two and isolate the dynamic operability problems associated with poor process design alone.

Until now development of a theoretical framework has been plagued by the classical feedback representation which does not allow separate analysis of the effects of the controller parameters and the design parameters. To circumvent this problem, recently a new representation called Internal Model Control (IMC) has been introduced by Morari (1983). As it turns out, IMC not only provides the right theoretical basis for the analysis of dynamic operability during process design but also constitutes a powerful framework for the synthesis of control systems.

We will now describe IMC only to the extent necessary for the development of the main results. For analysis purposes linear systems theory will be used while the effects of important nonlinearities will be captured by a modeling uncertainty description. Such a simplifying approach is quite acceptable in view of the fact that detailed nonlinear dynamic operability analysis during early stages of process design is usually not warranted. Also, a complete nonlinear analysis for later design stages still has to await new theoretical developments.

The IMC structure is shown in Fig. 2. Control system design consists of two steps: In the first a feedforward controller $G_I(s)$ is designed for the nominal system $\tilde{G}(s)$. In the second step a feedback controller, i.e., the filter F(s) is constructed so that the closed-loop stability is guaranteed for a given class of modeling errors and any deterioration from nominal performance due to model/plant mismatch (i.e., $G_p(s) \neq \tilde{G}(s)$) is minimized. Both controllers are assumed to be nonadaptive.

Assuming perfect model ($G_p = \tilde{G}$), the closed-loop relationship becomes

$$y = \tilde{G}G_I y^{sp} + (I-\tilde{G}G_I)d \tag{1}$$

If we choose for the controller the right inverse of $\tilde{G}$, $G_I = \tilde{G}^{-1}$, eqn. (1) reduces to

$$y = y^{sp} \tag{2}$$

Assuming the open-loop system and the controller to be stable, according to (2) we obtain "perfect dynamic operability" in the sense that we have perfect tracking of set-points and perfect attenuation of disturbances. However, this is only a theoretical limit, i.e., it is a target for dynamic operability which we can only try to <u>approach</u> by the right design.

In reality the inverse controller is not physically implementable due to the following <u>fundamental limitations</u> placed on $\tilde{G}(s)$ by the <u>underlying process design</u>:

1. Nonminimum phase (NMP) elements, i.e., time delays and right half plane (RHP) zeros which result in predictive and unstable controllers when $\tilde{G}$ is inverted.
2. Physical constraints on the manipulated variables. Inversion of $\tilde{G}$ often leads to improper controllers for which manipulated variables saturate at high frequencies.

Due to these constraints G_I has to depart from "perfect controller" and has to be designed so that it is the realizable (i.e., physically implementable) inverse of the invertible part of the model $\tilde{G}$. In doing so one can formulate and assess the effects of these limitations on dynamic operability in a quantitative manner.

Thus $\tilde{G}$ is first factored into

$$\tilde{G} = \tilde{G}_+ \tilde{G}_- \tag{3}$$

where $\tilde{G}_-$ is proper, stable, minimum phase, thus invertible and $\tilde{G}_+$ is strictly proper, nonminimum phase, thus noninvertible fraction of $\tilde{G}$. Different techniques are available to perform this structural factorization. Our research group used a technique based on Λ-generalized polynomial matrices which handles systems that have both time delays and RHP zeros. The method has been exclusively used for the control system design and for details the reader is referred to Arkun et al. (1985), Arkun and Manousiouthakis (1985) and Ravindranath (1983). Also, factorization of time delay systems has been discussed in detail by Holt and Morari (1985a) and the factorization of RHP zeros has been treated in Holt and Morari (1985b).

Above factorization, eqn. (3), allows us to determine the class of admissable controllers G_I and the corresponding set of achievable nominal performances.

Admissable Set of Controllers

$$G_I = \{G_I = \tilde{G}_-^{-1} M \text{ where M is proper and stable}\} \quad (4)$$

Achievable Nominal Performances

$$y = \tilde{G}_+ M y^{sp} + (I - \tilde{G}_+ M)d \triangleq H y^{sp} + Dd \quad (5)$$

It is clear that if $\tilde{G}$ contains RHP zeros and time delays, they can not be removed by any control system and they will appear in certain combinations in $\tilde{G}_+ M$. Thus the information gained through this development of achievable performances is very valuable. By studying H and D the designer can find out if a desired level of dynamic operability is feasible or not. During this assessment a certain degree of freedom exists since factorization and therefore $\tilde{G}_+$ and the selection of M is not unique. However, the designer can use this freedom to judiciously create the alternative nominal performances which are expected from a given process design. Holt and Morari (1985a,b) explored this freedom and showed how one can try to shift the effects of RHP zeros and time delays between different outputs and how this affects decoupling control and consequently dynamic operability. They derived several rules to generate $\tilde{G}_+$'s which are then evaluated either by inspection (e.g., magnitude of time delays, location of RHP zeros) or by a more quantitative performance index such as Integral Square Error. The method has been applied to distillation columns and heat exchanger networks to compare achievable dynamic performances of alternative designs, e.g., designs with bypass valves around different heat exchangers in the network.

Finally, the effect of physical constraints on dynamic operability can be measured by making use of Singular Value Decomposition techniques (Klema and Laub, 1980). Manipulated variables are less likely to be saturated by load changes for those process designs whose minimum singular values $\sigma_*(\tilde{G}(s))$ are large over a large frequency range (Morari, 1983). Note that $\sigma_*(\tilde{G})$ depends only on the process design, is independent of the particular controller used and is easy to compute and use as a dynamic operability measure.

Although significant progress has been made so far, dynamic process operability still remains to be a challenging research area with quite a few unsolved problems. Future research should concentrate on refining the existing techniques discussed above and developing new theories for large interconnected systems. Following research topics pertain to the fundamental limitations imposed by nonminimum phase elements and constraints on manipulated variables.

    a. The "best" factorization $\tilde{G}_+$ for systems with more than one RHP zero and for systems with both time delays and RHP zeros is an open question. For these systems rules to create alternative $\tilde{G}_+$'s are needed.

    b. Current results are derived for a centralized IMC structure, thus, they are not yet convenient for the study of large scale interconnected systems. There is a strong need for the development of decomposition techniques and the corresponding dynamic operability measures.

    c. Factorization and the analysis techniques should be incorporated into a computer aided software using reliable computational tools so that they can be easily tested on industrial processes.

    d. Most of the results in the literature focus on alternative process designs caused by variations in the control structure. Available theory is general enough to consider other changes as well, e.g., the structure of the flowsheet, the equipment sizes and the operating conditions. More studies should be undertaken to address this more difficult problem. Can the dynamic operability measures <u>predict</u> the design changes so that they guide the designer in a more constructive way?

e. Finally, existing theory should be extended to address open-loop unstable plants and hopefully include a nonlinear analysis. Is it possible to formulate the admissible set of controllers and the performance targets for the nonlinear IMC?

So far we have assumed <u>perfect</u> models in order to be able to illustrate the two fundamental limitations due to NMP elements and physical input constraints. However, it has to be noted that, the success of the controllers and hence the dynamic plant operability is further limited by the accuracy of the process models which are used to design the controllers. These models usually contain many assumptions and simplifications which arise as a natural consequence of the designer's limited knowledge about the process and/or are introduced to bring ease to the controller design. Depending on the sensitivity of the process design to such uncertainties, significant deviations from the nominal target operation can occur.

In order to minimize the effect of uncertainty on dynamic operability, this issue has to be addressed as early as possible during process design. In the sequel we will refer to <u>robust</u> dynamic operability in order to separate the uncertainty related operability problems from the other two aforementioned fundamental limitations. In the following section we will present new analysis tools and a design methodology which systematically develops plant designs with improved robust dynamic operability characteristics.

Design of Chemical Plants with Improved Dynamic Operability in an Environment of Uncertainty

Before establishing the fundamental link between robust dynamic operability and process design, we will focus on a motivating example.

Consider the two heat integrated reactors studied by Georgakis and Worthey (1978) which is shown in Fig. 3. The parameter η denotes the degree of heat integration. The control system regulates the reactors output temperatures by manipulating the temperatures of the auxiliary heating and cooling coils. The system is subject to uncertainty since the reaction mechanism and rate constants are poorly known and the feed properties can change during operation. We will consider two alternative designs I and II which differ in the values of the design parameters of the exothermic reactor (e.g., T_{f_2} and V_2). The objective is to compare their dynamic performances in the presence of uncertainty at a fixed degree of heat integration, $\eta = 0.2$.

As set-point responses in Fig. 4 indicate, assuming perfect models, both designs can achieve <u>equally</u> satisfactory performances. It should be noted that these are the best achievable performances obtained after implementing IMC so that the results are not biased by the controller design. However, in a more realistic environment such as when reaction rate constants are in error, design II's performance turns out to be much better. Stability of design I can be restored by detuning the controller but at the expense of a substantial performance loss. Having identified the more robust design II, let's see how it performs under different operating conditions, in this case, for different degrees of heat integration. We find out through simulation (see Fig. 4) that dynamic operation becomes more sensitive to uncertainty (i.e., less robust) as we try to accommodate more heat integration - a trend opposite to what steady-state economics would dictate. At this point the designer can either give up the heat integration or try to locate the source of the robustness problem and improve it by appropriate design changes.

This motivating example illustrates that sensitivity of dynamic operation to undertainty depends on process design parameters (i.e., equipment sizes and operating conditions) and raises two challenging questions:
1. Can we <u>predict</u> and rule out the designs with robustness problems without resorting to expensive dynamic simulations?
2. Can we <u>systematically</u> improve robust dynamic operability without seriously jeopardizing the steady-state feasibility and economics?

Similar views from industry (Rinard and Slaby, 1984) and academia (Grossmann and Morari, 1983) have repeatedly expressed the need for rigorous techniques to design chemical plants with <u>robust</u> dynamic operability characteristics. The key to the understanding of the design and operation of chemical plants in an environment of uncertainty lies in effectively characterizing the uncertainty. Only then, a meaningful setting can be established within which robust stability and performance problems can be tackled.

<u>Characterization of uncertainty</u>. The most popular description for uncertainty is an <u>unstructured</u> model uncertainty (Doyle and Stein, 1981) which has been widely used in the design of robust controllers. The only piece of information available about uncertainty is a bound on the magnitude of the modeling error which is characterized through its transfer function representation. No information about the mechanism that causes the model-error is assumed to be known. A typical example is the high frequency process dynamics which cannot be modelled accurately.

However, for chemical processes the model-error can be usually well-structured. The sources of structured modeling errors can be classified as:

(i) <u>Uncertainty in the parameters</u>. Examples include poorly known modeling parameters (e.g., reaction rate constants), disturbances, and uncertainty introduced in the parameters of linear models by linearization of nonlinear models for several different operating conditions.

(ii) <u>Uncertainty in the model structure</u> which stems from simplifications made either on transfer function models (e.g., model order reduction by dropping high-order dynamics and/or lumping similar time constants) or on state-space models (e.g., neglecting some secondary physical phenomena).

We will use a dual framework to mathematically represent the structured errors. In the <u>frequency-domain framework</u>, the true but unknown process transfer function model $G_p(s,\gamma,\alpha^*)$ will be assumed to belong to the set of possible parameterized models:

$$M_\alpha = \{G_p(s,\gamma,\alpha);\ \alpha \in B\} \tag{5}$$

The set is generated by the unknown parameters vector α which lies in a "ball" of uncertainty B centered at the known nominal parameters α^0 and containing the true parameters α^*. γ is the vector of process design parameters. For chemical plants B is usually expressed as a set of inequalities which define lower and upper bounds on uncertain α. We will assume that this is the only information available on α as this is most often the case in industrial practice.

Similarly in the <u>time-domain framework</u>, the true process state-space model will be assumed to belong to the set of possible nonlinear dynamic models:

$$M_\beta = \left\{ \begin{array}{l} \dot{x} = f(x,m,\gamma,\beta) \\ \\ y = g(x,m,\gamma,\beta) \end{array} , \beta \in D \right\} \quad (6)$$

where β is the vector of uncertain parameters for which again only upper and lower bounds are known through the set D. x is the vector of states; m is the vector of manipulated variables and y is the vector of output variables.

Thus, depending on the type of modeling information available, the designer will start either from M_α or M_β and develop robustness measures to analyze the effect of process design parameters γ on the dynamic operability of the plant. Before presenting these measures, the set descriptions M_α and M_β need to be cast into a more convenient mathematical form.

We can express $M_\alpha = \{G_p(s,\gamma,\alpha); \alpha \in B\}$ in the error multiplicative form:

$$G_p(s,\gamma,\alpha) = [I + L(s,\alpha)]\tilde{G}(s,\gamma,\alpha^0) \quad \forall \alpha \in B \quad (7)$$

where $\tilde{G}(s,\gamma,\alpha^0)$ is the known nominal model which can be reduced-order. The true plant model $G_p(s,\gamma,\alpha^*)$ thus lies somewhere in the neighborhood of $\tilde{G}(s,\gamma,\alpha^0)$ which contains the set of perturbed models generated by the error matrix $L(s,\alpha)$ for different α's $\in B$.

From (7) we can now derive an <u>estimate of the structured model-error</u> matrix as:

$$L(s,\alpha) = G_p(s,\gamma,\alpha)\tilde{G}^{-1}(s,\gamma,\alpha^0) - I \quad (8)$$

In the case of new chemical plant designs for which no previous data is given in the form of transfer functions (5) or (7), the designer will have first principle models as expressed by (6). Then, linear sensitivity analysis around any operating point gives the uncertain plant transfer function model:

$$G_p(s,\gamma,\beta) = \nabla_x g(\gamma,\beta)[sI - \nabla_x f(\gamma,\beta)]^{-1} \nabla_m f(\gamma,\beta) + \nabla_m g(\gamma,\beta) \quad \forall \beta \in D \quad (9)$$

Similar to equation (8) the corresponding structured model-error matrix can be computed from

$$L(s,\beta) = G_p(s,\gamma,\beta)\widetilde{G}^{-1}(s,\gamma,\beta^o) - I \quad (10)$$

<u>Remark 1</u>: $G_p(s,\gamma,\beta)$ depends on the design and uncertain parameters which later allows the designer to study the sensitivity of robust dynamic operability to process design changes in a direct fashion. Also in the above formulation $G_p(s,\gamma,\beta)$ is made explicitly dependent on the sustained input disturbances through β and thus certain effects of nonlinearities are accounted for.

<u>Remark 2</u>: Since an estimate of the structured error matrix is available (see (8) and (10)) an upper bound on its magnitude can be computed and later used by the robust operability measures. This requires the solution of the following optimization problem:

$$\underset{\beta}{\text{Max}} \; \sigma^*[L(j\omega,\beta)] \quad \forall \omega \in \Gamma \subset R^+ \quad (11)$$

$$\text{subject to } \beta \in D: \quad \beta_\ell \leq \beta \leq \beta^u$$

σ^* is the maximum singular value of L and Γ is the frequency band of interest for the system under consideration. We have developed efficient algorithms for handling such maximization problems (Guruswamy, 1983).

It should be emphasized that, when available, information on $L(s,\beta)$ and its magnitude is very valuable and can help reduce the conservatism inherent to the robustness tests to a certain extent. Sometimes such information may become crucial in favoring one process design over another.

<u>Remark 3</u>: Efficient computational methods are needed to carry the parameters γ,β of the nonlinear state-space model (6) to the transfer function model (9). We have developed and successfully implemented on various physical examples a computer software based on symbolic logic languages such as MACSYMA (see MACSYMA Reference Manual 1977) to perform the necessary parametric operations involving analytic differentiation and linearization with excellent accuracy.

<u>Remark 4</u>: Finally this type of uncertainty characterization constitutes the starting basis for the development of the robustness measures which will be used to analyze the robust dynamic operability of chemical plants. The next section dwells on this subject.

<u>Measures for assessing robust dynamic operability</u>. All the results are derived using the IMC structure for the reasons discussed earlier. Palazoglu et al (1985) has shown that the closed-loop system of Fig. 2 will be stable for all L satisfying $\sigma^*[L] < \ell_m$ if the following inequality holds:

$$(1-\alpha)\frac{1}{\varepsilon_1[\tilde{G}]\varepsilon_2[\tilde{G}]} + \alpha \frac{1}{(\varepsilon_1[\tilde{G}])^2} > (\ell_m(\omega)\sigma^*[\tilde{G}_+]\sigma^*[F])^2 \quad \forall \omega \in R^+ \quad (12)$$

where ε_1, ε_2 are defined as the <u>robustness indices</u> and are given by:

$$\varepsilon_1(\omega,\gamma) = \frac{\sigma^*[\tilde{G}(j\omega,\gamma,\beta^0)]}{\sigma_*[\tilde{G}(j\omega,\gamma,\beta^0)]} \qquad (13)$$

$$\varepsilon_2(\omega,\gamma) = \frac{\sigma^{*,2}[\tilde{G}(j\omega,\gamma,\beta^0)]}{\sigma_*[\tilde{G}(j\omega,\gamma,\beta^0)]} \qquad (14)$$

Here, $\sigma^*[.]$ and $\sigma_*[.]$ are the maximum and minimum singular values; $\sigma^{*,2}[.]$ is the second largest singular value.

Condition (12) is obtained under the assumption that the projection of the error matrix onto the "most sensitive direction" of the IMC structure is "small". The most sensitive direction is defined by the direction along which it is easiest to make the closed-loop system unstable. The parameter α denotes how small the error projection is. When $\alpha = 1$ (that is, the "fuzziest" case when no error structure information is available to make the projection small), (12) reduces to the result obtained by Morari (1983) for completely unstructured errors:

$$\frac{1}{\varepsilon_1} > \ell_m \sigma^*[\tilde{G}_+]\sigma^*[F] \qquad (15)$$

However, this measure inherently assumes that the uncertainty can occur in all directions and thus can become conservative if the error is more structured. In that case (12) is the appropriate measure since it incorporates the direction of the error into the analysis, and consequently guarantees stability for a larger magnitude of uncertainty than that allowed by (15). Also note that ideally one would try to achieve zero error projection ($\alpha=0$) either by an appropriate design of the filter F or as a result of the nature of the error structure L. In that case robust stability constraint (12) reads:

$$\frac{1}{\varepsilon_1[\tilde{G}]\varepsilon_2[\tilde{G}]} > (\ell_m(\omega)\sigma^*[\tilde{G}_+]\sigma^*[F])^2 \qquad (16)$$

It can be easily shown that the projections of the error onto the <u>other</u> "less sensitive directions" can be similarly incorporated. The important point to emphasize here is that as more projections are made smaller, more and more robustness indices ($\varepsilon_1, \varepsilon_2, \ldots \varepsilon_{n-1}$ with $\varepsilon_i = \frac{\sigma^{*,i}}{\sigma_*}$) corresponding to these projection subspaces will appear in the denominator of the left hand side of the stability constraints (12) and (16).

Following remarks elaborate on the above development and pave the way to the use of the robustness indices during process design.

<u>Remark 1</u>: Since the indices ε_i's depend only on the open-loop plant design (see (13) and (14)) and not on the controller nor the filter, we will explore the utility of these indices to assess the degree of robust dynamic operability of different chemical plant designs.

According to (12) and (16), the process designs with smaller robustness indices are less sensitive and can tolerate higher magnitudes of uncertainty and/or larger feedback filter gains to improve process response. Thus, for good robustness with respect to stability and performance, <u>all</u> the indices ε_i's of a process design should be as small as possible (close to 1) over a wide range of frequencies.

Remark 2: Since information about the error structure L is available through (10), the error bound ℓ_m for a given design can now be estimated using (11), and we do not have to assume equal ℓ_m's for alternative designs. Thus if different designs are shown to have significantly differing ℓ_m's, we then consider not only ε_i's but also ℓ_m's before selecting the final design (e.g., according to (16) $\varepsilon_1 \varepsilon_2 \ell_m^2$ should be minimized over all designs).

Remark 3: Finally it should be noted that ε_i's and ℓ_m are scaling dependent and this issue has been extensively covered in Palazoglu et al (1985). We scale $\tilde{G}$ to minimize the product of the robustness indices and the error bound. When no error information is available, only the robustness indices are minimized. This scaling procedure is maximizing the allowable filter gain and is seeking the highest performance possible subject to the robust stability constraints. The resulting optimal scaling problem is a min-max semi-infinite optimization problem for which computational techniques exist (Guruswamy, 1983; Arkun et al., 1985).

Above robust dynamic operability measures can be utilized basically in three different design strategies:

i. One can evaluate the operability characteristics of several given designs by comparing their robustness indices (see Morari et al., 1985).

ii. Starting from a given design, one can evolve to designs with better operability by conducting a sensitivity analysis in terms of the pertinent design variables, and making design modifications as suggested (see Palazoglu et al., 1985).

iii. The design problem can be conveniently pouched into an multi-objective optimization formulation which would yield the optimal solution with respect to steady-state economics and robust dynamic operability constraints (Palazoglu and Arkun, 1985a,b).

In the following section, the highlights of the sensitivity approach will be presented.

Evolutionary design using the sensitivity of robustness indices. A rigorous and systematic procedure to design robust chemical plants is lacking in the literature. When sensitivity data is not available, analyzing and improving isolated design decisions can not be efficient, simply because of the arbitrariness in such decisions and due to the difficulty of predicting the effects of design changes on robustness based on physical intuition alone.

The goal of our sensitivity analysis is to formulate in a direct fashion how the robustness properties of a chemical plant are affected by its design and operating conditions. The mathematical tools to be derived from this study can then be used in an evolutionary framework to systematically arrive at designs and operating regimes for which robust dynamic operability is improved.

The sensitivity of robustness indices with respect to any design parameter can be computed by the following formula (Palazoglu et al., 1985):

$$\nabla_{\gamma_i} \varepsilon_k[\tilde{G}] = \frac{1}{\sigma_*[\tilde{G}]} (\nabla_{\gamma_i} \sigma^{*,k}[\tilde{G}] - \varepsilon_k[\tilde{G}]\nabla_{\gamma_i} \sigma_*[\tilde{G}]) \qquad (17)$$

where

$$\varepsilon_k[\tilde{G}(j\omega,\gamma)] = \frac{\sigma^{*,k}[\tilde{G}(j\omega,\gamma)]}{\sigma_*[\tilde{G}(j\omega,\gamma)]} \qquad (18)$$

$$\nabla_{\gamma_i} \sigma_j[\tilde{G}(\gamma)] = \text{Re}[v_j^H(\gamma)\nabla_{\gamma_i}\tilde{G}(\gamma)u_j(\gamma)] \qquad (19)$$

$\sigma^{*,k}$ is the kth largest singular value, v_j and u_j are the left and right singular vectors of G corresponding to σ_j and $Re[A] = 1/2[A+A^H]$. Note that the sensitivities are to be calculated at each frequency ω. $\nabla_\gamma \tilde{G}(j\omega,\gamma)$ is calculated from $\tilde{G}(j\omega,\gamma)$ using the symbolic logic language MACSMYA. An interactive graphics software displays the sensitivities of different singular values and robustness indices as a function of frequency.

Typically, the designer will start with a given feasible plant design and calculate the sensitivities of the robustness indices. Based on these results, the <u>key</u> design parameters will be identified and appropriate changes be made in a direction in which robustness improves. Other design objectives can be also made part of this sensitivity-based evolutionary design procedure if they can be expressed in terms of the singular values (e.g. disturbance handling capacity and limitations on manipulated variables can be formulated using the minimum singular value $\sigma_*(\tilde{G})$).

Finally, we should again emphasize the fact that it is not an obvious and trivial task for the designer to use his physical intuition in determining how different design parameters affect the robustness of a chemical plant operation. In this respect the sensitivity study is extremely helpful to the designer.

<u>Example.</u> We will consider the reactor system of Fig. 3 discussed before. Let us first see if the robustness analysis will predict the dynamic operability problems which were earlier detected by simulation. The scaled robustness index ε_1 for design I is quite high (see Fig. 5) suggesting potential dynamic operability problems in the presence of uncertainty. Furthermore in Fig. 6 ε_1 is given as a function of the degree of heat integration. It is seen that ε_1 increases with η, thus predicting that heat integration has an adverse effect on the robust operation. When the sensitivity of the robustness index to a change in η for design I is evaluated at $\eta = 0.95$, the gradient directs the designer to lower η values for

all frequencies (see Fig. 7).

At this point we can check the sensitivity of ε, with respect to different parameters to identify the key ones. We consider the effects of the following design variables: $\gamma^T = [V_1 \; V_2 \; T_{f2}]$. The sensitivities are plotted in Fig. 8 for design I with $\eta = 0.2$. It is seen that T_{f2} has a pronounced effect on the robustness index, indicated by the large magnitude of its sensitivity. While V_2 has less impact on ε_1 as compared with T_{f2}, V_1 has no effect on ε_1 as its sensitivity is practically zero within the frequency range of interest.

Based on the predictions of this sensitivity analysis, two modifications of the original design I are proposed. In design II T_{f2} is increased by only 5% and V_2 is unchanged. In design III V_2 is increased by 18%. The new ε_1's are shown in Fig. 9. Increasing V_2 has improved the robustness but at the expense of higher investment and operating costs. Design II is the most robust system since the change in T_{f2} has suppressed ε_1 the most but again with an increase in the operating costs. Dynamic simulations, omitted here, confirm the prediction of the sensitivity study (see Palazoglu et al, 1985).

Before concluding with this example we should mention that doing detailed dynamic simulation to screen and improve many designs including many uncertainties and design parameters is almost impossible and not warranted. That's why robustness measures and sensitivity studies developed above are useful tools for the design of robust chemical plants.

Finally, while the significance of robust dynamic operability as a design objective is irrefutable, it has to be more rigorously complemented with economic and steady-state flexibility aspects to arrive at an efficient final design. This issue will not be addressed here but for details the reader is referred to our recent work (Palazoglu and Arkun, 1985a,b).

In the opinion of this author the issue of uncertainty will continue to be the focus of research on dynamic operability. As we better understand the problems of uncertainty, newer research areas will develop and more useful and efficient analysis and design techniques will emerge. Some of the important current research topics can be summarized as follows:

a. The key to the robustness analysis discussed above is that it introduces new ways for characterizing the structure of the model/plant mismatch, and estimating the associated model-error matrix and its magnitude bound. Using these results design conservatism is reduced to a certain extent and the effect of uncertainty on dynamic operability is analyzed more effectively. Recently Structured Singular Value (SSV) analysis has been proposed to study both stability and performance of control systems in the presence of uncertainty (Doyle, 1982). The method offers opportunities to treat several different uncertainty descriptions in a nonconservative way (Skogestad and Morari, 1985). It will be interesting to see how the uncertainty characterization discussed in this paper can be coupled with the SSV analysis and how the above results can be made less conservative.

b. Deterioration of the nominal achievable performance of a process design due to robust stability constraints such as (12) should be described more quantitatively.

c. Existing results should be extended to large-scale systems. New theories need to be established to formulate how the "robustness measures" of a given plant depend on the properties of individual subsystems of process units and their interconnections. Decomposition techniques and efficient numerical methods to quickly evaluate the robustness indices and their sensitivities should be developed. These computational techniques should be incorporated into equation-

oriented large-scale simulators so that they can be
effectively used to assess robust dynamic operability during
computer aided flowsheet analysis and synthesis.
d. Sensitivity study is primarily formulated in a parametric
framework. It is not clear how structural flowsheet
modifications can be made part of this study in an efficient
way.
e. Finally, assessment of robust dynamic operability of a given
process design depends on the type of process models used.
This interplay between modeling and robust operability brings
out an interesting and challenging problem: What should be
the level of sophistication of our process control models?

DYNAMIC OPERABILITY PROBLEMS RELATED TO CONTROL SYSTEM DESIGN

During the synthesis of process control systems, the designer
has to address three fundamental questions: (1) selection of
measurements and manipulated variables, (2) deciding on the
structure of interconnections between the measurements and the
manipulated variables, (3) choosing the control law and the tuning
parameters. Naturally, all these decisions will ultimately
influence the dynamic operation of the plant. In the subsequent
sections, we will briefly discuss the dynamic operability issues
pertaining to each of the above control system design activities.

Selection of the Control Structure

Within the last decade several attempts have been made to
develop heuristics, mathematical formulations, and algorithmic
procedures to synthesize control structures (Douglas, 1977, 1981;
Morari and Stephanopoulos, 1980; Govind and Powers, 1982). For a
detailed review, the reader is referred to Stephanopoulos (1983).
Douglas' approach is based on a steady-state operability analysis
which is carried out during preliminary process design stage. The

effects of disturbances on plant economics and design objectives are studied to identify the appropriate process design modifications as well as the preliminary selections of measurements and manipulated variables. Alternative control structures which are acceptable from a steady-state point of view can then be screened based on their dynamic operability characteristics. Govind and Powers (1982) uses a cause-and-effect graph to study the interrelationships among different process variables and thus they are able to identify alternative sets of measured and manipulated variables. The method of Morari and Stephanopoulos (1980) uses the concepts of controllability and observability to sytematically create all the alternative feasible control structures.

All the aforementioned techniques deal primarily with the creation of alternative control structures. Although several analysis tools based on heuristics, steady-state gains, time-delays, time-constants, dynamic interactions and simulations have been frequently mentioned, a concise screening strategy based on dynamic operability aspects has been missing for quite some time. With the introduction of the IMC structure, a powerful insight into the quality of control offered by alternative control structures is now possible. The <u>fundamental limitations</u> of alternative control structures can be studied by using the same theoretical framework which we have presented to tackle process design related dynamic operability problems. Alternative sets of measurements and manipulated variables will simply correspond to different $\tilde{G}(s)$'s for which performance limitations can be systematically identified. In the first level of screening those control structures whose achievable nominal performances (see (5)) are not acceptable due to their nonminimum phase elements are rejected. Control structures which are prone to excessive control actions (i.e., those with low $\sigma_*(\tilde{G})$ values) are screened as well. In the second level of screening the promising structures are compared based on how they cope with uncertainty. Again, each control structure has its own description in terms of $\tilde{G}$, L, ℓ_m and ε_i's. Therefore the robust dynamic operability characteristics of alternative control structures can be evaluated, compared and the best ones are selected using the robustness criteria presented earlier.

Finally it should be noted that nonsquare control structures (i.e., unequal number of measurements and manipulated variables) can be also included in this synthesis process. Morari et al. (1985), Arkun and Manousiouthakis (1985) give physical examples which illustrate how one can make use of additional manipulated variables to alleviate some of the dynamic operability problems.

Selection of the Controller Structure

Once a control structure is selected, the designer must decide on the interconnections between the measured and manipulated variables. One choice would be to interconnect all the measurements to all the manipulated variables to form a fully centralized controller structure which would give the best achievable performance if it were implemented in practice. However, the large scale nature of chemical plants has always forced the designers to search for simpler localized controller structures. The designer usually partitions the plant into subsystems (from an information flow point of view) and controls each subsystem individually in a decentralized fashion. Decentralized control approach, if formulated correctly, not only facilitates the design of the control system but also offers significant implementational advantages such as reduced communication cost, improved safety and fault tolerance, increased modularity and flexibility to update and expand the control system and finally easier monitoring and on-line tuning. If it were not for these considerations, one would have always adopted the centralized controller structure and there would be no selection task left to the designer.

Alternative controller structures arise due to alternative degrees of decentralization. In the one extreme we have the most classical single-input-single-output (SISO) controller structure which constitutes the highest degree of decentralization. In the other extreme we have completely centralized structure. Between these extremes lie all the other alternatives (see Fig. 10) whose performances need to be evaluated. Here it should be emphasized that decentralization of the controller structure imposes a new constraint on dynamic process operability.

The synthesis goal is to select those decentralized controller structures whose dynamic performances do not deviate significantly from that of the centralized controller which would in theory give the best achievable performance. A new methodology is described in the following sections to achieve this goal.

Block Relative Gain (BRG) concept. In order to select the best SISO decentralized controller structures, Bristol (1966) introduced the concept of relative gain. Because of its many useful properties the relative gain has found wide applications among practicing engineers. However, at the same time, its original development as a scalar limited its application to the synthesis of S.I.S.O. controller structures only. Reformulating and extending this concept from a scalar to a matrix (Block Relative Gain) creates a new synthesis framework which can address all forms of decentralized controller structures. In fact BRG can be used as a dynamic operability measure to effectively screen for the best controller structures. We will now present the block relative gain.

Consider a square (nxn) transfer function matrix G(s), (in the sequel we will drop s for convenience) partitioned as follows:

$$G = \begin{array}{c} \uparrow \\ m \\ \downarrow \\ \uparrow \\ n-m \\ \downarrow \end{array} \begin{bmatrix} \overset{\leftarrow \ m \ \rightarrow}{G_{11}} & \overset{\leftarrow \ n-m \ \rightarrow}{|} & G_{12} \\ - & + & - \\ G_{21} & | & G_{22} \end{bmatrix} \quad \text{with} \quad \begin{bmatrix} y_1 \\ y_2 \end{bmatrix} = G \begin{bmatrix} u_1 \\ u_2 \end{bmatrix} \qquad (20)$$

The plant is to be controlled by a decentralized control structure in which the first m outputs y_1 are interconnected with the first m inputs u_1 and the last n-m outputs y_2 are interconnected with the last n-m imputs u_2. The corresponding feedback configuration is shown in Fig. 11.

We can now define the <u>m-dimensional Block Relative Gain</u> (left and right) as:

$$BRG_\ell = \begin{bmatrix} \dfrac{\partial y_1}{\partial u_1} \Big|_{\substack{u_2=0 \\ F=0}} \end{bmatrix} \cdot \begin{bmatrix} \dfrac{\partial y_1}{\partial u_1} \Big|_{\substack{y_2=0 \\ F_1^2=0 \\ F_2^1=I}} \end{bmatrix}^{-1} = G_{11} \cdot [G^{-1}]_{11} \qquad (21)$$

$$BRG_r = \begin{bmatrix} \dfrac{\partial y_1}{\partial u_1} \Big|_{\substack{u_2=0 \\ F_1^2=0 \\ F_2^1=I}} \end{bmatrix}^{-1} \cdot \begin{bmatrix} \dfrac{\partial y_1}{\partial u_1} \Big|_{\substack{u_2=0 \\ F=0}} \end{bmatrix} = [G^{-1}]_{11} \cdot G_{11} \qquad (22)$$

where $[G^{-1}]_{11}$ is the first mxm block of G^{-1}:

$$G^{-1} = \begin{bmatrix} [G^{-1}]_{11} & [G^{-1}]_{12} \\ [G^{-1}]_{21} & [G^{-1}]_{22} \end{bmatrix} \qquad (23)$$

Note that in the case of 1-dimensional BRG which corresponds to SISO controller structure, left and right BRG's become identical (since G_{11} is scalar) and reduce to the classical Bristol's relative gain. In the case of n-dimensional BRG which corresponds to completely centralized controller, $BRG_\ell = BRG_r = I$ (identity matrix).

It can be proved that BRG is a measure of closed-loop dynamic interactions and as such it can be used to predict the dynamic performance of decentralized controller structures (Manousiouthakis et al., 1985a). In particular, it can be shown that the closed-loop performance of the subsystem corresponding to G_{11}, when the rest of the plant outputs are under perfect control, is a continuous function of BRG_ℓ:

$$\left. \frac{\partial y_1}{\partial y_1^{sp}} \right|_{\substack{y_2=0 \\ \text{All loops closed}}} = [I + (BRG_\ell^{-1} \cdot G_{11}K_1)^{-1}]^{-1} \qquad (24)$$

According to (24) if BRG_ℓ is equal to identity, the corresponding subsystem will not experience any interactions from other parts of the plant. <u>Thus during synthesis one should try to identify and select decentralized controller structures whose BRG's are close to identity</u>.

While defining BRG and deriving its relationship to the closed-loop dynamic performance, we have accepted the usual assumption of perfect control for the plant outputs y_2. Manousiouthakis (1985) has relaxed this assumption by defining a refined BRG:

$$\widetilde{BRG}_\ell = [I - G_{12}G_{22}^{-1}H_{22}G_{21}G_{11}^{-1}]^{-1} \qquad (25)$$

In (25) H_{22} is the nominal achievable performance for the outputs y_2 and contains all the fundamental limitations to perfect control (compare with (5)). Note that when $H_{22}(s) = I$ (i.e., perfect control), eqn. (25) reduces to (21) and $\widetilde{BRG}_\ell = BRG_\ell$ for all the frequencies.

At steady-state the requirement of perfect control is realistic and can be achieved by integral control and BRG_ℓ given by (21) can be safely used. If one wishes to investigate interactions not only at steady-state but over a range of frequencies, BRG_ℓ is a reliable frequency dependent interaction measure to compare alternative controller structures and need not be refined unless H_{22} is nonminimum phase. If H_{22} is nonminimum phase, one either uses BRG_ℓ at steady-state only, or, if interested in all the frequencies, uses $\widetilde{BRG}_\ell$ instead.

Properties of BRG. Manousiouthakis et al. (1985a) has developed a systematic screening procedure which utilizes various important properties of BRG. These properties effectively reduce the combinatorial problems and make the analysis of large scale systems feasible. Furthermore their appealing features such as scale independence, intimate relationship with the classical RGA, among many others, make BRG a very promising tool. At present these properties do not hold for the refined $\widetilde{BRG}$. In the following, we will present two of the properties which will be heavily utilized during analysis and synthesis.

Theorem:

Let BRG_ℓ and BRG_r be the m-dimensional block relative gains corresponding to the first m outputs and inputs of a given system G. Their <u>diagonal elements</u>

    a) are <u>independent of scaling</u> and

    b) can be expressed in terms of the 1-dimensional BRG's that correspond to these m outputs and m inputs (i.e., in terms of the corresponding RGA elements). Specifically

$$(BRG)_\ell(i,i) = \sum_{j=1}^{m} RGA(i,j) \qquad (26)$$

$$(BRG)_r(i,i) = \sum_{j=1}^{m} RGA(j,i) \qquad (27)$$

Proof. See Manousiouthakis et al. (1985a).

According to this theorem, by simply inspecting the relative gains, the designer can group inputs and outputs into attractive decentralized structures. This short-cut technique becomes especially helpful when decentralized controllers for large scale systems are to be selected. Next theorem establishes some important properties of the eigenvalues of BRG.

Theorem:

Let BRG_ℓ and BRG_r be the m-dimensional block relative gains corresponding to the first m outputs and inputs of a given system G. Then their eigenvalues $\lambda_i(BRG)_\ell$, $\lambda_i(BRG)_r$ i=1,...m
  a) are <u>independent of scaling</u>
  b) are the same for both BRG_ℓ and BRG_r, that is
  $\lambda_i(BRG)_\ell = \lambda_i(BRG)_r$ i=1,...,m

Proof. See Manousiouthakis et al. (1985a).

When both the eigenvalues and diagonal elements of either the "left" or "right" BRG are "close" to one, the designer can be fairly confident that both BRG's are approximately identity. Furthermore, by having to check only the diagonal elements and eigenvalues of the BRG's the designer utilizes only 2m terms instead of $2(m)^2$ terms for every m-dimensional BRG being considered. Thus, the eigenvalues and diagonal elements of the BRG's are very effective in speeding up the subsystem selection process for decentralized control systems.

Selection Procedure. Alternative decentralized controller structures result from partitioning G into blocks of different dimensions and assigning alternative sets of inputs and outputs into each of these blocks (see Fig. 12). A decentralized structure is said to be acceptable if all the BRG's corresponding to the partitioned blocks of different dimensions are "close" to the identity matrix. A BRG is considered to be "close" to the identity matrix if its diagonal elements and eigenvalues belong to the neighborhoods $B(1,\delta_1)$ and $B(1,\delta_2)$, centered at (1,0) in the complex plane, for all the frequencies. By adjusting the values of radii δ_1 and δ_2, the designer can moderate the number of alternatives he wishes to consider as well as the level of dynamic performance to be attained. Screening starts with the SISO structures and steady-state BRG's are inspected first. For the promising structures, analysis for other frequencies follows. If acceptable structures are discovered, the designer has the option to either stop or proceed to the higher dimensional BRG's in the hope of finding a more attractive decentralized control structure. At each level of screening, properties of diagonal elements and eigenvalues are utilized to facilitate the selection procedure. The method is guaranteed to conclude because eventually the n-dimensional BRG (i.e., centralized controller) will be reached if none of the decentralized structures is acceptable.

Process applications. We have implemented the method on several physical examples including a boiler furnace, heat integrated reactors and a heat exchanger network (Manousiouthakis et al., 1985b,c). Dynamic simulations have verified that BRG is a dependable tool to select high performance decentralized controller structures.

Selection of the Control Law and the Tuning Parameters

In this last stage of the synthesis of process control systems, decisions have to be made on the final design of the controller. The IMC structure provides the best framework to design and implement the control system. For a selected set of measured and manipulated variables, the design is performed in two stages: feedforward and feedback. Dynamic operability is an integral part of the feedforward design since the controller G_I is designed to realize the achievable nominal dynamic performance (see (4) and (5)). It is also part of the feedback design since the filter F is designed to achieve robust stability and dynamic performance in the presence of uncertainty. Thus all the important issues of stability, performance and robustness and the associated tradeoffs can be treated in a transparent and systematic way.

For the IMC design, we have recently developed a hybrid approach which combines algebraic design with frequency domain techniques. The algebraic approach is followed to construct the feedforward controller in an elegant and numerically efficient way. The frequency domain methods are utilized in the filter design so that robustness issues for the closed-loop system can be addressed. Details of the design method and its extension to decentralized control systems can be found in Arkun and Manousiouthakis (1985), and Manousiouthakis and Arkun (1985c).

Although dynamic operability can be rigorously assessed for new control system designs which will use the IMC structure, most of the existing systems in industry today are classical feedback loops. For these systems general and easy-to-use analysis tools are needed. Recently a new analysis procedure has been introduced by Arkun et al (1984) and Palazoglu and Arkun (1985c). By examining the singular values of certain frequency dependent operators, one can assess dynamic operability of a feedback control system in terms of its nominal performance, robustness with respect to stabilty, and robustness with respect to performance. Furthermore, if dynamic operability problems are detected, the key controller tuning parameters can be identified by a singular value sensitivity analysis (analogous to identifying key process design parameters) and they are retuned to alleviate the operability problems if possible. If simple tuning does not suffice, one can try to make changes in the control system as suggested by the singular value analysis (e.g., different decoupling schemes, manipulated variables, etc.). The analysis tools developed are very general and easy to implement since they use reliable and easily accessible computational tools like Singular Value Decomposition. For the description of the computer software which uses interactive graphics the reader is referred to Arkun et al (1985). As our powerful results on distillation columns indicate, the technique can be effectively used in making a comparative assessment of different controller designs in terms of how well they can operate in the presence of modeling errors which exist in a real industrial environment.

The main purpose of the above discussion is to convince the reader that the state of the art in process control system synthesis has significantly advanced over the last decade due to better formulation of dynamic operability problems and new analytical insights. Naturally this progress has created several new challenges:

a. In the selection of control structures, manipulated variables can be combined with each other in many different ways to create new sets of inputs (e.g., ratios of manipulated variables). Trevathan (1976) and Shinskey (1983) explored this additional degree of freedom in the synthesis of SISO control structures. At present there is no systematic procedure to create the best combinations of manipulated variables.

b. BRG has been developed for square (i.e., equal number of inputs and outputs) controller structures. In view of the fact that nonsquare controllers can offer substantial improvements in control quality, an extension of the BRG concept to nonsquare systems should be pursued.
c. Design of decentralized control laws for complete chemical plants within the framework of IMC is a challenging problem which rightly calls for further attention and creative research. Important issues include robust stability and performance of interconnected systems, dynamic operability tradeoffs among different subsystems and incorporation of feedforward control to account for some of the subsystem interactions.
d. Despite significant progress, the best way to design the IMC controller G_I and the filter F in the presence of model/plant mismatch is still an unsolved problem. Conservatism, uncertainty description and the simplicity of the design method are the key issues.
e. We have so far addressed nonadaptive and linear control systems only. Recently several adaptation schemes and nonlinear controllers are being incorporated into the IMC structure. It will be very valuable to develop methods to predict the achievable performances of such control systems. These techniques would then help the designer to decide whether there is any incentive to use adaptive or nonlinear control instead of fixed linear robust control for the class of modeling errors considered.

CONCLUSION

Although the subject of dynamic process operability has relentlessly occupied the minds of chemical engineers for quite some time, significant progress has been made only recently. Such advances came about after a better understanding and formulation of the problem, and the development of a creative theoretical foundation. Nevertheless, research in this area is still at its infancy. This paper has attempted to give a panorama of dynamic process operability by emphasizing some of the pertinent problems, describing recent theoretical developments and outlining new challenges for future research.

ACKNOWLEDGEMENTS

Acknowledgement is made to the National Science Foundation (CPE83-12341, CBT85-13213) and to A. Palazoglu for his comments and criticism.

REFERENCES

Arkun, Y., and V. Manousiouthakis (1985). A hybrid approach to design robust control systems. Int. J. of Contr., submitted for publicaton.

Arkun, Y., B. Manousiouthakis, and A. Palazoglu (1984). Robustness analysis of process control systems. A case study of decoupling control in distillation. Ind. Eng. Chem. Proc. Des. Dev., 23, pp. 93-101.

Arkun, Y., B. Manousiouthakis, A. Palazoglu, V. Guruswamy, and P. Putz (1985). Computer-aided analysis and design of robust multivariable control systems for chemical processes. Comp. and Chem. Eng., 9, No. 1, pp. 27-59.

Bristol, E. H. (1966). On a new measure of interaction for multivariable process control. IEEE Trans. Aut. Contr., AC-11, No. 1, 133.

Douglas, J. M. (1977). Preliminary process design. Quick estimates of control economics. 69-th Annual AIChE Mtg., New York.

Douglas, J. M. (1981). Process operability and control of preliminary designs. CPCII. Proc. of the 1981 Eng. Found. Conf., Sea Island, Georgia.

Doyle, J. C., and G. Stein (1981). Multivariable feedback design: concepts for a classical modern synthesis. IEEE Trans. Aut. Contr., AC-26, pp. 4-16.

Doyle, J. C. (1982). Analysis of feedback systems with structured uncertainties. IEE Proc., 129, D(6), 242.

Georgakis, C., and D. J. Worthey (1978). On dynamical methods of heat integration design. AIChE J., 24, (6), pp. 976-984.

Govind, R., and G. J. Powers (1982). Control system synthesis strategies. AIChE J., 28, (1), 60.

Grossmann, I. E., and M. Morari (1983). Operability, resiliency and flexibility. Process design objectives for a changing world. Proc. of the Second Int. Conf. on Found. of Comp. Aid. Proc. Des. (Snowmass, Colorado), pp. 931-1010.

Guruswamy, V. (1983). Use of singular values in process design and control: new computational techniques and applications in chemical engineering. M.S. Thesis, RPI, Troy, New York.

Holt, B. R., and M. Morari (1985a). Design of resilient processing plants V. The effect of deadtime on dynamic resilience. Chem. Eng. Sci., in press.

Holt, B. R., and M. Morari (1985b). Design of resilient processing plants VI. The effect of right-half-plane zeros on dynamic resilience. Chem. Eng. Sci., 40, 59-74.

Klema, V. C., and A. J. Laub (1980). The singular value decomposition: its computation and some applications. IEEE Trans. Autom. Contr., AC-25, 164.

MACSYMA Reference Manual (1977), The Math Lab group, Laboratory for Computer Science, MIT, Cambridge, Mass., Version 9.

Manousiouthakis, V. (1985). Design of control systems for uncertain chemical plants. Ph.D. Dissertation, RPI, Troy, New York.

Manousiouthakis, V., R. Savage, and Y. Arkun (1985a). Synthesis of decentralized process control structures using the concept of Block Relative Gain. AIChE J., in press.

Manousiouthakis, V., and Y. Arkun (1985b). Synthesis of control structures for large systems using Block Relative Gain. Annual AIChE Mtg, Chicago, Ill.

Manousiouthakis, V., and Y. Arkun (1985c). Design of robust decentralized process control systems. Ame. Contr. Conf. Proc., Boston, MA, pp 1304-1309.

Morari, M., and G. Stephanopoulos (1980). Studies in the synthesis of control structures for chemical processes. Part II. AIChE J., 26, (2), 232.

Morari, M. (1983). Design of resilient processing plants - III. A general framework for the assessment of dynamic resilience. Chem. Eng. Sci., 38, pp 1881-1891.

Morari, M., M. J. Oglesby and I. D. Prosser (1985). Design of resilient processing plants-VII. Design of Energy management system for unstable reactors - new insights. Chem. Eng. Sci, 40, 187-198.

Palazoglu, A., B. Manousiouthakis, and Y. Arkun (1985). Design of chemical plants with improved dynamic operability in an environment of uncertainty, <u>Ind. Eng. Chem. Proc. Des. Dev.</u>, <u>24</u>, pp 802-813.

Palazoglu, A., and Y. Arkun (1985a). Design of chemical plants in the presence of process uncertainty: a multiobjective approach. <u>Comp. and Chem. Eng.</u>, submitted for publication.

Palazoglu, A., and Y. Arkun (1985b). Design of chemical plants with multiregime capabilities and robust dynamic operability characteristics. <u>Comp. and Chem. Eng.</u>, submitted for publication.

Palazoglu, A., and Y. Arkun (1985c). Robust tuning of process control systems using singular values and their sensitivities. <u>Chem. Eng. Comm.</u>, <u>37</u>, pp 315-331.

Ravindranath, T. K. (1983). Design of internal model controllers for chemical processes using algebraic theory. M.S. Thesis, RPI, Troy, New York.

Rinard, I. H., and J. Slaby (1984). A prototype multipurpose simulator for process operability analysis. AIChE Annual Mtg, Anaheim, CA.

Shinskey, F. G. (1983). From process knowledge to computer control: a methodology. <u>Proc. of PACHEC '83</u>, III, pp 85-89.

Skogestad, S., and M. Morari (1985). Design of resilient process systems. Effect of model uncertainty on dynamic resilience. <u>Chem. Eng. Sci.</u>, submitted for publication.

Stephanopoulos, G. (1983). Synthesis of control systems for chemical plants - A challenge for creativity. <u>Comp. and Chem. Eng.</u>, <u>7</u>, pp 331-365.

Trevathan, V. L. (1976). Process controls in chemical industry. In Foss, A. S., M. M. Denn (Ed.), <u>AIChE Sym. Series</u>, <u>159</u>.

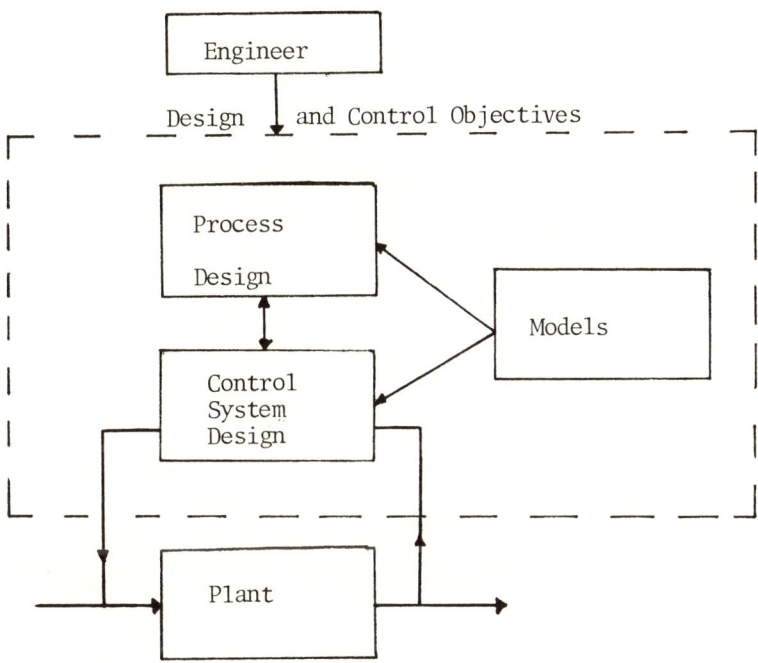

Fig. 1. Conceptual representation of the design and operation of a chemical plant.

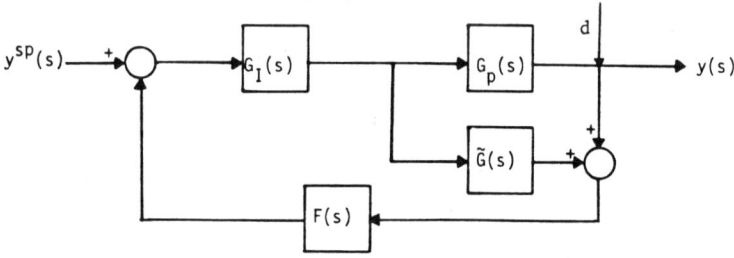

Fig. 2. Internal Model Control (IMC) structure.

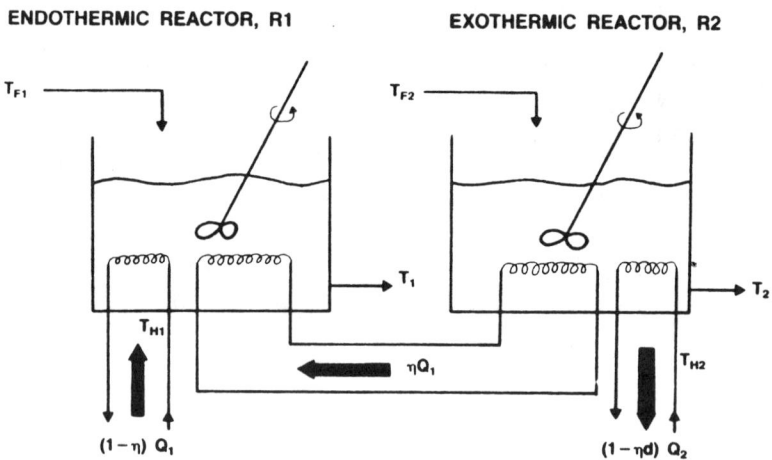

Fig. 3. **Heat Integration scheme between two CSTRs. Here** $Q_1 < Q_2$ **and** $0 < \eta < 1$.

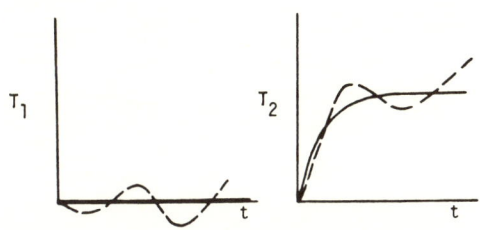

(i) Design I. ———: Model, - - -: Plant

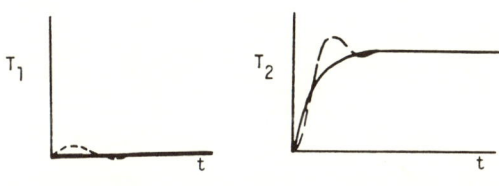

(ii) Design II. ———: Model, - - -: Plant

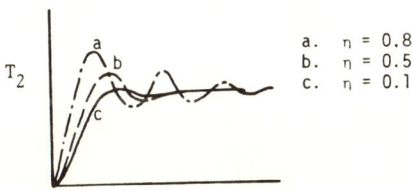

a. η = 0.8
b. η = 0.5
c. η = 0.1

(iii) Design II

Fig. 4. Set-point responses for different designs and for various degrees of heat integration

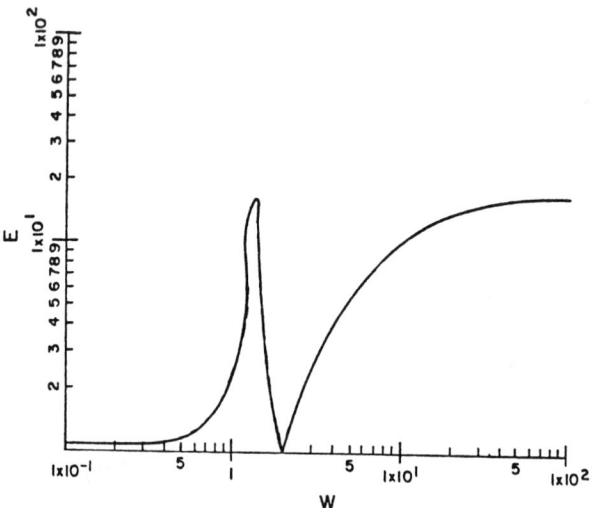

Fig. 5. ε_1 for design I after scaling.

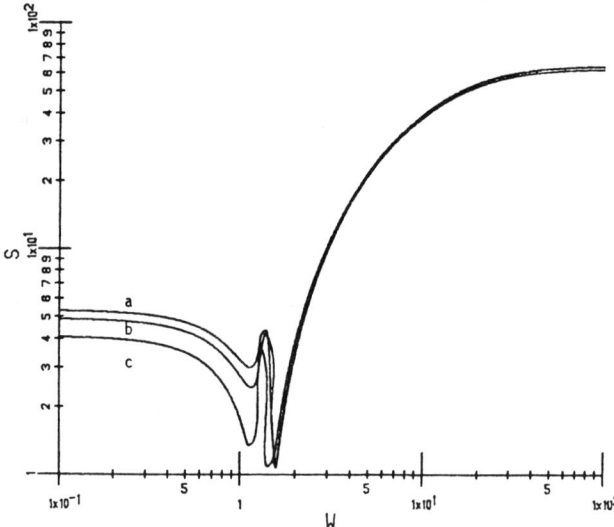

Fig. 6. ε_1 for design I when (a) $\eta = 1$, (b) $\eta = 0.8$, (c) $\eta = 0$

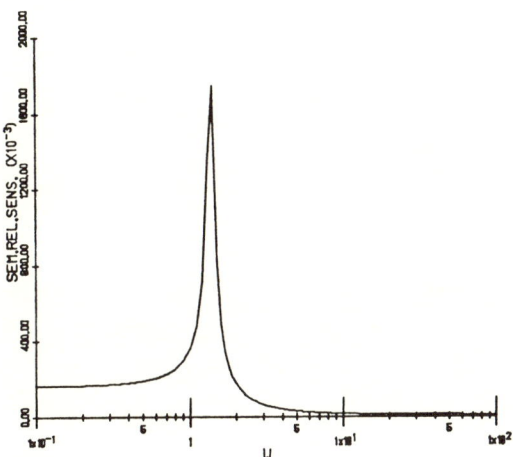

Fig. 7. Plot of $\partial\epsilon_1/\partial\eta$ for Design I at $\eta = 0.95$

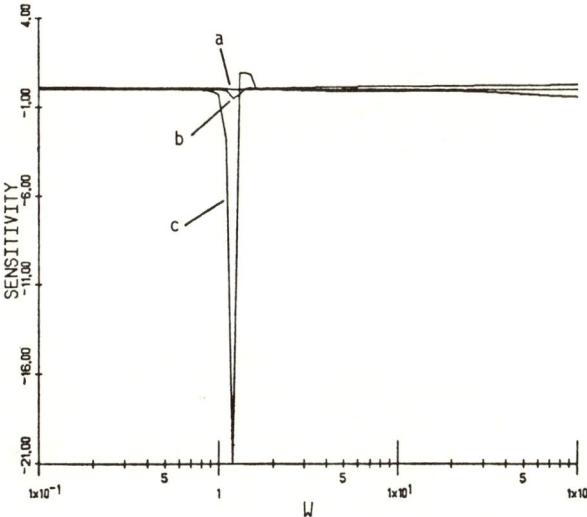

Fig. 8. Semi-relative sensitivities a) $\partial\epsilon_1/\partial\tilde{v}_1$ b) $\partial\epsilon_1/\partial\tilde{v}_2$ c) $\partial\epsilon_1/\partial\tilde{T}_{f2}$ when $\eta = 0.2$ for design I. $\tilde{\gamma} = \gamma/\gamma_0$ where γ_0 is the nominal design value.

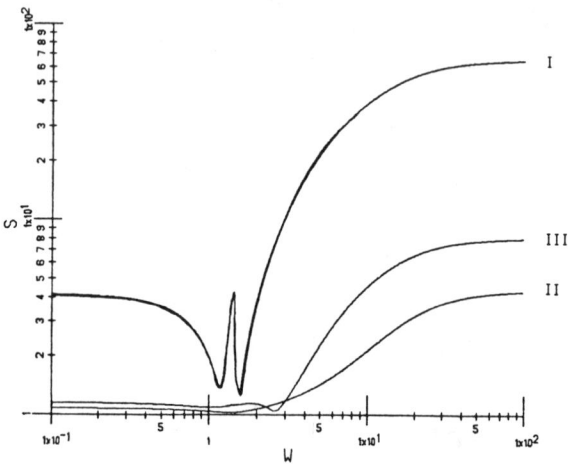

Fig. 9. ε_1 for three different designs when $\eta = 0.2$

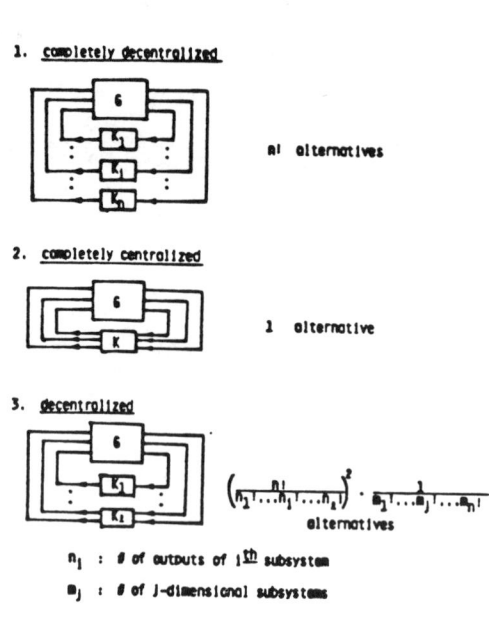

1. **completely decentralized**

   n! alternatives

2. **completely centralized**

   1 alternative

3. **decentralized**

   $\left(\frac{n!}{n_1!\ldots n_i!\ldots n_q!}\right)^2 \cdot \frac{1}{m_1!\ldots m_j!\ldots m_n!}$ alternatives

   n_i : # of outputs of ith subsystem
   m_j : # of j-dimensional subsystems

Fig. 10. Interconnection Modes for an n×n System

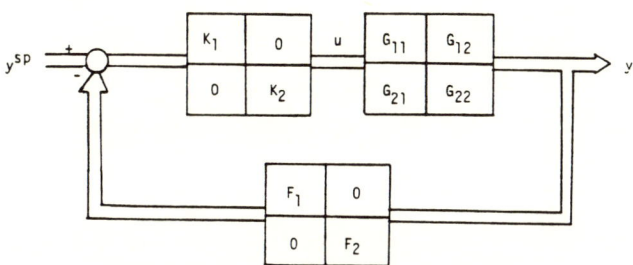

Fig. 11. Decentralized Feedback System

Fig. 12. Partitioning of G into blocks of different dimensions

A SHORT-CUT METHOD FOR PROCESS CONTROL AND OPERABILITY ANALYSIS

Thomas E. Marlin
Stone & Webster Engineering Corporation, Boston, MA 02107

Thomas J. McAvoy
Maria Marino-Galarraga
Naveen Kapoor
University of Maryland, College Park, MD 20742

Abstract. A method for process control and operability analysis which requires limited engineering effort is presented in this paper. The method is used to design control schemes for multivariable systems based on predicted control performance; design decisions involve selecting control schemes and loop pairing and determining when decoupling is appropriate. The method involves three approaches of increasing complexity. The first approach uses only steady-state gains to estimate the effects of interaction on control performance. The second uses further design data to develop approximate dynamic models and better estimates of control performance. The third uses a simplified, block-oriented simulation to give more information on the transient reponse. The overall method is useful for industrial problem solving; several examples of distillation control are presented. In addition, the method yields new, general insights by predicting favorable and unfavorable interaction in multivariable control schemes and by elucidating the effects of process configuration on dynamic response.

Keywords. Controller Tuning; Decoupling; Distillation Control: Dynamic Modelling; Dynamic Simulation; Interaction; Multivariable Control; Operability.

INTRODUCTION

A plant that is difficult to control will often be operated far from its peak efficiency. Excessive energy will be consumed, product yields will not be optimum, and off-specification material may often be produced. A key objective is to develop process and control designs which are operable, i.e., which can be maintained for extended periods of time near their best operating conditions. Operability is a broad term, encompassing a wide range of process and control topics from efficient normal operation through startup and shutdown. Each practitioner and researcher has a slightly different definition of operability. For the purposes of this paper, operability is limited to control performance. Predicting control performance and relating it to process and control design are the main topics of this paper.

Practicing engineers are often faced with the challenge of designing process and control systems simultaneously. Typically, limited data is available, and time is of the essence. The engineer can often meet this challenge by relying on past experience with similar processes to make good design decisions. However, operating experience with similar processes is not always available, and the engineer must perform a more thorough analysis.

The choices available to the engineer in performing this analysis are shown schematically in Fig. 1. Steady-state information on the process must be available for process design. The most accurate transient information can be determined through detailed nonlinear dynamic simulation of the system. This option is costly both in engineering time and delays in finalizing the design. However, it is a necessary step in some cases. The other option involves some form of short-cut analysis. A short-cut method maximizes the benefit of the engineering effort expended. General goals of short-cut methods are to 1) completely solve many problems, 2) reduce the initially large number of possible designs to a few for further analysis and 3) develop general insights into how process and control design decisions influence operability.

Most experienced control practitioners have a short-cut "tool kit" that may range from rules of thumb to quickie calculations to standard design guidelines. The short-cut method proposed in this paper provides an integrated approach that complements and extends existing techniques. A newly developed approach uses only steady-state information to predict control performance. An extension of the short-cut method makes use of approximate dynamic models developed by an approach presented in this paper. These models enable the engineer to more accurately predict the control performance. The most demanding analysis in the short-cut method is simplified dynamic simulation. The engineer can employ only those parts of the method required for a specific problem.

The three approaches are presented in the paper in the order shown in Fig. 1. Prior to covering the method the goals of short-cut analysis are further defined with respect to the example process that is adressed throughout the paper. The paper concludes with a discussion of extensions and limitations of the short-cut method.

DISTILLATION CASE STUDY

The short-cut method is applicable to a wide variety of processes and controllers. To facilitate the presentation, the paper addresses distillation in all of its examples. Distillation is selected because it is industrially important and presents the important difficulties in controlling nonlinear, multivariable process. These case studies highlight the strengths and weaknesses of the short-cut method.

The general distillation control problem is shown schematically in Fig. 2. The key aspects of the method are summarized below.

* Control performance is determined in the <u>time domain</u>. This enables the engineer to directly compare the peformance of many potential process and control designs. Also, it enables the engineer to compare the performance to criteria determined by process engineers.

* Disturbances must be specified. They can be upsets or set point changes. Specifying disturbances is <u>necessary</u> <u>for</u> <u>any</u> <u>control</u> <u>evaluation</u> <u>method</u> since the performance of control schemes is strongly influenced by the type and magnitude of the disturbances.

* Multivariable control is addressed. The complexities of interactions between control loops and of designing multivariable controllers is an integral aspect of the short-cut method. The most challenging problem with tight control of both composition control loops is emphasized.

In addition to narrowing the presentation to distillation, two further limitations are observed in the paper. First, the multivariable control problem involves only two controllers. Second, the feedback control design is the standard proportional-integral controller with and without decoupling. These limitations are not central to the application of the method but serve to limit the discussion to a manageable range of topics. Extensions of the short-cut method are discussed briefly later in the paper.

SHORT-CUT ANALYSIS USING STEADY-STATE INFORMATION

Useful predictions of the control performance can be made with the use of steady-state information. These predictions are introduced by first developing the general transfer function for the two-by-two control problem. The structure of the transfer function is analyzed, fundamental dimensionless parameters are identified, and the prediction of control performance is derived. Several examples are analyzed. Finally, a few useful generalizations are given.

Prediction of Control Performance

The measure of control performance used in this method is integral error. It measures the ability of the control system to maintain the controlled variable at its desired value as disturbances occur. The disturbances can be either upsets or set point changes. Naturally, a small integral error is desired, and a large integral error indicates opportunity for improved control. While not a comprehensive measure of all aspects of control performance, integral error captures the essence of performance in one number.

The control performance calculation is introduced for a single input system as shown in Fig. 3. For any system, the closed-loop transfer function relating the error to a disturbance can be developed. The integral error can be calculated directly from the LaPlace transform of the error (Marlin, 1983). This LaPlace transform is determined directly from the transfer function and the disturbance as:

$$\int e(t) \, dt = E(s)_{s=0} \qquad (1)$$

Application of the general equation to the single loop control system yields the equation below.

$$\int e_{sv} = K_D T_I / K_C K_P \qquad (2)$$

The integral error is a function of the disturbance magnitude, K_D, and the process dynamics which affect the controller tuning, $K_c K_p$ and T_I. The controller tuning can be determined from the process dynamics by using methods such as the Ziegler-Nichols rules, tuning charts, or other approaches. Tuning rules ensure a stable system and prevent the (foolhardy) engineer from increasing the closed-loop gain to a very high value so that the integral error calculation predicts a small value. This integral error equation can be used to calculate the control performance of single loop controllers.

Similar analysis can be used to calculate the control performance of a multivariable system (Marino-Galarraga et al, 1985; Stanley et al, 1985). Again, the transfer function is derived; in this case, the transfer function includes parameters from both loops because of interactions. The general block diagram of the two-variable control problem is given in Fig. 4, and the transfer function is given below.

$$\frac{E_1(s)}{D(s)} = - \frac{F_1 \, GG_2 \, RDG(s)}{RGA(s) \{ 1 + GG_1 + GG_2 \} + GG_1 GG_2} \qquad (3)$$

with

$$GG_i = G_{ii} \, G_{ci} \qquad (4)$$

$$RDG(s) = \frac{1}{(1-G_{12}G_{21}/G_{11}G_{22})} \left\{ 1 - \frac{F_2 G_{12}}{F_1 G_{22}} - \frac{1}{GG_2} \right\} \qquad (5)$$

$$RGA(s) = \frac{1}{(1-G_{12}G_{21}/G_{12}G_{22})} \qquad (6)$$

The terms GG_i are essentially properties of each single loop and do not involve interaction. All interaction effects are included in two dimensionless parameters. One appears in the denominator of the transfer function and thus affects the stability of the controller. It is the Relative Gain, here designated as $RGA(s)$. The other appears in the numerator and affects the transient response. It is a newly identified parameter termed the Relative Disturbance Gain, $RDG(s)$ (Stanley et al, 1985; Marino-Galarraga et al 1985). When referring to the steady-state values of these parameters the LaPlace variable, s, is dropped, and they are signified by RGA and RDG, respectively. The steady-state Relative Gain is the same term introduced twenty years ago by Bristol (1966).

The integral error is determined by applying the principles given in equation (1) to the transfer function in equation (3). The result is given below.

$$\int e_{mv} = (\text{ RDG })(\text{ Tuning Factor })\int e_{sv} \qquad (7)$$

$$\text{Tuning Factor} = \frac{(K_C/T_I)_{sv}}{(K_C/T_I)_{mv}} \qquad (8)$$

The Tuning Factor accounts for how much the controller under consideration must be detuned due to interaction.

The ratio of multivariable to single variable integral errors is a key indicator of the effect of interaction on control performance. Its absolute value indicates the change in control performance. The ratio can be calculated by rearranging equation (7). This ratio is referred to as the "integral error ratio" throughout the paper. The calculation of the integral error ratio involves the evaluation of two parameters, the Relative Disturbance Gain and the Tuning Factor. How they are evaluated is now presented.

The steady-state Relative Disturbance Gain is related to the changes in the manipulated variable, which is assumed equal to the controller output, for single and multivariable control when an upset occurs (Stanley et al 1985). The relationship is shown in Fig. 5 and given below.

$$RDG_1 = \frac{(\partial m_1/\partial d)|_{x_1,x_2}}{(\partial m_1/\partial d)|_{x_1,m_2}} \quad \begin{array}{l}\text{(mv control)}\\ \text{(sv control)}\end{array} \quad (9)$$

Importantly, the Relative Disturbance Gain can be calculated directly from steady-state gains as given in the following equation for a two-variable system.

$$RDG_1 = \frac{1}{(1-K_{12}K_{21}/K_{11}K_{22})} \{1-K_{F2}K_{12}/K_{F1}K_{22}\} \quad (10)$$

The first term covers inherent interaction in the process and is the Relative Gain. The second term covers the interaction resulting from the specific upset being considered. Interaction can reduce the Relative Disturbance Gain when the second term is small, that is, when the term $(1-K_{F2}K_{12}/K_{F1}K_{22})$ is close to zero. The interaction can be considered favorable when the effect of the other control loops tends to return the control variable being considered to its set point. This favorable interaction results in a small Relative Disturbance Gain and a small integral error. Later in the paper, a few sample distillation systems that exhibit favorable interaction are discussed. Naturally, the converse also occurs with unfavorable interaction leading to large integral errors.

The prediction of favorable and unfavorable interaction and its quantitative effect on integral error is an important feature of the short-cut method.

To complete the calculation of integral error, a method is needed to calculate the multivariable tuning so that the Tuning Factor can be estimated. It is emphasized that the tuning is to be used in the prediction of integral error. Therefore, the tuning need not be "optimal" but must provide good transient responses and be representative of the tuning used in plant implementations. The multivariable tuning differs from the single variable because of the process interaction. From equation (3), it is apparent that the interaction parameter that influences stability is the Relative Gain. A general approach has been developed to calculate the controller tuning constants for multivariable PI systems with a wide range of process dynamics (Marino-Galarraga et al, 1984, 1986a).

The approach can be simplified for use in the short-cut method. The simplifications are presented in three categories. In the first, at least one of the "interaction terms" (G_{12} or G_{21}) is much slower than the "diagonal terms" (G_{11} and G_{22}). In this case, the interaction influences the control loops slowly, and the two controllers can be tuned according to single loop tuning rules. Therefore, the Tuning Factor is equal to one for this category.

In the second category, the dynamics of the interaction and diagonal paths are similar. This is often the case since the physiochemical processes for the paths are similar. In this category, the steady-state Relative Gain can be used to determine the Tuning Factor.

The calculation of the Tuning Factor is summarized below.

$$\text{for } \frac{T_{11}T_{22}}{T_{12}T_{21}} = 1.0 \quad (\text{ about } 0.50 \text{ to } 2.0) \quad (11)$$

$$\text{and } \theta_{ij}/T_{ij} < 2.0$$

for RGA $\geq$ 1.0

$$\text{Tuning Factor} = \frac{1}{\text{RGA} - (\text{RGA}^2 - \text{RGA})^{.5}}$$

for RGA < 1.0

$$\text{Tuning Factor} = \frac{1}{AA * BB}$$

$$AA = 1 - 1.5/pi\ [\tan^{-1}((1/RGA-1)^{.5})]$$
$$BB = RGA^{.5}\{1 - 1/pi\ [\tan^{-1}((1/RGA-1)^{.5})]\}$$

The resulting relationship between the Tuning Factor and the Relative Gain is shown graphically in Fig. 6. The Tuning Factor is about one in the region of little interaction, where the Relative Gain is nearly one. The Tuning Factor is also bounded below two for large relative gains and below about three for small relative gains (0.5<RGA<1.0).

For the third category, the interactive dynamics in both directions are faster than the diagonal paths. In these few cases, the steady-state method of calculating the Tuning Factor can yield results that are too small by a factor of two or more. A frequency domain analysis is required. Several techniques are available, and an approach has been developed as part of the short-cut method that gives good estimates for multivariable systems (Marino-Galarraga et al 1984, 1986a). It is easier to use than others available since it does not require trial and error calculations or interactive graphics.

With simple methods for calculating the Relative Disturbance Gain and The Tuning Factor, the effect of interaction on the control performance can be directly calculated from the integral error ratio. Before proceeding with further developement of the short-cut method, some examples of the steady-state method are presented.

Examples of Steady-State Analysis

Two distillation control examples are presented here to demonstrate the power of the steady-state method in selecting control designs. The first short-cut example is a typical analysis of two alternative control schemes and loop pairings. The column considered separates a feed mixture of benzene and toluene with an approximate relative volatility of 2.4 into a top composition of 98 percent benzene and a bottom composition of 2 percent benzene. Further details on the column design and nonlinear simulation method are available (Weischedel and McAvoy 1980, Tower B).

The schemes are a direct material balance scheme and an energy (indirect material) balance scheme. The material balance scheme pairs the top composition with the distillate flow and the bottom composition with the reboiler heat input. The energy balance pairs the top composition with the reflux flow and the bottom composition with the reboiler heat input. The key disturbance is chosen to be an upset in feed composition. The evaluation of only two schemes is done to simplify the discussion; any number of pairings and disturbances can be analyzed. The top composition control performance calculations for the first example are summarized below.

CONTROL	RGA	TUN FACTOR	RDG	INTEGRAL ERROR RATIO
Material Balance	0.6	2.5	1.1	2.7
Energy Balance	5.3	2.0	0.25	0.5

The Relative Disturbance Gain and the Tuning Factor for the top composition are calculated according to equations (10) and (11), respectively. They are used to calculate the integral error ratios. The results show clearly that interaction is favorable for the energy balance scheme since the integral error ratio is less than one (0.5). Also, the interaction is unfavorable for the material balance control scheme since the ratio is much greater than one (2.7). Typically, the top composition open-loop dynamic responses are similar for the material and energy balance controls. Therefore, the much smaller integral error ratio for energy balance control indicates that energy balance control provides better control performance for feed composition upsets. The indication is confirmed by nonlinear simulation of the two control schemes. As shown in Fig. 7, energy balance provides better control, giving a smaller integral error and a smaller peak deviation from set point.

The second short-cut example addresses the selection of appropriate multivariable decoupling. The column considered was first analyzed by Wood and Berry (1973). The control scheme is energy balance, and the upset is in feedrate. The decoupling design procedure is the simplified approach as defined by Luyben (1970) and McAvoy (1983). Decoupling is appropriate when unfavorable interaction degrades control performance and is not appropriate when favorable interaction improves control performance. The integral error ratio is useful in predicting when decoupling should be used.

The integral error ratio of a decoupled system usually has a value close to one. This result follows from the fact that the decoupler is designed to break interactions and achieve a control performance close to the single loop performance. Decoupling will improve the control performance of a system that has a large integral error ratio with two single-loop controllers. This conclusion is demonstrated below.

$$\frac{\frac{(\int e_{mv})_{2L}}{\int e_{sv}}}{\frac{(\int e_{mv})_{Dec}}{\int e_{sv}}} = \frac{(RDG)(Tun\ Factor)}{1.0} = \frac{(\int e_{mv})_{2L}}{(\int e_{mv})_{Dec}} \quad (12)$$

with

    2L = control system with two single-loop controllers
    Dec = decoupled control system

Thus, when the product of the Relative Disturbance Gain times the Tuning Factor is much greater than one, decoupling improves control performance by decoupling unfavorable interactions. When the product is less than one, decoupling should not be used since it would increase the integral error by decoupling favorable interaction.

The control scheme is evaluated according to equation (12) to determine if one-way decoupling is beneficial. The calculations are summarized below.

CONTROLLED VARIABLE	RGA	TUN FACTOR	RDG	TUN FACTOR*RDG
Top	2.1	1.55	-0.5	-0.77
Bottom	2.1	1.55	1.2	1.85

Decoupling from the top loop to the bottom loop should be beneficial because the control with two single-loop controllers is predicted to have a larger integral error than the decoupled system (by a factor of 1.85). Decoupling from the bottom to the top loop is not appropriate because the two single-loop controllers have a smaller integral error than the decoupled system (by a factor of 0.77). These predictions are substantiated by simulations shown in Figs. 8 and 9. One-way decoupling from top to bottom loop significantly improves the bottom control performance without degrading the bottom performance. One-way decoupling from the bottom to the top does not improve, and actually slightly degrades, the top control performance, as predicted.

These two examples demonstrate clearly the usefulness of the short-cut method. Key design decisions can be made based on the control performance predictions. Before extending the analysis to include dynamics, a few of the more interesting aspects of the the steady-state analysis are presented.

Additional Results From Steady-State Analysis

Several additional insights are presented here that apply to all multivariable control schemes.

Set point disturbances. The first is that the Relative Disturbance Gain for a set point disturbance is always equal to the Relative Gain. Thus, the Relative Gain is the fundamental control performance parameter for set point changes (Stanley et al, 1985).

Disturbances considered. The second is that the integral error should be calculated for several common disturbances before selecting a control design. Note that even if integral errors are small for the disturbances considered, control designs with a large Relative Disturbance Gain for other disturbances or a large Relative Gain are to be avoided when possible. The analysis of dynamic responses based on the Relative Disturbance Gain and the Relative Gain is explained further (Marino-Galarraga et al, 1986b).

Manipulated Variable disturbances. The third addresses disturbances that enter through a manipulated variable. An example is a heating medium disturbance that alters the heat exchange duty in a distillation reboiler. For the case of a disturbance entering through a manipulated variable (m_i), it can be shown that the multivariable control performance for the loop (x_i) is not much different from its single variable control performance (Marino-Galarraga et al, 1985).

RDG-RGA relationship. The fourth demonstrates the relationship betwen the Relative Gain and the Disturbance Relative Gain (Stanley et al, 1985). Straightforward algebraic manipulation leads to the following equation which gives the relationship for a two-variable control system.

$$RDG_2 = \frac{(1 - RDG_1) \, RGA}{(RGA - RDG_1)} \qquad (13)$$

Summary of Steady-State Analysis with a Caveat

In summary, the control performance analysis based on steady-state information provides guidance in the selection of good control schemes. Each loop can be analyzed separately. The results are obtained by a small engineering effort and require limited data. While the results are not definitive (the integral error ratio is calculated), important design decisions can often be based on the method. In addition, considerable new insight is generated into the existance and causes of favorable and unfavorable interaction.

The great simplification in the steady-state approach is not achieved without debits. The engineer must be aware of the debits, which arise from assumptions, to use the approach correctly. A key assumption in comparing control systems is that the relative integral errors of the single loop control systems are known. For example, the loop pairing example covered in this section assumes that the top quality control loops for both energy and material balance have similar single loop integral errors. This is generally a good assumption since they typically have similar open-loop dynamics. In the course of our research, we have encountered a few cases in which alternate control schemes have had substantially different single loop integral errors, and the differences have been large enough to change the multivariable control design. The cases involved distillation energy and material balance control, and the loop affected was the loop that manipulated the reboiler energy input to control bottom composition. The performance of this loop for the material balance scheme was inferior to the performance for the energy balance scheme.

Therefore, the engineer should understand the open-loop dynamics of the system to employ the steady-state approach. The engineer often, but not always, has this information. The next section presents a technique for quickly estimating the dynamics of complex processes.

DYNAMIC ANALYSIS USING APPROXIMATE MODELS

The analysis of steady-state data developed the ratio of integral errors to be used to predict control performance. The denominator of the ratio, the single variable integral error, can be calculated according to equation (2). The terms $K_c K_p$ and T_I require a knowledge of the open loop process dynamics. If the open loop dynamics are not known from prior experience, they must be estimated as part of the short-cut method. An approach is presented in this section that facilitates the development of approximate models. The approach is first explained; then, examples are presented.

Approximate Modelling Approach

The most straightforward technique for developing simplified models is to directly linearize a fundamental, nonlinear model. Many times this technique provides a model adequate for short-cut operability analysis. However, the technique has two shortcomings. First, the resulting linear models may be too complex for short-cut analysis, and further simplification is required. Second, direct linearization can lead to models which do not accurately represent the actual dynamic response. These two shortcomings are addressed in this section by presenting a complementary modelling approach.

The modelling approach develops approximate models with time domain dynamic responses similar to the exact process response (Paynter and Takahashi 1956; Gibilaro and Lees, 1969; Marlin 1983). The moments of any response can be easily determined from its LaPlace Transform. Models with several moments that are the same will give similar dynamic responses. The approach forces the exact and approximate models to have several moments that are the same. Each moment gives an independent equation which is used to determine one parameter in the approximate model. Thus, a few moment equations provide sufficient information to fit an exact (complex) model with an approximate (simple) model.

The moment method can be used for any linear model structure. For the cases considered in this paper, a first order with deadtime approximate model is sufficient. This model will also be sufficient for the vast majority of practically important problems, but extensions to more complex forms, if needed, is straightforward.

Approximate Model for Recycle Processes

An approximate model for recycle processes is worthwhile to develop because such process configurations are common in industry. Also, the development lends insight into models of more complex processes. To explain the effect of recycle on process dynamics the simple system shown in Fig. 10 is analyzed. Then, the analysis is extended to more complex systems, including distillation.

The system in Fig. 10 can be approximated by a first order process (Kapoor et al, 1986a). The constants in the approximate model can be determined by the method of moments. The resulting equations are given below.

$$K_p = \frac{K_{FF}}{1 - K_{RR}K_{FF}} \tag{14}$$

$$T = \frac{T_{FF} + K_{RR}K_{FF}T_{RR}}{1 - K_{RR}K_{FF}} \tag{15}$$

This approximate model predicts a transient response to a step disturbance that is very close to the exact model as shown in Fig. 11. Also apparent from the figure is the long response time compared to the forward and recycle time constants. This occurs as the product of the forward and recycle gains ($K_{RR} K_{FF}$), referred to as the loop gain, approaches one. The example just analyzed is typical of systems with positive feedback recycle structure. As the loop gain increases, both the overall gain and the dominant time constant of the system increase. This increase is more pronounced as the loop gain approaches one.

This recycle example shows the importance of process structure on dynamic response. It also demonstrates the power of the method of moments in easily developing accurate approximate models. These findings are applied to distillation in the next subsection.

Approximate Distillation Model

Distillation is the physical separation method of choice in the petroleum and chemical industries, and its control is central to the operation of these plants. Naturally, the design of distillation controls requires a good understanding of the dynamics. Considerable effort over the last 15 years has been invested in developing approximate models of distillation (Wahl and Harriott 1970; Weigand et al 1973). The goals were to improve controls and gain insight into the dynamics. None of these previous approaches were successful in developing models useful over a wide range of design conditions. Typical results, taken from a published study (Fuentes and Luyben, 1983), are given below.

| | TIME CONSTANTS (MIN) | |
COLUMN	LINEARIZED	N-L MODEL
Moderate Purity	30	35
High Purity	170000	30

The moderate purity column has 27 theoretical trays, a relative volatility of 2.0, and distillate and bottom impurities of 5 percent. The high purity column has 90 theoretical trays, a relative volatility of 2.0, and distillate and bottom impurities of 0.001 percent. Direct linearization of the exact, tray-to-tray equations gives good results when compared to a detailed, nonlinear simulation for the moderate purity tower. However, the linearization is in extreme error for the high purity tower.

The structure of the distillation process is exploited in developing an approximate model. Distillation is a complex countercurrent process; it can be analyzed as a set of recycle processes. For example, a feed composition upset propagates from the feed tray down the column to the reboiler, and the composition upset is partially vaporized and recycled back to the feed tray. This one recycle path is combined with many others to form the column dynamic response. These recycle paths contribute positive feedback paths that can significantly influence distillation dynamics.

An approximate model has been developed based on the multiple recycle structures and the method of moments (Kapoor et al 1986a, 1986b). The block diagram in Figure 12 is useful in developing an approximate model. Each block represents a major dynamic element in the column, e.g., feed tray, reboiler and so forth. In each block, a balance is made on the light component in and outflow. The transfer function relating any input to any output can be developed from the block diagram.

A typical transfer function, the one between the feed composition and top product, is analyzed to demonstrate the key effects of structure on dynamic response. The net transfer function is given below.

$$\frac{DX_D}{FX_f} = \frac{H_S}{(1 - H_S H_E)} \frac{G_9 G_{10}}{(1 - G_{10} G_{11})} \qquad (16)$$

with

$$H_E = G_7 + \frac{G_8 G_9 G_{11}}{(1 - G_{10} G_{11})}$$

$$H_S = G_1 + \frac{G_2 G_3 G_5}{(1 - G_4 G_5)}$$

Since the flows are constant for a feed quality disturbance, they can be factored out of the equations above.

The transfer function contains several positive feedback terms. These terms are $(1-H_S H_E)$, $(1-G_{10}G_{11})$, and $(1-G_4 G_5)$. There are more than these three; in fact, the countercurrent flows between adjacent trays gives numerous positive feedback terms. However, the lumping used in this development helps elucidate the underlying cause of the large distillation time constants calculated by direct linearization. As with the recycle examples, the dynamic response of the system changes rapidly as a positive feedback term approaches a value of zero since it appears in the denominator. It has been shown that the only term that typically approaches zero for distillation columns is the term involving H_e and H_s (Kapoor el al 1986). Gains for the two cases considered are given below.

COLUMN	GAINS		
	$G_4 G_5$	$G_{10} G_{11}$	$H_S H_E$
Moderate Purity	.77	.41	.72
High Purity	.56	.38	.99986

It is clear that the positive feedback term is not near zero for the moderate purity base case but is very near zero for the high purity base case. By analogy with the simple recycle systems analyzed above, this high gain indicates that the linearized time constant is also large.

A more detailed investigation of the H_S H_E term give further insight into the causes of the large time constant and suggests an approach for estimating more accurate values. The investigation involves evaluating the gain of the H_S H_E term at various feed compositions. The calculations were performed by changing the feed composition from the base case without altering the number of trays or the flows. The calculations were done with a slightly modified version of Luyben's stepping technique (1973). This technique efficiently evaluates the frequency response of selected distillation transfer functions. The results are given in Fig. 13. Both the gain of the positive feedback term and the dominant column time constant have high values at the base case. Their values decrease rapidly in a small region about the base case as the feed composition is altered. These results suggest that the tower dynamics during a disturbance may be much faster since the column will then operate at perturbed conditions.

Before exploiting this insight to develop improved time constants, it is worthwhile to reflect on the high purity column results. The mathematical analysis shows that the linear approximation is good in only a small region about the base case. Also, the unusual circumstance exists that the gain and time constant change rapidly at the boundary of the region and are nearly constant at lower values on each side of the region. A physical explanation is suggested by these results and has been verified by simulation studies. In a very small region around the base case a small change in feed composition requires a large change in tower concentration profile. This large change is because the tower is altered from a condition in which each component is separated into pure products to a condition in which one product becomes relatively impure. Nearly all incremental feed component goes into the impure product. After the transition, one product is essentially pure, and the other is impure. Further changes in feed composition require only small changes in tower composition profile. The linear analysis indicates a long time constant because of the large composition changes occurring for the small incremental feed composition around the base case. The fallacy is that this real effect is only true for a very narrow region around the base case; any significant disturbance shifts the column to a perturbed operation with a smaller time constant.

Studies using a perturbed base case provided much improved estimates for the column time constants. Perturbed cases were taken for the feed composition ten percent higher and ten percent lower that the original, high purity base case. The calculated dynamic model parameters are much more reasonable than the base case values. A comparison of the high purity case prediction with a nonlinear simulation is shown in Fig. 14. For this comparison, the gain of the linear model was altered slightly (about 10 percent) to agree with the nonlinear simulation. Similar good results are obtained for other moderate and high purity columns (Kapoor et al 1986a). In addition, the analysis method can be extended to other changes such as reboiler and reflux changes (Kapoor et al 1986b).

The approximate distillation models can be used in the calculation of integral error. When the open-loop dynamics are known, the single-loop integral error can be calculated from equation (2). This value, along with the Relative Disturbance Gain and Tuning Factor, provide all of the information required in equation (7).

The approximate distillation models discussed to this point have been developed with the use of a small computer program that evaluates the frequency response of the linear model. A purely analytical approach is possible by matching moments of the tray-to-tray linearized model, equation (16), with a first order with deadtime model. The approach is similar to the approach presented for the recycle system, although much more complicated to carry out. The resulting approximate model must also use values from a perturbed operation when applied to high purity distillation columns. The analytical models provide good dynamic responses for the columns studied; in fact, they are nearly as accurate as the frequency response method. The approach is beyond the scope of this paper, but a full exposition is available (Kapoor et al 1986b).

Approximate Modelling Summary

The dramatic effect of process design on dynamic response has been demonstrated and quantified for recycle processes. The insight has been used to develop approximate distillation models for a wide range of operation, including high purity. These results demonstrate that the method of moments is a useful adjunct to standard linearization methods for developing approximate models of complex process configurations. The approximate models can be used for many purposes such as frequency response analysis. Most relevant here is their use in calculating the single variable integral error, thus enabling the value of the multivariable integral error to be calculated from equation (7). In the next section, the use of approximate models in simplified simulation is discussed.

SIMPLIFIED SIMULATION

In some cases, simulation is required to supplement the integral error analysis. The additional factor that usually must be estimated is the maximum error, i.e., the maximum difference between the controlled variable and its desired value. The maximum error should be maintained below a specified value to prevent undesirable consequences like diverting products to lower value streams, activating safety shutdown systems, or upsetting downstream processes. Naturally, the few cases chosen for simulation should be indicated by the short-cut integral error analysis.

Much has been written about dynamic simulation; so, this brief discussion is limited to the special features needed within the context of short-cut analysis. The simulation system should enable the engineer to construct, execute, and analyze a simulation quickly. To meet these needs, the system should be easy to use, have good ergonomics, and be able to efficiently solve simplified dynamic models.

A system that meets these needs is shown schematically in Fig. 15. Naturally, the first step involving problem formulation, initial analysis, and model development are the tasks of the engineer; they are addressed in the preceeding two sections of this paper.

A key feature of this system is the support software that relieves the engineer of the programming burden. The solution of most problems can be achieved by configuring standard blocks, each of which performs a specific algorithm. The blocks are used to simulate the process, upsets, and controllers and to calculate control performance. The engineer inputs all configuration information and adjustable parameters through an interactive dialogue with the system. Naturally, the program allows the engineer to link with user-written programs, but nearly all short-cut analysis can be performed with the standard features.

Execution of the simulation is interactive. The engineer can introduce disturbances, upsets or setpoint changes on a prearranged schedule or interactively via the console. The engineer can observe the progress of the simulation through various graphic and schematic displays which are updated in realtime. Cases can be stopped, altered, initialized, and restarted with standard interactive commands.

Studies like the linear simulations in this paper can be performed quickly with such a simulation system. The availability of the system in a personal computer brings this capability to each engineer in a cost-effective manner.

EXTENSIONS AVAILABLE THROUGH THE SHORT-CUT METHOD

The short-cut method is presented in this paper through application to distillation control. The method is much more general: a few of the more prominent extensions are presented in this section.

Controllers

The distillation examples used standard proportional-integral controllers exclusively. The short-cut method is applicable to any linear controller. For example, control performance analysis via integral error has been applied to a predictive controller, the Internal Model Controller (IMC) design. New design methods were developed that take advantage of favorable process interactions when they exist (Marino-Galarraga et al 1986b).

Processes

The modelling and analysis methods are general for any linear system. Heat exchangers, boilers, and steam systems have been analyzed (Marlin 1983, 1985). A study indicates that approximate models are adequate for boilers and high pressure steam systems (Bertrand and McAvoy 1986).

Sensitivity

The analytical results on integral error are especially useful in determining the sensitivity of control performance to parameter changes. Of special interest is the sensitivity of a PI-decoupled system to errors in the decoupler. It can be analytically demonstrated that the integral error is proportional to a term involving the Relative Gain. Thus, Relative Gain is the fundamental sensitivity parameter of the system. This result has been demonstrated previously several times by simulation case studies.

LIMITATIONS

Naturally, the short-cut method has limitations that the engineer must consider. The first is that the entire method is based on linear analysis. It has been successfully applied to nonlinear distillation cases. Success of the method for highly nonlinear systems like strong acid-base neutralization is problematical. Such systems are best simulated.

A second limitation is that the integral error can be small due to the cancellation of large positive and negative components. Experience to date has not encountered such cases that lead to incorrect control design, but the engineer should recognize the possibility.

To date, the method has been applied only to two-variable control systems. The Relative Disturbance Gain has been extended to any order system. The approach for calculating the tuning factor has also been extended to any order system. However, these calculations are cumbersome and further work is required to determine valid, shortcut methods. The extension to very high orders may give a method too complicated for short-cut procedures.

CONCLUSIONS

An integrated short-cut method has been developed. A time domain control performance measure, integral error, is predicted and used to select control designs. The Disturbance Relative Gain and the Relative Gain were found to be the important parameters, and their steady-state values are often sufficient to predict control performance. Additional accuracy can be obtained by developing approximate dynamic models. Should further analysis of the transient response be necessary, a simulation system like the one described in the paper is appropriate for simplified dynamic studies. The entire approach has been demonstrated by several case studies.

The method was developed as a problem solving technique for use by control practitioners. Extensive theoretical studies have been consolidated into straightforward calculations using information typically available to the engineer. The engineer can integrate previous process experience, combine results from other short-cut techniques, and use only the rigor required for the problem at hand. For example, the use of steady-state data is sufficient if the engineer can predict the open-loop process dynamics from experience. As demonstrated, it also yields important insight into the interaction between process and control design, especially into the parameters that contribute to favorable and unfavorable interaction.

We believe that the features of this method earn the method a place in the control engineer's "tool kit". A short-cut analysis at the beginning of each study will provide satisfactory final results in some cases and better direct futher analysis in many other cases.

Appendices to this paper that provide a brief survey of operability analysis and describe the method of moments are available. A copy can be obtained by writing the authors.

REFERENCES

Bertrand, C. and T. McAvoy (1986). Short-Cut Analysis of Pressure Control in Steam Headers, Automatic Control Conference, Seattle, WA

Bristol, E. (1966). On a New Measure of Interaction for Multivariable Process Control, IEEE Trans Autom Control, AC-II, 133-134

Fuentes, C. and W. Luyben, (1983). Control of High Purity Distillation Columns, IECPPD, 22, 361-366

Gibilaro, L. and F. Lees (1969). Reduction of Complex Transfer Function Models to Simple Models Using the Method of Moments, Chem Eng Sci, 24, 85-93

Kapoor, N., T. J. McAvoy, and T. E. Marlin (1986a). Effect of Recycle Structure on Distillation Tower Time Constants, accepted for publication in the AIChEJ

Luyben, W. (1970). Distillation Decoupling, AICHEJ, 16, 2, 198-203

Luyben, W. L. (1973). Process Modelling, Simulation, and Control of Chemical Processes. McGraw Hill, New York

Marino-Galarraga, M., T. J. McAvoy, and T. E. Marlin, (1984). The Effect of Interaction on Tuning and Operability of Classically Controlled Systems, American Control Conference, San Diego, CA, June 1984

Marino-Galarraga, M., T. E. Marlin, and T. J. McAvoy, (1985). Using Relative Disturbance Gain to Analyze Process Operability, American Control Conference, Boston, MA, June 1985

Marino-Galarraga, M., T. J. McAvoy, and T. E. Marlin, (1986a). Short-Cut Operability Analysis, Part II - Estimation of f_i Tuning Parameter for Classical Control Schemes, submitted to IECPPD

Marino-Galarraga, M., T. J. McAvoy, and T. E. Marlin, (1986b). Short-Cut Operability Analysis, Part III - Short-Cut Methodology for the Assessment of Process Control Designs, submitted to the IECPPD

Marlin, T. E. (1983). Short-Cut Operability Analysis, Second Biennial Short Course on Control in the Chemical Process Industries, University of Maryland

Marlin, T. E. (1985). Short-Cut Operability Analysis, Third Biennial Short Course on Contral in the Chemical Process Industries, University of Maryland

McAvoy, T. (1983), *Interaction Analysis*. Instrument Society of America, Research Triangle Park, NC

Paynter, H. and Y. Takahashi (1956). A New Method for Evaluating Dynamic Response of Counterflow and Parallel-Flow Heat Exchangers, *Trans ASME*, 78, 749-758

Stanley, G., M. Marino-Galarraga, and T. J. McAvoy, (1985). Short-Cut Operability Analysis, Part I - The Relative Disturbance Gain, *IEC PDD*, 24, 1181-1188

Wahl, E. and P. Harriott (1970). Understanding and Prediction of Dynamic Behavior of Distillation Columns, *IEC PDD*, 9, 3, 396-407

Weigand, et al (1972). *AICHEJ*, 18, 6, 1243-1252

Weischedel, K. and T. McAvoy, (1980). Feasibility of Decoupling Conventionally Controlled Distillation Columns, *IEC Fund*, 19, 379-384

Wood, R. and M. Berry (1973). *CES*, 28, 1707

NOMENCLATURE

d	=	disturbance
D	=	distillate flow
$D(s)$	=	LaPlace transform of the disturbance
e	=	error, difference between the measured variable and set point
$E(s)$	=	LaPlace transform of error
F_i	=	transfer function of the disturbance to the ith output variable

G_{ij} = ij^{th} transfer function in multivariable system
G_c = controller transfer function
GG_i = see equation (4)
K_C = controller gain
K_D = disturbance magnitude (single loop control)
K_{Fi} = disturbance magnitude (i^{th} output in multivariable system)
K_{FF} = forward path gain in recycle structure
K_{ij} = gain of ij^{th} element in multivarable control
K_P = process gain (single loop)
K_{RR} = return gain in recycle structure
m_i = i^{th} input variable in multivariable system
pi = 22/7
RDG = relative disturbance gain, equation (5), (10)
RGA = relative gain
s = LaPlace transform variable
t = time
T_{FF} = forward path time constant in recycle structure
T_I = controller integral time
T_{ij} = time constant of ij^{th} transfer function
T_{RR} = reutrn path time constant in recycle structure
V = vapor flow
x_i = i^{th} output (measured) variable in time domain
X_i = LaPlace transform of x_i
y = vapor composition

Subscripts

Dec = decoupled control system
sv = single variable
mv = multivariable
2L = control system with two single-loop PI controllers

Greek

θ_{ij} = deadtime of the ijth transfer function

FIGURE 1

SHORT-CUT ANALYSIS MAXIMIZES RESULTS FOR ENGINEERING EFFORT

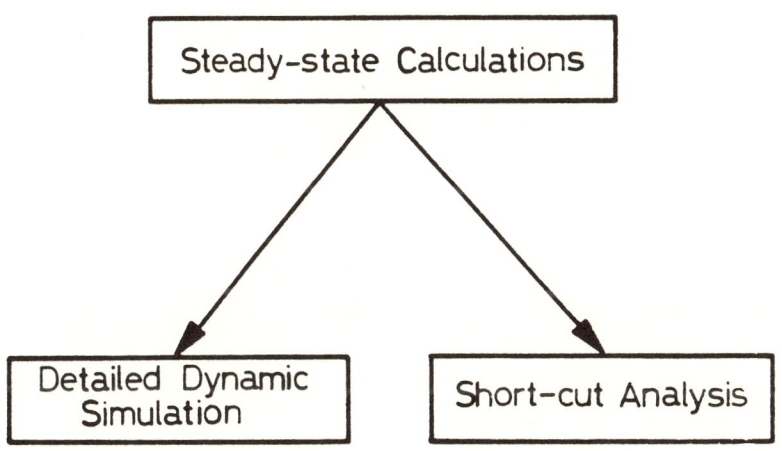

- Steady-state Analysis
- Approximate Dynamic Models
- Simplified Dynamic Simulation

FIGURE 2

SHORT-CUT ANALYSIS IS BASED ON CONTROL PERFORMANCE

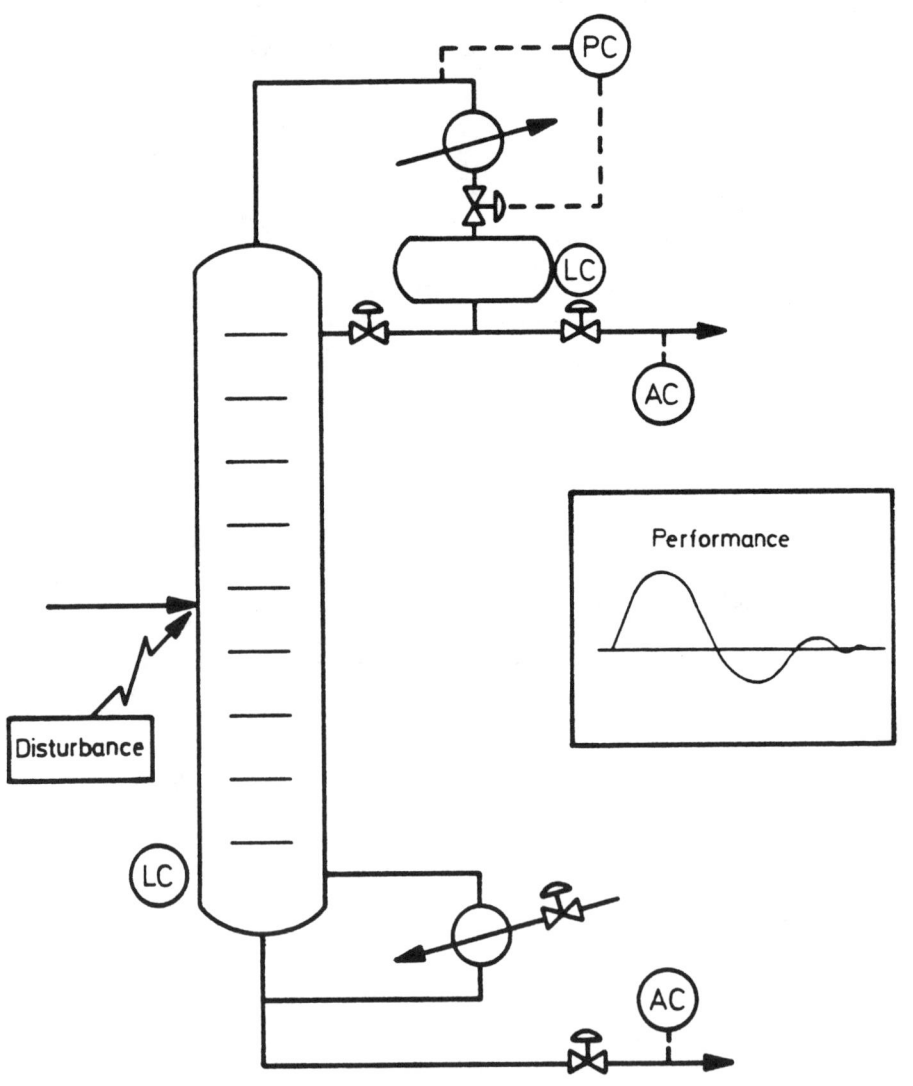

FIGURE 3

INTEGRAL ERROR CAN BE CALCULATED FROM STEADY-STATE PARAMETERS

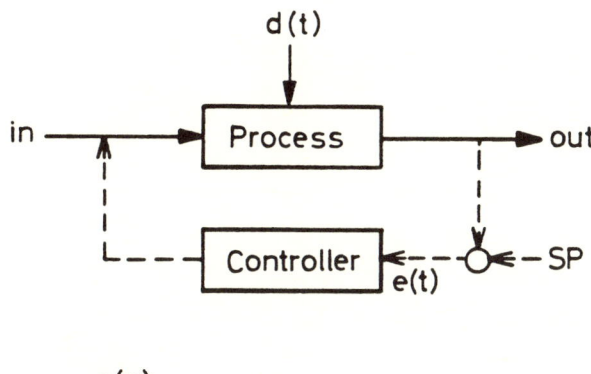

$$\frac{e(s)}{d(s)} = P(s)$$

$$\int_0^\infty e\, dt = P(s)d(s)\Big|_{s=0} = E(o)$$

Single Variable PI

$$\int e_{sv} = \frac{K_D T_I}{K_C K_P}$$

Two Variable PI

$$\int e_{mv} = (RDG)(\text{Tuning Factor})\left(\int e_{sv}\right)$$

FIGURE 4

GENERAL TWO-VARIABLE CONTROL BLOCK DIAGRAM

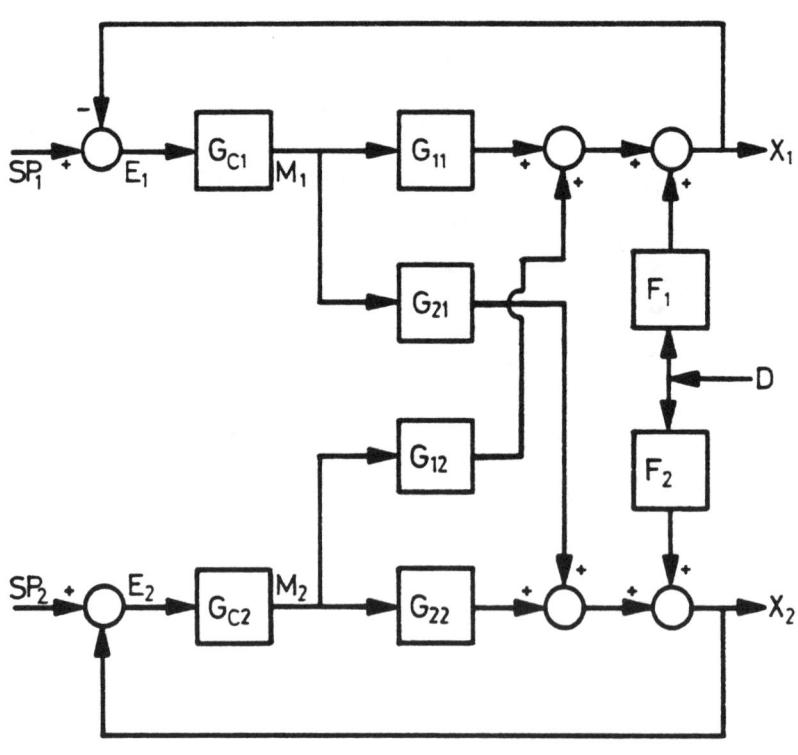

FIGURE 5

THE RELATIVE DISTURBANCE GAIN IS RELATED TO MANIPULATED VARIABLE CHANGE FOR A SPECIFIED UPSET

$$RDG = \frac{(\Delta \text{manipulated variable})|mv}{(\Delta \text{manipulated variable})|sv} = \frac{B}{A}$$

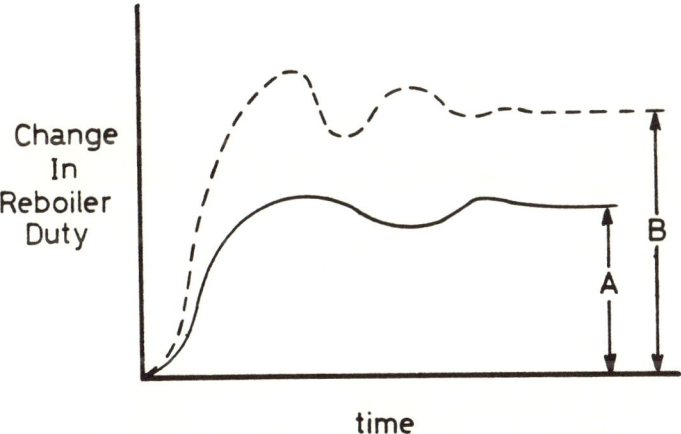

$$RDG_1 = \left(\frac{1}{1 - \frac{K_{12} K_{21}}{K_{11} K_{22}}} \right) \left(1 - \frac{K_{F2} K_{12}}{K_{F1} K_{22}} \right)$$

FIGURE 6

THE TUNING FACTOR IS RELATED TO INTERACTION

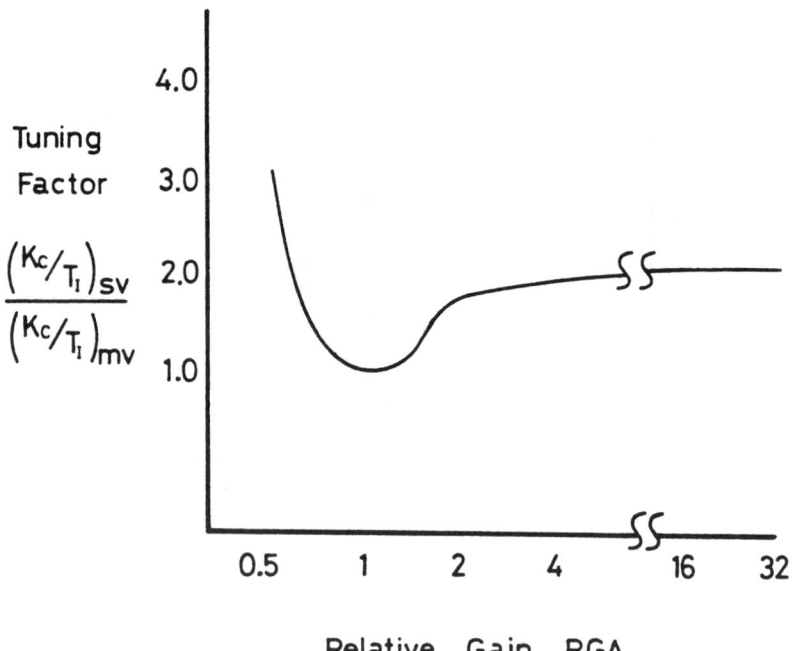

(Diagonal and Interactive Dynamics Similar)

FIGURE 7

SIMULATION CONFIRMS BETTER PERFORMANCE BY ENERGY BALANCE CONTROL

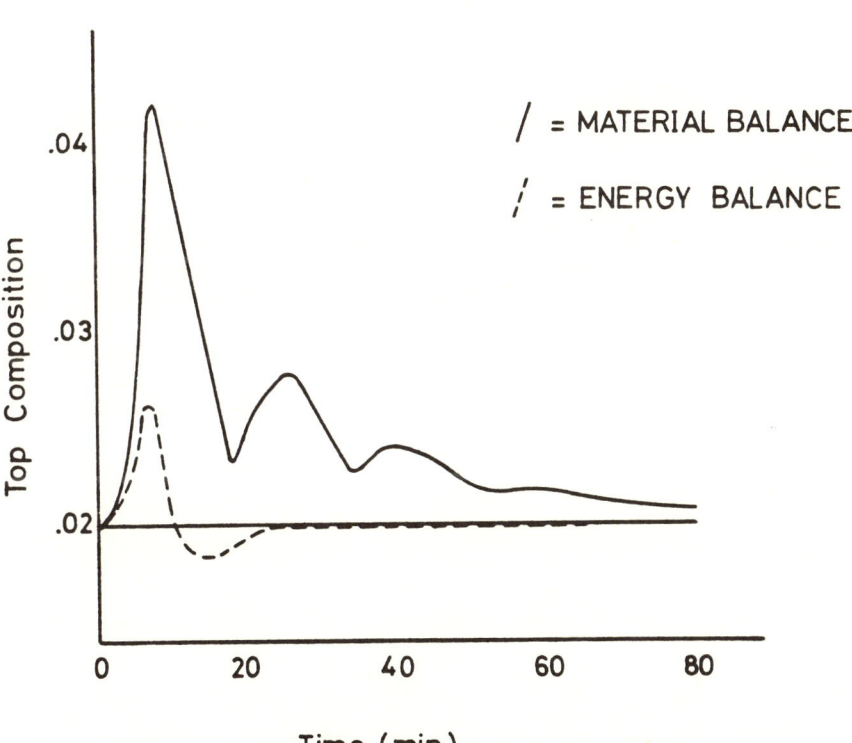

FIGURE 8

SIMULATION CONFIRMS IMPROVED CONTROL PERFORMANCE FOR TOP TO BOTTOM DECOUPLING

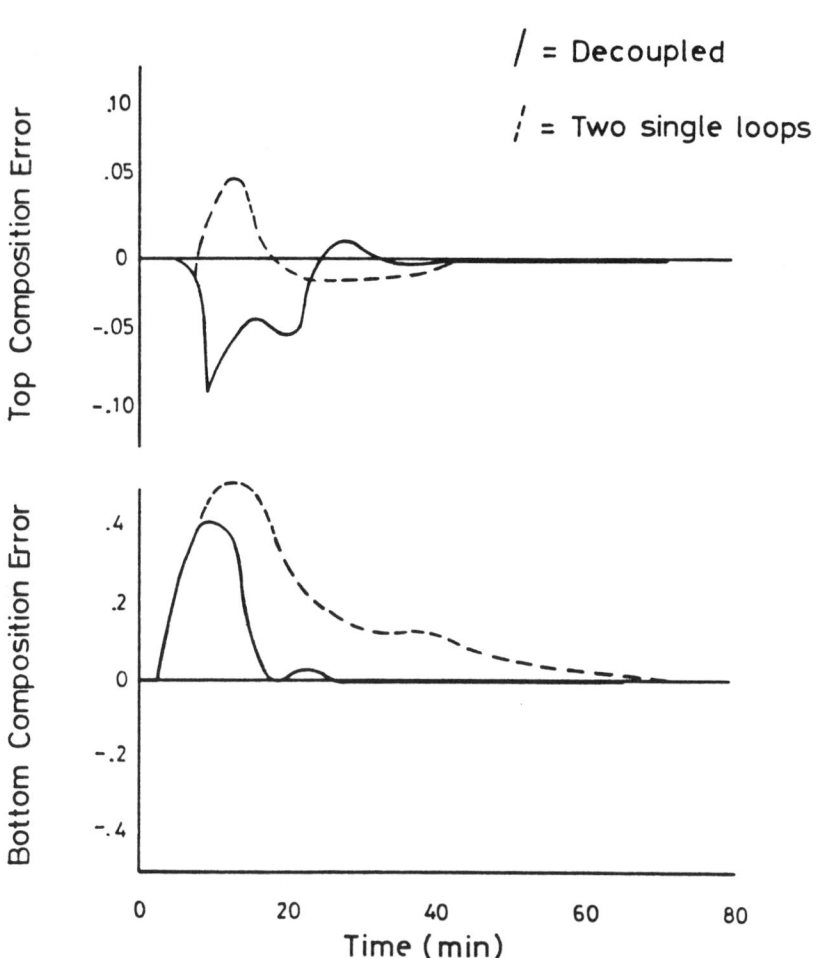

FIGURE 9

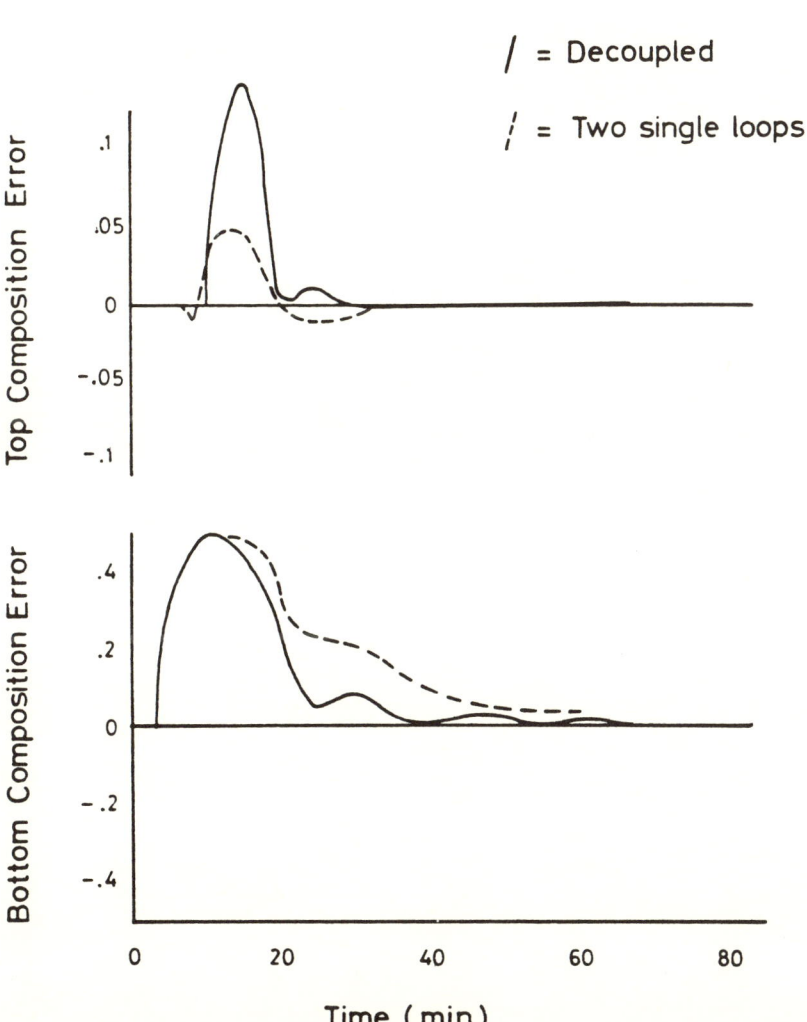

FIGURE 10

RECYCLE SYSTEM BLOCK DIAGRAM

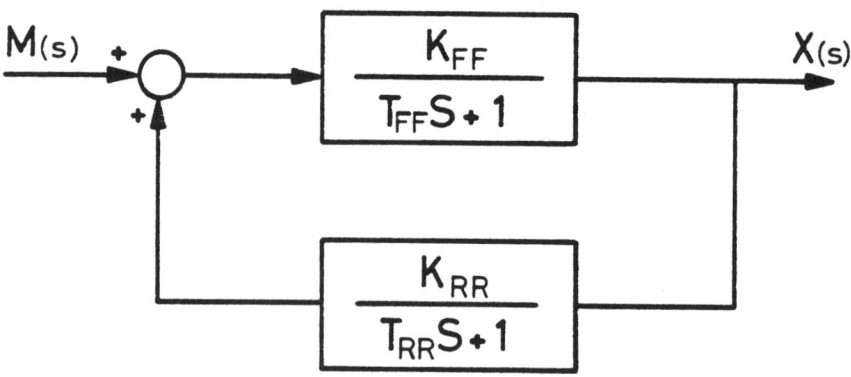

FIGURE 11

RECYCLE SYSTEM DYNAMIC RESPONSE

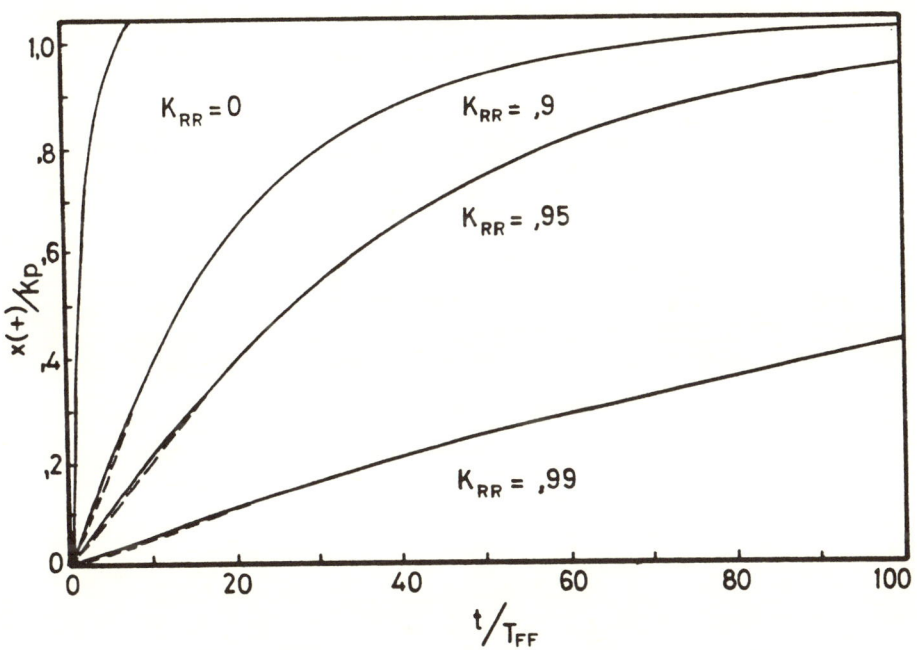

FIGURE 12

DISTILLATION BLOCK DIAGRAM

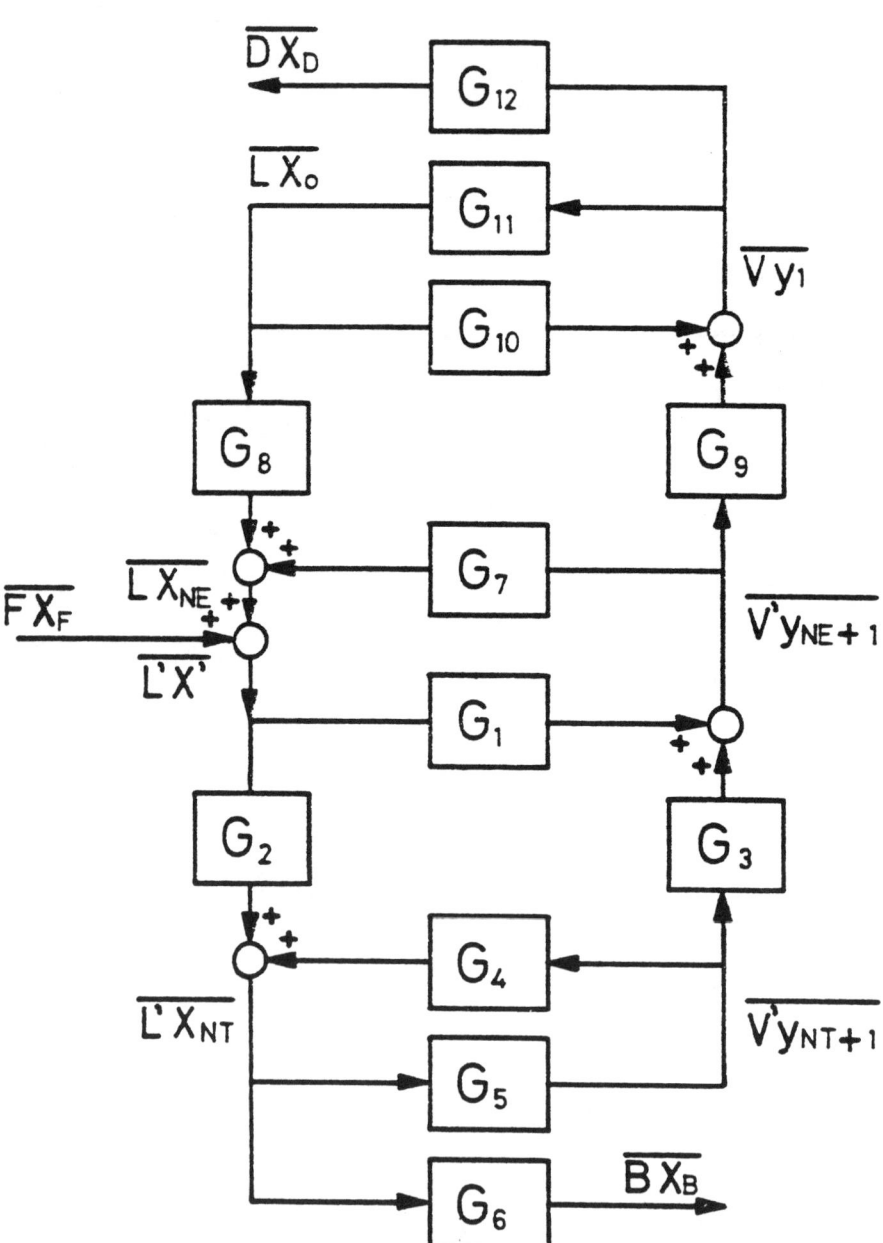

FIGURE 13

LINEARIZED DOMINANT TIME CONSTANT AND POSITIVE FEEDBACK GAIN ARE LARGE AT THE HIGH PURITY STEADY STATE

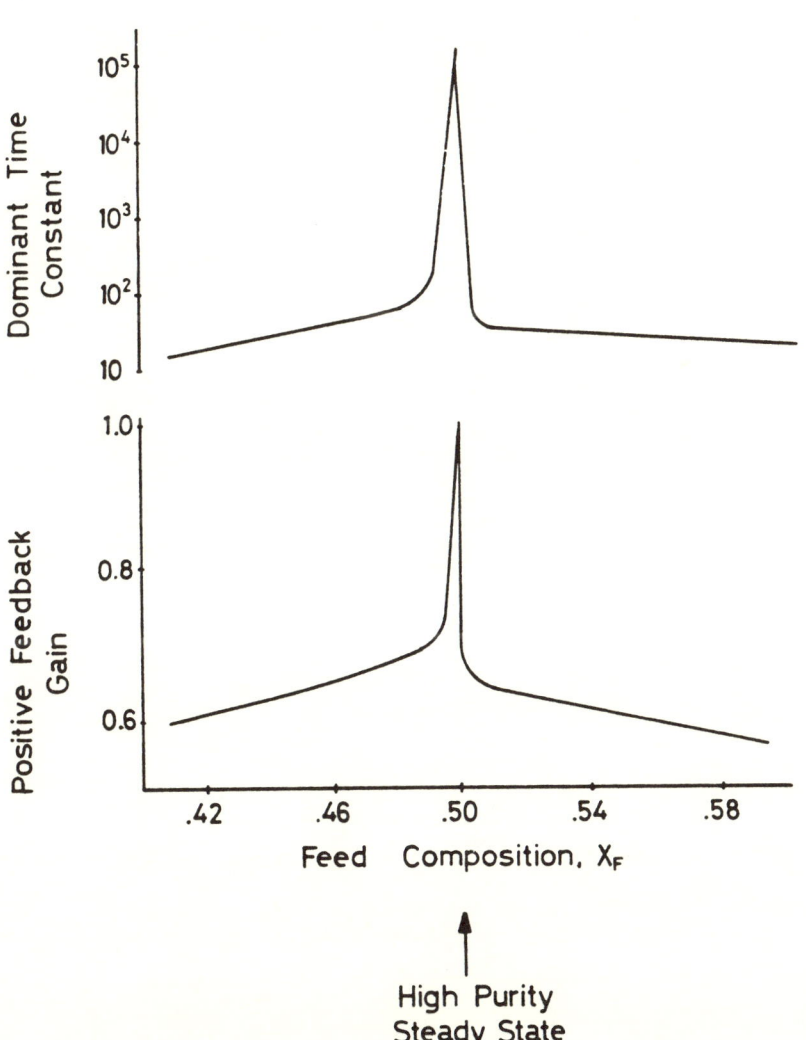

FIGURE 14

MODEL BASED ON PERTURBED BASE CASE
GIVES ACCURATE OPEN-LOOP RESPONSE
(FEED COMPOSITION CHANGE)

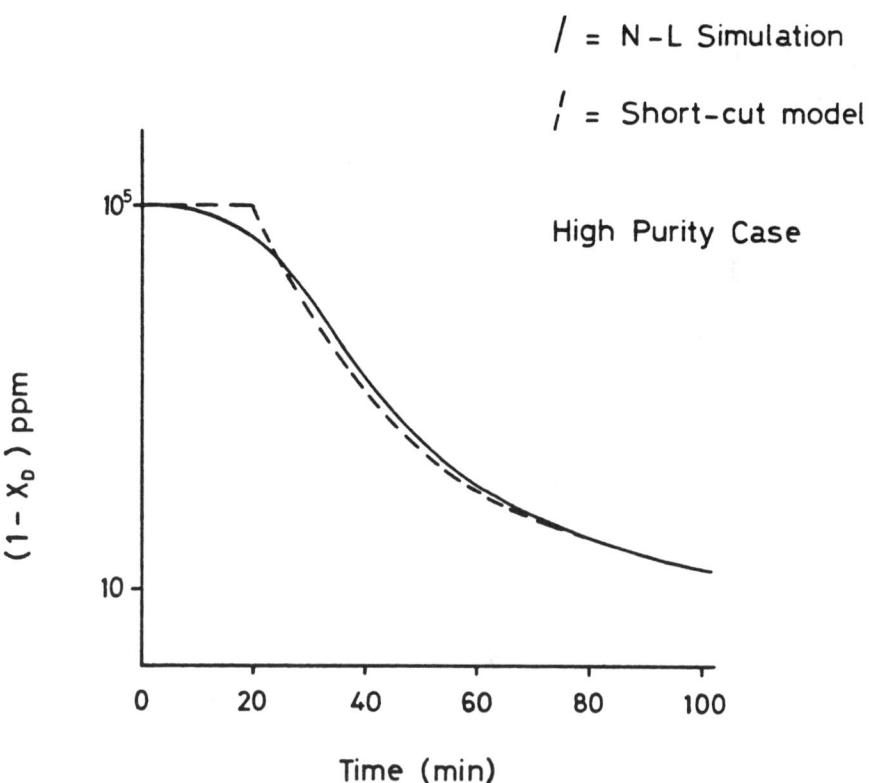

FIGURE 15

SIMPLIFIED SIMULATION FACILITATED BY SIMULATION SYSTEM

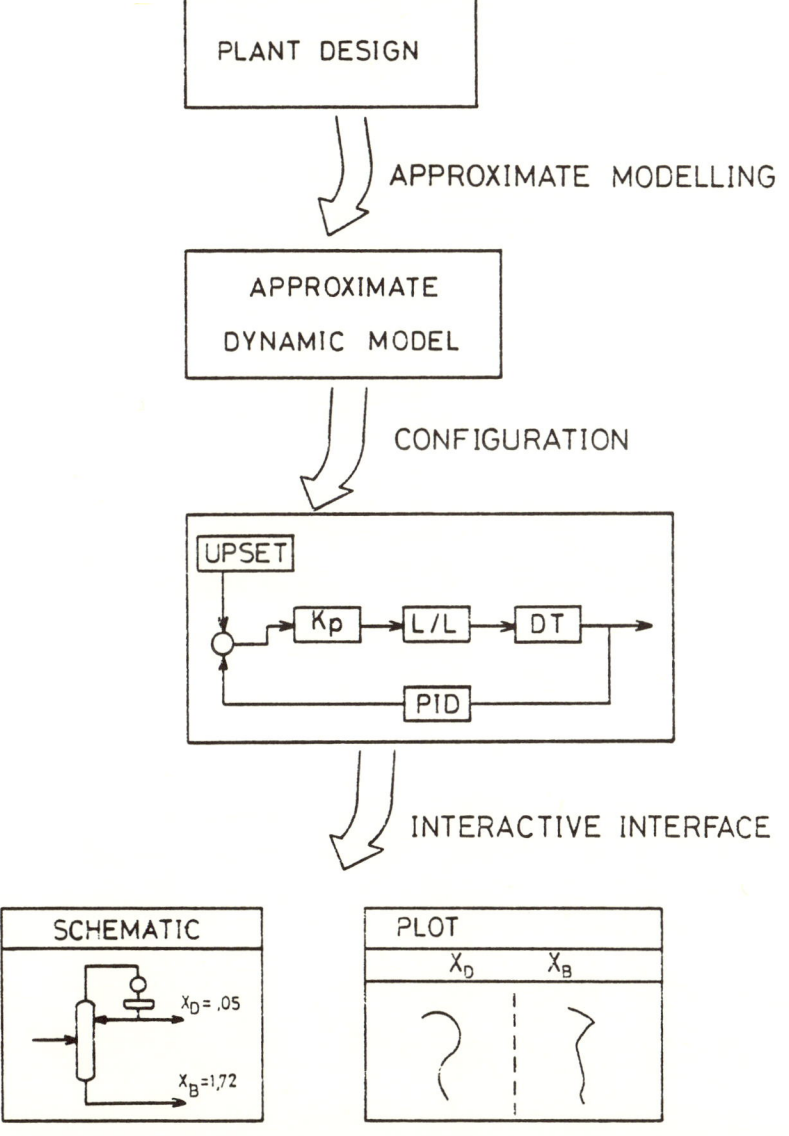

ADAPTIVE CONTROL

B. Erik Ydstie
The University of Massachusetts, Amherst, MA 01003

A SIMPLE PROBLEM WITH COMPLEX DYNAMICS
Consider a model reference adaptive controller

$$u(t) = (r - z(t)y(t))/K, \quad K \neq 0,$$

with one parameter $z(t)$ updated so that

$$z(t) = z(t-1) + py(t-1)(y(t)-r), \quad 0 < p < \infty, \ t = 1,2,3\ldots$$

Here $u(t)$ is the manipulated variable, $y(t)$ is the process output, K is a fixed estimate of the process gain, r is the setpoint and p is the adaptation gain. The closed loop system exhibits a benign behaviour when we apply this simple nonlinear feedback strategy to a first order process. There is no model mismatch and strong results with global implications can be obtained using simple analysis. This situation is unlikely to occur in practice, however.

To illustrate the effect of model mismatch we apply the same feedback strategy to a second order process

$$y(t) + a_1 y(t-1) + a_2 y(t-2) = bu(t-1).$$

Now, the situation with respect to simple analysis and achievable results is entirely different. The associated discrete map $F_\phi(y,z)$: $R3 \to R3$, $\phi = (a_1, a_2, b/K, r, p)$, can exhibit exceedingly complicated behaviour. One can bring to bear the whole gamut of nonlinear analysis on this apparently simple problem without being able to completely characterize the global characteriscs of $F_\phi(y,z)$. Some phase plane plots of (y,z) are shown in Fig. 1 below. The map exhibits limit cycles with different periods, strange attractors and chaotic motion.

THE APPARENT SOLUTION
The presentation provided by Prof. K. J. Aström and Dr. G. A. Dumont as well as the discussion provided by Dr. M. Fjeld at the CPC-III meeting underscored the point illustrated above. Namely: Current adaptive control theory suffers from the fact that few general statements can be made about the global behaviour of adaptive controllers when they are applied to real processes. Thus, there exist a multitude of

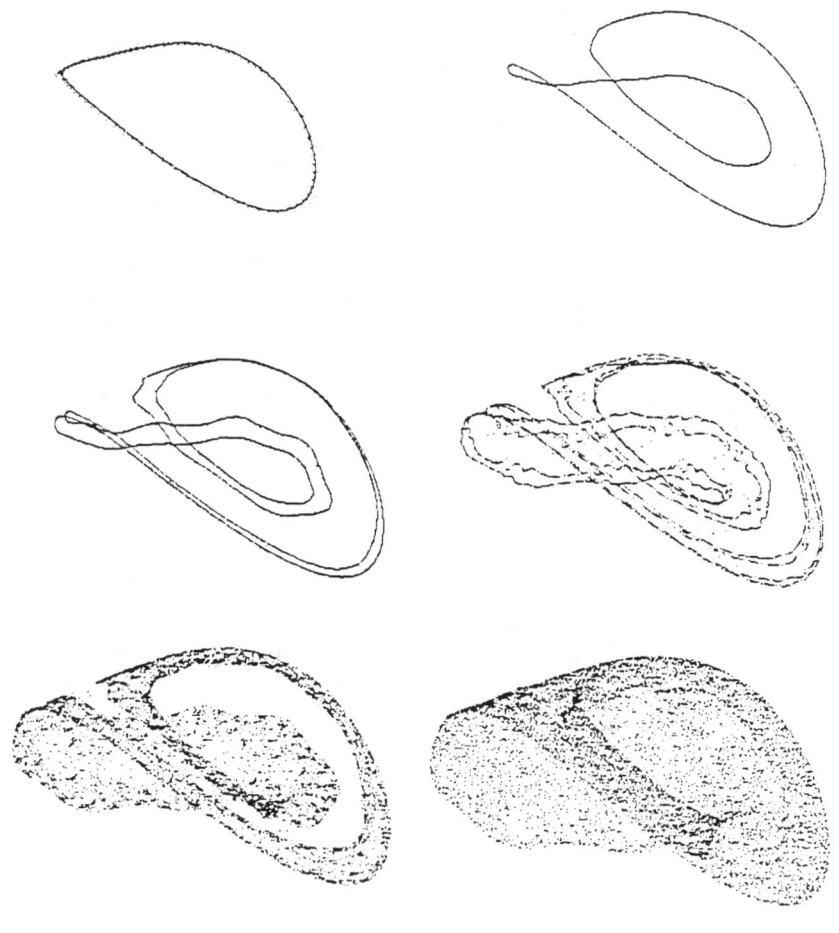

FIGURE 1. Complex dynamics in a simple adaptive control system $\phi = (-0.9, a_2, 0.311642, 10, 0.01169)$. The parameter a_2 varies so that $a_2 = 0.653, 0.7, 0.71, 0.74, 0.76, 0.78$, starting at the top left corner.

algorithms that all have slightly different characteriscs, but they are difficult to compare since some work well in some cases but in other cases they do not. These complex algorithms are not amenable to simple analvsis and a "magic box" like the one shown in Fig. 2, is not likely to appear on the market, or in the control journals in the near future. However, this does not mean that all is lost and that adaptive control algorithms cannot find a useful and important role to play in the chemical processing industries. But, the emphasis has to be placed on localized analysis and engineering developments aimed towards specific process applications. Great strides are currently being made in this direction. The engineering commitment involved in making the adaptive theory work may appear overwhelming at first sight. (Fig. 3)

Drs. Dumont and Fjeld pointed out that adaptive control algorithms have been applied to solve important control problems in the paper industries. One of the reasons for this successful development is that the processes in question, though complicated, often are standardized and well-characterized. Often it is clear what can be gained from applying more advanced control policies to let's say a Fourdrinier paper machine. The engineering effort required to make the adaptive control system work can then be justified based on economic judgements. Similar justifications can be made in a number of chemical process applications, however. One example is the distillation column control problem. It has been demonstrated that multivariable adaptive control algorithms can be made to work reliably in an industrial environment, and that they can be standardized. An algorithm that has been developed for use on one distillation column can be applied to other columns even if they have different characteristics. Substantial improvements in the process performance have been reported [1]. Other adaptive algorithms have been applied to the control of nuclear power plants [2]. The common theme in these, and the many other successful applications mentioned in the adaptive control session at the CPC-III meeting, is that a pleasing outcome always has come as a direct result of substantial research and development effort. The most important part of the final algorithm often is not the adaptive strategy itself, but a rather large number of monitoring and safety features surrounding the adaptive policy. These ensure proper performance under normal operating conditions. They also guarantee that the process "fails gracefully" if abnormalities occur.

Prof. Åstrøm, in his presentation, pointed out that the

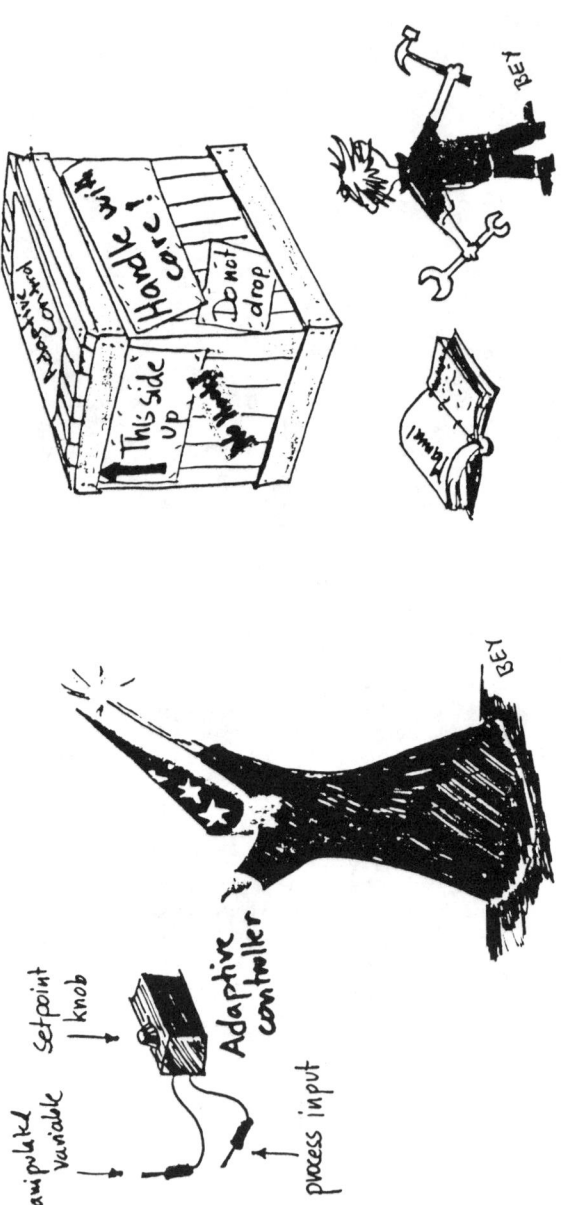

FIGURE 2. The Adaptive Control Theoretician doing his mysterious work.

FIGURE 3. The Adaptive Control Engineer trying to fathom the results and apply the ideas.

adaptive algorithms and their surrounding features efficiently can be developed within the framework of an "Expert System". An integral part of this expert system may be a number of different algorithms for signal processing, auto tuning and adaptive control as well as different control algorithms. These are combined and used as the need arises. Again, the needs relevant to a process control engineer are dictated by specific process applications and process economics.

PERSPECTIVE FOR THE FUTURE
Recent research in the area of robust adaptive control has focused on the local rather than the global behaviour of the adaptive control algorithms [3,4]. This has led up to linearization procedures around tuned trajectories, and the determination of what is meant by "slow adaptation" and small model mismatch. The results provide a framework for further theoretical development, and important results on robust adaptive control no doubt are just around the corner.

The theoretical results referred to above apply to recursive algorithms. The recursive approach is computationally efficient and pleasing from the point of view of analysis. But, the recursive approach also poses a number of theoretical and practical problems that have yet to be satisfactorly resolved. This is especially true for multivariable and feed forward control systems where the signal conditions often are more difficult to monitor. However, it is in the area of multivariable and feed forward control systems that the greatest gains from applying adaptive control theory can be expected in the chemical process industries. A number of problems associated with the recursive approach can be avoided or at least made easier to solve by applying batch estimation and adaptive calibration techniques.

The bandwidth requirements for most process control systems are not strict; and, the majority of process control problems do not need continual updating of parameters. A low bandwidth controller like a PID type controller with adaptive calibration of gains and time constants combined with gain scheduling is often all that is required. The application of high performance pole-zero cancellation policies is usually not merited and the problems associated with the use of these policies usually far outweigh their advantages. Examples of the application of simple adaptive calibration systems are discussed in Refs. [5,6]. The theory for adaptive calibration systems has not been studied yet and it is an open area both for research and application.

Dr. Dumont pointed out that the current adaptive control literature is difficult to read and understand. Nevertheless, a great number of individuals in the chemical processing industries have embraced the concept of adaptive control and many companies have commited resources to the evaluation of turnkey adaptive control systems as well as developing their own. An informal poll taken during the meeting revealed that about 50% of the audience at the CPC-III had been working with, or were familiar with the application of adaptive controllers. About 75% of the audience believed that adaptive control in the future can play an important role in solving industrial process control problems. Whether that is going to be so or not depends largely on the practicing control engineers in industry. Their ability and willingness to <u>make</u> adaptive control work for particular process applications is going to determine the outcome.

Experiences reported from the european countries (Belgium, Scandinavia, UK, France and Germany) strongly suggest that in the short run, the successful development of reliable algorithms is most likely to occur where there is a close University-Industry cooperation.

[1] Unbehauen, H. and P. Wiemer. "Application of Multivariable Adaptive Control Schemes to Distillation Columns." Paper presented at the 4th Yale Workshop on Applications of Adaptive Systems Theory. Yale University, New Haven, May 29-31, (1985).

[2] Irving, E. "Commande Adaptative, Partie 1," ESE, Paris, France, (1984). Private Communication.

[3] Kosut, R. L. and C. R. Johnson, Jr. "An Input-Output View of Robustness in Adaptive Control." Automatica, pp. 569-581, Vol. 20, No. 5, (1984).

[4] Riedle, B. D. and Kokotovic, P. V. "Integral Manifolds of Slow Adaptation." IEEE, T-AC-31, pp. 316-224, No. 4, (1986).

[5] Bartman, R. V. "Dual Composition Control in a C_3/C_4 Splitter." CEP, pp. 58-62, September, (1981).

[6] Whatley, M. J. and Pott, D. C. "Adaptive Gain Improves Reactor Control." Hydrocarbon Processing, pp. 75-78, May, (1984).

ADAPTATION, AUTO-TUNING AND SMART CONTROLS

Karl Johan Astrom
Lund Institute of Technology, S-221 00 Lund, Sweden

Abstract. This paper reviews some advances in adaptive control that have occured since CPCII. This includes theoretical development, auto-tuning and industrial use. The possibility to use a collection of different algorithms for estimation, control design and monitoring which are coordinated by an expert system is a new emerging concept which is beginning to be explored.

Keywords. Adaptive Control; Robustness; Automatic Tuning; Expert Control; Knowledge Based Systems.

1. INTRODUCTION

There have been significant advances in adaptive control after the CPCII which was held in January of 1981. Theory and algorithm have been improved. More important however is the emergence of several industrial products for industrial process control. Leeds and Northrup announced their Electromax V which is a single loop controller with a self-tuning option in 1981. The swedish Company ASEA announced their Novatune, which is a small DDC system with several adaptive modules, in 1982. Three adaptive controllers were announced in 1984. The British company Turnbull Controls introduced their TCS 6355 auto-tuning controller, which is a single loop regulator with adaptive and auto-tuning facilities. The swedish company NAF Controls announced their Autotuner which is based on a novel scheme to tune PID regulators. Foxboro announced the adaptive single loop regulator Exact. There are also several other adaptive controllers which have been announced or which are about to appear.

Today there are several thousand loops under adaptive control. The practical experiences from their operation is naturally accumulating. A description of some experiences are given in the paper (Dumont, 1986).

The products are based on different concepts and different regulator structures. The demands on the user are also quite different both with respect to operational issues and in the effort required to understand how they work. Most products are based on the PID algorithm but there are a few that uses other types of algorithms. Some use the traditional approach to adaptive control based on recursive parameter estimation and automatic control design, but others are using nonconventional methods for estimation and control design. The Foxboro Exact uses an heuristic design method which mimics the tuning procedure used by an operator. The NAF Autotuner uses a novel method to determine the process dynamics based on relay oscillation.

Most adaptive schemes currently used can be characterized as local gradient algorithms. This means that given good initial values they will drive the system towards a very good performance. The effort required to obtain the initial values or the prior knowledge may be substantial. Several adaptive systems therefore have what is called a "pretune mode" which typically uses a pulse test to obtain the required prior knowledge. The autotuner is different because it requires very little prior knowledge. It also generates the test signals automatically. There is also a growing awareness of the need for safeguards to ensure that the adaptive regulators work well under all possible operating conditions.

The purpose of this paper is to look at some of the approaches to adaptive control their strengths and weaknesses. In doing so it is found that systems with very attractive properties can be obtained by combining several different approaches. An autotuner can be used to arrive at a simple control law in a robust way. The information gathered by the autotuner can also be used to derive the prior information required by more sophisticated adaptive schemes. We will thus arrive at a system which contains several different algorithms. To monitor their operation it is then useful to introduce algorithms which supervise the operation of the system and which can initiate switching between algorithms. It is clear that a system of this type will involve a substantial amount of heuristic logic. Expert system methodologies provide a systematic approach for dealing with this logic. The term expert control is therefore coined to describe systems of this type. Once the expert system approach is taken it is also possible to obtain control systems with learning functions.

The purpose of this paper is to pinpoint some of these interesting developments that have taken place. The paper is organized as follows. The auto-tuner which is a simple and robust way to design systems with "push button tuning" is described in Section 2. The technique can also be used as a pre-tune mode for more complicated adaptive regulators. Conventional adaptive control based on recursive parameter estimation and control design is discussed in Section 3. The focus of the presentation is on algorithmic development. Some advances in adaptive control theory are presented in Section 4. This includes stability, convergence, robustness and universal stabilizers. Practical aspects on implementation of auto-tuning and adaptive systems are presented in Section 5. This is based on some published material on the commercial products and on my own experience. The discussion clearly indicates that there is a considerable amount of heuristics in current implementations. This serves as a motivation for Section 6 where it is attempted to combine algorithms and heuristics in an organized fashion by merging the fields of automatic control and expert systems. Some speculations on the future development of the field are given in the conclusions.

2. AUTOTUNING

For a long time the efforts in adaptive control were concentrated to comparatively complicated control systems. Only moderate interest were given to adaptation of simple controllers of the PID type. My own interest in this field started around 1980 when trying to respond to questions like the one posed by Ray Ash at CPCII: "Why don't you just provide an ordinary PID regulator with a tuning button?" A novel approach which solves this problem will be discussed in this section. This approach was originally presented in Åström and Hägglund (1983, 1984 a,b,c) and in Hägglund and Åström (1985 a,b,c). The approach was motivated by a desire to develop a simple robust tuning scheme which requires very little prior information. The approach is based on a special technique for

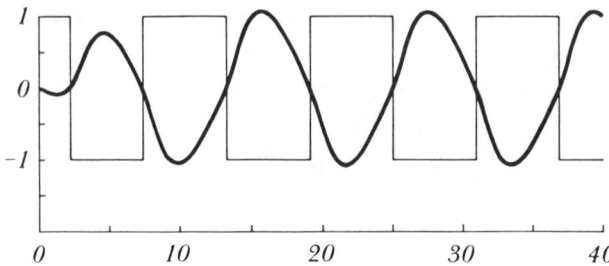

Fig. 1. Input and output signals for a system under relay feedback.

system identification which automatically generates an appropriate test signal and a variation of the the classical Ziegler-Nichols (1943) method for control design.

The Basic Idea

The Ziegler-Nichols method is based on the observation that the regulator parameters can be determined from knowledge of one point on the Nyquist curve of the open loop system. This point is the intersection of the Nyquist curve with the negative real axis. It is traditionally described in terms of the ultimate gain k_c and the ultimate period T_c. In the original scheme, described in Ziegler and Nichols (1943), the ultimate gain and period are determined in the following way: A proportional regulator is connected to the system. The gain is gradually increased until an oscillation is obtained. The gain k_c when this occurs is the critical gain and the oscillation has the critical period. It is difficult to perform this experiment automatically in such a way that the amplitude of the oscillation is kept under control.

The autotuner is based on the idea that the ultimate gain and the ultimate frequency can be determined by introducing relay feedback. A periodic oscillation is then obtained. The ultimate period T_c is simply the period of the oscillation and the critical gain can be determined from the relay amplitude and the amplitude of the oscillation, see Fig. 1.

If the process attenuates high frequencies so that the first harmonic component dominates the response it follows that the input and the output are out of phase. Furthermore if the relay amplitude is d it follows from a Fourier series expansion that the first harmonic of the input is $4d/\pi$. If the amplitude of the output is a the process gain is thus $\pi a/4d$ and the ultimate gain becomes

$$k_c = \frac{4d}{\pi a} \qquad (1)$$

Exact analyses of relay oscillations are also available. See Hamel (1949), Tsypkin (1958) and Åström and Hägglund (1984a). The period of an oscillation can be determined by measuring the times between zero-crossings. The amplitude may be determined from the peak-to-peak values of the output. These estimation methods are easy to implement because they are based on counting and comparison only. Simulations and extensive experiments on industrial processes have shown that the simple estimation method works well in comparison with the more sophisticated estimation methods. The simple methods also have some additional advantages, see Åström (1982).

Control Design

When the critical gain k_c and the critical period are known the parameters of a PID regulator can be determined by the Ziegler-Nichols rule which can be expressed as

$$k = \frac{k_c}{2} \qquad T_i = \frac{T_c}{2} \qquad T_d = \frac{T_c}{8} \qquad (2)$$

This rule gives a closed loop system which is sometimes too poorly damped. There are therefore many modifications of the basic Ziegler Nichols rule.

A block diagram of a control system with auto-tuning is shown in Fig. 2. The system can operate in two modes. In the tuning mode a relay feedback is generated as was discussed above. When a stable limit cycle is established its amplitude and period are determined as described above and the system is then switched to the automatic control mode where a conventional PID control law is used.

The tuner is very easy to use. The process is simply brought to an equilibrium by setting a constant control signal in manual mode. The tuning is then activated by pushing the tuning switch. Simplicity is the major advantage of the auto-tuner. It is very easy for the operator to use it. It is also easy to

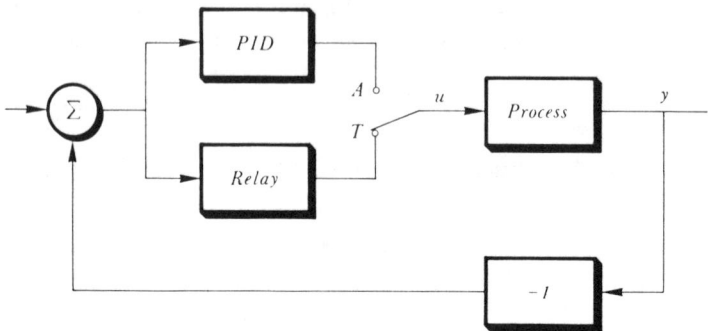

Fig. 2. Block diagram of an auto-tuner.
The system operates as a relay controller in the tuning mode (T) and as an ordinary PID regulator in the automatic control mode (A).

explain the auto-tuner to the instrument engineers. The properties of the autotuner are illustrated in Fig. 3, which shows an application to level control in three cascaded tanks. After bringing the system to an equilibrium the auto-tuner is initiated. The relay oscillation then appears. The amplitude measured in the first half-period indicates that the relay amplitude is too high. The relay amplitude is therefore reduced. When the oscillation has stabilized so that the amplitudes of two consequtive half periods are sufficiently close the critical gain and the critical period are determined and the regulator is switched to normal PID control. A set point change is later introduced manually. This shows that the tuning has resulted in a system with good transient behavior.

Prior Information

A major advantage of the autotuner is that it requires little prior information. Only two parameters the relay amplitude and the hysteresis width are required. In the NAF autotuner these parameters are set automatically. The relay amplitude is initially set to fixed proportion of the output range. The amplitude is adjusted after one half period to give an output oscillation of specified amplitude. The modified relay amplitude is stored for the next tuning. The hysteresis width is set automatically based on measurements of the measurement noise.

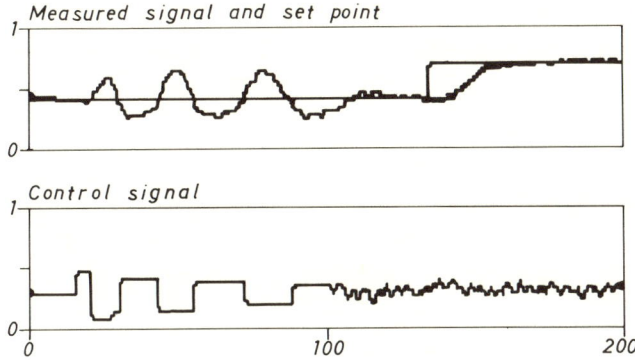

Fig. 3. Results obtained applying an auto-tuner to level control of three cascaded tanks.

Practical Aspects

There are several practical problems which must be solved in order to implement an auto-tuner. It is e.g. necessary to account for measurement noise, level adjustment, saturation of actuators and automatic adjustment of the amplitude of the oscillation. It may be advantageous to use other nonlinearities than the pure relay. A relay with hysteresis gives a system which is less sensitive to measurement noise. Measurement noise may give errors in detection of peaks and zero crossings. A hysteresis in the relay is a simple way to reduce the influence of measurement noise. Filtering is another possibility. The estimation schemes based on least squares and extended Kalman filtering can be made less sensitive to noise. Simple detection of peaks and zero crossings in combination with an hysteresis in the relay has worked very well in practice. See e.g. Åström (1982).

The process output may be far from the desired equilibrium condition when the regulator is switched on. In such cases it would be desirable to have the system reach its equilibrium automatically. For a process with finite low-frequency gain there is no guarantee that the desired steady state will be achieved with relay control unless the relay amplitude is sufficiently large. To guarantee that the output actually reaches the reference value, it may be necessary to introduce manual or automatic reset. It is also desirable to adjust the relay amplitude automatically. A reasonable approach is to require that the oscillation is a given percentage of the admissible swing in the output signal.

An Industrial Application

The concept of autotuning has been incorporated into a commercial regulator manufactured by NAF Controls in Sweden (Bååth and Hägglund, 1985). Figure 4 shows an application of this regulator to temperature control in a distillation column. The control loop considered had been behaving poorly for a long time. It was oscillating with the settings normally used ($K = 8$, $T_i = 2000$, and $T_d = 0$). At time 11.30 the regulator was switched to manual. Two hours later the output had settled reasonably well and the tuning was initiated at time 14.00. The logic for automatic selection of the noise limits and the relay amplitude took about an hour to settle. The measurement of the period and the amplitude of the oscillation started about time 15.00. The measurement was completed at time 20.00 and the regulator automatically switched to automatic control mode. Notice that the whole procedure was fully automatic from the time 14.00 when the tuning was initiated. Also notice that the severe disturbances at time 17.00 - 18.00 did not pose difficulties because of the robustness facilities built into the system. Finally observe the good performance of the regulator when the tuning was complete.

Extensions

There are several extensions of the simple auto-tuner. More information about the process characteristics can be extracted by analysing the waveform obtained under relay control. Improved design methods can also be obtained by measuring several points on the Nyquist curve. It is also easy to determine several points on the Nyquist curve by making relay feedback experiments with relays having modified characteristics. A relay with hysteresis has the describing function shown in Fig. 5A. A relay experiment with such a relay gives determines the intersection of the Nyquist curve with the describing function shown in Fig. 5A. The describing function can be translated vertically by changing the hysteresis width. By modifying the relay characteristics we can also obtain the characteristics shown in Fig. 5B. Several points can be

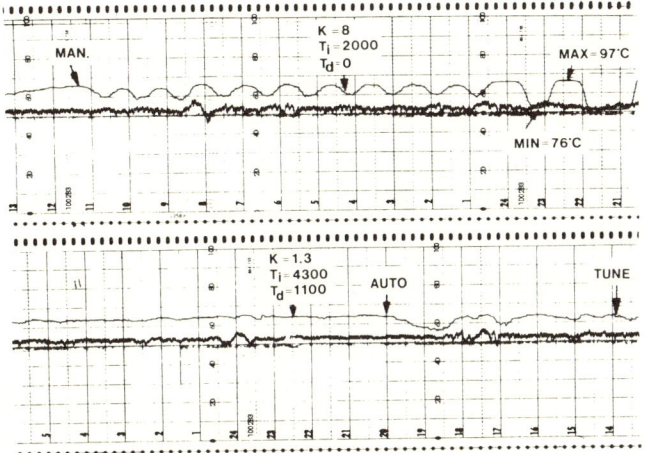

Fig. 4. Application of the auto-tuner to temperature control in a distillation column.

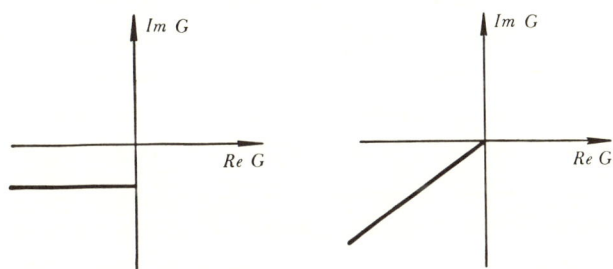

Fig. 5. Describing functions of relay with hysteresis and a relay with a modified hysteresis characteristics.

determined by changing the angle φ. Design techniques based on several points on the Nyquist curve are discussed in Åström and Hägglund (1984c), Hägglund and Åström (1985).

Auto-Tuning with Scheduling

Auto-tuning is a simple way to reduce uncertainty by experimentation. In many cases the characteristics of a process may depend on the operating conditions. If it is possible to measure some variable which correlates well with the changing process dynamics it is possible to obtain a system with interesting characteristics by combining the auto-tuner with a table look-up function. When the operating condition changes a new tuning is performed on demand from the operator. The resulting parameters are stored in a table together with the variable which characterizes the operating condition. When the process has been operated over a range covering the operating conditions the regulator parameters can be obtained from the table. A new tuning is then required only when other conditions change. A system of this type is semi-automatic because the decision to tune rests with the operator. The system will, however, continue to reduce the plant uncertainty.

3. ADAPTIVE CONTROL

A block-diagram of a conventional adaptive regulator is shown in Fig. 6. The adaptive regulator can be thought of as composed of two loops. The inner loop consists of the process and an ordinary linear feedback regulator. The parameters of the regulator are adjusted by the outer loop, which performs recursive parameter estimation and control design calculations. To obtain good estimates it may also be necessary to introduce perturbation signals. This function is not shown in Fig. 6 in order to keep the figure simple. Notice that the system may be viewed as automated modeling and design.

The block labeled "regulator design" in Fig. 6 represents an on-line solution to a design problem for a system with known parameters. This is called the underlying design problem. It is useful to consider this problem because it gives the characteristics of the system under the ideal conditions when the parameters are known exactly.

The adaptive regulator shown in Fig. 6 is very flexible. Both model reference adaptive system and self-tuning regulators can be represented by it. Many different design methods and many different parameter estimation schemes can be used. There are adaptive regulators based on phase- and amplitude margin design methods, pole-placement, minimum variance control, linear quadratic gaussian control and optimization methods. An interesting avenue which have not yet been pursued is to use robust design techniques which

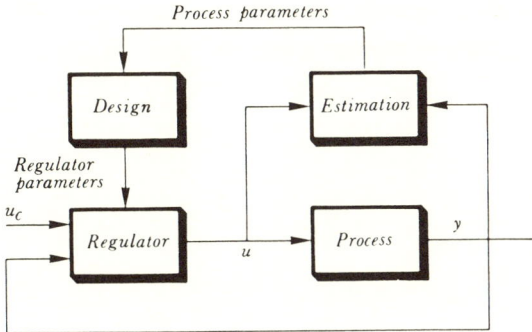

Fig. 6. Block diagram of a conventional adaptive regulator.

inherently will offer some insensitivity to modeling errors. Many different parameter estimation schemes have also been used, for example stochastic approximation, least squares, extended and generalized least squares, instrumental variables, extended Kalman filtering and the maximum likelihood method. See Åström (1983a) which gives an overview and many references. An example illustrates a typical case.

Example 1. Estimate the parameters of the second order model

$$y(t) + a_1 y(t-h) + a_2 y(t-2h) = b_1 u(t-h) + b_2 u(t-2h) \tag{3}$$

recursively. Let $\hat{a}_i$ and $\hat{b}_i$ denote the parameter estimates. The control law

$$u(t) = t_0 r(t) - s_0 y(t) - s_1 y(t-h) - r_1 u(t-h)$$

where

$$t_0 = (1+p_1+p_2)/(\hat{b}_1+\hat{b}_2)$$

$$r_1 = [(p_1-\hat{a}_1)\hat{b}_2^2 - (p_2-\hat{a}_2)\hat{b}_1\hat{b}_2]/N$$

$$s_0 = [(p_1-\hat{a}_1)(\hat{a}_2\hat{b}_1-\hat{a}_1\hat{b}_2) + (p_2-\hat{a}_2)\hat{b}_2]/N$$

$$s_1 = -\hat{a}_2 r_1/\hat{b}_2$$

$$N = \hat{b}_2^2 - \hat{a}_1\hat{b}_1\hat{b}_2 + \hat{a}_2\hat{b}_1^2$$

gives a closed loop system whose pulse transfer function from the command signal to the output is given by

$$H_m(z) = \frac{1 + p_1 + p_2}{b_1 + b_2} \cdot \frac{b_1 z + b_2}{z^2 + p_1 z + p_2}$$

where

$$p_1 = -2 e^{-\zeta \omega h} \cos \omega h \sqrt{1 - \zeta^2}$$

and

$$p_2 = e^{-2\zeta \omega h}$$

The closed loop system will thus retain the open loop zero and the closed loop poles correspond to a sampled second order system with bandwidth ω and relative damping ζ. □

Some minor modifications of the control law in the example are needed to handle bias and integral action. A detailed discussion of these factors is given in Åström (1979). The commercial regulators, Electromax V and TCS 6355 are based on estimation of parameters in the model (3). They do, however, use control design methods which are different from the one used in the example.

The self-tuner shown in Fig. 6 is called an indirect selftuner or an STR based on estimation of an explicit process model. It is sometimes possible to reparameterize the process so that it can be expressed in terms of the regulator parameters. This gives a significant simplification of the algorithm because the design calculations are eliminated. In terms of Fig. 5 the block labelled design calculations disappears and the regulator parameters are updated directly. This idea was used in the self-tuning regulator which is based on minimum variance control and least squares parameter estimation given in Åström and Wittenmark (1973). An example illustrates the idea which is also used in the ASEA Novatune.

<u>Example 2</u>. The self-tuner discussed in Åström and Wittenmark (1973) is based on the mathematical model

$$y(k+d) = s_0 y(k) + s_1 y(k-1) + \ldots + s_{n_s} y(k-n_s)$$
$$+ r_0 u(k) + \ldots + r_{n_r} u(k-n_r) + \varepsilon(k+d) \qquad (4)$$

where u is the control variable, y the measured output and ϵ is a disturbance. If ϵ is independent of the other terms on the right hand side the minimum variance control law for the plant (4) is simply

$$u(k) = -[s_0 y(k) + s_1 y(k-1) + \ldots + s_{n_s} y(k-n_s)$$
$$+ r_1 u(k-1) + \ldots + r_{n_r} u(k-n_r)]/r_0 \qquad (5)$$

The basic self-tuning algorithm can be described as follows:

Algorithm. Repeat the following steps at each sampling period:

Step 1. Update the estimates of the parameters of the model (4), so that a weighted sum of squares of the errors ϵ are minimal.

Step 2. Compute the control signal u(k) from past data y(k), y(k-1),...,u(k-1),... using (5) with the estimates obtained from Step 1. □

Notice that when least squares estimation is used the error $\epsilon(k+d)$ will be uncorrelated with the other terms in the right hand side of (4). Also notice that no design calculations are required since the parameters of the regulator (5) are obtained directly from the model parameters because of the special model structure used in (4).

Direct and Indirect Adaptive Control

An advantage of indirect adaptive control is that many different design methods can be used. The key issue in analysis of the indirect schemes is to show that the parameter estimates converge. This will in general require that the model structure used is appropriate and that the input signal is persistently exciting. To ensure this it may be necessary to introduce perturbation signals. The direct adaptive control schemes are simpler than the direct schemes. They may also work well even if the model structure used is not correct. The direct schemes will, however, require other assumptions.

Prior Knowledge

The parameter estimation step is a crucial part in all adaptive schemes. The sampling period is a critical parameter when discrete time models are fitted to data. The parameter estimation is insensitive to the sampling period if the true system is actually governed by a low order model. The sampling period is however critical when a low order model is fitted to a high order process. A low order model can be a very good approximation of a high order system if the sampling period is reasonably long. Results for short sampling periods can, however, be very poor because the parameters b_1 and b_2 will be underestimated, the computed gain becomes too high and the closed loop unstable. Experience indicates that it is not possible to obtain a good model (3) unless the order of magnitude of the sampling period is known. This means that it is not possible to construct a universal regulator for process control based on (3) unless some device for finding the sampling period is devised. For the regulator in Example 1 this can be achieved by relating the sampling period to the desired bandwidth and letting the operator choose it. The adaptive systems Electromax V and TCS 6355 both require prior knowledge of a time scale which among others is used to set the sampling period. A fairly elaborate "pretune" scheme is provided to determine the time scale by experimentation in both systems.

The self-tuning regulator given in Example 2 also requires prior knowledge. The following data is needed:

- h   sampling period
- d   delay in number of sampling periods
- n_r degree of the polynomial R
- n_s degree of the polynomial S
- λ forgetting factor
- θ_o initial estimate
- P_o initial covariance
- uh   high control limits
- ul   low control limits

The sampling period is critical as was discussed above. The integer d is also crucial. The closed loop system will become unstable if h and d are underestimated. The parameters are particularly important. Since the self-tuner is based on minimum variance control they will directly determine the closed loop bandwidth. The parameters n_r and n_s are not particularly critical. A calculation of covariances of inputs and outputs will show if they are too small, see Åström (1970). The parameter λ determines the trade-off between the tracking ability and the steady state variance of the recursive parameter estimator. The parameters θ_0 and P_0 determine the initial transient of the estimator but are otherwise unessential.

In control system design it is frequently necessary to make a trade-off between the response time and the size of the control signal. In minimum variance control this trade-off is made indirectly via selection of the sampling period. The regulator gain decreases and the response time increases with increasing sampling period. The minimum variance control law cannot handle nonminimum phase system because the process zeros are canceled by the controller. By increasing the sampling period and the delay d used in the adaptive control law the problems with nonminimum phase systems will, however, disappear. See Åström and Wittenmark (1985). Sampling of a stable system, with nonzero steady state gain, always gives a minimum phase sampled system provided the sampling period is sufficiently long. See Åström et al. (1984). This is also true for unstable systems provided that the unstability is caused by a single pole. The quality of the approximation by a low order system will also be improved when the sampling period is increased. The drawbacks with a long sampling period are slow responses to disturbances and changes in the set point. Notice that a sampled data system runs open loop between the sampling instants.

Predictive Control

There have recently been a considerable interest in adaptive regulators based on predictive control. Such regulators are based on estimation of models of the type

$$y(k+d) = s_0 y(k) + s_1 y(k-1) + \ldots + s_{n_s} y(k-n_s)$$
$$+ r_{-d} u(k+d) + \ldots + r_{-1} u(k+1)$$
$$+ r_0 u(k) + \ldots + r_{n_r} u(k-n_r) + \varepsilon(k+d) \quad (6)$$

The specifications are often expressed in terms of the desired step response of the closed loop system which is easy to describe to the operator. There are many different algorithms of this type e.g. the extended horizon minimum variance control (Ydstie, 1984) and extended prediction self-adaptive controls (de Keyser and Van Cauvenberghe, 1982, 1985; de Keyser et al., 1985). There are also variations based on linear quadratic optimization criteria. See Peterka (1984), the Musmar algorithm Mosca et al. (1982) and Lemos and Mosca (1985). These algorithms are also related to dynamic matrix control (Cutler and Ramaker, 1980) and model predictive control (Richalet et al., 1978), which is dealt with at length in Session III of this meeting. There are also multivariable extensions of the algorithms (Rouhani and Mehra, 1982).

4. THEORY

Theory has different roles in analysis and design of adaptive control systems. Analysis aimed at understanding specific algorithms is one goal. Creation of new adaptive control laws is another role. Adaptive systems are complex and difficult to analyse because they are inherently nonlinear. Progress in theory has been slow and much work remains before a reasonably complete coherent theory is available.

Because of the complex behavior of adaptive systems it is necessary to consider them from several points of view. Theories of nonlinear systems, stability, system identification, recursive estimation, convergence of stochastic algorithms and optimal stochastic control all contribute to the understanding of adaptive systems.

Generic Problems

A considerable effort has been devoted to construction of models which can serve as prototypes for general adaptive problems. The early work concentrated on systems where there was only a variation in the process gain. Much attention was later devoted to single-input single-output systems described by the equation

$$A(q)y(t) = B(q)u(t) + v(t) \qquad (7)$$

In this model u is the control variable, y is the measured output and v is a disturbance. A and B are polynomials in the forward shift operator i.e.

$$A(q) = q^n + a_1 q^{n-1} + \ldots + a_n \quad \text{and} \quad B(q) = b_0 q^n + \ldots + b_m$$

Multivariable systems where u and y are vectors and A and B are matrix polynomials have also been explored.

The model (7) represents a system where the system dynamics is totally unknown. In many applications the situation is quite different because the system is partially known. This situation has not been investigated much because each problem has a special structure.

It is customary to separate the tuning and the adaptation problem. In the tuning problem it is assumed that the process to be controlled has constant but unknown parameters. In the adaptation problem it is assumed that the parameters are changing. Many issues are much easier to handle in the tuning problem. The convergence problem is to investigate if the parameters converge to their true values. The corresponding problem is much more difficult in the adaptive case because the targets are moving. The estimation algorithms are the same in tuning and adaptation. They can be described by

$$\theta(t+1) = \theta(t) + P(t)\psi(t)[y(t+1) - \varphi(t)\theta(t)] \tag{8}$$

The gain matrix P behaves, however, very differently in the two cases. It goes to zero in the tuning case as t increases but it does not converge to zero in the adaptive case.

Stability

Stability is a basic requirement on a control system. Much effort has also been devoted to analysis of stability of adaptive systems. It is important to keep in mind that the stability concepts for nonlinear differential equations refer to stability of a particular solution. It is thus often the case that one solution is stable and another one unstable.

Stability theory has been the major inspiration for the development of model reference adaptive systems. Many attempts were made to provide stability proofs during the seventies. Several crucial issues were however overlooked and it was not until 1980 that correct stability proofs appeared. See Egardt (1979), Fuchs (1979), Goodwin et al. (1980), Gawthrop (1980), de Larminat

(1979), Morse (1980), and Narendra et al. (1980). An elegant formalism for the proof has recently been published by Narendra and Annaswamy (1984).

<u>Assumptions for stability proof</u>. The following assumptions are essential for the stability proof.

(A1) the relative degree d = deg A - deg B is known,
(A2) the sign of the leading coefficient b_0 of the polynomial B(q) is known,
(A3) the estimated model is at least of the same order as the process,
(A4) the polynomial B has all zeros inside the unit disc.

The stability theorems are important because they give simple and rigorous analysis of a reasonable adaptive problems. The assumptions required are, however, very restrictive.

The assumption A1 means for discrete systems that the time delay is known with a resolution of one sampling period. This is not unreasonable. For continuous time systems the assumption means that the slope of the high frequency asymptote of the Bode diagram is known. Together with assumption (A2) it also means that the phase is known at high frequencies. If this is the case, it is possible to design a robust high gain regulator for the problem, see Horowitz (1963), Horowitz and Sidi (1973). For many systems like flexible aircraft, electromechanical servos and flexible robots, the main difficulty in control is the uncertainty of the dynamics at high frequencies, see Stein (1980).

The assumption A2 was believed necessary for a while. A clever demonstration that this was not the case was published by Nussbaum (1983). Further exploration of Nussbaums results have given rise to the notion of universal stabilizers which will be discussed in more detail below.

Assumption A3 is very restrictive, since it implies that the estimated model must be at least as complex as the true system, which may be nonlinear with distributed parameters. Almost all control systems are in fact designed based on strongly simplified models. High frequency dynamics are often neglected in the simplified models. It is therefore very important that a design method can cope with model uncertainty, see Horowitz (1963). It was demonstrated by Rohrs et al. (1982) that instabilities could easily be generated if the assumption A3 is violated. This has generated a lot of research into the <u>robustness</u> of adaptive control. Several modifications of the algorithms have been proposed to improve robustness. One idea is to introduce a term $-a\theta$ to the right hand side of (8). This is called "leakage". Another idea is to filter the error and the regression vector φ in (8). A third idea is to introduce nonlinear modifications of the estimation algorithm. These issues are discussed at length in the monograph (Kosut et al., 1986).

Assumption A4 is also crucial. It arises from the necessity to have a model, which is linear in the parameters. It follows from the discussion in the Appendix that this is possible only if $B^- = b_0$. In other words the underlying design method is based on cancellation of all process zeros. Such a design will not work even for systems with known constant parameters if the system has an unstable inverse.

The analysis by Egardt (1979) also applies to the case when there are disturbances. Egardt has given counterexamples which show that modifications of the algorithms or additional assumptions are necessary if there are disturbances. One possibility is to bound the parameter estimates a priori for example by introducing a saturation in the estimator. Another possibility is to introduce a dead zone in the estimator which keeps the estimates constant if the residuals are small. These results also hold for continuous time systems as has also been shown by Peterson and Narendra (1982).

Instability Mechanisms

Apart from the stability proofs it is also useful to have an understanding of the mechanisms that may create instability. To develop this insight we will consider a simple model reference adaptive control problem which is described by the equations

$$\begin{aligned} y &= G(p)u \\ u &= \theta^T \varphi \\ \frac{d\theta}{dt} &= -k\varphi e \\ e &= y - y_m \end{aligned} \qquad (9)$$

where u is the process input, y the process output, y_m the desired model output, e the error and θ a vector of adjustable parameters. The transfer function of the process is G and $p = d/dt$ denotes the differential operator. The components of the vector φ are functions of the command signal, the system input and output. It follows from (9) that

$$\frac{d\theta}{dt} + k\varphi[G(p)\varphi^T\theta] = k\varphi y_m \qquad (10)$$

This equation gives insight into the behavior of the system.

<u>Slow adaptation</u>. Assume first that the adaptation loop is much slower than the process dynamics. The parameters then change much slower than the regressive vector and the term $G(p)\varphi^T\theta$ in (9) can then be approximated by its average

$$G(p)\varphi^T\theta \approx \overline{[G(p)\varphi^T(\theta)]}\theta \qquad (11)$$

Notice that the regression vector depends on the parameters. The following approximation to (10) is then obtained

$$\frac{d\theta}{dt} + k\varphi(\theta)\overline{[G(p)\varphi^T(\theta)]}\theta \approx k\varphi y_m \qquad (12)$$

This is the normal situation because the adaptive algorithm is motivated by the fact the parameters change slower than the other variables in the system under this assumption. Notice, however, that it is not easy to guarantee this.

Equation (12) is stable if $k\varphi[G(p)\varphi^T]$ is positive. This is true e.g. if G is SPR and if the input signal is persistently exciting.

<u>Fast Adaptation</u>. The approximation (12) is based on the assumption that the parameters θ change much slower than the other system variables. If the parameters θ change faster than φ then (10) can be approximated by

$$\frac{d\theta}{dt} + k\varphi\varphi^T G(p)\theta \approx k\varphi y_m \qquad (13)$$

A linearization for constant φ_0 shows that the stability is governed by the algebraic equation

$$\det[sI + k\varphi_0\varphi_0^T G(s)] = s^{n-1}[s+KG(s)] = 0 \qquad (14)$$

where I is the identity matrix and K is given by

$$K = k\varphi_0^T \varphi_0$$

is the equivalent adaptive loop gain. The stability can then be determined by a simple root-locus argument.

For sufficiently large $k\varphi_0^T\varphi_0$ the system will always be unstable if the pole-excess of G(s) is larger than or equal to 2. Also notice that the equivalent gain K is proportional to $\varphi_0^T\varphi_0$. The equivalent gain can thus be made arbitrarily large by choosing the command signal large enough. It thus seems intuitively clear that the adaptive system can be made unstable by making the command signal large enough.

Once the source of the difficulty is recognized it is easy to find a remedy. Since the equivalent gain K in the adaptive loop is too large because of its signal dependence, one possibility is simply to modify the parameter updating law to

$$\frac{d\theta}{dt} = -\frac{k}{1+\varphi^T\varphi}\varphi e$$

Equation (13) then holds with

$$K = k\frac{\varphi\varphi^T}{1+\varphi^T\varphi}$$

The equivalent gain in the adaptation loop is then bounded and the parameters θ will change arbitrarily slow for all signal levels. The actual value of the k can be chosen based on a simple root-locus argument for (14).

The modification of the parameter updating law has been used by many authors e.g. Narendra and Lin (1980). It is also worthwhile to note that a law of this type is obtained automatically when adaptive laws are derived from recursive estimation, see Åström (1983b). The high gain instability mechanism is the same as the one discussed in Cyr et al. (1983).

An Example

Many of the robustness issues can be illustrated by a simple example. Consider model reference adaptive control of a first order system. Let the

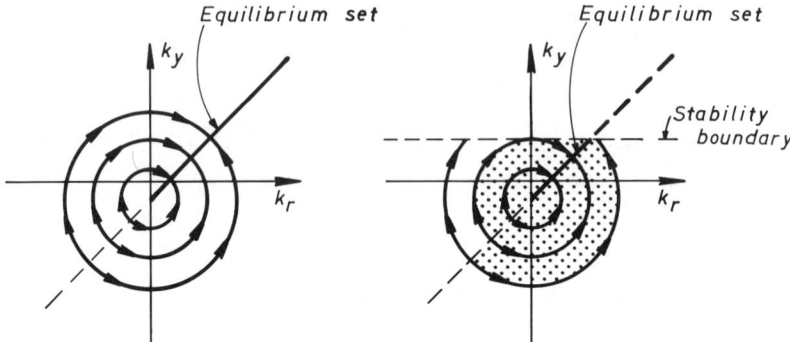

Fig. 7. Parameter trajectories for a model reference adaptive system with two parameters. Figure 7A shows the nominal case where the plant is of first order. Figure 7B shows the case when the plant is of higher order.

nominal plant be characterized by

$$G(s) = \frac{b}{s + a} \tag{15}$$

and the model by

$$G_m(s) = \frac{b_m}{s + a_m}$$

The closed loop system is then described by (9) with

$$\varphi = [r \quad -y]^T$$
$$\theta = [k_r \quad k_y]^T$$

If the command signal r is a step and if the adaptation is sufficiently small it was shown in Åström (1984) that the parameters follow the trajectories shown in Fig. 7A. A characteristic feature is that the equilibrium is not unique. The parameters can settle anywhere on the half-line shown in Fig. 7A. The reason for this is that the command signal is a step, which gives reliable information about the steady state gain. The parameters moves towards the equilibrium along arcs which are approximately circular. Notice also that the feedback gain k_y can

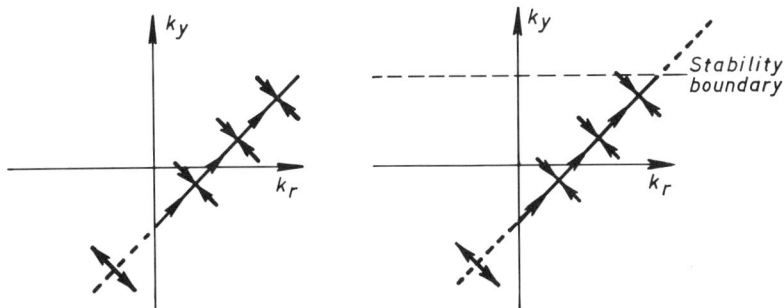

Fig. 8. Parameter trajectories of a model reference adaptive system with measurement noise.

be arbitrarily high if the initial conditions are chosen appropriately. This does not give rise to any problems in the nominal case. If the process to be controlled has additional dynamics which is not modeled by (15) like timedelays it may, however, be unstable when the feedback gain is sufficiently high. The closed loop system will then be unstable for sufficiently large initial values of the parameters as is shown in Fig. 7B. We can thus conclude that the adaptive system designed for a first order plant may be unstable when applied to a system with more complicated dynamics.

The case when there is measurement noise is shown in Fig. 8. The effect of measurement noise is that there will be a drift along the equilibrium line. The feedback gain will thus increase continuously . This will not give rise to difficulties in the nominal case. A plant with more complicated dynamics may, however, become unstable for high gains.

Having described the instability mechanisms we can now also discuss various measures used to improve the robustness. In Egardt (1979) it is shown that stability can, roughly speaking, be guaranteed even in the presence of disturbances by imposing one of the conditions

    a) Parameters are bounded.
    b) The parameters are not updated if the errors are small.

It seems intuitively reasonable that these conditions will help in the example discussed by keeping the parameters bounded we can avoid that the feedback gain k_y becomes too high. A device, which has been proposed to keep the parameters bounded, is to modify the equation for updating the parameters from

$$\frac{d\theta}{dt} = - k\varphi e$$

as in (9) to

$$\frac{d\theta}{dt} = - k\varphi e - \alpha\theta$$

This is referred to as introducing "leakage" in the estimator, see Ioannou and Kokotovic (1983). By stopping the updating when the error is small the drift of the parameters along the equilibrium line will also be eliminated. This is also referred to as a "dead zone". It was introduced in Egardt (1979) and has later been explored in Narendra and Petersen (1981). A technique of making the dead-zone adaptive is discussed in Goodwin (1986). All practical adaptive regulators have used some device of this nature to switch off the adaptation when there is little information to be gained from the process inputs and outputs.

System identification theory gives another way to explain the difficulty illustrated in Fig. 8. A step input is only persistently exciting of order 1. This means that only one parameter can be determined reliably and that any attempt to determine more parameters is futile. This can be used for diagnosis as discussed in Wittenmark and Åström (1984). It also suggests that the problem can be avoided by introducing perturbations which will allow all parameters to be reliably determined. This is discussed in Åström (1984). The usefulness of perturbations to gain useful information about the parameters is also suggested by dual control theory, see Åström (1983a).

Another interesting fact that has emerged from recent analysis is that there is a difference between the case of continuous time and discrete time regulators. In Rohrs et al. (1985) it is shown that unmodeled continuous dynamics is significantly reduced by the operation of sampling.

Universal Stabilizers

Adaptive control systems are nonlinear systems with a special structure. They are often designed based on the idea of automating modeling and design. It is natural to ask if there are other types of nonlinear controls which also can deal with uncertainties in the process model. A special class of systems were generated as attempts of solving the following problem which was proposed by Morse (1983). Consider the system

$$\frac{dy}{dt} = ay + bu$$

where a and b are unknown constants. Find a feedback law of the form

$$u = f(\theta,y)$$
$$\frac{d\theta}{dt} = g(\theta,y)$$

which stabilizes the system for all a and b. Morse conjectured that there are no rational f and g which stabilize the system. Morse's conjecture was proven by Nussbaum (1983) who also showed that there exist nonrational f and g which stabilize the system, e.g. the following functions

$$f(\theta,y) = (y-r)\,\theta^2 \cos\theta$$
$$g(\theta,y) = (y-r)^2$$

This correspond to proportional feedback with the gain

$$k = \theta^2 \cos\theta$$

The behavior of Nussbaum's regulator can be described as follows: Sweep the regulator gain k over positive and negative values. Find a way to stop the sweep rapidly if a stable system is obtained. Figure 9 shows a simulation of this control law applied to an integrator with unknown gain. Notice that the regulator is initialized so that the gain has the wrong sign. In spite of this the regulator recovers and changes the gain appropriately. Nussbaum's regulator is of considerable principal interest because it shows that the assumption A2 is not necessary. The control law is, however, not necessarily a good control law in a practical situation because it may generate quite violent control actions. The initial conditions for the simulation shown in Fig. 9 were in fact chosen quite carefully.

Nussbaums work has created a lot of interest. A clever multivariable extension is given by Mårtensson (1985 a,b).

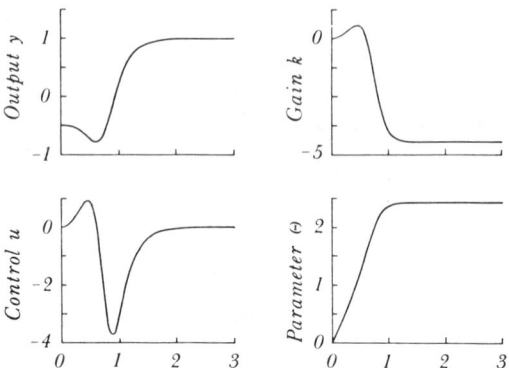

Fig. 9. Simulation of an integrator with Nussbaum's control law.

Convergence

The behavior of the parameters is an important issue in adaptive control. This is also a problem which has been the subject of much theoretical investigation. A typical approach is to assume that the system to be controlled is known and to investigate the behavior of the estimated parameters. The key problems have been investigated. Problems of this type have also been investigated in connection with determination of convergence conditions, possible convergence points and convergence rates. These problems have also been investigated in connection with system identification, see e.g. Åström and Eykhoff (1971). The result depends in complicated way on the process model, the disturbances and the estimation algorithm. There is, however, one complication in the adaptive case because the input to the process is generated by feedback. It is then more difficult to establish persistency of excitation. The feedback also makes the process input correlated with the disturbances.

A few simple observations can be made. If there are no disturbances, if the process input is persistently exciting and if the model structure is appropriate then the parameters can be determined exactly in a finite number of sampling periods. A recursive estimation algorithm which achieves this has time-variable gain. A constant gain algorithm will give exponential convergence. The situation is quite different when there are random disturbances. It is then necessary to have algorithms with decreasing gain in order to obtain estimates that converge.

The gain will typically decrease as 1/t. For algorithms whose gains do not go to zero the estimates will fluctuate. The magnitude of the fluctuations decreases with decreasing gain. Selection of suitable gains in adaptive control algorithms is thus a compromize between tracking rate and precision. When discussing convergence rates it is also important to keep in mind that performance measures are approximately quadratic functions of the parameter errors.

Parameterization

Parameterization is an important issue which enters many aspects of the adaptive control problem. The number of parameters is important. With fewer parameters to estimate less requirements are imposed on the input signal to achieve persistent excitation. For direct adaptive control it is also important to have a model which is linear in the parameters was also emphasized. Different parameterizations will thus lead to systems having different characteristics.

Finally it is worthwhile to observe that the formulation of a generic model like (7) with all parameters unknown is often a poor model because in practice it often happens that part of the dynamics is known.

5. PRACTICAL ASPECTS

Some practical aspects on the implementation of adaptive regulators will be given in this section. An ordinary PID-regulator is first discussed to provide some perspective. This regulator is ideally described by

$$u(t) = \left[e(t) + \frac{1}{T_i} \int^t e(s)ds + T_d \frac{de(t)}{dt} \right] \qquad (16)$$

The linear behavior of PID-control can be understood very well from this equation. Suitable values of the parameters can be determined. The performance of the closed loop system can be predicted etc. The actual operation of a PID regulator must however take nonlinear behavior into account. It is thus necessary to consider switching between manual and automatic operation and transients due to parameter changes. The actuators will saturate for some period in virtually all applications. This gives rise to problems with windup of the integrator. It is also becoming increasingly more common to connect PID regulators with logic selectors which brings up additional nonlinear problems. An operational industrial PID regulator thus consists of an implementation of the Equation (16) and some heuristic logic that takes care of the problems mentioned above. Although these heuristic factors are of extreme importance for good control they have not attracted much interest from theoreticians. They are instead hidden in practical designs and rarely discussed in the control literature. One reason for this is commercial secrecy, another is that most control engineers, being thoroughly indoctrinated by linear system theory, are poorly equipped to understand nonlinear phenomena. We can thus conclude that practical PID control is not solved by linear theory alone, but that nonlinearities plays an important role. They are typically handled by logic that surrounds the linear control law given by Equation (16). The logic is often designed heuristically.

Heuristic logic is even more important in adaptive control. The fundamental control law is much more complicated in this case. Windup can occur not only in the integrator but also in the estimator. Since there is a parameter estimator in the loop it is also necessary to safeguard against poor performance of the estimator due to poor data e.g. during an instrument failure. The adaptive algorithms also require some amount of apriori information. An example of the information needed to apply a general adaptive regulator was given in Section 3. To acquire this information it may be necessary to carry out a preliminary system identification phase. An empirical evidence of this is the pre-tune phase which exist in several commercial systems. To obtain a well functioning adaptive control system it is necessary to provide it with a considerable amount of heuristic logic. This goes under many names like safety nets or safety jackets. Experience has shown that it is quite time consuming to design and test this heuristic logic. Some practical issues are discussed in Wittenmark and Åström (1984). It is difficult to get information about what is actually done in practical systems because the manufacturers of adaptive systems are therefore understandably reluctant to disclose their tricks.

The key issues to get a robust controller are good data and an appropriate model structure. It is important that the model is accurate at the cross-over frequency. To obtain a good reduced order model it is essential that the input signal has sufficient energy content around the cross-over frequency and that it is so rich in frequency that it is persistently exciting. To guarantee a good model it is thus necessary to monitor the excitation and the energy of the input signal in the relevant frequency bands. A more detailed discussion is found in Åström (1984).

6. EXPERT CONTROL

The properties of auto-tuners and adaptive regulators are complimentary. The auto-tuner requires little prior information. It is very robust and it can generate good parameters for a simple control law. Adaptive regulators like model reference adaptive controllers or self-tuning regulators can use more complex control laws with potentially better performance. The self-tuners have local gradient procedures. Starting from reasonably good a priori guesses of system order, sampling period, and parameters, the algorithms can adjust the regulator parameters to give a closed loop system with very good performance. The algorithms will however not work if the prior guesses are too far off. With poor prior data they may even give unstable closed loop systems. This has led to the development of the safety jackets mentioned previously. The adaptive algorithms are also capable of tracking a system provided that the parameters do not change too quickly. It thus seems natural to try to combine auto-tuners and adaptive control algorithm. In Åström and Anton (1984) and Åström et al. (1986) it was proposed to use an expert system for this purpose.

Expert Systems

One objective for expert systems is to develop computer-based models for problem solving which are different from physical modeling and parameter estimation. See Barr and Feigenbaum (1982), Davis (1982), and Hayes-Roth et al. (1983) It attempts to model the knowledge and procedures used by a human expert in solving problems within a well-defined domain. Knowledge representation is a key issue in expert systems. Many different approaches have been attempted such as first order predicate calculus (logic), procedural representations, semantic networks, production systems or rules, and frames. The architecture of a knowledge-based is shown in Fig. 10. It consists of a knowledge

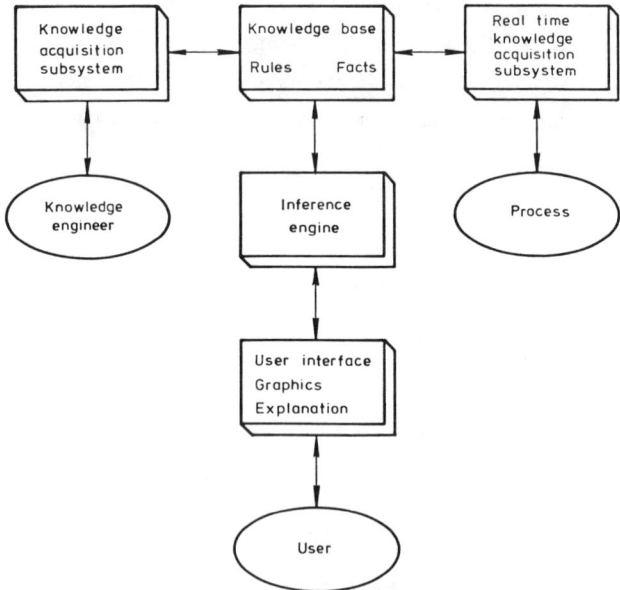

Fig. 10. A knowledge based expert system.

base, an inference engine and a user interface.

The Knowledge Base

The knowledge base consists of data and rules. The data can be separated into facts, and goals. Examples of facts are statements like "the control variable is in the range 0 to 50", "there is hysteresis in the actuator", "the system appears to be stable", "PI control is adequate", "deviations are normal". Typical examples of goals are "minimize the variations of the output", "maintain steady state control at specified limit", "find out if gain scheduling is necessary" or "find a scheduling table". Data is introduced into the database by the user or via the real time knowledge acquisition system. New facts can also be created by the rules.

The rulebase contains the production rules of the type: "if <premise> then <conclusion> do <action>". The <premise> represents facts or goals from the database. The <conclusion> can result in a new fact being added to the data base or a modification of an existing fact. The <action> can be to activate an algorithm for diagnosis, control or estimation. These actions are different from those found in conventional expert systems. The rulebase is often structured in groups or knowledge sources that contain rules about the same subject. This simplifies the search.

In the control application the rules represent the skills about the control and estimation problem that we want to build into the system. This includes the appropriate characterization of the algorithms, judgemental knowledge on when to apply them and supervision and diagnosis of the system. The rules are introduced by the knowledge engineer via the knowledge acquisition system, which assists in writing and testing rules.

Inference Engine

The inference engine processes the rules to arrive at conclusions or to satisfy goals. It scans the rules according to a strategy which decides from the context (current data base of facts and goals) which production rules to select next. This can be done according to different strategies. In forward chaining it is attempted to find all conclusions from a given set of premises. This is typical for a data driven operation. In backward chaining the rules are traced backward from a given goal to see if it can be supported by the current premises. This is typical for a diagnosis problem. The search can be organized in many different ways depth first or breadth first. There are also strategies that use the complexity of the rules to decide the order in which they are searched. To devise efficient search procedures it is often convenient to decompose the rule base into pieces dealing with related chunks of knowledge. If the rules are organized in that way it is also possible for a system to focus its attention on a collection of rules in certain situations. This can make the search more efficient.

User Interface

The user interface of a production system can be divided into two parts. The first part is the development support that the system gives. This contains tools such as rule editor and rule browser for development of the system knowledge base. The other part is the runtime user interface. This contains explanation facilities that makes it possible to question how a certain fact was concluded, why a certain estimation algorithm is executing etc. It is also possible to trace the execution of the rules. The user interface wan also contain facilities to deal with natural language. Fancy graphics can also be helpful.

Planning

Expert control contains an element of planning. Consider for example the actions to be taken at ank on-line fault, or when it is desired to change operating conditions. The development of a suitable plan of actions may be viewed as a search through a large network to reach the desired goal. This searching and planning in a complex environment is a fundamental activity in AI systems.

Real Time Expert System

Expert systems normally interact via an operator who gives premises and goals. An interesting aspect of the expert control systems is that they can acquire knowledge automatically from the environment by injecting signals into a system and observing responses. Premises can also be generated automatically by signals from the sensors. It may take a long time to search through a large rule base. In an expert control system it may also happen that premises change with time. This poses significant problems.

Expert Control

The idea of expert control is to have a collection of algorithms for control, supervision and adaptation which are all supervised by an expert system. This offers several interesting possibilities. It was mentioned in Section 4 that heuristic logic is important for ordinary PID regulators and even more so for adaptive regulators. The logic shows up as if-then-else or case statements in the regulator code. In many cases the code for the logic is larger than the code for the control algorithm. The debugging, modification, and testing of the control logic can be very time consuming. An expert system is a very convenient way to implement this logic even if it is an overkill for PID control. In Åström

(1983b) it is shown that the logic for an auto-tuner is very conveniently implemented using an expert system.

An expert system has the interesting ability to explain its reasoning. This offers interesting possibilities for the control problem. We can thus get answers to questions like. What control law is beeing used? Why was this control law chosen? What is the current knowledge of the process and its environment? Are the fluctuations in the process output normal? The word 'expert control' has also been used in other contexts. Moore et al. (1984 a,b) have proposed to use the expert system in a supervisory mode as control advisors and alarm advisors. Other applications are given in Trankle and Markosian (1985) and Sanoff and Wellstead (1985).

An Example

The notion of expert control is illustrated by an example. Consider a simple regulation loop where the goal is to keep the process output close to a set point for a wide range of operating conditions. A list of the major operations in the system is given below.

```
MainMonitor:
    StabilitySupervisor
    ComputeMeansAndVariances
AutoTuning
    Tune
    KcTcEstimator
    DeterminePidStructure
    EstimateTimeDelay
BackUpControl:
    PidControl
    PidSupervisor
FixedGainMinimumVarianceControl:
    MinimumVarianceControl
    MinimumVarianceSupervisor
        RingingDetector
        DegreeSupervisor
Estimation:
    ParameterEstimation
    EstimationSupervisor
        ExcitationSupervisor
        PerturbationSignalGenerator
        JumpDetector
```

>   DriftDetector
> SelfTuning:
>   SelfTuningRegulation
>   SelfTuningSupervisor
> Learning:
>   GetRegulatorParameters
>   SmoothAndStoreRegulatorParameters
>   TestSchedulingHypothesis

The following discussion explains some of the operators or actions that are used in the system. The "action" MinimumVarianceControl is a primary function of the regulator. The preconditions for this action include knowledge of an appropriate sampling period and models for the process and the disturbances. The process zeros are cancelled in minimum variance control. This may lead to ringing if the cancelled zeros are not sufficiently well damped. To detect ringing and to take the appropriate actions it is useful to include a RingingDetector. Ringing can be avoided by increasing the parameter d or by increasing the sampling period h, see Åström and Wittenmark (1985). There is a convenient way to find out if a process is under minimum variance control simply by calculating the autocorrelation of the process output, see Åström (1970). This can be used in the MinimumVarianceSupervisor.

If the process model required for minimum variance control is not available a self-tuning regulator may be used. This requires certain preconditions as was discussed in Section 3. If the prior information for a self-tuner is not available it can be attempted to use an auto-tuner, which requires less prior information. The data obtained from the auto-tuning experiment can be used to generate initial conditions for the self-tuner. The performance of a self-tuner depends critically on the process being properly excited. An ExcitationSupervisor can check this. If there is not enough excitation there are two options. Either to stop the updating or to introduce perturbation signals. using a PerturbationSignalGenerator. Other functions may also be provided. Assume that it is known that the process dynamics changes with a few parameters like production. Gainscheduling and learning may then be considered. This is done by storing control parameters for different operating conditions in tables.

TABLE 1   Main Monitoring Table

An entry is made whenever there is a mode switch or a set-point change.

#	Time	u	σ_u	y	σ_y	Stable	Regulator type

Process data is stored in lists in the system data base. It is convenient to have event lists associated which each of the knowledge sources listed above. There will thus be a main monitoring table a minimum variance control table an auto-tuning table etc. A typical example of such a table is given in Table 1. An entry is made in this table when there is a major event in the system e.g. a set point change, a tuning, a switching of control modes etc.

It may be useful to add a few entries in the table such as max and min values or percentile values. From the data shown in Table 1 it is possible to make deductions like: What are the relations between the mean values of u and y? Do these relations change with time? Are there any relations between the standard deviations and the mean value of the control signal? What are the patterns of the mode switches? Does the system go to tuning mode after large set point changes? What control modes are used for most of the time? Are these drastic variations in performance with time and modes? The answers to these questions will allow us to make inference about the characteristics of the process.

A prototype system of the type outlined above has been implemented by Årzén using a VAX 11/780 running under VMS is used. The expert system is implemented in Lisp with the algorithms written in Pascal. Parallel processes are implemented using the VMS mail box facility. The expert system framework OPS4 is used. The design and some experiments are described in Årzén (1986).

7. CONCLUSIONS

Control of systems with unknown parameters has been approached from two points of view automatic tuning and adaptive control. It has been demonstrated that both approaches lead to controllers which contain numerical algorithms as well as heuristic logic. The approaches are also complementary with respect to the prior information needed. It has been suggested to use an expert system to coordinate the different techniques and to add facilities like monitoring and tables for storing information about the process and its control system. The approach which clearly can be applied to a wide variety of problems seems to offer interesting possibilities to combine analytical and heuristic approaches. The incorporation of heuristics through AI structures results in systems that are far more flexible and transparent than selector and safety-jacket logic. Experience from building expert systems for real applications has shown that their power is most apparent when the problem considered is sufficiently complex. This paper has pointed out that an expert system can provide a framework for blending numerical algorithms with this detailed knowledge of dynamics and process control. This results in a feedback system with many interesting features which includes learning, store of increased process knowledge and explanatory power.

Acknowledgement

This work was partially supported by the National Swedish Board of Technical Development (STU) under contract no. 85-3225.

REFERENCES

Årzén, K.E. (1986). Experiments with expert systems for process control. To appear in Proc. 1st Int. Conf. on Applications of Artificial Intelligence to Engineering Practice, Southampton, UK, 1986.

Åström, K.J. (1970). Introduction to Stochastic Control Theory. Academic Press, New York.

Åström, K.J. (1979). Simple self-tuners 1. Report CODEN: LUTFD2/TFRT-7184, Department of Automatic Control, Lund Institute of Technology, Lund, Sweden.

Åström, K.J. (1982). Ziegler-Nichols auto-tuners. Report CODEN: LUTFD2/ TFRT-3167, Dept. of Automatic Control, Lund Institute of Technology, Lund, Sweden.

Åström, K.J. (1983a). Theory and applications of adaptive control - A survey. Automatica, 19, 471-486.

Åström, K.J. (1983b). Implementation of an auto-tuner using expert system ideas. Report CODEN: LUTFD2/TFRT-7256. Dept. of Automatic Control, Lund Institute of Technology, Lund, Sweden.

Åström, K.J. (1984). Interactions between persistent excitation and unmodeled dynamics in adaptive control. Proc. 23rd IEEE Conf. on Decision and Control.

Åström, K.J., and J.J. Anton (1984). Expert Control. IFAC World Congress, Budapest.

Åström, K.J., and P. Eykhoff (1971). System identification - A survey. Automatica, 7, 123-162.

Åström, K.J., and T. Hägglund (1983). Automatic tuning of simple regulators. Preprints IFAC workshop on Adaptive systems in control and signal Processing, San Francisco, CA.

Åström, K.J., and T. Hägglund (1984a). Automatic tuning of simple regulators. Proc. IFAC 9th World Congress, Budapest, Hungary.

Åström, K.J., and T. Hägglund (1984b). Automatic tuning of simple regulators with specifications on phase and amplitude margins. Automatica, 20, 645-651.

Åström, K.J., and T. Hägglund (1984c). Automatic tuning of simple regulators. In I.D. Landau et al. (Eds.). Adaptive Systems in Control and Signal Processing 1983. Proc. IFAC Workshop, San Francisco, USA. Pergamon Press, 1984.

Åström, K.J., and B. Wittenmark (1973). On self-tuning regulators. Automatica, 9, 185-199.

Åström, K.J., and B. Wittenmark (1985). The self-tuning regulators revisited. Proc. 7th IFAC Symp. Identification and System Parameter Estimation, York, UK.

Åström, K.J., P. Hagander, and J. Sternby (1984). Zeros of sampled systems. Automatica, 20, 31-38.

Åström, K.J., J.J. Anton, and K.E. Årzén (1986). Expert control. Automatica, 22:3.

Bååth, L., and T. Hägglund (1985). NAF Controls autotuner. Report NAF Controls AB, Solna, Sweden.

Barr, A., and E.A. Feigenmaum (Eds.) (1982). The Handbook of Artificial Intelligence, Volume 2. William Kaufmann, Inc., Los Altos, California.

Cutler, C.R., and B.C. Ramaker (1980). Dynamic matrix control - A computer control algorithm. Joint Automatic Control Conference, San Francisco, paper WP5-B.

Cyr, B., B. Riedle, and P. Kokotovic (1983). Hopf bifurcation in an adaptive system with unmodeled dynamics. Proc. IFAC Workshop on Adaptive Systems in Control and Signal Processing, San Francisco, pp. 20-22.

Davis, R. (1982). Expert systems. Where are we? And where do we go from here?. The AI Magazine, Spring 1982, 3-22.

De Keyser, R.M.C., and A.R. Van Cauwenberghe (1982). Simple self-tuning multistep predictors. In G.A. Bekey, G.N. Saridis (Eds.), Identification and System Parameter Estimation - 6th IFAC Symposium, Oxford, Pergamon Press, 1558-1563.

De Keyser, R.M.C., and A.R. Van Cauwenberghe (1985). Extended prediction self-adaptive control. Invited session on Applications of Adaptive and Self-tuning Control, Proc 7th IFAC/IFORS Symp on Identification and System Parameter Estimation, York, UK.

De Keyser, R.M.C., and Ph.G.A. Van de Velde, and F.A.G. Dumortier (1985). A comparative study of self-adaptive long-range predictive control methods. Proc 7th IFAC/IFORS Symp on Identification and System Parameter Estimation, York, UK.

Dumont, G. (1986). A survey of the applications of adaptive control in the process industries. This conference.

Egardt, B. (1979). Stability of Adaptive Controllers. Lecture notes in Control and Information Sciences, Vol. 20, Springer-Verlag, Berlin.

Fuchs, J.J. (1979). Commande adaptative directe des systemes linéaires discrets. Theses D.E., Univ. de Rennes, France.

Gawthrop, P.J. (1980). On the stability and convergence of a self-tuning controller. Int. J. Control, 31, 973-998.

Goodwin, G.C. (1986). Digital adaptive control - A robust approach. This conference.

Goodwin, G.C., P.J. Ramadge, and P.E. Caines (1980). Discrete-time multivariable adaptive control. IEEE Trans. Aut. Control, AC-25, 449-456.

Hägglund, T., and K.J. Åström (1985a). Automatic tuning of PID controllers based on dominant pole design. IFAC Workshop on Adaptive Control of Chemical Processes, Frankfurt/Main, FRG.

Hägglund, T., and K.J. Åström (1985b). "Förfarande för inställning av en PID regulator för en process", Swedish patent 8104989-2.

Hägglund, T., and K.J. Åström (1985c). "Method and an apparatus in tuning a PID regulator", American patent 4549123.

Hamel, B. (1949). Contribution a l'etude mathematique des systemes de reglage par tout-ou-rien. C.E.M.V. Service Technique Aeronautique, 17.

Hayes-Roth, F., D. Waterman and D. Lenat (1983). Building Expert Systems. Addison-Wesley, Reading, Mass.

Horowitz, I.M. (1963). Synthesis of Feedback Systems. Academic Press, New York.

Horowitz, I.M., and M. Sidi (1973). Synthesis of cascaded multipleloop feedback systems with large plant parameter ignorance. Automatica, 9, 589-600.

Ioannou, P.A., and P.V. Kokotovic (1983). Adaptive Systems with Reduced Models. Lecture Notes in Control and Information Sciences, vol. 47, Springer-Verlag.

Kosut, R.L., P.V. Kokotovic, B.D.O. Anderson, R.R. Bitmead, B. Riedle, C.R. Johnson Jr., L. Praly, and I. Mareels (1986). Stability of Adaptive Systems: Passitivity and Averaging Analysis. MIT Press, to appear.

Larminat, Ph. de (1979). On overall stability of certain adaptive control systems. 5th IFAC Symp. on Identification and System Parameter Estimation, Darmstadt, FRG.

Lemos, J.M., and E. Mosca (1985). A multipredictor-based LQ self-tuning controller. Proc 7th IFAC/IFORS Symp on Identification and System Parameter

Mårtensson, B. (1985a). The order of any stabilizing regulator is sufficient a priori information for adaptive stabilization. System & Control Letters, 6, 87-91.

Mårtensson, B. (1985b). Adaptive stabilization of general multivariable, continous- or discrete time systems. 7th Int. Symp. on Mathematical Theory of Networks and Systems (MTNS-85), Stockholm, Sweden.

Moore, R.L., L.B. Hawkinson, and C.G. Knickerbocker (1984a). Expert systems applications in industry. Proc. of the ISA International Conference and Exhibit, Houston, Texas, vol. 39, part 2, Oct 22-25.

Moore, R.L, L.B. Hawkinson, C.G. Knickerbocker and L.M. Churchman (1984b). A Real-Time Expert System for Process Control. LISP Machine, Inc.

Morse, A.S. (1980). Global stability of parameter-adaptive control systems. IEEE Trans. Aut. Control, AC-25, 433-439.

Morse, A.S. (1983). Recent problems in parameter adaptive control. In I.D. Landau, Ed., Outils et Modeles Mathematiques pour l'Automatique, l'Analyse de Systemes et le Traitement du Signal. Editions NRS, vol. 3, pp. 733-740.

Mosca, E., G. Zappa, and C. Manfredi (1982). Progress on multistep horizon self-tuners: the MUSMAR approach. Ricerche di Automatica, XIII, 85-105.

Narendra, K.S., and A.M. Annaswamy (1984). A general approach to the stability analysis of adaptive systems. Proc. 23rd IEEE Conf. Decision and Control, Las Vegas, Nevada, pp. 1298-1303.

Narendra, K.S., and Y.H. Lin (1980). Design of stable model reference adaptive controllers. Applications of Adaptive Control. Academic Press, New York.

Narendra, K.S., and B.B. Peterson (1981). Adaptive control in the presence of bounded disturbances. Proc. 2nd Yale Workshop on Applications of Adaptive Control.

Narendra, K.S., Y.-H. Lin, and L.S. Valavani (1980). Stable adaptive controller design - Part II: Proof of Stability. IEEE Trans. Aut. Control, AC-25, pp. 440-448.

Nussbaum, R.D. (1983). Some remarks on a conjecture in parameter adaptive control. Systems & Control Letters, 3, 243-246.

Peterka, V. (1984). Predictor-based self-tuning control. Automatica, 20, 39-50.

Peterson, B.B., and K.S. Narendra (1982). Bounded error adaptive control. IEEE Trans. Aut. Control, AC-27, 1161-1168.

Richalet, J., A. Rault, J.L. Testud, and J. Papon (1978). Model predictive heuristic control: Applications to industrial processes. Automatica, 14, 413-428.

Rohrs, C.E., G. Stein, and K.J. Åström (1985). Uncertainty in sampled systems. Proc. American Control Conference, Boston, Massachusetts.

Rohrs, C.E., L. Valavani, M. Athans, and G. Stein (1982). Robustness of adaptive control algorithms in the presence of unmodeled dynamics. Proc. 21st IEEE Conf. Decision and Control, Orlando, Florida.

Rouhani, R., and R.K. Mehra (1982). Model algorithmic control - Basic properties. Automatica, 18, 401-414.

Sanoff, S.P., and P.E. Wellstead (1985). Expert identification and control. Preprints 7th IFAC/IFORS Symp. on Identification and System Parameter Estimation, York, UK.

Stein, G. (1980). Adaptive flight control: A pragmatic view. In K.S. Narendra and R.V. Monopoli, Eds., Applications of Adaptive Control. Academic Press, New York, pp. 291-312.

Trankle, T.L., and L.Z. Markosian (1985). An expert system for control system design. Proc. IEE Int. Conf. Control 85, Cambridge, UK.

Tsypkin, J.A. (1958). Theorie der Relais Systeme der Automatischen Regelung. R. Oldenburg, Munich.

Wittenmark, B., and K.J. Åström (1984). Practical issues in the implementation of self-tuning control. Automatica, 20, 595-605.

Ydstie, B. (1984). Extended horizon adaptive control. Proc. 9th IFAC World Congress, Budapest, Hungary, paper 14.4/E-4.

Ziegler, J.G., and N.B. Nichols (1943). Optimum settings for automatic controllers. Trans. ASME, 65, 433-444.

ON THE USE OF ADAPTIVE CONTROL IN THE PROCESS INDUSTRIES

Guy A. Dumont
University of British Columbia, Vancouver, BC, Canada

Abstract. In 1981, it appeared that adaptive control was on the verge of becoming widely applied in the process industries. The purpose of this paper is to find, five years later the extent to which this prediction has been fulfilled. The answer is, by and large: to a very limited extent. Three sectors, pulp and paper, chemical and petrochemical have been surveyed for industrial applications of adaptive control. Sectoral and geographical differences are evident in the results and are tentatively explained. Although not used much, adaptive control is perceived by the industrials as useful, but still too complicated to use. Closer cooperation between industrials, vendors and academia is needed. Researchers should, wherever required unify the theory and address such issues as use of a-priori knowledge and choice of time-scale. Only then will adaptive control find a niche as a standard industrial tool.

Keywords. Adaptive Control; Chemical Industry; Control Applications; Industrial Control; Paper Industry; Petrochemical Control; Process Control; Pulp Industry; Self-tuning Regulators

INTRODUCTION

"As an overall statement, notwithstanding some good demonstrations, process control has missed the modern control bandwagon. Now it is time to catch it." (Fjeld, 1981)

"The theoretical advances, together with the practical lore coming out of further pilot applications, should make adaptive control a fairly common tool within a few years." (Bélanger, 1981)

Those statements were made at the last Chemical Process Control Conference, in Sea Island. Now, 5 years later, what is the status of adaptive control applications in the process industries? Have the pilot plant applications really led to full scale industrial applications? Has the industry caught the adaptive control bandwagon?

The purpose of this paper is to bring answer to these questions. First and foremost it is the result of a fact finding mission, this is not an exhaustive survey. For this, the reader is referred to the excellent survey paper by Seborg et al. (1986), which , in addition to a tutorial on adaptive control, gives rather complete lists of reported applications of adaptive control to pilot and industrial plants. This paper is limited to three industries, pulp and paper, chemical and petrochemical, primarily North American. As is well known, some companies within these sectors of activity tend to keep many of their activities proprietary, and thus this survey cannot claim to be exhaustive. However, I feel that it gives a rather accurate picture of the present level of applications of adaptive control. Various methods were used to gather that information. First, a questionnaire was sent to about sixty industrial users, including vendors, in North-America and Europe. Replies were received from about half of those. Through my daily activities, I already had a rather good idea of the situation in the pulp and paper industry. To compensate for this lack of day-to-day rapport with the chemical and petrochemical sectors, I paid a visit to the control engineering groups of two major US chemical companies and a major US oil company.

The paper is organized as follows. The next section attempts to draw a picture of the extent of application of adaptive control in the three sectors. In the discussion, we then try to summarize the industrial perception of adaptive control and we follow with a critical review of the past five years of adaptive control literature. Finally, some recommendations are made to the three parties involved, industry, vendors and academia, as well as some suggestions for research aimed at improving the current adaptive control methods, some of which is already taking place.

SURVEY RESULTS

Pulp and Paper Industry

Of the three industries covered in this survey, the pulp and paper industry has the largest number of reported applications of adaptive control. There is obviously the pioneering work done in Sweden a decade ago, see Borisson and Wittenmark (1974) and Cegrell and Hedqvist (1975). However, this did not lead to continuous industrial applications. In fact, it is only recently that the industry has manifested interest to apply self-tuning control to a paper machine. The first application reported by a supplier of turnkey control systems to the industry can be found in Fjeld and Wilhelm (1981), Wilhelm (1982) and Kelly (1982), where application to machine direction control of moisture on a fine paper machine is presented. As in the pioneering Swedish work, feedforward terms are included to decouple the moisture from the stock flow. The Clarke-Gawthrop algorithm, in its velocity form with fixed forgetting factor is used along with undisclosed proprietary modifications. This scheme is said to further reduce the moisture variations by 50% compared to conventional control. This industrial scale trial has now led to two industrial applications. One is for the control of moisture and basis weight on a tissue machine with steam heated Yankee drier and gas fired hood. The other is for the control of moisture on a multiple headbox paperboard machine. The previous adaptive scheme has been slightly enhanced by the use of a variable forgetting factor. For this application, the main advantage of using adaptive control is the ability of tracking plant and noise dynamics that change significantly from grade to grade. Both systems have been in continuous operation for about a year. This control scheme is however not yet available as an off-the-shelf product.

Another application from the same group is reported in Wilhelm and Fjeld (1983) and Wilkinson (1983). It concerns the cross-direction distribution of fibers to control basis weight and moisture profiles on fourdrinier paper machines. The spatial response of profile to headbox slice motion is a complex function of the behavior of the mechanical system made up by the actuator and the slice and of the hydrodynamic behavior of the stock on the wire. This function is strongly affected by the operating point of many variables, the machine condition and stock properties. Because of the slow control sampling rate on this loop, this is a steady-state control problem where the purpose is to solve the decoupling problem between the numerous actuators. The experimental system characterized the process by a large band diagonal matrix with three estimated parameters, and then used a generalized minimum-variance control law to compute the actuator movements required to obtain the desired profile. This system has been commercially available from Accuray for over two years, as an off-the-shelf product known as the Profile-Manager^{TM}. In a newer version, called Smooth-Profiler^{TM}, an indirect adaptive control scheme is used, where the 8 coefficients of a non-causal, spatial ARMAX model are identified. The current system now also incorporates constraints on bending moments of the slice lip and hard constraints on actuator movement, as well as relative weighing of basis-weight and moisture, for more details see Fjeld (1986). Many applications of both systems are reported around the world.

An application of STC to machine direction moisture control on a linerboard machine is reported in Sikora (1983) and Sikora and Bialkowski (1984). The design is based on the Clarke-Gawthrop algorithm with fixed forgetting factor, with feedforward from the couch vacuum, machine speed and moisture setpoint. This latter feature, used in conjunction with a complicated logic greatly added to the complexity of the controller (up to 28 parameters had to be estimated!) and was probably responsible for the eventual failure of the project. Industrial trials for this application took over a year of time-consuming tuning and painstaking patching of the original scheme. After that period, the STC was judged to be not robust enough for industrial operation and the potential reduction in moisture variance too small and the project was abandonned. This is a typical case of overkill, where the designer wants the STC to achieve too many things. This project failed due to the lack of identifiability caused by the large number of parameters estimated on-line as well as by the lack of persistent excitation of most of the feedforward variables. For example, it should have been realized that, whereas on lighter grades the couch vacuum contains information, on a thick grade like linerboard, it is basically flat and should not have been used. It should be noted that simple techniques, such as a dead zone on the prediction error or simple monitoring of the inputs are available to avert this kind of problem.

Other applications of adaptive control in papermaking have been reported, all involving the ASEA Novatune system, which has now been commercially available for a number of years. The most popular application of this system is for the control of reel density and web tension on roll winders, see Eriksson et al. (1983). Because the process dynamics change continuously during the winding operation, the self-tuning controller outperforms conventional PID control. About a dozen applications of this system have been reported worldwide. It must be noted that vendor field engineers spend several weeks commissioning and tuning the system, with very little involvement from the customer. It is thus hard to say how much of the improvement in performance is due to the adaptive control law and how much is due to improved sensors and actuators or the general strategy. The signal used for density feedback is computed assuming constant basis weight of the parent reel. However, the actual basis weight is variable and thus the density calculation is erroneous. In conventional feedback control this is troublesome, but in adaptive control, where this signal is used for on-line identification, this can become an acute problem. Some of our recent work indicates that density measurement variations may be due for the most part to basis weight variations. A possible but costly improvement would be an additional beta gauge on the winder for feedforward control. Other potential applications of the Novatune to the pulp and paper industry such as consistency control, Yankee drier moisture control or lime kiln are described in Sinner (1983) and in the Asea documentation, however, I could not trace any of those in the industry.

Pulp mills with their poorly understood processes, poor measurements, long deadtimes and highly variable raw materials seem ideal places for applications of adaptive control, and yet reported applications are not legion. A major US paper company reports in-house development of a diffusion washer control system using adaptive Kalman filtering, as well as a digester control system using variable deadtime compensation. Those schemes are reported to result in improved process stability; unfortunately no details are given.

In the pulping area, applications have gravitated around the continuous digester, the first work being reported by Cegrell and Hedqvist (1974). The problems in controlling continuous digesters include long and variable deadtimes, variable wood chips characteristics and acute lack of on-line sensors. A MISO STR is used to control chip level by means of the chipmeter and the blow flow. Cooking temperature is manipulated to control the predicted degree of delignification, as indicated by a predicted Kappa number. The parameters of the model, which is linear in the coefficients, are updated every time a new Kappa number test is available. This system was tested on an industrial Kamyr digester in Sweden, however there is no indication that it was followed by a permanent implementation. A later in-house development is described by Sastry (1978). He develops a minimum-variance STR to control the chip level by manipulating the blow flow. Results of plant trials on an industrial Kamyr digester are presented. However, a few weeks later the system was abandoned due to lack of reliability and robustness of the STR. Further inspection of the design in question indicates several potentially fatal flaws in the scheme, such as a numerically non-robust implementation of the least-squares estimator as well as inconsistencies between the assumed deadtime and the order of the implicit model used.

A more recent approach to this problem is found in Dumont et al. (1984) and Bélanger et al. (1986). This time the Clarke-Gawthrop STC with estimation of a bias and a fixed forgetting factor is used. During industrial trials, four different strategies were evaluated, with various combinations of manipulated variables. All resulted in similar performance and the choice of the strategy to be implemented for the permanent system was essentially guided by the comments and desires of the operators. As a result of this scheme, the chip level variations were reduced in half and the number of feed interruptions due to high level conditions reduced by a factor of three. This last factor contributes to the high level of confidence that the operators put in the system, as it relieves them of one of their main concerns. As it is estimated that about 50% of the Kappa number variations are related to level variations, the scheme has a direct impact on pulp quality. This scheme has been in continuous operation since March 1983 in a Canadian pulp mill. As a result of its success and of the demand of the industry, a stand-alone, off-the-shelf version has been developed at PAPRICAN.

In Dumont (1982), a STR is used to control the motor load on a chip refiner in a thermo-mechanical pulping plant. The motor load control problem is interesting since the incremental gain of the process is subject to a slow drift due to plate wear and, occasionally can change sign due to catastrophic behavior. To solve that problem, a recursive least-squares estimator with variable-forgetting factor is used in conjunction with a Dahlin algorithm. This system was successfully used for a 6-month period on an industrial production unit. A similar system has reportedly been in operation for two years at the M. Peterson & Son Company in Norway, for control of the motor load on 3 plate refiners in a beater room. Prior to the use of adaptive control, the process whose gain can change by a factor of 30 could not be controlled. The system was developed in cooperation with the Norwegian Pulp and Paper Institute.

In Martin-Sanchez and Dumont (1985), an Adaptive-Predictive Control Scheme is used to control the pre-tower chlorine residual on the chlorination stage of a bleach plant. However, results of a few days of tests indicate that the adaptive scheme does not clearly outperform a well-tuned PID controller.

Reports have also been received of applications of the FOXBORO EXACT^{TM} self-tuning PID controller. In particular, Ducey (1985) reports its application to the control of the pH of the stock out of the refiner chest. The control problem is complicated by a variable process time delay. After a day of operation, the tuning had improved considerably. This is said to result in significant savings. Other applications have been reported on waste treatment pH control and chlorination tower residual control.

The recovery cycle in a kraft mill also appears as an ideal candidate for adaptive control. Of particular interest is the control of the green liquor density out the smelt dissolving tank. Although not a final product, a green liquor of constant quality is important as it affects strongly the quality of the white liquor used in the digester and thus the quality of the pulp. Difficulties in this loop include irregular smelt flow from the recovery boiler as well as variable strength for the weak wash used as dissolving agent. A US pulp mill reports that, compared to a manually tuned PID controller, an EXACT^{TM} controller reduces the density variations by a factor of four. A Swedish mill reports applying the Novatune on a similar loop. Although it outperformed a PID controller, the reduction of 18% in density standard deviation was insufficient to warrant permanent installation.

Automatic tuning has also been applied in various instances in the pulp and paper industry. For example, work has been performed at PAPRICAN, McGill University and UBC to develop a PID autotuner. The step response of the closed-loop system is identified as a series of Laguerre functions and the regulator is then tuned by minimization of a quadratic performance index. The scheme has been successfully tested on various loops in three pulp mills, including a chlorination residual control loop and various pH control loops, for more details see Dumont et al. (1985) and Zervos et al. (1985). The method is presently being evaluated for potential marketing by a distributed control vendor. Measurex also recently introduced a tuning package for use on their turnkey control systems, for tuning of the Dahlin regulator, see Kunde et al. (1983). A recursive least-squares algorithm is used to identify the plant while the manipulated variable is changed in a stepwise manner. The Dahlin regulator is then designed based on the certainty-equivalence principle. Åström (1980) reveals that a Swedish pulp and paper mill uses the STR as a tuner on paper machines, a recovery boiler and various low level loops. A major US paper company reports difficulties when experimenting on simulations and pilot plant with various self-tuning control schemes. This led to the choice of an explicit scheme used as a tuning tool. An external signal is sent to identify the plant and an optimization routine is then used to tune a specified controller structure, generally a PID and sometimes a Dahlin algorithm. This scheme has been in use for two years in three mills as a tuner for periodical retuning of loops on recovery boilers, power boilers and headboxes. It has proved particularly useful on noisy or inverse response loops.

Chemical Industry

One of the US chemical companies that I visited reports to have two adaptive controllers in operation in its facilities, an ASEA Novatune and a FOXBORO EXACT^{TM}. The first one controls the pH out of a three tank neutralization system. The major control difficulties are changing deadtime and a gain that can change by a factor of 10 due to variations in the waste stream characteristics. Because it was one of the early applications of the Novatune, the company spent some 6 man-months of engineering to evaluate it and to make it work reliably for this application, with the addition of many ad-hoc features. Now, it is said to work better than a previous PI controller. The FOXBORO self-tuning PID controller is also used to control the pH of a waste stream, this time in a ditch. It is working remarkably well, and is said to result in an annual saving of $250,000. An interesting comment concerning the generally good results reported when applying the EXACT controller to pH control; is that it is ideally suited for it, as it is looking for sinusoidal perturbations, obviously present here. To determine the applicability of that controller to other loops the same company is testing it on simulated processes with drifts and noisy signals. It is also worth to note that until these controllers came on the market, that company was working on two similar products. It reportedly spent close to $1 million developing a Novatune like self-tuning controller which, when properly used, was said to perform remarkably well. Unfortunately, it had up to 64 parameters to be set up by the user and was too complex to use in an industrial environment. Until recently, the group had access to a remarkably versatile pilot plant to test its designs. Another in-house development is a PID controller tuner, based on RLS identification of a FOPDT model and pole placement. Despite good results obtained on simulation, tests on real processes were disappointing. However, experience with tuners found in the literature was equally disappointing on real processes.

The other US chemical company interviewed reported one application of adaptive control among its various facilities. This time, it is an in-house development. An adaptive Kalman filter is used to estimate the polyester inherent viscosity out of a polycondensation reactor, by means of auxiliary measurement. The filter provides an estimate every five minutes. A lab measurement, available every few hours is used to update two parameters of the Kalman filter, using RLS with variable forgetting factor scheme. The estimate of the inherent viscosity is then controlled by feedback, using a prediction model controller that manipulates the polycondensation pressure. This system has been in use for two years and has resulted in a significant improvement in the quality of the polyester for the customer. The same group is presently investigating with on-line estimation of the impulse response for adaptive Dynamic-Matrix-Control (DMC).

Brooks and Nazer (1985) report an application of adaptive control to a distillation column at Esso Chemical Canada. During 8 weeks, a generalized Clarke-Gawthrop scheme, with the structure of an Analytical Predictor controlled the stripping section tray temperature by manipulating the reboiler heat duty. Compared to a Smith predictor, this resulted in a variance reduction of 30%. Then, the estimator was taken off and the information gathered during the project was used to build a look-up table for gain scheduling of the analytical predictor, according to the operating conditions.

Although the applications reported in the US chemical companies are few, the situation seems to be different in Europe. A striking example is the case of Bayer AG in Leverkusen, Germany, where since 1981, 15 adaptive controllers have been implemented on production plants. Krahl et al. (1985) describe the application of an implicit multivariable self-tuning controller (Koivo, 1981) to the control of a distillation column. Steam supply and reflux ratio are manipulated to control top and bottom temperatures. Production runs were preceded by extensive simulation and pilot plant tests. The scheme is now in use on five distillation columns, none of them ultra-high purity. A 20% energy savings, relative to conventional SISO control is claimed. An amazing observation about this application is that the Koivo scheme, which requires the same delay in all loops, could be used successfully.

One of the longest running self-tuners in the industry is described in Dumont and Bélanger (1978; 1981). This system has been operating on two TiO_2 rotary kilns continuously since 1977. It controls the most crucial property of the final product, the rutile content. Key aspects are the structure of the scheme (STC around fixed controller), and its simplicity (only two estimated parameters). This system has resulted in increased production and improved quality. Academia was the major force in the project and expertise was present in the plant for the commissioning period.

Albert and Kurz (1985) describe an application to pH control in a waste water neutralization plant. The influent, with a pH between 4 and 11, is neutralized by a combination of hypochloric acid and caustic soda to produce an effluent with pH between 6.5 and 9.5. A PID controller with variable gain is used to control the pH. The gain is adjusted by a simple model-reference scheme. Because of its simplicity, the scheme could be implemented on a low-cost programmable digital controller. The system has been in use since October 1984 and its performance has been so good that the pH setpoint could be raised. This reduces the environmental burden by separating the heavy metals while meeting the pH specifications.

Niederlinski et al. (1985) report successful application of the Clarke-Gawthrop algorithm, with variable forgetting factor to an industrial PVC reactor. Preliminary maximum-likelihood identification was performed to chose the model structure used in the self-tuner.
For bulk temperature control, the STC performs slightly better than a well-tuned PID controller. However, the latter could not constantly perform well, due to the reactors non-stationary and non-linear dynamics. Another application of adaptive control to an industrial exothermal batch reactor is reported by Åkesson (1985). The Novatune is used to control the reaction rate in Latex and PVC reactors. Control difficulties include long and variable deadtime, sudden

changes as well as drifts due to wear and environmental conditions and a multiplicity of grades.

Najim et al. (1982) report successful application of a generalized Clarke-Gawthrop STC to moisture control in an industrial phosphate rotary dryer. The problem is to keep the moisture content constant despite flow rate and feed moisture content variations. Several weeks of continuous operation showed that, compared to a well-tuned conventional multi-loop PID control scheme, the STC reduces the moisture standard deviation by 2.5 while reducing the fuel consumption by 4.3%. This project was performed by an academic research group over a period of several years.

Petrochemical Industry

The only US oil company interviewed had but one industrial application of adaptive control in operation in its facilities to report. Again, this is an EXACT controller on a pH control loop. The company is however working on several developments that may lead to industrial applications. First, adaptive DMC is being investigated for potential application to pH control. Second, an adaptive PID controller is being developed, based on the identification of a pseudo ARMAX model and use of Internal Model Control (IMC) for the design of the PID controller. In addition, a CAD package is being developed for use in the plants.

Other oil companies that responded to the questionnaire had practically no application to report. Some applications of model-based gain scheduling were reported for control of surge drum level, of light ends tower product quality and crude distillation units. All these are in-house developments. One company reports successful testing on simulation of an adaptive controller, in cooperation with a university, unfortunately followed by failure on the real process attributed to lack of persistent excitation. A Canadian oil company reports experimenting with adaptive control on an unspecified process.

Although simulation results were encouraging, the industrial application failed due to lack of persistent excitation and thus of identifiability. The project was then abandoned.

Control System and Equipment Vendors

The market for PID autotuners and selftuners is a very active one. Introduced a few years ago, the Electromax V from Leeds and Northrup was one of the first self-tuning PID controllers on the market (Hoopes et al., 1983). The FOXBORO self-tuning PID controller has already been mentioned several times. It is unique in that it is based on expert systems technology, using a tuning knowledge base of about 200 production rules. It automatically updates its tuning constants when required. NAF Controls, from Sweden recently introduced an autotuner based on the work of Åström and Hägglund (1984). This autotuner is available as an option on their Unic S computer control system. Contrary to the EXACT controller, it has to enter a tuning sequence to compute new settings. An attractive feature of this system is its ability to automatically build its own gain scheduling. The Honeywell UDC 5000 controller has also an auto-tuning mode. The frequency response of the plant is identified by means of a Fast-Fourier-Transform algorithm and the PID tuning constants are then computed using frequency domain techniques. The user specifies whether slow, medium or fast tuning is desired. A system developed at Brown Boveri in Germany is claimed to have been used to automatically tune 160 loops in a coke gas plant (Hildenbrand and Wilhelm, 1985). The automatic tuner identifies a second-order plus deadtime model of the plant using least-squares. For this, a PRBS is sent to the open-loop plant input. Then, using Ziegler-Nichols-like rules, the PID controller is tuned. Auto-tuned PID controllers have also been announced by Turnbull Control Systems, Doric Scientific while many vendors are reportedly developing similar devices.

However, the ASEA Novatune still seems to be the only "general-purpose", adaptive controller not limited to the PID case. Indeed, it can handle processes with dead time, and perform minimum-variance control and its extended form as well as pole placement. Because it requires a working knowledge of stochastic control theory, it is more difficult to use than the above PID controllers but it is more versatile. ASEA reports several hundreds of Novatune systems in use worldwide, the bulk of them in Europe. For more details on the Novatune see Bengtsson and Egardt (1984). Based on the same methodology as the Novatune but process specific there is the paper machine cross-direction control system available off-the-shelf from Accuray.

DISCUSSION

Industrial Perception of Adaptive Control

Included with the questionnaire was a list of statements about adaptive control. The respondents were asked to rate their level of agreement with these on a scale from one to ten. The results are summarized on Fig. 1. Most respondents tend to agree that adaptive control is no longer only an academic research topic, that there is a need for it in their industries and that there are definitely some loops where it can bring some improvement. Although most tend to agree that PID auto-tuners should be included in distributed control systems, they also say that it is not sufficient. Respondents were also asked to list the processes where they feel adaptive control would be most beneficial. The list would be too long to reproduce here, but the answers indicate that the lack of use of adaptive control is not due to a lack of potential applications. Most applications cited involve control of product quality and processes which display non-linear or time-varying dynamics and have long dead-times. We have seen above actual realizations of some of these potential applications in the pulp and paper and chemical industries, however none in the petrochemical sector.

It is not easy to explain the differences noted between the three sectors. Two factors may play in favor of the pulp and paper industry, its openness and the fact that the equipment is rather standard. Thus, despite the relative small size of pulp and paper mills, a novel control scheme has generally a rather wide market to justify the initial development work and is easily transferable. The pulp industry could soon loose that advantage as it will move away from the commodity market to produce specialized pulps in small quantities using proprietary processes. It is ironic that it is then, because of strict specifications and the large number of grades produced, that adaptive control would be most useful. The petrochemical industry, due to its size and the size of its plants, does not need to be as open as the pulp and paper industry to benefit from a much larger economy of scale. But in that industry, the emphasis is on production costs more than on quality, whence the interest in optimization techniques. Moreover, control engineers in the chemical and petrochemical sectors tend to see their control problems as being primarily multivariable. Certainly, multivariable adaptive control is much more difficult and currently available methods are not entirely satisfactory. At the other end of the spectrum, some chemical companies use proprietary, unique processes to produce small batches of expensive chemicals with tight specifications. One of the companies interviewed manufactures 500 different products in rather small quantities using such processes. Their approach is to integrate process and control design by evaluating flowsheets for dynamic operability. Thus, they claim to eliminate the need for adaptive control by clever process design.

During my survey, I came across several criticisms of the work performed by the academic community. Despite the plethora of papers on adaptive control of pilot distillation columns, very few industrial columns (none in North America) are controlled in such a manner. Most of the academic work is directed at low purity columns, whereas many industrial columns are high purity and have a highly nonlinear behavior. However, if academia was addressing that problem, would the North-American industry be willing to perform industrial tests, in view of the high stakes involved?

Very rarely in the literature is the benefit of using adaptive control quantified. Because adaptive control still requires substantial development, many companies hesitate to embark on such a project without a clear idea of the benefits. Maybe academia working on applications should try harder to quantify those benefits. Few industrials realize the cost of poorly tuned loops. In a plant where hundreds of loops behave poorly and a significant proportion of which stay in manual mode, the hidden cost may be large. In a recent informal survey, it was found that in Canada, only 15% of the paper machine moisture control systems perform significantly better than in manual control. For 50% of the systems, there is only a marginal improvement compared to manual, and for 35% there is a significant performance degradation! This is profoundly disturbing, considering that this is a critical loop for the final product quality.

Another often heard criticism is that most of the adaptation effort in the present adaptive controllers is simply spent in trying to fit a linear dynamic model to a nonlinear process. This is probably true in some cases. To control a time invariant plant with a well-defined nonlinearity, the obvious solution is to use model-based nonlinear control. More often than not, however the nonlinearity is poorly known or may be time-varying. If the form of the nonlinearity is known from first principles, then it could be included in the adaptive control model with some parameters estimated on-line. Nice applications of that idea are demonstrated on simulation for the control of waste water treatment and pH neutralization by Goodwin et al. (1982), and for the control of microbial growth systems by Dochain and Bastin (1984). Another interesting idea is the use of chemical invariant theory to design adaptive pH control systems (Gustafsson, 1982). Also, by nature many processes are bilinear, an area of adaptive control rather undeveloped.

A related criticism is that current adaptive controllers are black-boxes that do not make use of process knowledge. Actually,

PID controllers are black boxes that do not make much use of process knowledge. In present adaptive control schemes, a-priori plant knowledge is required to choose the structure of the model and for initialization. Thus, self-tuning controllers actually require more knowledge about the plant than a PID does. Many applications, that were justified on the basis of poor process knowledge, failed because of lack of perception of that point. Adaptive control cannot replace process knowledge and good engineering. If anything, good engineering is certainly more important in an adaptive control application than in a standard PID loop.

Many industrial users are more interested in eliminating oscillations than in minimizing variance. Robustness is thus seen as much more important than performance, then simple controllers are generally sufficient. In case of troublesome control loops, many industrial control engineers would contemplate process changes first. As put by one of them, "I would rather move a pipe than start developing an adaptive controller". There are nevertheless those critical loops where consistent performance is required and for which I believe adaptive control is an attractive solution.

Thus, there are definitely some loops which are perceived as potentially benefitting from adaptive control. Why then, is adaptive control not currently applied to those? We shall see in the next section the technical reasons due to the limitations of the current adaptive schemes. There are also non-technical obstacles that the industry might do something about.

An obvious problem is the lack of commercially available adaptive controllers, apart from the increasing number of auto-tuning and self-tuning PID controllers such as the EXACT controller. As for more general-purpose adaptive controllers, the Novatune is there, but because it would mean bringing an additional supplier in the plant, of communication problems, and of its reputation for not being easy to use, North American companies have given it a rather cool reception. More traditional North American suppliers of control systems have yet to include such an adaptive controller as an option on their systems.

Until such a low-cost product appears, either as a stand alone, compatible unit or as a software option, fully supported by the vendor, I see little chance of increased application of adaptive control, apart for the PID case.

Presently, if a company wants to develop a non-PID-based adaptive controller, a prerequisite is the presence of qualified control personnel in the plant. Although some large companies have excellent control staff in their central engineering department, who can develop adaptive control schemes, few plants have control staff to maintain and update those schemes. Any non-standard control scheme (adaptive or not) requires a commitment to support it for the future in face of changes in the plant, sensors, etc. Many of the successful applications reported here have involved academia at the design stage but were implemented and are supported by plant control staff. Obviously, the availability of low-cost adaptive controllers should lessen that problem.

Finally, when comparing Europe and North America it is obvious that in general European plants have better educated staff, which is as a rule more open to new concepts. A recent survey (Bialkowski, 1983) found that about 25 qualified control engineers were employed by the entire Canadian pulp and paper industry, most of them in five companies! This is about the same number found in **one** of the largest Swedish pulp and paper companies. Some Japanese pulp and paper companies have as many as fifteen control engineers (Snider, 1986). Maybe comparisons like these explain the greater use of adaptive control in Europe relative to North America.

A Critical Review of Current Adaptive Control Research

A few years ago, K.J. Åström called the adaptive control field a "fiddlers paradise". It is still true today, as at each control conference, "new" adaptive control schemes seem to appear. It is particularly true at control conferences within the chemical engineering community, where the audience is bombarded with new acronyms, that rarely correspond to a really new scheme but most often to a slight variant of an existing one. I do not mean to be controversial or to throw a stone at the chemical engineering community, but I think this only adds to the confusion. We should strive for unification and clarity.

When looking at the progress in adaptive control over the past five years, two highlights immediately come to mind, convergence analysis and robustness. Certainly, the most significant theoretical result in the adaptive control field up to now is the proof of convergence for various types of adaptive controllers. This remarkable achievement came as the result of several years of painstaking mathematical work by leading researchers in the field. Most schemes for which convergence has been formally proven use the stochastic approximation identification, not used in practice because of its slow convergence. The recursive least-squares identification scheme has to be made more robust in order to prove convergence in the stochastic case. The proofs of convergence usually require a semi-positive realness (SPR) condition, persistent excitation, as well as knowledge of the plant order. However remarkable this result is, it nevertheless has limited practical implications. In practice, the conditions under which the theorems are valid are rarely met. Exact knowledge of the plant order is unrealistic, and because the plant is unknown to start with the SPR condition cannot be checked. In any case, as simulations have shown, that condition is probably too restrictive. Industrial users are also reluctant to send an external perturbation, in order to ensure persistency of excitation, although simple practical solutions exist to handle that situation.

A more fundamental question is the meaning of convergence in practice, as most applications of adaptive control are justified by time-varying dynamics, which were a major motivation for the early work in the field. Ironically, most current adaptive controllers are fundamentally designed for the time-invariant case, for which auto-tuning is a more appropriate technique. In that case, it is much easier to guarantee convergence as the reluctance to send a perturbation is less since it is done infrequently and for short periods. However, as noted by Goodwin et al. (1984), convergence analysis studies help discriminating between good and bad adaptive control methods and show the way for new research directions.

Shortly after these theoretical results were obtained, the robustness issue was raised by Rohrs and et al. (1982), and "spoiled the party" so to speak. They investigated what happens when the plant order is underestimated. By showing that a particular adaptive control algorithm could be destabilized in the presence of unmodelled dynamics or noise, some extremely serious doubts about the practicality of adaptive control were expressed. In particular, it was concluded that "existing adaptive algorithms cannot be considered as serious practical alternatives to other methods of control". This had the effect of a bombshell. On the positive side, it forced the adaptive control community to increase its understanding of these methods and it triggered various efforts to try to explain the problem described. Today, robustness of adaptive control is a very active research field and our understanding of the phenomenon has improved tremendously. There were however several negative consequences of the robustness issue. Although it made the adaptive control sessions at control conferences lively and entertaining, it probably gave the newcomer to the field the impression of an immature theory, of extremely divided experts, and certainly of a field not ready for applications. Surely, after attending such a session, a newcomer would not have gone back home eager to apply adaptive control on a real plant. All this despite the fact that there already were numerous documented successful industrial applications. In 1983, Åström proposed his own explanation of the phenomenon observed by Rohrs et al. and even proposed simple remedies such as anti-aliasing filters and careful choice of the sampling interval (both standard features of digital control systems), to filter out high frequency signals. As opposed to all applications, the scheme studied by Rohrs et al. was not discrete but continuous. This fundamental difference has far-reaching consequences. Indeed, as shown later by Rohrs et al. (1985), even if the unstructured uncertainty may be large in the continuous time domain, in practice it can be made as small as desired by lowering the sampling rate. This latter work gives much insight about the crucial role of the sampling interval and the care with which it must be chosen.

Early implicit adaptive control schemes had difficulties with non-minimum characteristics. Explicit adaptive LQG control schemes handle that situation very well. However, in case of over-parametrization pole-zero cancellation causes numerical problems when solving the Diophantine equation. Recently, Gohneim (1985) proposed an iterative way to solve the Diophantine equation that does not require preliminary elimination of common factors. Grimble (1984) proposes new implicit and explicit LQG self-tuning controllers that appear more robust than previous schemes. In Åström and Wittenmark (1985), it is shown that the basic STR can handle non-minimum phase systems by extending the prediction horizon. Related adaptive schemes based on the idea of extended horizon predictive control had been previously developed (Clarke et al., 1985; De Keyser and Van Cauwenberghe, 1985; Martin-Sanchez, 1976; Ydstie, 1984). With proper choice of the horizon, they can handle non-minimum phase plants with unknown deadtime. The prediction horizon can also be adjusted to find the best compromise between output and input variances. This class of adaptive schemes appears to be robust and simple to use and is thus of significant practical importance.

A related problem is that of varying deadtime. Early implicit adaptive schemes required a good knowledge of the process deadtime. This is now routinely solved by estimating an extended B-polynomial, although the above approaches are more attractive.

Two issues of great practical importance are still outstanding, those of the choice of the time scale (sampling period) and of the choice of the model structure (number of poles and zeros). The relay method of Åström and Hägglund (1984) could be used to automatically determine the sampling period. As for the model structure, most successful applications use models with about 6 or 7 parameters. As shown by Rohrs et al. (1982), under-parameterization may lead to robustness problems. On the other hand, over-parameterization may lead to identifiability and numerical problems due to pole-zero cancellations. Most practical applications are preceded by a batch identification experiment for structure determination. There are, however, no general guidelines. The same problem exists for the choice of the initial conditions, the use of a fixed- or variable-forgetting factor, the so-called safety net and associated parameters that have to be chosen by the designer. The eventual success of an application is as much a function of these factors as of the adaptive control method used. This engineering aspect of adaptive control projects is very much an art and is rarely addressed in the literature. The expert systems technology seems a promising way of automating the design of adaptive (or for that matter, any) control systems. Preliminary ideas in this direction are exposed in Åström and Anton (1984), who propose an expert system for coordinating an auto-tuner and a self-tuner, and in Sanoff and Wellstead (1985), who propose an expert system to configure self-tuners as well as a run-time expert for on-line performance monitoring and improvement.

The problem of applying adaptive control to time-varying systems has received little attention. An interesting approach is proposed in Hägglund (1983). A new estimation scheme based on fault detection techniques deals with large parameter changes. To handle slow parameter changes, the P-matrix is updated such that it converges toward a specified diagonal matrix.

Research on multivariable systems has been oriented toward the problem of plants with unstable inverse, i.e. multiple deadtimes. For this problem, it seems more promising to perform the control design using state-space techniques than polynomial ones. However, a polynomial representation seems more appropriate for the identification phase. Schemes that identify a matrix ARMAX model which is then put in state-space form for control design have been developed (Bortolotto and Jorgensen, 1985) and seem promising on pilot plant applications. Another promising approach is that proposed by Jones and Porter (1985). A matrix ARMAX model is identified using a RLS scheme and the B_1 matrix, which represents the step response matrix is used to design a multivariable, robust PI controller. Needless to say, the problem of structure determination is even more acute in the multivariable case. As noted by Sargent (1985), physically based models will have to be used to reduce the number of estimated parameters in multivariable adaptive control. Problems that have been neglected and that could have great practical significance are those of adaptive decoupling and of combination of MISO self-tuners to solve the multivariable adaptive control problem, see Gawthrop (1984) for some preliminary work in that direction.

For a long time the industry had been expressing the need for better ways to tune the thousands of PID controllers in operation. It is only in the past five years that practical devices that automatically tune PID controllers have been developed and commercialized. Judging by their success, these devices were long overdue and may eventually become standard equipment. Hopefully they will pave the way for more sophisticated adaptive controllers.

RECOMMENDATIONS

My recommendations will be aimed at the three separate players in the adaptive control game, the industrial users, the control systems vendors and the academic researchers. Opportunities for progress will exist only where close cooperation between these three parties has been established.

Industrials

Industrials should stop seeing the computer control systems installed in their plants as alien and hire adequate control staff to maintain their performance. In view of the large number of loops to be tuned, automatic tuners should be used as much as possible. Critical loops, that can potentially benefit from adaptive control should be identified. Adaptive feedforward control applications should not be overlooked as they are simple, and more robust. Adaptive control should also be used to build gain scheduling look-up tables (e.g. on a fine paper machine). Pressure should be put on the consulting engineering firms and the vendors to provide adaptive control systems. Finally, if control theory is to stay relevant to industrial process control, industries should share their control problems with researchers and open their plants to them for carefully planned experimentation.

Vendors

Commercial control systems now make use of state-of-the-art hardware, but for the most part still use the same control methods as 30 years ago. A comforting sign is the increasing activity in the market for auto-tuning and self-tuning PID controllers. Widespread use of adaptive control will not happen unless it becomes available on commercial systems. Adaptive control theory has now evolved to the point to be ready for industrial use. Vendors should now make a major effort to include some of that theory as options on their systems. They should take advantage of their experience with turnkey systems for developing commercial adaptive controllers that are easy to use and with which users feel confident.

Researchers

Researchers in adaptive control have to be praised for the theoretical advances of the past five years. Thanks to this work, adaptive control theory is on more solid grounds. However, when it comes to the design of adaptive control, as noted in Ljung and Anderson (1984), a general framework for design engineeers is still lacking. For instance, what class of systems should adaptive control be applied to? Applications fall mainly in two categories non-linear, time-invariant systems and time-varying systems. For the first category, non-linear adaptive control schemes, that make use of process knowledge should be developed. For the second category, present schemes are not entirely adequate as they are ad-hoc modifications of algorithms for the time-invariant case. Better algorithms for the time-varying case need to be developed. Asynchronous adaptive control schemes where the parameters may be estimated every sample but the control parameters are updated less frequently are very interesting from a practical point of view and should be studied.

A major reason for the lack of use of present adaptive control schemes is that they are not easy to use. The user has to choose between a myriad of explicit, implicit schemes based on stochastic control, pole-placement, etc. Although to a good engineer, the basis for control design should be clear, the choice between explicit and implicit schemes is more difficult. The present literature does not address that issue, it should.

Another problem is the need for structural knowledge. Elimination of the need for structural knowledge requires development of adaptive control methods based on other parameterizations than transfer functions, see for instance Dumont and Zervos (1986). The other outstanding problem is that of the choice of time scale. More research should be aimed at automating that choice, most likely via a pre-tune mode. Safety nets, or practical tips are an integral part of any control scheme and have been used in all successful applications of adaptive control but are rarely described in the literature (Dumont, 1982; Isermann, 1982; Seborg et al., 1986; Wittenmark and Åström, 1984). Ideally, the user should only have to choose **performance related** parameters instead of **process related** ones. Schemes with that characteristic have yet to appear.

Finally, the current work on combined use of auto-tuning, self-tuning and expert systems is very promising. Åström (1985) proposes an expert system that monitors the performance of a control system, decides the type of control to be used, when to use auto-tuning or self-tuning, automatically builds gain scheduling tables, etc. Such a system is very ambitious but could become widely applied by the next decade. This is a very promising research field.

CONCLUSIONS

In conclusion, one may say that adaptive control is still not a common tool in the process industries despite some impressive applications. My prediction is that it will change only if the following happens:

Industrials must hire trained control staff for their plants, must realize the costs of poorly tuned control loops and must open their plants to researchers.

Control system vendors must develop inexpensive, easy-to-use adaptive controllers either as stand-alone systems or as software options.

Researchers must unify the theory where required, reduce the need for a-priori knowledge and automate the choice of the time scale.

As any prediction, it will be shown wrong by the person standing in my place five years from now. I only hope that I will be shown to have been too pessimistic.

Acknowledgements. I wish to thank Prof. T. McAvoy for helping me establish contact with major US chemical and petrochemical companies. Partial support for this survey came from them, I thank them very much (they know who they are). I also wish to thank all the others who replied to my questionnaire. Finally, thanks to M. Fjeld for his helpful comments on the early version of this paper.

REFERENCES

Åkesson, (1985). Adaptive automatic control of reaction speed in exothermal batch processes. IFAC Workshop on Adaptive Control of Chemical Processes. Frankfurt-am-Main, Germany.

Albert, W., and H. Kurz (1985). Adaptive control of a waste water neutralization process - Control concept, implementation and practical experiences. IFAC Workshop on Adaptive Control of Chemical Processes. Frankfurt-am-Main, Germany.

Åström, K. J. (1980). Self-tuning regulators - design principles and applications. In Narendra and Monopoli (Eds.), Applications of adaptive control. Academic Press, New York, pp 1-68.

Åström, K. J. (1983). Analysis of Rohrs counterexamples to adaptive control. 22nd IEEE CDC, San-Antonio, Texas.

Åström, K. J. (1985). Auto-tuning, adaptation and expert control. 1985 ACC, Boston, Mass.

Åström, K. J., and J. J. Anton (1984). Expert control. IFAC 9th World Congress. Budapest, Hungary.

Åström, K. J. and T. Hägglund (1984). Automatic tuning of simple regulators with specifications on phase and amplitude margins. Automatica, 20, 645-651.

Bélanger, P. R. (1981). A review of some adaptive control schemes for process control. 2nd International Conference on Chemical Process Control. Sea Island, Georgia.

Bélanger, P. R., L. Rochon, G. A. Dumont, and S. Gendron (1986). Self-tuning control of chip level in a Kamyr digester. AIChE Journal, 32, 65-74.

Bengtsson, G., and B. Egardt (1984). Experiences with self-tuning control in the process industries. IFAC 9th World Congress. Budapest, Hungary.

Bialkowski, W. L. (1983). Computer systems engineering staff in the Canadian pulp and paper industry. Pulp and Paper Canada, 84, T190-194.

Borisson, U., and B. Wittenmark (1974). An industrial application of a self-tuning regulator. 4th IFAC Symposium on Digital Computer Applications to Process Control. Zurich.

Bortolotto, G. E., and S. B. Jorgensen, (1985).Adaptive control
of a recycle reactor.IFAC Workshop on Adaptive Control of
Chemical Processes. Frankfurt-am-Main, Germany.

Brooks, and Nazer (1985). Self-tuning control applied to an industrial
distillation column. IFAC Workshop on Adaptive Control of Chemic
al Processes. Frankfurt-am-Main, Germany.

Cegrell, T., and T. Hedqvist (1973). A new approach to continuous
digester control. 4th IFAC Symposium on Digital Computer
Applications to Process Control. Zurich.

Cegrell, T., and T. Hedqvist (1975). Successful adaptive control
of paper machines. Automatica, 11, 53-60.

Clarke, D. W., P. S. Tuffs, and C. Mohtadi (1985). Self-tuning
control of a difficult process. 7th IFAC/IFORS Symposium.
York, UK.

De Keyser, R. M. C., and A. R. Van Cauwenberghe. Extended prediction
self-adaptive control. 7th IFAC/IFORS Symposium.
York, UK.

Dochain, D., and G. Bastin (1984). Adaptive identification and control
algorithms for nonlinear bacterial growth systems. Automatica,
20, 621-634.

Ducey, M. J. (1985). Self-tuning PID controller reduces refiner
chest pH stock variations. Pulp and Paper, 59, 6, 94-95.

Dumont, G. A., and P. R. Bélanger (1978). Self-tuning control of a
TiO_2 kiln. IEEE Trans. on Autom. Control, 23, 521-531.

Dumont, G. A., and P. R. Bélanger (1981). Successful industrial
application of advanced control theory to a chemical process.
IEEE Control Syst. Mag., 1, 1, 12-16.

Dumont, G. A. (1982a). A practical guide to self-tuning regulators. ISA 1982 Joint Symposium, Columbus, Ohio.

Dumont, G. A. (1982b). Self-tuning control of a chip refiner motor
load. Automatica, 18, 307-314.

Dumont, G. A., S. Gendron, and P. R. Bélanger (1984). Practical
experience with Kamyr digester self-tuning control. 1984 ACC,
San Diego, Calif.

Dumont, G. A., C. Zervos, and P. R. Bélanger (1985). Automatic
tuning of industrial PID controllers. 1985 ACC, Boston, Mass.

Dumont, G. A., and C. C. Zervos (1986). Adaptive Controllers Based on Orthonormal Series Representation. 2nd IFAC Workshop on Adaptive Syst. in Control and Signal Proc., Lund, Sweden.

Eriksson, L. G., L. Lind, and I. Olsson (1983). Computerised control of reel winding improves quality. Svensk Papperstidning, 86, 4, 24-28.

Fjeld, M. (1981). On closing the gap with modern control theory that works. 2nd International Conference on Chemical Process Control. Sea Island, Georgia.

Fjeld, M., and R. G. Wilhelm (1981). Self-tuning control - the software way. Control Eng.,28, October.

Fjeld, M. (1986). Adaptive control - Perspective from industry. This Volume.

Gawthrop, P. J. (1984). Multi-loop self-tuning control: Cascade systems. 9th IFAC World Congress. Budapest, Hungary.

Gohneim, Y. (1985). Indirect adaptive using the linear quadratic solution. Ph. D. Thesis, E. E. Dept., McGill Univ., Montreal, Canada.

Goodwin, G. C., D. J. Hill, and M. Palaniswami (1984). A perspective on convergence of adaptive control algorithms. Automatica, 20, 519-531.

Goodwin, G. C., B. Mcinnis,and R. S. Long (1982). Adaptive control algorithms for waste water treatment and pH neutralization. Optimal Control Applications and Methods, 3, 443-459.

Gustafsson, T. K. (1982). A simulation study of a class of algorithms for adaptive pH control. Report 82-13, Process Control Laboratory Abo Akademi, Finland.

Hägglund, T. (1983). New estimation techniques for adaptive control. Ph. D. Thesis, Lund Institute of Technology, Lund, Sweden.

Hildenbrand, P., and H. Wilhelm (1985). Adaptive tuning of regulators in a bus-oriented process control system. IFAC Workshop on Adaptive Control of Chemical Processes. Frankfurt-am-Main, Germany.

Hoopes, H. S., W. M. Hawk, and R. C. Lewis (1983). A self-tuning controller. ISA Trans., 22, 3, 48-58.

Isermann, R. (1982). Parameter adaptive control algorithms - A tutorial. Automatica, 18, 513-528.

Jones, A. H. and B. Porter (1985). Design of adaptive digital setpoint tracking PID controllers incorporating recursive step-response matrix identifiers for multivariable plants. 24th IEEE CDC. Fort Lauderdale, Florida.

Kelly, W. H. (1982). Self-tuning control strategies: modern solutions to papermaking control problems. Can. Industr.Comp. Society Conf.. Hamilton, Ontario.

Keviczki, L., I. Vajik, and J. Hetthessy. Intellicon: An industrial multiloop adaptive regulator. 7th IFAC/IFORS Symposium. York, UK.

Krahl, F., W.Litz, and F. Rhiel (1985). Experiments with self-tuning controllers in chemical plants. IFAC Workshop on Adaptive Control of Chemical Processes. Frankfurt-am-Main, Germany.

Koivo, H. N. (1980). A multivariable self-tuning controller. Automatica, 16, 351-366.

Kunde, K., T. S. Cheng, and J. Mathre (1983). Tuning weight and moisture control automatically. Tappi, 66, 7, 35-38.

Ljung, L., and B. D. O. Anderson (1984). Adaptive control, where are we?. Automatica, 20, 499-500.

Martin-Sanchez, J. M. (1976). A new solution to adaptive control. Proc. IEEE, 64, 1209-1218.

Martin-Sanchez, J. M., and G. A. Dumont (1985). Industrial comparison of an auto-tuned PID regulator and an adaptive predictive control system. IFAC Workshop on Adaptive Control of Chemical Processes. Frankfurt-am-Main, Germany.

Najim, K., M. Najim, and H. Youlal (1982). Self-tuning control of an industrial phosphate dry process. Optimal Control Applications and Methods, 3, 435-442.

Niederlinski,A., J. Moscinski, J. Kasprzyk, K. Arnold, and P. Warczykowski (1985).The application of self-tuning control to an industrial PVC batch reactor. IFAC Workshop on Adaptive Control of Chemical Processes. Frankfurt-am-Main, Germany.

Rohrs, C., L. Valavani, M. Athans, and G. Stein (1982). Robustness of adaptive control algorithms in presence of unmodelled dynamics. 21st IEEE CDC. Orlando, Florida.

Rohrs, C., G. Stein, and K. J. Åström (1985). A practical robustness theorem for adaptive control. 1985 ACC, Boston, Mass.

Sanoff, S. P., and P. E. Wellstead (1985). Expert identification and control. 7th IFAC/IFORS Symposium. York, UK.

Sargent, R. W. H. (1985). Trends in the development of process control systems. Chem. Eng. Res. Des., 63, 349-352.

Sastry, V. (1978). Self-tuning control of Kamyr digester chip level. Pulp and Paper Canada, 79, T160-T163.

Seborg, D. E., T. F. Edgar, and S. L. Shah (1986). Adaptive Control Strategies for Process Control: A Survey. AIChE Journal, 32, in press.

Sikora, R. F. (1983). Self-tuning control of moisture content. 3rd Int. Pulp and Paper Proc. Contr. Symp.. Vancouver, B.C.

Sikora, R. F., and W. L. Bialkowski (1984). A self-tuning strategy for moisture control in papermaking. 1984 ACC.. San Diego, Calif.

Sinner, B. (1983). Advantages with advances control systems in papermaking processes. 3rd Int. Pulp and Paper Proc. Contr. Symp.. Vancouver, B.C.

Snider, E. H. (1986). Tokyo technology conference and tour highlight Japanese concentration on systems engineering . Pulp and Paper Canada, 87, 1, 15-17.

Wilhelm, R. G. (1982). Self-tuning control strategies: multifaceted solutions to paper machine control problems. ISA Spring Symp.. Colombus, Ohio.

Wilhelm, R. G., and M. Fjeld (1983). Control algorithms for cross directional control: tha state of the art. 5th IFAC PRP Conf.. Antwerp.

Wilkinson, A. J. (1983). A new control technique for the cross-machine control of basis weight. 5th IFAC PRP Conf.. Antwerp.

Wittenmark, B., and K. J. Åström (1984). Practical issues in the implementation of self-tuning control. Automatica, 20, 595-605.

Ydstie, B. E. (1984). Extended horizon adaptive control. 9th IFAC World Congress. Budapest, Hungary.

Zervos, C., P. R. Bélanger, and G. A. Dumont (1985). On PID controller tuning using orthonormal series identification. IFAC Workshop on Adaptive Control of Chemical Processes. Frankfurt-am-Main, Germany.

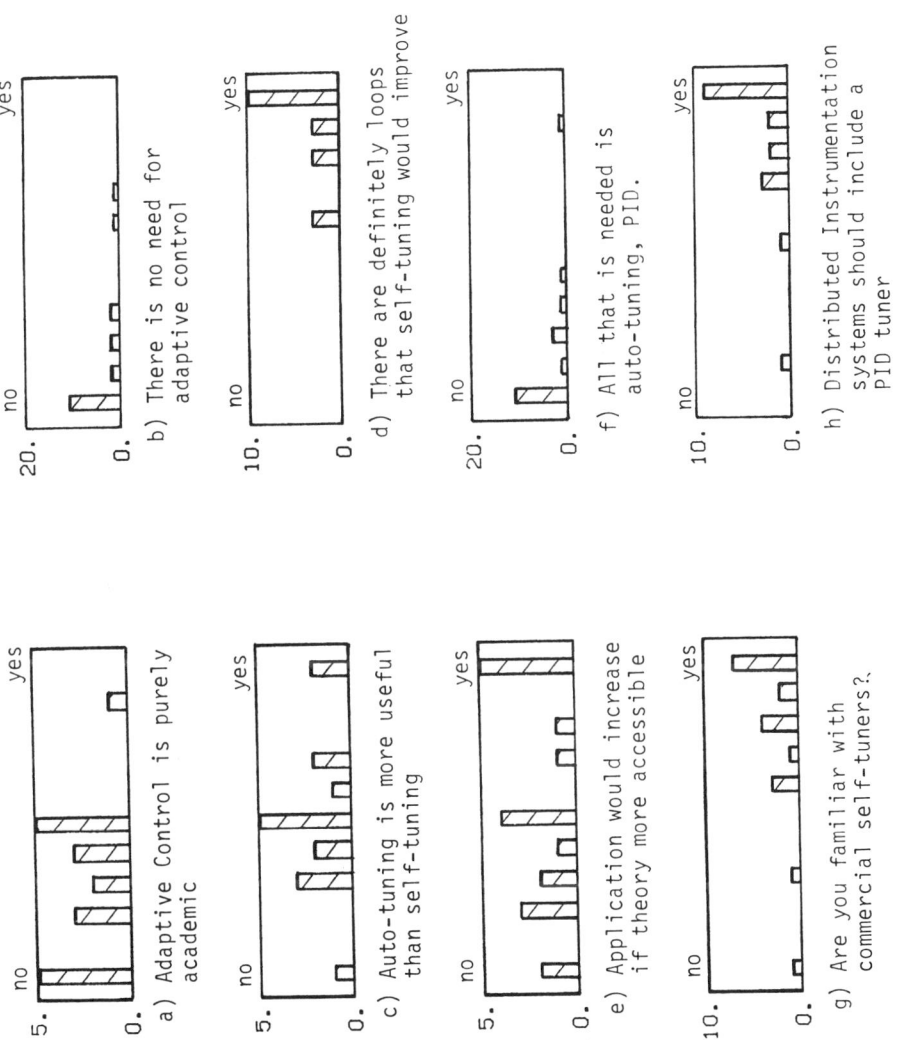

Figure 1: Distribution of answers to the questionnaire.

INDUSTRIAL PERSPECTIVE OF THE USE OF ADAPTIVE PROCESS CONTROL

Magne Fjeld

Statoil, Klaebuvn, N-7000 Trondheim, Norway

Abstract. A brief explanation of the theoretical basis for steady-state cross directional control is given. Some additional applications in Norway are briefly mentioned. Some of the pitfalls existing in the engineering and application of adaptive controls are outlined, as well as what makes applications successful. A discussion of the use of adaptive control in the process industries, as presented in a paper by Dr. Guy A. Dumont, is included as well.

Keywords. Adaptive Control; Cross Machine Control; Paper Machines; Optimal Control; Profile Control; Robust Estimation; LQG Adaptive Control; Parametrisations in Adaptive Control

INTRODUCTION

The purpose of this paper is twofold. Firstly, to review critical issues in application of adaptive control, and to present an adaptive profile control concept which is now a commercial system. Secondly, the session chairman has asked this author to discuss Dr. Guy A. Dumont's paper "On the use of adaptive control in the process industries" as it was presented at the conference, and I shall also give some of my own experience on the use of such controls.

INDUSTRIAL SURVEY

First, some comments regarding application of adaptive control to paper machine control.

Important reasons why turnkey suppliers of control systems did not take an interest in A.C./S.T. (Adaptive Control/Self-Tuning) until lately, after the pioneering work in Sweden more than 10 years ago, are:

- It has been uncertain whether customers are willing to pay the price for the A.C. option. Also, whether it pays off, will be dependent on the actual machine and disturbance conditions and the variety of grades.

- The techniques, plus their basic weaknesses, and possible pitfalls in application engineering and commissioning were not sufficiently understood or developed until around 1980, such that A.C. could be perceived to be applicable on a wide and *continuous* basis in the industry. Now I would like to think that the situation is different.

On the cross machine (CM) control problem for basis weight and moisture content:

Many people ask if adaptive control is needed, also at this conference. Well, here is an application where the adaptive control development for CM control was stimulated by the market pull. Typically, for paper machines producing fine and kraft papers with a variety of grades, the market, and need for cross machine control is almost as big as for machine direction control. The same well-known economical benefits from machine direction control also apply to CM control, and possibly other benefits, such as a CM uniformity requirement for print paper. Since the introduction of the "Profile Manager" a few years ago, approx. 120 systems have been sold, and around 80 systems installed. The average price tag is approx. USD 300.000,-, and will in the most comprehensive installations encompass such items as installation of slice screw actuators, including screw position measurements, a traversing robotic tool for performing slice screw control adjustments, and of course in any case, a computer system. The number of jack screws may typically be between 40 and 100.

Since the solution of such a control problem may not be well known, let me briefly describe the principles. Earlier references may be found in Wilhelm and Fjeld (1983) and Wilhelm (1982). Spatially, the control problem is strongly multivariable, since there are strong interactions between a response at a certain point, and the corresponding actuator as well as its neighbouring actuators. The character of these interactions may change, dependent of a multitude of factors, such as when grade and speed changes occur. See Fig. 1.

One of the first things to observe for this problem, is that the dominant part of the CM profiles in moisture and basis weight are changing very *slowly* due to disturbances with origin a.o. in the felts, the driers, in the wire drainage, and sometimes possibly due to changes in the hydrodynamics through the headbox. This leads to an approach where CM control is decoupled from machine direction control, considering the first problem to be a very low-frequency control problem as far as time domain is concerned. More rapid changes occur during grade changes, speed changes and other manipulations on the machine. Therefore, for most CM problems during stationary situations, it suffices with a correspondingly slow sampling rate both in control and in the recursive model update.

For the control of this process, only one control agent - the headbox slice in this case - is at disposal, whereas both basis weight and moisture profiles are affected by the slice positions. We could therefore say that it is a "vectorial one input/two outputs"-control problem (see Fig. 2).

The control problem may be formulated as a deterministic linear-quadratic one on a *spatial* domain, using the criterion

$$J = \text{Min} \int_{-L}^{L} \left[q\, \Delta y(v)^2 + r\, \frac{d^2 u(v)}{dv^2} \right] dv \qquad (1)$$

where the factor r puts weight on slice *bending*.

The "output" deviation $\Delta y(\ell)$ from the setpoint profile for $\ell \varepsilon [-L, L]$ can be taken as any quadratic form in ΔBW and ΔM, such as

$$\Delta y^2 \triangleq \left[a \cdot s^2 \cdot \Delta BW^2 + (1-a) \cdot \Delta M^2 \right] \qquad (2)$$

where $s = \partial M / \partial BW$ = sensitivity in moisture due to basis weight variations.

The user may, by changing the relative weighting between moisture and dry basis weight control errors in the performance criterion, direct how much of the effect of control will be directed to the benefit of reducing basis weight profile variations versus moisture profile variations. For the model, we assume a symmetric (steady-state) spatial impulse response

$$\phi(\ell - v) = \phi(v - \ell) \qquad (3)$$

(Green's function), such that a response variable $x(\ell)$ can be expressed as

$$x(\ell) \simeq \int_{-L}^{L} \phi(\ell - v)\, \Delta u(v)\, dv + x_o(\ell) \qquad (4)$$

$x_o(\ell)$ is an initial profile, before application of Δu. [-L, L] is the width of the machine, such that L is assumed "large". This means that the effect of possible boundary conditions at the edges are assumed not to influence the response at any point. (This is a good assumption for "well-behaved" headboxes, at least at some short distance from the edges.)

The control problem above may be expressed in its (approximate) spatial-discrete form. The description eq. (2) then has an input/output equivalent rational transfer function, represented by a finite number of *poles* and *zeros* in general.

The approximate finite-dimensional system description is:

$$\sum_{i=-n}^{n} a_i z^{-i} \Delta x(\ell) = \sum_{i=-m}^{m} b_i z^{-i} \Delta u(\ell) \qquad (5)$$

The discretization of eq. (1) is straightforward, approximating the 2nd derivative by the usual 3-point discrete approximation.

Note:

Symmetry has been assumed, i.e.

$a_{-i} \equiv a_{+i}$
$b_{-i} \equiv b_{+i}$

This simplifies the parameter estimation. It is straightforward to show that $p \equiv r \equiv o$ in the criterion J will result in a *decoupling feedback controller*.

Assuming stochastic stationarity in the spatial direction of CM disturbances, and symmetry of their autocorrelation function, the model description eq. (5) may be extended to the full ARMAX form, extending the r.h.s. with a polynomial

$$\sum_i c_i z^{-i} v(\ell)$$

on the basis of the assumptions, $c_{-i} \equiv c_{+i}$

Since the formulation of the problem is not strictly valid at the boundaries(edges) of the wire, care must be taken in the application engineering.

The pole-zero locations for an arbitrary, dynamic (causal) system may look as in
Fig.3a. For an arbitrary, symmetric static spatial system, the approximate discrete form may have pole/zero locations as depicted in Fig. 3b: For any pole or zero, there exists a conjugate pole or zero.

The problem now has been described as a spatially stochastic system, and the optimum control criterion may either be expressed as a generalized minimal variance criterion

$$J = \text{Min } E\left\{q \cdot \Delta y(\ell_k)^2 + r \cdot \left.\frac{d^2 \Delta u(\ell)}{d\ell^2}\right|_{\ell=\ell_k}\right\} \tag{6}$$

or with the full LQG criterion

$$J = \text{Min } E\left\{\sum_{k=-K}^{k=+K}\left[q\,\Delta y(\ell_k)^2 + r \cdot \left.\frac{d^2 \Delta u(\ell)}{d\ell^2}\right|_{\ell=\ell_k}\right]\right\} \tag{7}$$

where one may use the approximation

$$\left.\frac{d^2 u(\ell)}{d\ell^2}\right|_{\ell=\ell_k} \simeq \frac{u(\ell_{k-1}) - 2u(\ell_k) + u(\ell_{k+1})}{4\Delta^2} \tag{8}$$

where $\Delta = |\ell_k - \ell_{k-1}|$, which is the distance between two adjacent actuators on the slice.

Performing control iterativety at instants t, t+1, t+2, ..., it can be shown that Δu^*, the optimal control vector increment, can be expressed as

$$\Delta u^*(t+1, \ell_k) = g \cdot \frac{G(z)\,\overline{G(z)}}{H(z)\,\overline{H(z)}}\, e(t, \ell_k) \tag{9}$$

for a particular point ℓ_k, and where ( ) denotes conjugation, and g is a gain factor <1.. It is hence seen that the Δu^* at a particular point is implicitly expressed through the error profile $\{e(\ell_k)\}$, and must be eliminated from a set of linear equations. Both LQG and the GMV control will in principle result in the form eq. (9), but G and H will be different for these two types of control strategies.

Application areas for this kind of control strategy include *sheet processes,* such as on a paper machine, (cross machine control of the sheet formation process on the wire, caliper control, sheet drying by sectionalized hoods, moisture control by rewetting), sheet extruders in the plastic industry, distributed paint spraying processes, etc.

There is a lot more to be said both theoretically and in particular with regard to engineering(edge effects, wind-up problem, actuator constraints, etc.) of such a control package, but this should give the main ideas. However, the reader should also appreciate the fact that process *profile control* is a genuine class of process control in general.

On some of the fact-findings in the survey chapter of Dr. Dumont, one can certainly state that a number of things may go wrong in AC/ST. We do know that AC/ST usually relies upon:

(a) a model
(b) one-line identification
(c) computes a control law on the basis of an updated model

The *choice* of (a) has an effect on both (b) and (c). The *result* of (b) has an effect on (c). However, do not let the computer grind on this, and the application engineer blindly accept the results. Basic filtering principles and basic control engineering sense must be followed! *If a control law structure cannot work well for a particular problem, certainly it will not work well if it is made adaptive in the parameters.*

There are examples in the industries where lack of appreciation of this fact has cost dearly.

Dr. Dumont's list of applications are certainly not exhaustive, since it covers only those who received and responded to the survey invitation. In Norway, for instance, immediately two process industry examples come to my mind:

- An STR, tuning one parameter only, to control the specific refiner energy in pulp refining, was developed at PFI (The Pulp and Paper Research Institute in Oslo, Norway), and installed on PM5 with M. Peterson & Son. The problem arises, because the gain between the pressure applied to the plates and the actual power dissipated (as measured on the refiner motor), is depending on many other variables that are difficult to model and measure. The adaptive estimation of the gain makes possible a precise control (without overshoots) of the specific energy usage (kWhrs/ ton). The system has been running successfully for a couple of years. Attempts to develop an adaptive master controller to control paper porosity, cascading into the refiner control loop, failed for unspecified reasons.

- AC in the electrolysis of aluminium with the Norwegian company "Årdal og Sunndal Verk".

The curve exhibiting both resistance (R) versus % Al_2O_3 (x) is named the so-called bath-tub curve. Of crucial interest is the parameter b_1, which is the derivative of R w.r.t. x.

Three coefficients, including b_1, are estimated in a MA-model, adding colored noise to the process model. Changes in R are taken as measurements y. The control u is the feeding of x, i.e. the time derivative of x.

The bath is controlled, applying the control u as a certain function of b_1. In this way, the process operates almost without the periodic arc ignitions which are so common elsewhere, when x is getting too low. Energy usage is decreased, and productivity is increased.

This system has been running successfully for a couple of years. However, as is typical in difficult processes, it is a kernel only in an extensive supervisory and correction control software, that a.o. is based on mass balance supervision. Many special precautions had to be taken in the engineering of the algorithm, in order to deal with a number of discrete events occurring in this type of process. The system has also been installed successfully in plants in Saudi Arabia.

INDUSTRIAL PERCEPTION OF ADAPTIVE CONTROL

It is true that in some cases processes perform in spite of the system trying to control it. However, when making surveys of this kind, one must be careful with the data gathering. A fair picture may only be possible if recording data over several weeks or even a few months. Also, the manual control base case (for comparisons) is not easy (even unrealistic) to obtain for extensive time durations.

Non-adaptive paper machine moisture controls may work well for some grades (hopefully for thebulk of the grades), and may degrade the performance for other grades. This may be so because tuning has been made for a dominant grade, and remains fixed (without retuning). Also, tuning in such cases is done quite conservatively.

According to the basic principle found in theory of cybernetics, a good controller must contain a model of the process to be controlled. It may well be then, that there may exist no naive control solution that will work well for the control of a difficult process. The processes that can give significant payoff with AC, are usually also requiringmore careful and tailored designs, in particular w.r.t. supervisory software interfacing to the basic algorithm, because of the environmental conditions and requirements in operation of the process.

It is likely that many of the trials made which have not materialized into a reliable adaptive control system, and therefore have been removed after the trial period, are due to either neglect or ignorance of these requirements, or no funds or personnel available to attack the problems in full.

RECOMMENDATIONS

AC theory and experiments have demonstrated good ways to achieve robust AC systems:

Good approach to adaptive control/tuning:

The basic control algorithm must be robust
+
The parameter estimation method must be robust
+
Make sure that the estimation scheme and the feedback control are sufficiently "decoupled"
=
Adaptive robust control

An area which has received little attention in the adaptive control literature, is application of robust estimation principles, based on the literature of mathematical statistics. *Poulsen* and *Holst* (1982) have done some work to handle *outliers* and *load changes* (step disturbances). There exist possibilities for hypothesis testing, before data are actually used to update estimates. The reason for this is, that if non-stationarities are tested for and detected, updating of parameters may be delayed provisionally for several samples, without detrimental effects. This time frame can be used to perform hypothesis testing. If, say, a step disturbance is detected, then appropriate changes in the calculation of the innovation can be made to avoid the resulting detrimental transient effects in the parameter values.

Good adaptive methods to track time-varying parameters should be more researched. For "packaged controllers", methods should not require additional parameters to be chosen at commissioning time. If AC algorithms are to be useful for vendors manufacturing turnkey systems, they cannot rely on any extensive knowledge of control or identification theory. Even "cookbook recipies" for setting parameters are undesirable if they require too much attention, reasoning and time for the engineer commissioning the system. Remember that field engineers are all-round systems engineers (electronic/software/measurements) who have a lot of other things to pay attention to.

The Fortesque et al. method of variable forgetting factor $\lambda(t)$ (Fortesque et al., 1981) was introduced a.o. to avoid blow-up in the covariance matrix. However, the original Fortesque et al.'s way of computing $\lambda(t)$ does not prevent blow-up if measurement noise is present. This is because the information measure used is not a good measure of information content unless the process is nearly deterministic, as has been pointed out by Fortesque et al. themselves. There exist, however, remedies to this, such as P-matrix regularisation, or eventual reinitialisation (which has proved not to be so attractive), the method of Hägglund, and some others. A method that does not require *any* choice of additional parameters (even not a memory length, or an "information content"), is a method attributed to S. Sælid and B. Foss (1983), where a $\lambda_i(t)$ is computed for each particular parameter θ_i such that exactly the same amount of information, supplied by the measurement at a given time, is forgotten, when the initialization period is passed. The result is that the diagonal terms of P will remain constant for all times, whereas the off-diagonal terms will change according to the variations in the regression vector. This effectively eliminates blow-up.

When embarking on identification or AC projects, using ARMAX models, I do feel that it is all too common that we unconditionally accept parametrisations like Ay = Bu + Cv, or Ay = Bu + v . However, some thought should be given as to where process noise is dominantly entering the process. Often, it is very unlikely that v is filtered through the same poles (A) as the control variable. Such a priori knowledge will speed up parameter estimation (because of reduced parameter covariance and fewer parameters), with additional benefit to robustness of the control loop. There are many possible parametrisations, and in important applications, a few of these should be thought about and tested in the pre-engineering phase. See three examples in figure 4a-c.

In autopilots, navigation problems, dynamic positioning, etc. state space models have often proved to give effective parametrisations, since some parameters then can be plugged into the models on the basis of equipment/vessel data, etc. at commissioning time. Similar possibilities should also be kept in mind for process control, although we know that even simplified, low order state space modelling based on 1st principles is usually much harder in the process industries.

CONCLUSIONS

Cross machine control of some sheet processes asks for more than naive control structures, and when running a variety of grades, adaptive control may be necessary. In fact, early attempts in the industry to do moisture profile control on paper machines using a large number of simple controllers (feedback from dry end moisture at a certain point to the corresponding actuator) failed. The reason for this was that the cross-coupling would dictate such a low controller gain, to avoid instability (which would result in travelling "waves" back and forth across the machine).

In the successful application of adaptive process control, it is very important to observe the requirements for good identification, and take measures to avoid processing garbage data. Further, the model parametrisation should be sensible, if not, odds to fail in implementation will increase. Finally, the controller structure must be appropriate for the process to be controlled. Adaption will not help us if the basic approach to control is no good.

REFERENCES

Fjeld, M. and R. G. Wilhelm (1981). Self-tuning control - the software way. Control Eng., 28, October.

Fortesque, T.R., L.S. Kershenbaum and B.E. Ydstie (1981). Implementation of self-tuning regulators with variable forgetting factors. Automatica, 17, 6, 831-835.

Grimble, M. (1984). Implicit and explicit LQG self-tuning controllers. IFAC 9th Triennal World Congress, IFAC Proceedings Series 1985 no. 2.

Poulsen, N.K. and J. Holst (1982). Robust self-tuning controllers in nonstationary situations. Richerche di Automatica, 13, 197-217.

Sælid, S. and B. Foss (1983). Adaptive controllers with a vector variable forgetting factor. Conference on Decision and Control, San Antonio.

Wilhelm, R.G. (1982). Self-tuning control strategies: Multifaceted solutions to paper-machine control problems. ISA Spring Symp., Columbus, Ohio.

Wilhelm, R.G. and M. Fjeld (1983). Control algorithms for cross directional control: the state of the art. 5th IFAC PRP Conf., Antwerp.

Westerlund, T., H. Toivonen and K.-E. Nyman (1980). Stochastic modelling and self-tuning control of a continuous cement raw material mixing system. Modeling, Identification and Control, 1, 1, 17-37.

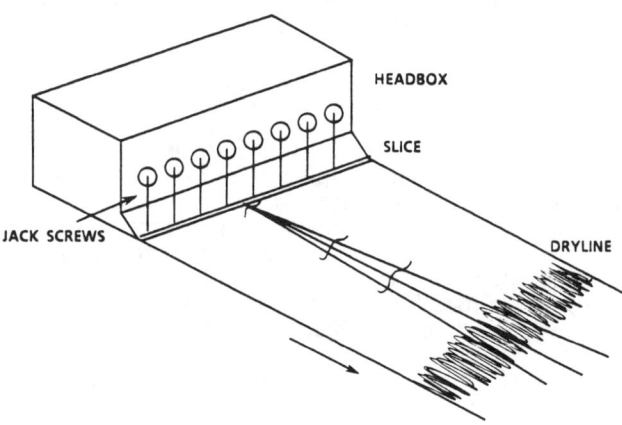

Fig. 1 Headbox, slice, and forming section of paper machine

Industrial Perspective on Adaptive Process Control   511

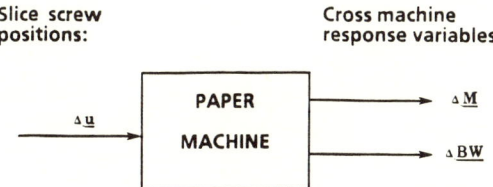

Fig. 2 Cross machine control model

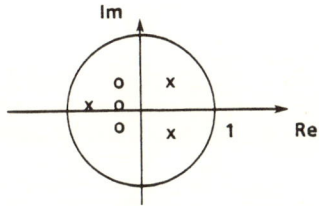

Fig. 3a Pole-zero locations for a causal, dynamic system

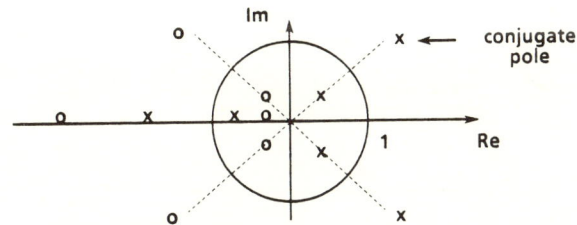

Fig. 3b Pole-zero locations for a spatial, static ("noncausal") system

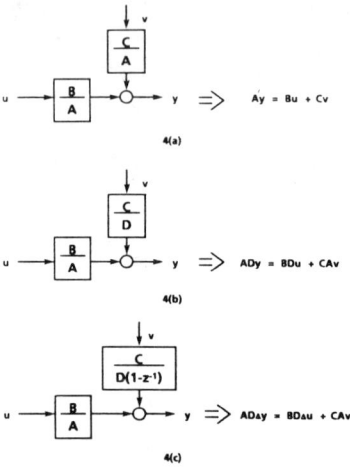

Fig. 4(a) - (c)

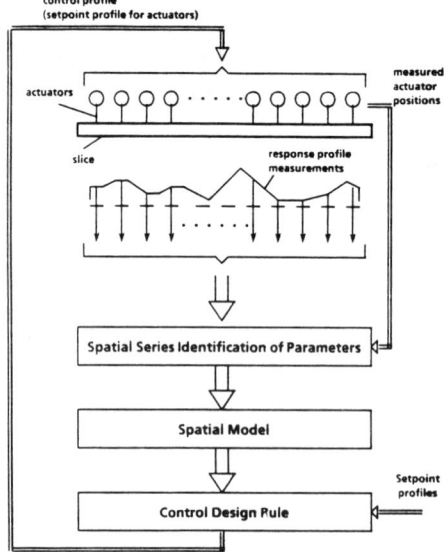

Fig. 5 Block diagram structure of adaptive cross machine control

ON-LINE IDENTIFICATION AND OPTIMIZATION

Thomas F. Edgar
University of Texas, Austin, TX 78712

The papers in this session were intended to inform process control engineers about recent advances in identification and optimization algorithms and to describe how these advances are being used in the chemical process industries.

The first paper was an appropriate bridge to the previous session on adaptive control. Graham Goodwin has been developing a series of very promising and significant results, including deriving algorithm convergence properties and dealing with the model-mismatch problem. Professor Goodwin gave a status report on the development of robust estimation algorithms, where the controller must be designed to compensate for errors in model structure as well as in the parameters. In this sense, it is an extension of the recent robustness ideas of Doyle and Morari, with the advantage that the model parameters can be adjusted on-line.

Goodwin's paper contains several other valuable tidbits of information for general design of digital controllers. The new version of the discrete transform (δ transform) has the advantage in that it reduces to the operator (d/dt) as the sampling time (Δ) approaches zero. Hopefully, this operator may help clear up the murky interface between continuous and discrete time controller designs. Goodwin also offered some new insights on the selection of Δ for computer control.

The second part of the session dealt with the more traditional application of linear and non-linear programming (NLP) to chemical processes. The paper by Mike Morshedi of Shell Development dealt with the solution of large-scale optimization problems which result when an optimal control problem is solved as an NLP. Morshedi mentioned that successful optimization of refinery product inventories represents major profit opportunities for Shell Oil in the future. The method employed by Morshedi resembles the DMC algorithm in some respects; hence, it has been given the name "Universal DMC." The technique is essentially a successive quadratic program which uses a Newton iteration to solve the two-point boundary value problem. The use of NLP to treat an optimal control problem with linear state and control constraints is not new; however, Morshedi's algorithm is quite different from those proposed in the past, and it aroused a great deal of discussion and controversy among some of the conference participants.

Linear programming (LP) is a much more familiar method to control engineers. Dr. Donald Shobrys of Chesapeake Decision Sciences presented the final paper of the session on the use of LP and NLP to solve scheduling problems in the process industries. This paper was authored by Leon Lasdon of the University of Texas and Tom Baker of Chesapeake Decision Sciences, neither of whom were able to attend the conference. Shobrys pointed out the problems of communicating optimized set points from the management level to the process control (loop) level. Due to a lack of feedback from lower levels to upper levels, much of the scheduling in many companies is done by manual introduction of data into process control computers rather than by direct communication. This can lead to suboptimal results and lack of coordination. Perhaps one of the most pertinent comments regarding this paper was made by John Haydel (Shell). He suggested that, given the current nature of multi-level decision making in U.S. companies, the control engineer should be skeptical about any set point sent verbally to the process level. Haydel stated that the control engineer should ask the origin of the information and inquire about the basis of the decision prior to implementation.

DIGITAL ADAPTIVE CONTROL
(A ROBUST APPROACH)

Graham C. Goodwin
University of Newcastle, N.S.W. 2308, Australia

Abstract. Adaptive control has been a topic of interest to control engineers for at least four decades. However, initial attempts to use these methods in practice were largely unsuccessful due to inadequate computer hardware and to lack of a suitable supporting theory. In recent years this situation has changed somewhat due to advances in microprocessor technology and due to the development of a better theoretical understanding of adaptive control. The aim of this paper will be to survey some of the recent results in digital adaptive control and to relate them to practical design issues where possible.

Keywords. Adaptive Control; Robustness; Self Tuning Control; Discrete Time Systems; Delta Operators.

INTRODUCTION

The performance of industrial control systems is frequently limited by a number of factors including

- poorly known models
- time varying parameters
- external disturbances and noise
- nonlinear effects
- time delays, and
- multivariable interactions

In spite of these difficulties there are   benefits to be
gained from improved control. These benefits include

- greater energy efficiency
- higher product quality
- improved safety margin
- reduction of out-of-tolerance product
- reduction of pollution, and
- higher plant throughput

Of course, one's ability to design a high performance control system is limited by the available knowledge about the plant and its environment. Adaptive control is an interesting idea in this regard since this form of controller includes a procedure for on-line estimation of a model for both the plant and its environment (Astrom (1983a)).

For this reason there has been considerable interest in the development of simple and robust adaptive controllers which are suitable for both practical use. The history of adaptive control can be traced back at least 4 decades. However, early attempts to implement adaptive algorithms were frustrated by inadequate computer hardware and due to the absence of suitable supporting theory. However, in recent years these two difficulties have been addressed.   Using current theory and technology it seems that one should be able to implement adaptive control laws which outperform traditional control algorithms. Adaptive control algorithms are certainly an order of magnitude more complex than traditional algorithms but current computer technology means that this increase in complexity can be handled for little marginal increase in cost.

SURVEY OF CONVERGENCE THEORY

The theory of adaptive control is important since it helps delineate the best algorithm for a given application and since it highlights potential failure modes. Thus the convergence theory of adaptive control plays a role somewhat similar to that played by Nyquist's stability theorem in classical control.

Early work in the theory of adaptive control was restricted to the case when the plant order was exactly known. Within this framework, the

first theoretical stability results were obtained in the late 1970's.
At this time several groups (Feuer and Morse (1978), Goodwin,
Ramadge and Caines (1980), Narendra, Lin, Valavani (1980), Morse (1980),
Egardt (1979), Narendra and Lin (1980) published results which
established global convergence for a class of model reference adaptive control laws applied to linear minimum phase plants. It was
later realized (Goodwin and Sin (1984), Kreisselmeier (1985)) that
essentially the same proof worked for non-minimum phase plants
provided the parameters could be projected into a known convex set
in which no unstable pole-zero cancellations could occur. More
recently, it has been shown (Elliott, Christi, Das (1985), Goodwin and
Teoh (1985)) that global convergence can be achieved provided a
suitably rich external input can be applied.

More recently, attention has shifted to the so-called robustness
problem in adaptive control. The issue here is whether or not an
adaptive control algorithm can tolerate unmodelled dynamics. Rhors
et al (1983) showed that the basic adaptive control algorithms
could indeed fail if they were applied blindly to systems having
even small unmodelled dynamics or disturbances. On the other hand,
there is some practical evidence (Astrom (1983b), Wittenmark
and Astrom (1985)) which suggests that, at least some, adaptive control
algorithms are robust provided fairly simple practical precautions
are taken.

There has been considerable research effort aimed at gaining a
better understanding of adaptive control algorithms under non-ideal
conditions. For example, the adaptive control problem has been cast
into an input-output stability framework by Gawthrop and Lim (1982),
Kosut and Friedlander (1985), and Ortega and Landau (1983). Then
passivity or sector constraints are imposed to identify allowable
unmodelled dynamics. Kosut and Anderson (1984) have considered local-
output stability conditions using the same methods. Ioannou and
Kokotovic (1984) use singular perturbation techniques to suggest a
modification to the usual parameter update to ensure tracking to within

a uniform bound. Persistency of excitation guarantees the exponential stability of the homogeneous part of the adaptive system and hence ensures a measure of robustness. Anderson and Johnstone (1983), Chen and Cook (1984), Bodson and Sastry (1984), Narendra and Annaswamy (1983) and Kreisselmeier (1982) have explored this approach. Another line of study has been the method of averaging. Its application in adaptive control was first suggested by Astrom (1983b). The basic idea is that if the adaptation loop runs more slowly than the process dynamics then the unmodelled high frequency parts are less likely to be excited. This idea has been taken up in many papers including (Riedle and Kokotovic (1984), (1985), Fu, Bodson and Sastry (1985), Kokotovic, Riedle and Praly (1985) and Bitmead and Johnson (1985)).

Adaptive control algorithms are sometimes capable of quite complicated characteristics. For example, Mareels and Bitmead (1985) have recently shown that the parameter estimation trajectories for a very simple algorithm can fluctuate widely whereas the plant can be satisfactorily stabilized. For this reason, many authors have tried to ensure that the properties of the parameter estimator remain sensible irrespective of the nature of the control law and despite the presence of unmodelled errors. One way of achieving this is to introduce a deadzone into the parameter estimator - see for example Goodwin and Sin (1984), Egardt (1979), Samson (1983), Peterson and Narendra (1982), Martin-Sanches (1984), Goodwin, Hill and Palaniswami 1984) and (1985). The idea has recently been extended to cover unmodelled dynamics by using a relative deadzone (see for example Kriesselmeier and Anderson (1985),Ioannou and Tsakalis (1985), Goodwin, Leal, Mayne and Middleton (1986) and Goodwin, Hill, Mayne, Middleton (1985)). Extensions to time varying plants are explored in Anderson (1982) and Goodwin, Hill, Palaniswami (1985).

The culmination of the above theory is that it is now known that adaptive control algorithms are capable of robustly stabilizing those plants which can be approximately modelled by linear models of a given order. For the case of time-invariant plants this seems to be about as good a theoretical result as one could hope for and lends credibility to the application of adaptive control to practical problems.

To be concrete, we shall give emphasis in this paper to a particular adaptive control algorithm. This algorithm has been studied extensively in theory but probably requires further practical experience to verify its robustness properties. However, initial results seem encouraging.

THE ROLE OF THE DIGITAL COMPUTER

It perhaps goes without saying that adaptive control algorithms will be implemented digitally. It is therefore relevant to consider the effect of digital implementation of algorithms. There are two main issues here, namely the effect of the sampling rate and of finite wordlengths. These two effects combine to give a strict relationship between the desirable working frequency range and the sampling rate. The upper frequency is constrained by the fact that if the bandwidth approaches the sampling rate then intersample ripple problems can occur especially for lightly damped systems (de Souza and Goodwin (1983)). Thus it is desirable, if possible, to ensure that the sampling rate is 5 to 10 times the maximum frequency of interest.

At the other end of the spectrum, the finite wordlength and finite accuracy of A/D converters places a limit on the lowest frequency that can be sensibly distinguished from d.c. To illustrate the idea, consider a simple first order continuous system:

$$\frac{d}{dt} y = -\frac{1}{\tau} y + \frac{1}{\tau} u \qquad (3.1)$$

The corresponding discrete model at sampling interval Δ is

$$y(t+1) = ay(t) + bu(t) \text{ where } a = e^{-\Delta/\tau}, b = 1 - a. \qquad (3.2)$$

Since we are interested in the low frequency limit, it is reasonable to approximate $e^{-\Delta/\tau}$ as $1 - \Delta/\tau$. Then the model (3.2) becomes:

$$y(t+1) \simeq y(t) - \frac{\Delta}{\tau}[y(t)-u(t)] \qquad (3.3)$$

Assuming $y(t)$, $u(t)$ are scaled to the range $(-1,1)$ then the model (3.3) reduces to the useless model: $y(t+1) = y(t)$ whenever Δ/τ is less than the least significant bit, i.e. 2^{-b} where b is the number of binary digits in the A/D converter or worldlength. Thus for acceptable model accuracy we need

$$\frac{\tau}{\Delta} \ll 2^b. \qquad (3.4)$$

Noting that the frequency domain breakpoint is $f_\ell = 1/2\pi\tau$ and defining $f_s = 1/\Delta$, equation (3.4) suggests the following lower bound on the frequency range:

$$f_\ell > \frac{f_s}{2^b}. \qquad (3.5)$$

With 10 bits accuracy (which incidentally is roughly what can be achieved with standard analogue components), then

$$f_\ell > \frac{f_s}{1000}. \qquad (3.6)$$

The key factor in determining this limit is the rate of change of output and thus the same conclusion holds for pure integrators. We simply replace the time constant τ by the time taken (T_I) to integrate a 1 unit input to 1 unit output. Then the integrator will only work effectively if

$$\frac{T_I}{\Delta} \ll 2^b \qquad (3.7)$$

and the corresponding frequency limit can be calculated from $f_\ell = 1/2\pi T_I$.

Combining the upper and lower frequency constraints, we find that the desirable bandwidth in digital estimation and control is ideally

$$\frac{f_s}{1000} < f < \frac{f_s}{10} ; \quad f_s = \frac{1}{\Delta} \qquad (3.8)$$

4. NUMERICAL CONSIDERATIONS

This topic has been central to studies in digital filtering (see for example, Rabiner and Gold (1975)) but has been less extensively studied in the context of estimation and control. The problem has been studied by some authors including Belanger and Ghorein (1980), Moroney, Willsky and Houpt (1980), Tan and McInnes (1982), Williamson and Sridharan (1984) and Ahmed and Belanger (1984). Important insights have been gained especially with respect to the topic of scaling and numerical robustness. However, work to date has generally been concerned with shift operator or ARMA type models. We argue below that there seems to be a strong case to consider alternative model structures. One possibility would be lattice forms (Itakura and Saito, 1971). However, we advocate here a different form (Delta) which has the advantage of retaining intuitions based on continuous time transfer function ideas and also allows continuous and discrete methods to be brought together.

Let q denote the usual forward shift operator used in the analysis of discrete time systems. Then we define the discrete Incremental Difference Operator or Delta operator by

$$\delta = \frac{q-1}{\Delta} \qquad (4.1)$$

where Δ is the sampling interval.

The notion of the Delta operator is not new. It has been suggested in the digital filtering literature as a way of improving numerical properties (see for example Agarwal and Burrows, (1975) and Orlandi and Martinelli, (1984)).

Given a polynomial of any order n in the shift operator q, this will be exactly equivalent to some polynomial in δ of order n. Thus the usual transfer function discrete models of a linear system as a rational function in q will be exactly equivalent to another rational function in δ (with numerator and denominator of the same degree as the numerator and denominator of the rational function in q). Also a filter or controller expressed as a rational function in δ may be implemented without converting back to shift form by simply using an observer (or controller) state space model implemented using the Delta operator.

A new basic building block is necessary to implement delta models; namely the function (δ^{-1}). This building block has properties similar to a continuous time integrator and replaces the unit delay (q^{-1}) used in the standard discrete time models. The required computation to implement δ^{-1} is set out below:

If
$$\alpha_t = \delta^{-1} \beta_t \qquad (4.2)$$
then
$$\alpha_{t+1} = \alpha_t + \Delta \beta_t \qquad (4.3)$$

The hardware, or software, necessary to implement this building block is only marginally more complex than that necessary to implement the more usual shift operator, q^{-1}. Delta blocks can be manipulated in much the same way as integrators are manipulated in the implementation of continuous time models.

Of course, δ^{-1} as given in (4.2), (4.3) looks like a Euler approximation to an integral. This implies that δ (in discrete time) and $D = d/dt$ (in continuous time) have a similar (though not identical) physical interpretation. This means that insights obtained in continuous designs can be mapped rather easily into the discrete domain. For example, the stability boundary for a δ-model is a restricted left half plane consisting of a circle of radius $1/\Delta$ centered on the point $(- 1/\Delta, 0)$. We also see that as the sampling rate increases (i.e. $\Delta \to 0$) then the discrete case reduces to the continuous case.

In fact, one can replace all discrete design methods in the Z-domain by corresponding design methods using Delta operators. For example, Z-transforms can be replaced by the following Γ-transform which is better matched to the Delta model

$$Y_\Gamma(\gamma) = \Delta \sum_{k=0}^{\infty} y(k\Delta)(1 + \gamma\Delta)^{-k} \qquad (4.4)$$

The properties of this transform are more akin to Laplace Transforms. Also the Γ transform converges to the Laplace transform as $\Delta \to 0$. We believe that this link between continuous and discrete design methods has considerable appeal in the teaching of digital design methods (see Goodwin and Middleton (1986a)).

A bonus arising from the use of delta operators is that the numerical properties of digital control laws are improved relative to the usual shift operator models. The difficulty with shift models can be traced to the fact that, as the sampling rate increases, all of the relevant poles and zeros cluster around the point $(1 + j0)$ in the complex plane. Thus, much of the available wordlength is wasted in continually representing this off-set. The delta operator, on the other hand, effectively subtracts 1 from the pole-zero locations and this greatly improves the numerical properties.

These issues are explored in more detail in Middleton and Goodwin (1985), Goodwin (1985) and Goodwin and Middleton (1986b). In these papers it is established that the delta operator has superior numerical properties relative to shift operators with respect to:

- coefficient representation accuracy
- round off noise
- on-line design law calculations
- on-line parameter estimation.

For these reasons we will use delta operators (rather than shift operators) in the remainder of this paper. More generally, we commend delta operators as an interesting alternative discrete time model.

In the next two sections we shall describe robust parameter estimation and robust control law synthesis based on delta operators. Then in Section 7 we will show how these can be combined to yield a possible digital adaptive control law.

5. ROBUST PARAMETER ESTIMATION

For simplicity we assume that the nominal plant can be approximately modelled in the following form:

$$y = H_0 u + H_1 z + d + \xi \qquad (5.1)$$

where u, y, z, d, ξ denote the plant input, output, external perturbations, deterministic disturbances and plant modelling error respectively. The nominal plant transfer function, H_0, and perturbation transfer function, H_1, are assumed to be rational functions of the form

$$H_0 = \frac{B}{A} \quad ; \quad H_1 = \frac{F}{A} \tag{5.2}$$

where we have assumed a common denominator for simplicity. In (5.2) A,B are coprime polynomials of degree ∂_A and ∂_B ($\leq \partial_A - 1$) respectively in the δ operator with A monic, F is a polynomial of degree $\delta_F \leq \delta_A$ in δ.

The purely deterministic disturbance d will typically be a d.c. offset. However, more generally, disturbances can be accommodated. In fact, it is possible in parable to treat any disturbance which can be modelled in the form:

$$Od = 0 \tag{5.3}$$

where O is a monic polynomial in δ of degree ∂_O having non-repeated zeros on the stability boundary. (If d is constant then $O = \delta$ suffices).

Multiplying both sides of (5.1) by AO gives

$$AOy = BOu + FOz + AO\xi \tag{5.4}$$

The above equation is unsuitable for parameter estimation due to the fact that the "equation error" term $AO\xi$ involves near differentiation of the errors including noise. (Note that this is true irrespective of the operator used in the model). We shall therefore introduce a filter EQ where E is of degree ∂_A and Q is stable and 'close' to O, (i.e. if $O = \delta$ then $Q = \delta+\varepsilon$, $\varepsilon > 0$ suffices).

Operating on (5.4) by 1/EQ gives

$$Ay_f = Bu_f + Fz_f + \eta_f \tag{5.5}$$

where

$$y_f \triangleq \frac{O}{EQ} y \tag{5.6}$$

$$u_f \triangleq \frac{O}{EQ} u \tag{5.7}$$

$$z_f \triangleq \frac{O}{EQ} z \tag{5.8}$$

$$\eta_f \triangleq \frac{AO}{EQ} \xi \tag{5.9}$$

Equation (5.5) can also be written as

$$\bar{y} = (E-A)y_f + Bu_f + Fz_f + \eta_f \qquad (5.10)$$

where

$$\bar{y} \triangleq Ey_f = \frac{O}{Q} y$$

Equation (5.10) can then be expressed in regression form as

$$\bar{y}(k) = \phi(k-1)^T \theta_* + \eta_f(k) \qquad (5.11)$$

where

$$\phi(k-1)^T = [y_f(k),\ldots,\delta^{n-1}y_f(k),u_f(k),\ldots,\delta^m u_f(k),z_f(k),\ldots\delta^r z_f(k)]$$

$$\ldots\ldots (5.12)$$

$$\theta_* = [e_0 - a_0, \ldots e_{n-1} - a_{n-1}, b_0, \ldots, b_m, f_0, \ldots f_r] \qquad (5.13)$$

where

$$A(\delta) \triangleq \delta^n + a_{n-1}\delta^{n-1} \ldots + a_0 \;;\; n = \partial_A \qquad (5.14)$$

$$B(\delta) \triangleq b_m \delta^m + b_{m-1}\delta^{m-1} + \ldots + b_0 \;;\; m = \partial_B \qquad (5.15)$$

$$E(\delta) \triangleq \delta^n + e_{n-1}\delta^{n-1} + \ldots + e_0 \qquad (5.16)$$

$$F(\delta) \triangleq f_r \delta^r + f_{r-1}\delta^{r-1} + \ldots + f_0 \;;\; r = \partial_F. \qquad (5.17)$$

Note that we have used the notation:

$\bar{y}(k)$, $\phi(k-1)$ since $\bar{y}(k)$ depends on $\{y(t): t \leq k\}$
and $\phi(k-1)$ depends on $\{y(t), u(\tau): t \leq k-1\;;\; \tau \leq k-n+m\}$.

The regression model (5.11) suggests the following least squares parameter estimator:

$$\delta\theta(k-1) = \frac{a(k)}{\Delta} \frac{P(k-2)\phi(k-1)}{1 + \phi(k-1)^T P(k-2)\phi(k-1)} e(k) \qquad (5.18)$$

$$\delta P(k-2) = -\frac{a(k)}{\Delta} \frac{P(k-2)\phi(k-1)\phi(k-1)^T P(k-2)}{1 + \phi(k-1)^T P(k-2)\phi(k-1)} \qquad (5.19)$$

where $e(k)$ is the prediction error

$$e(k) \triangleq \bar{y}(k) - \phi(k-1)^T \hat{\theta}(k-1) \qquad (5.2)$$

The term $a(k)$ in equations (5.18), (5.19) implements a deadzone to account for modelling errors (see Section 2). The key idea is to set $a(k) = 0$ whenever the prediction error falls below some pre-specified bound. This ensures that the parameter estimator is not "thrown off course" by small errors. We have found this idea to be very helpful in getting on-line parameter estimators to work in practice.

<u>Remark 5.1</u>

In practice, it is desirable to slightly modify the algorithm (5.18), (5.19) to prevent the matrix P from going to zero. Many possible modifications have been proposed in the literature including

- exponential forgetting
- covariance modification
- covariance resetting, and
- constant trace algorithms

The convergence properties of the algorithm (5.18), (5.19) apply equally well to all of these variants provided only that P increases when it is modified and provided an upper bound is placed on trace P (see Goodwin, Hill and Palaniswami (1985)).

An algorithm that we have found to work quite well in practical cases is the following Regularized Constant Trace Algorithm:

$$P'(k-1) = P(k-1) - a(k) \frac{P(k-2)\phi(k-1)\phi(k-1)^T P(k-2)}{1 + \phi(k-1)^T P(k-2)\phi(k-1)} \quad (5.21)$$

$$P(k-1) = c_1 \frac{P'(k-1)}{\text{Trace}[P'(k-1)]} + c_2 I \quad ; \quad c_1 > 0, \; c_2 > 0 \quad (5.22)$$

<u>Remark 5.2</u>

In setting up equation (5.12), (5.13) we have assumed that all parameters are unknown. However, in some cases, it is known a-priori that the plant numerator or denominator contains a particular factor. For example, servo systems usually have an integrator in the plant. These known factors can be removed from the parameter estimator by simply multiplying the corresponding factor into the appropriate terms in the regression vector.

<u>Remark 5.3</u>

If prior information is available specifying a known region in

which the parameters lie then the parameter estimator can be
improved by projecting the estimates to lie inside this region. This
is readily achieved - see Goodwin and Sin (1984, p. 91). For example,
it is often desirable to ensure that the estimated gain of the plant
($\hat{b}_0$ in δ form) does not get too small.

Remark 5.4

It sometimes happens that measurement of an auxiliary variable
gives valuable information about the parameter estimates. Well known
examples of these are the dynamic pressure in aircraft and speed in
ships. This information can be used to initialize the parameter
estimator with good values. Thus "gain scheduling" and feedback
adaptive control can be combined.

Remark 5.5

An important function of the filter $\frac{O}{EQ}$ is to reduce the effects
of noise dynamics. For example, if one adopts a purely random model
for the noise, then 1/E would correspond to the optimal steady state
Kalman Filter for the process. The additional filter O/Q arises due
to the presence of deterministic disturbances. This filter can be
thought of as a rational approximation to a p-point robuts filter
(Kassam and Poor (1985) and Vastola and Poor (1984)) for removing
the deterministic component d, without otherwise distorting the
spectrum of the noise. If d = constant then we can use $O/Q = \frac{\delta}{\delta+\varepsilon}$
which, as expected, is a high pass filter.

The composite filter $\frac{O}{EQ}$ can also be derived in the stochastic
case by Kalman Filtering Theory for systems having purely
deterministic disturbances (Goodwin and Sin (1984) and Chan,
Goodwin and Sin (1984)).

Remark 5.6

Another function of the filter $\frac{O}{EQ}$ is to ensure that unmodelled
(high frequency) response data is not passed to the parameter estimator.
Thus the bandwidth of 1/E should be roughly comparable to the desired
closed loop bandwidth.

Remark 5.7

In the above discussion, we have assumed that the filter

polynomials E, O, and Q are fixed and given. This covers most
cases of practical interest. However, it is possible, if desired,
to estimate these polynomials also but this leads to a much more complicated
algorithm (see Goodwin and Sin (1984) Part B and Goodwin (1985a)).

6. ROBUST CONTROL SYSTEM DESIGN

When the model for the plant and the environment are known,
then there exist many methods for designing a robust control system,
e.g. frequency domain algorithms, optimal control, model reference
control, pole assignment etc. Here we will use pole-assignment since
it is relatively simple and since many other design methods turn out
to be special cases. We will also use the Internal Model Principle
(Francis and Wonham (1976)) so that the classical three term control
law will turn out to be a special case.

We assume that the desired plant output, y^*, satisfies a model
of the form

$$Sy^* = 0 \qquad (6.1)$$

where S is a monic polynomial in δ of degree ∂_S having non-repeated
roots on the stability boundary. For example, if y^* = constant then
$S = \delta$ suffices. However, we can treat arbitrary periodic signals
in the same way.

We first factor B as $B = B_\alpha B_\beta$ where B_α is stable and well-damped
and B_β is not necessarily so. We will cancel B_α in the closed loop.
The input is then generated from

$$B_\alpha LOSu' = -Py + Gy^* \qquad (6.2)$$
$$u = u' - Hz - \gamma \qquad (6.3)$$

where G,H are rational transfer functions whose purpose is to
provide feedforward action and where L and P are polynomials in δ
determined from the following pole assignment equation:

$$ALOS + B_\beta P = A^* \qquad (6.4)$$

where A^* is a monic stable polynomial of degree $\partial_{A^*} = 2\partial_A + \partial_0 + \partial_S - 1$.
The degree of L and P are respectively $\partial_L = \partial_A - 1$ and $\partial_P = \partial_A + \partial_0 + \partial_S - 1$.
When substituted into (6.1) this gives a proper but not strictly

proper control law. A proper control law can be obtained by increasing the degree of A* and L by 1. The control law is shown diagrammatically in Figure 6.1.

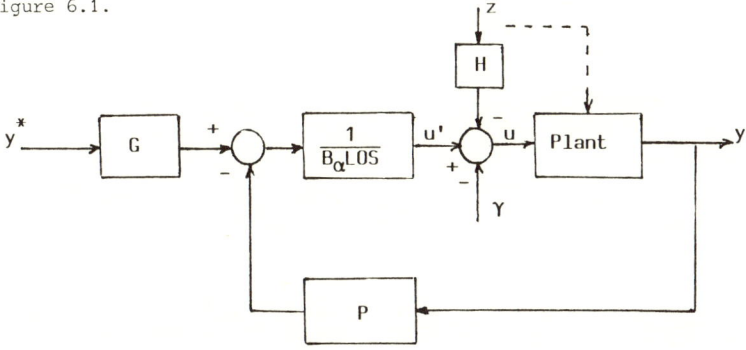

Figure 6.1 Structure of Feedback Control Law

Let V be any monic stable polynomial of degree $\partial_L + \partial_O + \partial_S$. Then equation 6.2 can also be expressed as

$$u' = K\left(\frac{u'}{V}\right) - P\left(\frac{y}{V}\right) + G\left(\frac{y^*}{V}\right) \qquad (6.5)$$

where $K = V - LOS$.

In equation (6.5), the operator $1/V$ plays the role of an observer in the usual state space terminology. In the case of a strictly proper controller, then a suitable choice for V is C_s where C_s is a monic stable polynomial close to SOE, i.e. $C_s = S'QE$ suffices where S' is stable and is close to S. Then equation 6.5 simplifies to

$$u'' = K'\left(\frac{u''}{E}\right) - P\left(\frac{y}{EQS'}\right) + G\left(\frac{y^*}{EQS'}\right) \qquad (6.6)$$

and where $[SO]u' = [S'Q]u''$ and $K' = E-L$ (of degree ∂_A).

To ensure that (6.4) is solvable for L,P we require that AOS and

B be relatively prime.

Substituting (6.2), (6.3) into (5.4) gives the following closed loop equation:

$$A^*y = B_\beta G y^* + LOS[F - B_\alpha B_\beta H]z + EQLS\ \eta_f + LOS[Ad - B_\alpha B_\beta \gamma] \quad (6.7)$$

We now briefly discuss the user choices in the above control law.

(a) **Choice of A^***

Ideally A^* should be chosen so that $1/A^*$ has as wide a bandwidth as possible. However, large feedback gains must be avoided since this reduces robustness in the presence of unmodelled dynamics. Hence the additional poles introduced in S and O cannot be shifted very far. (This is the usual classical result that integral action should be sufficiently slow that it does not destabilize the plant.) Also, disturbance rejection is enhanced if E is included as part of the closed loop characteristic polynomial. Thus a suitable choice for A^* is

$$A^* = \bar{A}^* E Q S' \quad (6.8)$$

where E, Q, S' are as defined earlier and $\bar{A}^*$ is an appropriate stable polynomial of degree equal to $\partial_{\bar{A}}$. Using this choice in (6.7), the tracking error due to the noise term η_f is seen to be approximately $(L\eta_f)/\bar{A}^*$. This is arbitrarily close to what can be achieved with non-robust control laws such as minimum variance control.

(b) **Determination of H**

It is clear from (6.7) that the effect of z on the closed loop is reduced if $F \simeq B_\alpha B_\beta H$. A suitable choice for H then is

$$H = \frac{F}{B_\alpha B_\beta' D_1} \quad (6.9)$$

where B_β' is stable and is 'close to' B_β in some sense, D_1 is any stable polynomial making H proper. Sensible choices are $B_\beta' = 1$ and $1/D_1$ having wide bandwidth such that the d.c. gain of F equals that of $B_\alpha B_\beta H$. This gives perfect cancellation at d.c. and near cancellation up to the 'breakpoints' in B_β and D_1.

(c) Determination of γ

Since $Od = 0$, it is clear from (6.7) that $\gamma = 0$ is sufficient to give zero steady state tracking error due to the deterministic disturbance term d. However d typically represents a load disturbance and this can undergo infrequent step changes. The term γ can then be used to provide feedforward action. Unlike the term z, the disturbance term, d, is not directly measured. However, from (5.1) and (5.2) we see that (neglecting the random noise term ξ) the value of Ad is approximately given by $Ay - Bu - Fz$. Hence from (6.7), the influence of d can be reduced if we choose γ so that $B_\alpha B_\beta \gamma$ is approximately $Ay - Bu - Fz$. A suitable choice for γ is thus

$$\gamma = \frac{Ay - Bu - Fz}{B_\alpha B_\beta' D_2}$$

where $B_\beta' D_2$ is stable and $B_\alpha B_\beta' D_2$ has degree $\geq \partial_A$. The d.c. gain of $B_\alpha B_\beta' D_2$ should be approximately the same as that of B. However, one must check that closed loop stability is maintained especially if estimates of A,B,F are used in forming γ. D_2 is important in this regard, since it limits the bandwidth over which the model is used to estimate d.

Note that if A is stable, then the above expression for γ can be rewritten as

$$\gamma = \frac{A}{B_\alpha B_\beta' D_2} \hat{d} \text{ where } \hat{d} = y - \frac{Bu}{A} - \frac{Fz}{A}$$

In this case the estimate $\hat{d}$ is simply the difference between the model output and the plant output. This is essentially the same idea as is used in the well known Smith Predictor (Smith (1959)). We have successfully used a generalization of the idea (as described above) to eliminate sinewave like eccentricity disturbances in steel rolling mills.

(d) Determination of G

If $1/\bar{A}^*$ can be made to have a wide bandwidth, then simple unity feedback suffices, i.e. we can choose $G = P$. With this choice it can be easily seen that (6.7) simplifies to (in steady state)

$$A^*(y-y^*) = LOS[F - B_\alpha B_\beta H]z + EQLS \, \eta_f \qquad (6.10)$$

Thus, y asymptotically tracks y^* to within the error induced by z and η_f alone.

In some cases, e.g. when the system has a significant pure time delay, it is not possible to robustly achieve a wide closed loop bandwidth. In these cases, the transfer function G can be used to give feedforward "forcing" of the plant ensuring improved transient response to changes in y^*. In essence, one uses G as an inverse of the closed loop plant over some frequency range.

For simplicity, we consider the common practical case when $y^* = $ constant and $SO = \delta$. Thus, we shall consider the case of simple integral action (more general cases can be treated similarly). Choosing $G = p_0 G'$ where p_0 is the coefficient of δ^0 in P and G' has d.c. gain of 1, then retains zero steady state tracking error from y^* to y. G' can then be chosen to 'cancel' the closed loop poles and zeros, i.e. the following choice suffices

$$G' = \frac{A^*}{\bar{A}^* B_\beta'} \cdot \frac{\bar{a}_0^* (b_\beta')_0}{a_0^*} \qquad (6.11)$$

or, more simply $G' = \dfrac{A^*}{\bar{A}^*} \cdot \dfrac{\bar{a}_0^*}{a_0^*}$ where $1/\bar{A}^*$ has wide bandwidth and B_β' is stable and 'close to' B_β.

Remark 6.1

Note that the operator $\dfrac{A^*}{\bar{A}^*} y^*$ gives a bandlimited "differentiation" of y^* and hence one must be careful with noise problems. The bandwidth of $1/\bar{A}^*$ is fixed by the range of validity of the model and the saturation limits on the input.

As a simple example, say $\frac{B_\beta}{A^*} = \frac{1}{s+1}$ and we use $G' = \frac{5(s+1)}{s+5}$, then for a unit step input, the various signals of interest are as shown below

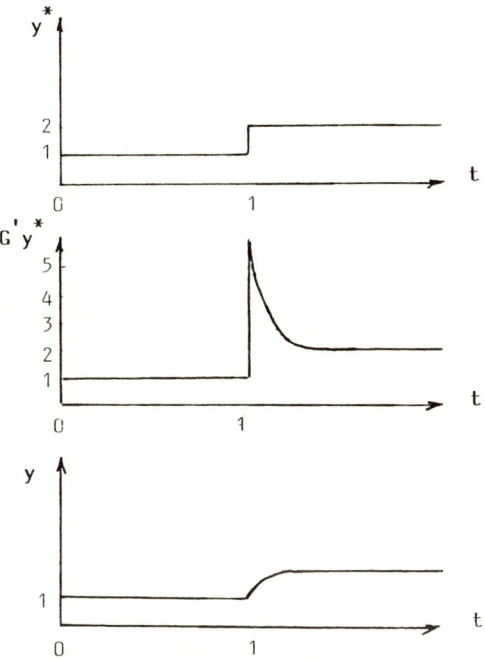

Of course, if a disturbance occurs at the output then this decays with a time constant of 1 second unless feedforward via γ is also incorporated.

Remark 6.2

A special case of (6.11) occurs when the open loop plant is stable and the feeback terms are small. In this case A^* approximately contains A as a factor. G' can then be taken as $\frac{A}{A^*}\left(\frac{\bar{a}_0^*}{a_0}\right)$. This then becomes equivalent to using a "Smith Predictor" (Smith (1959), Palmer and Shinnar (1979)). The latter control law is usually implemented as in Fig. 6.2

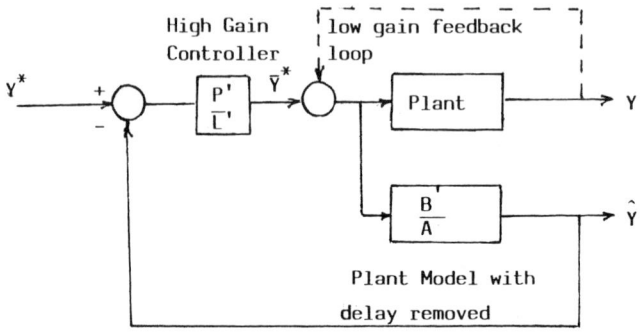

Figure 6.2 Structure of Smith Predictor

The transfer function from Y* to $\bar{Y}$* is readily seen to be $\left[\frac{P'A}{AL' + P'B'}\right]$. Thus the feedback structure in Fig. 6.2 simply uses feedback from $\hat{Y}$ to place $\frac{A}{\bar{A}^*}$ in series with y* where $\bar{A}^*$ = AL' + P'B'. This is more directly achieved in (6.11).

The result is also equivalent to minimum variance control for systems having long deadtimes - see Goodwin and Sin (1984, p. 415) and Palmer and Shinnar (1979).

7. ADAPTIVE CONTROL

An adaptive control law can, in principle, be obtained by combining any robust parameter estimator with any robust control law synthesis procedure as shown in Fig. 7.2. For example, if the least squares algorithm of Section 5 is combined with the pole-assignment method of Section 6, then one obtains the adaptive control law studied more fully in Goodwin, Hill, Mayne, Middleton (1985). We have found this controller to work quite well in practice (at least for simple servo systems). Typical results for a 20:1 change in system gain are shown in Fig. 7.1. Under these circumstances, the system goes unstable when a fixed parameter (non-adaptive) control law is used.

Digital Adaptive Control

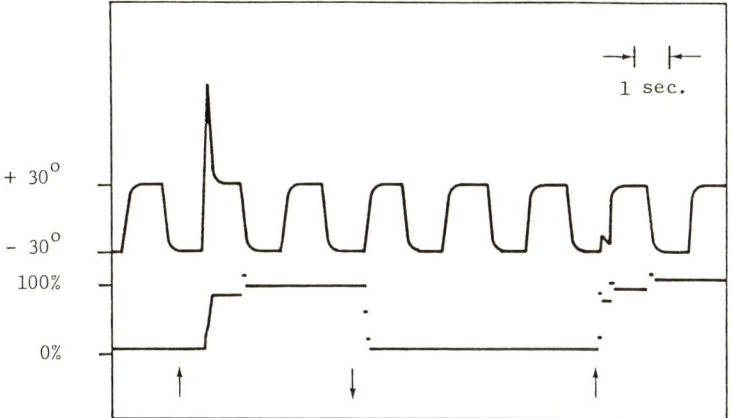

Figure 7.1

Adaptive Control of Servo System
Upper trace: system response
Lower trace: estimate of system gain ($\hat{b}_0$)
↑ 20:1 gain increase
↓ 20:1 gain decrease

Potential advantages of adaptive control relative to conventional control are:

- 'tight feedback control can be maintained even if the plant parameters changes,

- feedforward action depends on open-loop cancellation and hence on-line parameter estimation is particularly helpful since it ensures that the feedforward terms are appropriate,

- if the adaptation mechanism is removed, then the controller simply reverts to a classical robust fixed parameter control law.

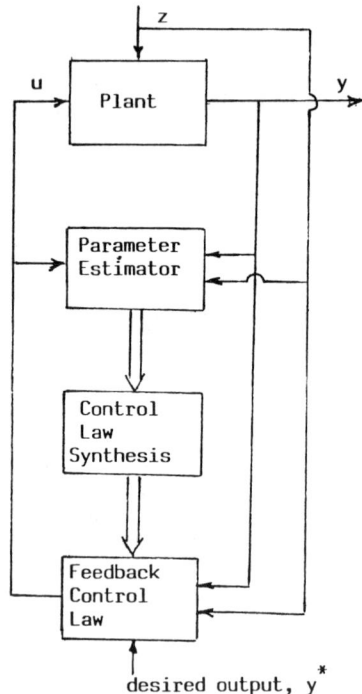

Figure 7.2 Structure of Adaptive Control Law

8. PLANTS WITH VARIABLE TIME DELAY

In many cases of practical interest the plant can have a variable time delay. This can be incorporated into the design procedure outlined earlier. A key observation in this regard is that if the time delay is relatively small compared with the dominant plant time constants then a _very_ high level of approximation can be achieved with a simple rational function (e.g. second order Padé). On the other hand, if the delay is large compared with dominant plant time constants then tight feedback is impossible and again there is no point having a highly accurate model for the time delay. Thus, it seems that in many cases, it suffices to use an approximate rational model for the delay, say

$$D_r(\delta) = \left[\frac{1 - \frac{\delta T}{2r}}{1 + \frac{\delta T}{2r}} \right]^r \qquad (8.1)$$

where T is the time delay and r is the order of the approximation. For example, with rapid sampling then with $r = 2$, for frequencies up to $1/3T$ the percentage error in phase is of the order of 5%. This is clearly insignificant for robust control law synthesis.

Since the function given in (8.1) is rational it can be estimated along with the other plant parameters. The filter $1/E$ now plays a key role in limiting the bandwidth of signals reaching the parameter estimator to those for which the rational approximation (8.1) is realistic. (Clearly, if very high frequency signals are passed to the parameter estimator then a rational approximation to the delay will no longer be sensible - see also Remark 5.6.) The control system synthesis procedure of Section 6 also applies to the case of a time delay. Further details of the implication of these ideas in adaptive control is contained in de Souza, Goodwin and Palaniswani (1985). Preliminary experimental results using the idea seem quite encouraging.

9. CONCLUSIONS

This paper has described recent results on robust adaptive control.

The emphasis has admittedly been on the theory. However, our aim has been to point to some different ideas in the hope that this might lead to some discussion and reflection.

Further practical work is necessary but it is, at least, feasible adaptive control laws could replace existing control algorithms especially where high performance is critical. This claim is made more plausible by the fact that it is possible to ensure that traditional fixed control laws, such as the ubiquitous PID controller, are obtained as special cases of the adaptive control laws.

10. ACKNOWLEDGEMENTS

The author gratefully acknowledges helpful interaction with Professor David Mayne of Imperial College during the preparation of this paper.

11. REFERENCES

Agarwal, R.C. and C.S. Burrows (1975). New recursive digital filter structures having very low sensitivity and roundoff noise, I.E.E.E. Trans. Circuits and Systems, CAS22, 12.

Ahmed, M.E. and P.R. Belanger (1984). Scaling and roundoff in fixed point implementation of control algorithms, I.E.E.E Trans. Indust. Electronics, August.

Anderson, B.D.O. (1982). Exponential convergence and persistant excitation, Proc. 21st I.E.E.E. Conf. on Decision and Control, Orlando, Florida.

Anderson, B.D.O. and R.M. Johnstone (1983). Adaptive systems and timevarying plants, Int. J. of Control, 37, 367-377.

Astrom, K.J. (1983a). Theory and applications of adaptive control - a survey, Automatica, 19, 5, 471-487.

Astrom, K.J. (1983b). Analysis of Rhors counter examples to adaptive control, Proc. 22nd Conf. on Decision and Control, San Antonio, Texas.

Belanger, P.R. and Y. Ghoreim (1980). Fixed point arithmetic microprocessor implementation of self tuning regulators. Joint Automatic Control Conf., San Francisco.

Bitmead, R. and C.R. Johnson Jr. (1985). Discrete averaging and robust identification, Control and Dynamic Systems, XXXIV, Ed. C.T. Leondes.

Bodson, M. and S. Sastry (1984). Exponential convergence and robustness margins in adaptive control, Proc. 23rd CDC, Las Vegas, December.

Chan, S.W., G.C. Goodwin, K.S. Sin (1984). Convergence properties of the Riccati difference equation in optimal filtering of nonstabilizable systems, I.E.E.E. Trans. Auto. Control, AC-29, 2, 110-119.

Chen, Z.J. and P.A. Cook (1984). Robustness of model reference adaptive control systems with unmodelled dynamics, Int. J. Control, 39, 1, 201-214.

De Souza, C.E. and G.C. Goodwin (1983). Robustness effects of sampling in model reference and minimum variance control, 22nd CDC Conference, San Antonio, Texas.

De Souza, C.E., G.C. Goodwin and M. Palaniswami (1985). An adaptive control algorithm for linear systems having variable time delay, Tech. Report, Dept. of Elec. and Com. Eng., University of Newcastle, Australia.

Egardt, B. (1979). Stability of adapative controllers, Lecture notes in Control and Information Series, Springer, New York.

Elliott, H., R Christi, M. Das (1985). Global stability of adaptive pole placement algorithms, I.E.E.E. Trans. on Auto. Control, AC-30, 4, 348-357.

Feuer, A. and A.S. Morse (1978). Adaptive control of a single input single output linear system, I.E.E.E. Trans. Auto. Control, AC-23, 557-570.

Francis, B. and W.M. Wonham (1976). The internal model principle of control theory, Automatica, 12, 457-465.

Fu, L.C., M. Bodson, S.S. Sastry (1985). New stability theorems for averaging and their applications to the convergence analysis of adaptive identification and control schemes, Memorandum No. UCB/ER1 M8J/21, Electronics Research Laboratory, University of California, Berkeley, USA.

Gawthrop, P.J. and K.W. Lim (1982). Robustness of self tuning controllers, Proc. I.E.E., 129, 21-29.

Goodwin, G.C. (1985a). Some observations on robust estimation and control, Plenary address, 7th IFAC Symposium on Identification and System Parameter Estimation, York, UK.

Goodwin, G.C. (1985b). UNAC - The University of Newcastle Adaptive Controller, Tech. Report, University of Newcastle, Australia.

Goodwin, G.C., D.J. Hill, D.Q. Mayne, R.H. Middleton (1985). Adaptive robust control (convergence, stability and performance), Department of Electrical Engineering, Tech. Report, Imperial College, UK.

Goodwin, G.C, D.J. Hill, M. Palaniswami (1984). A perspective on convergence of adaptive control algorithms, Automatica, 20, 5, 519-532.

Goodwin, G.C., D.J. Hill, M. Palaniswami (1985). Towards an adaptive robust controller, 7th IFAC Symposium on Identification and System Parameter Estimation, York, UK.

Goodwin, G.C., R. Lozano Leal, D.Q. Mayne, R.H. Middleton (1986). Rapprochement between continuous and discrete model reference adaptive control. To appear, Automatica.

Goodwin, G.C. and R.H. Middleton (1986b). Continuous and discrete adaptive control, Control and Dynamic Systems, XXXIV, Ed. C.T. Leondes.

Goodwin, G.C. and R. Middleton (1986). Digital Estimation and Control: a unified approach, book (in preparation).

Goodwin, G.C., P.J. Ramadge, P.E. Caines (1980). Discrete time multivariable adaptive control, I.E.E.E. Trans. Aut. Con., AC-25, 449-456.

Goodwin, G.C. and K.S. Sin (1984). Adaptive Filtering, Prediction and Control, Prentice Hall.

Goodwin, G.C. and E.K. Tooh (1985). Presistency of excitation in the presence of possibly unbounded signals, I.E.E.E. Trans. Auto. Con., AC-30, 6, 595-597.

Ioannou, P. and P.V. Kokotovic (1984). Robust redesign of adaptive control, I.E.E.E. Trans. Aut. Con., AC-29, 202-211.

Ioannou, P.A. and K. Tsakalis (1985). Robust discrete-time adaptive control, report 85-10-1, Elec. Eng. Systems, University of Southern California, USA.

Itakura, F. and S. Saito (1971). Digital filtering for speech analysis and synthesis, Proc. 7th Int. Conf. Acoust., Budapest, 25-C-1, 261-264.

Kassam, S.A. and H.V. Poor (1985). Robust techniques for signal processing: a survey, Proc. I.E.E.E., 73, 3, 433-481.

Kokotovic, P., B. Riedle and L. Praly (1985). On a stability criterion for continuous slow adaptation, Systems and Control Letters, 6, 7-14.

Kosut, R.L. and B.D.O. Anderson (1984). Robust adaptive control: conditions for local stability, Proc. 23rd CDC, Las Vegas,

Kosut, R.L. and B. Friedlander (1985). Robust adaptive control: conditions for global stability, I.E.E.E. Trans. Aut. Con., AC-30, 7, 610-624.

Kreisselmeier, G. (1985). An approach to stable indirect adaptive control, Automatica, 21, 4, 425-433.

Kreisselmeier, G. (1982). On adaptive state regulation, I.E.E.E. Trans. Aut. Con., AC-27, 3-17.

Kreisselmeier, G. and B.D.O. Anderson (1985). Robust model reference adaptive control, DFVLR Report, March, to appear I.E.E.E. Trans. Aut. Con.

Mareels, I.M.Y. and R.R. Bitmead (1985). Nonlinear dynamics in adaptive control; chaotic and periodic stabilization, Australian National University Tech. Report, August.

Martin-Sanches, J.M. (1984). A globally stable APCS in the presence of bounded noise and disturbances, I.E.E.E. Trans. Aut. Con., AC-29, 461-464.

Middleton, R.H. and G.C. Goodwin (1985). Improved finite wordlength characteristics in digital control using delta operators, to appear in I.E.E.E. Trans. Aut. Con.

Moroney, P., A.S. Willsky, P.K. Houpt (1980). The digital implementation of control compensators: the coefficient wordlength issue, I.E.E.E. Trans. Aut. Con., AC-25, pp 621-630.

Morse, A.S. (1980). Global stability of parameter adaptive control systems, I.E.E.E. Trans. Aut. Con., AC-25, 3, 433-439.

Narendra, K.S. and A.M. Annaswamy (1983). Persistent excitation and robust adaptive algorithms, Proc. 3rd Yale Workshop on Applications of Adaptive Systems Theory, Yale, USA

Narendra, K.S. and Y.H. Lin (1980). Stable discrete adaptive control, I.E.E.E. Trans. Aut. Con., AC-25, 3, 456-461.

Narendra, K.S., Y.H. Lin, L.S. Valavani (1980). Stable adaptive controller design, part II: proof of stability, I.E.E.E. Trans. Aut. Con., AC-25, 3, 440-449.

Orlandi, G. and G. Martinelli (1984). Low sensitivity recursive digital filters obtained via the delay replacement, I.E.E.E. Trans. Circuits and Systems, CAS-31, 7, July.

Ortega, R. and I.D. Landau (1983). On the model mismatch tolerance of various parameter adaptation algorithms in direct control schemes, a sectoricity approach, Proc. IFAC Workshop on Adaptive Control, San Francisco, USA, June.

Palmer, Z.J. and R. Shinnar (1979). Design of sampled data controllers, Ind. Eng. Chem. Process Des. Dev, 18, 1.

Peterson, B.B. and K.S. Narendra (1982). Bounded error adaptive control, I.E.E.E. Trans. Aut. Con., AC-27, 1161-1168.

Rabiner, L.R. and B. Gold (1975). Theory and Application of Digital Signal Processing, Prentice-Hall, New Jersey.

Rhors, C., L. Valavani, M. Athans, G. Stein (1983). Robustness of adaptive control algorithms in the presence of unmodelled dynamics, 21st I.E.E.E. Conf. on Decision and Control., Orlando, 1983. Also, I.E.E.E. Trans. Aut. Con., AC-30, 9, 881-889 (1985).

Riedle, B.D. and P.V. Kokotovic (1984). A stability instability boundary for disturbance free slow adaptation and unmodelled dynamics, Proc. 23rd I.E.E.E. CDC Conf., Las Vegas.

Riedle, B.D. and P.V. Kokotovic (1985). Integral manifolds and slow adaptation, University of Illinois technical report, DC-87, October.

Samson, C. (1983). Stability analysis of adaptively controlled systems subject to bounded disturbances, Automatica, 19, pp 81-86.

Smith, O.J.M. (1959). ISA, J., Vol 6(2), 28.

Tan, C.I. and B. McInnes (1982). Adaptive digital control implemented using residue number systems, I.E.E.E Trans. Aut. Con., AC-27, 499-502, April.

Vastola, K.S. and H.V. Poor (1984). On the p-point uncertainty class, I.E.E.E. Trans. Inf. Theory, IT-30, 2, 374-376, March.

Williamson, D. and S. Sridhavan (1984). Coefficient and state wordlength correction in digital Kalman filters, Technical Report,
University of New South Wales, Australia.

Wittenmark, B. and K.J. Astrom (1985). Practical issues in the implementation of self-tuning control, Automatica, 20, 5, pp 595-606,
September.

UNIVERSAL DYNAMIC MATRIX CONTROL

A. M. Morshedi

Shell Development Company, Houston, TX 77001

Abstract. Universal Dynamic Matrix Control (UDMC) is an algorithm for optimization and control of multiple-input - multiple-output nonlinear processes. UDMC is the generalization of the previously developed QDMC algorithm; it extends the optimization of the quadratic objective function subject to linear constraints to the solution of a nonlinear optimization problem subject to nonlinear differential equations. The classical approach for solving such an optimization problem leads to a difficult nonlinear two-point boundary value problem. UDMC simplifies the task of solving the two-point boundary value problem by solving the original optimization problem via a nonlinear programming approach. This nonlinear optimization problem is formulated by approximating the continuous input function by a finite sequence of piecewise continuous functions.

Keywords. Linear Programming (LP); Quadratic Programming (QP); Generalized Reduced Gradient (GRG); Simple functions; Piecewise Continuous; Dynamic Matrix Control (DMC); Quadratic Programming solution of Dynamic Matrix Control (QDMC); Linear programming solution of Dynamic Matrix Control (LDMC); Nonlinear Optimization; Calculus of Variation; Lagrange function; Euler-Lagrange equations; Jacobian Matrix; Perturbation; Sensitivity equation.

INTRODUCTION

Functional optimization subject to differential equation constraints is not a new subject to the literature. This optimization problem is a classical problem of the calculus of variations. Variational optimization differs from the conventional optimization problem. Conventional optimization is optimization subject to constraints without the presence of differential equations. Such optimization problems are treated extensively in the literature. The most notable are the well known methods of linear programming (LP) for problems where both objective function and constraints are linear, generalized reduced gradient (GRG); Abadie and Carpentier (1969) and most recently the quasi-Newton approach to the solution of the nonlinear optimization problem (NLP) known as successive quadratic programming (SQP); Powell (1977a, 1977b). SQP utilizes information on the curvature of the Lagrangian of the associated nonlinear optimization problem. Since the second order information exists, the whole optimization algorithm becomes like a classical method for the solution of a set of nonlinear algebraic equations using a quasi-Newton method. Here the nonlinear set of equations comes from the gradient of the Lagrangian set to zero; Fletcher (1980, 1981).

The variational optimization problem, however, differs from optimization in that, in the former, the independent input is not a finite number of variables, but consists of a so-called function of functions. The optimum is described by the behavior of the argument functions.

General attention was first drawn to problems of this type in 1696 by John Bernoulli's treatise on the "Brachistocrone Problem" which deals with a minimum-time trajectory of a particle in a gravitational field. Studies in classical calculus of variations focused on the well-known Euler-Lagrange equations; Courant and Hilbert (1953). Application of the Euler-Lagrange equations to variational optimization problems leads to a two-point boundary value problem. The solutions are obtained as functions over the interval; Keller (1968); Ray (1981); Kubecek and Hlavacek (1983).

Another work in the area of dynamic optimization is that of Hasdorff (1976), where he described in detail the use of the conjugate gradient method for solving optimal control problems. His method requires iterative intergration of the original model equation and adjoint sets of differential equations (two point boundary value problem) utilizing piecewise constant input functions. The method as outlined by Hasdorff is impractical (due to compulation and storage requirements) for most engineering applications. Also, these methods cannot handle general inequality constants because no type of active set strategy is employed.

Sargent and Sullivan (1978) describe a method along the lines of Hasdorff by implementing piecewise continuous basis functions for the independent variables, but still have to integrate the model and the adjoint equation in a forward and backward manner, respectively, to obtain the gradient information. They solve the resulting nonlinear programming problem with a projected variable metric method. A major advantage of this method is ability to solve much larger optimal control problems than demonstrated before by employing an integration package that exploits sparsity in the model equations.

Ray (1981) proposed another solution technique to the dynamic optimization problem. In his method, he assumed a continuous functional form (ploynomials) for the input function and then solved the resulting optimization problem using a gradient search. An advantage of Ray's method is that he reduced the classical two point boundary value problem into the standard nonlinear optimization problem. A disadvantage of the method is that a prior knowledge of continuous functional forms for the input variables must be available. These type of functions are not usually easy to write down, especially for the input functions which are nonsmooth. This is the case when the true optimal trajectory for the independent variable is a combination of continuous and piecewise constant functions. Even if one is able to describe the input function by a continuous function, it must be discretized afterward for digital computer implementation.

In this work a different approach to the solution of optimization subject to differential equation constraints is presented. The underlying concept is that of Dynamic Matrix Control (DMC) developed at the Shell Development Company in the late 1960's to solve a linear multivariable control problem in real time. The DMC formulation allows for computation of future finite discrete control moves to replace the continuous control function while approaching a desired setpoint; Cutler and Ramaker (1979); Cutler, Haydel and Morshedi (1983); Garcia and Morshedi (1984); Morshedi, Cutler and Skrovanek (1985, 1985); Garcia and Morshedi (1985).

DMC is structured such that the entire control algorithm can be put into a linearly constrained optimization problem. The objective function for this optimization problem can be formulated in two ways i) sum of the error-squared known as QDMC and ii) sum of the absolute values of the errors known as LDMC. In both of these formulations the constraints are the same and are linear, and the underlying QP and LP solutions are the optimum solutions to the control problem.

However, when the process models are nonlinear, the problem structure and its solution are not as simple as that obtained for the linear dynamic models. For linear models all the columns of the Dynamic Matrix are obtained from the first column and stay constant during the entire optimization calculation so long as the parameters in the underlying linear models do not change. In nonlinear control problems using the DMC concept, the nonzero values of the columns of the computed Dynamic Matrix have no relation to each other. The simple QP or LP (depending on the objective function) may no longer be a one-shot solution to the problem. A nonlinear optimization technique, such as GRG or SQP, should be used to obtain the optimal solution. At each optimization iteration, the Dynamic Matrix must be recomputed and utilized by the algorithm in order to obtain information with regard to the curvature (assuming SQP is used). In this work, a new algorithm, UDMC, is presented for the solution of among other problems, the nonlinear control problem subject to nonlinear equality and inequality constraints.

Universal Dynamic Matrix Control is a generalization of the DMC concept. UDMC is designed to solve problems having nonlinear as well as linear models (in the case of a linear model, the UDMC becomes QDMC or LDMC). The objective function is not restricted to quadratic or other special forms. Although UDMC's formulation leads to a conventional constrained optimization problem, it is independent of the type of optimization algorithm used. In the development of UDMC, SQP was used. SQP or any optimization method based on the second order information has shown to be superior to other nonlinear optimization methods for typical nonlinear problems encountered in practice; Schittcowski (1980); Chamberlain, et al. (1982). For parametric optimization, which is essential for a fast on-line application, second order information (the model's curvature) can be updated, saved, and used for subsequent optimization problems. These repeated optimization problems possess the same process model structures with different starting points.

In the following sections, the structure and the mathematical formulation of the UDMC algorithm will be presented. The idea of using simple functions to describe process input functions will be introduced.

Universal Dynamic Matrix Control relies on accurate computation of the Jacobian matrix of the nonlinear dynamic model representing the process. This Jacobian matrix must be computed rapidly with maximum computational efficiency. An efficient computational method based on the sensitivity equations has been developed and will be presented. Finally, we will close by giving two complete algorithms and discussing some practical considerations for efficient implementation of the UDMC algorithm.

Problem Statement

The general variational optimization can be stated as:

$$\text{Minimize}_{y,u} \int_0^{x_f} \phi [\, y(x), u(x) \,] \, dx \qquad (1a)$$

subject to

$$\frac{dy}{dx} = g(y,u) \qquad (1b)$$

$$h(y,u) \geq 0 \qquad (1c)$$

$$y(0) = y(x = 0) \text{ given} \qquad (1d)$$

$$u \in R^n \,;\, y \in R^m \,;\, x, \phi \in R^1$$

$$g : R^{m+n} \longrightarrow R^n$$

$$h : R^{m+n} \longrightarrow R^p$$

Problem (1) which is the general variational problem appears in a variety of industrial problems. In particular the gasoline blending optimization was the basic motivation behind the development of the Universal Dynamic Matrix Control algorithm.

In the gasoline blending optimization (1a) is the cost function integrated over time, equation (1b) represents the inventory balance given by a set of differential equations and (1c) is the nonlinear algebraic property model and hydraulic constraints involved in the overall optimization problem. The initial condition given by (1d) is the starting inventory. In the gasoline blending optimization, equation (1b) is a set of linear differential equations written around the tanks. All nonlinearities are encountered in the property model given by (1c).

Classical Approach

The Euler-Lagrange equations of problem (1) can be written by first forming the Lagrange function of problem (1). The Lagrange function is given by:

$$\underset{(y,u,\lambda,\mu)}{\text{Minimize}} \int_0^{x_f} \{\phi - \lambda^T [\frac{dy}{dx} - g(y,u)] - \mu^T h \} dx \qquad (2)$$

where λ's and μ's are the corresponding Lagrange multipliers for the equality and inequality constraints, respectively.

The Euler-Lagrange equations of the above integral equation can be given by the following system of equations:

$$\frac{\partial \phi}{\partial y} + \left(\frac{\partial g}{\partial y}\right)^T \lambda - \left(\frac{\partial h}{\partial y}\right)^T \mu + \frac{\partial \lambda}{\partial x} = 0 \qquad (3a)$$

$$\frac{\partial \phi}{\partial u} + \left(\frac{\partial g}{\partial u}\right)^T \lambda - \left(\frac{\partial h}{\partial u}\right)^T \mu = 0 \qquad (3b)$$

$$\frac{dy}{dx} = g(y,u) \qquad (3c)$$

$$h(y,u) \geq 0 \qquad (3d)$$

$$\mu^T h = 0 \qquad (3e)$$

$$\mu \geq 0 \qquad (3f)$$

$$\lambda(x_f) = 0 \tag{3g}$$

$$y(0) \quad \text{given}$$

The above system of equations is a nonlinear two-point boundary value problem with the mixture of differential/algebraic equations. This set of equations when solved would give the optimum functions for all the variables with respect to x. However, the solution of (3) is not trivial. Nonlinear two-point boundary value problems are inherently time-consuming and tough depending on the degrees of nonlinearities. The problem becomes much more difficult when the complementary conditions (equations 3e-f) are present and must be satisfied at each point. This condition means if an inequality constraint is active its associated multiplier is nonzero and visa-versa.

It is possible to pose the problem such that the original optimization problem (1) is free of inequality constraints. This is done through the introduction of squared slack variables. By doing so, more nonlinearities would be added to the problem formulation (1) which in turn makes the problem more difficult to solve. Such an elimination of inequalities could lead to convergence difficulties as well; Fiacco and McCormick (1968); Bard (1972); Fletcher (1980, 1981); Tapia (1980).

UDMC does not attempt to solve the nonlinear two-point boundary value problem. Construction of a Dynamic Matrix through the use of piecewise constant functions eliminates the need to revert to the Euler-Lagrange equations. Each continuous input function u is replaced by a finite-dimensional vector. The replacement reduces problem (1) to an ordinary nonlinear optimization problem which can be solved with any nonlinear optimization algorithm. The complementary conditions (equations 3e-f) are automatically satisfied during each optimization iteration. Since the problem is a straightforward nonlinear optimization problem with a finite number of discrete values for the inputs, a suitable algorithm based on the structure and nonlinearity of the process models can be selected to accomplish the computational task.

Piecewise Constant Functions

In this section the use of piecewise constant functions and its application to the input discretization will be described.

Let the input function u(x) be given. Then u(x) may be approximated by a finite number of discrete input values as follows:

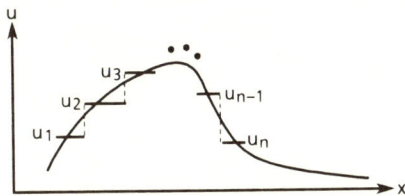

Figure 1. Discrete Representation of Input Function

where

$$u(x) = u_i \quad \text{for} \quad x_{i-1} \leq x \leq x_i$$

Define the characteristic function X_I such that

$$X_I(x) = \begin{cases} 1 & x \in I \\ 0 & x \notin I \end{cases} \quad (4)$$

where I is an interval, u(x) can be written as

$$u(x) = \sum_{i=1}^{n} u_i X_{I_i}(x) \quad (5)$$

where
$$I_i = \{x : u(x) = u_i\}$$

The above linear combination expressing a piecewise constant function is called a simple function; Royden (1963). Equation (5) is written in terms of the value of the input function u_i in an interval. In practice, however, it is convenient to work with the changes in the input rather than its value. The input change in the jth interval is given by: $\Delta u_j = u_j - u_{j-1}$; the step change from interval I_{j-1} to I_j. Graphically Δu_j can be shown as:

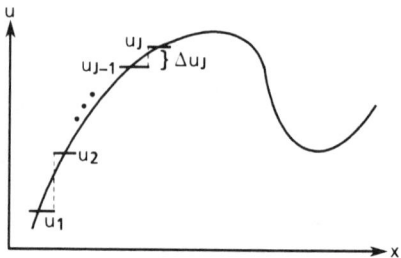

Figure 2. Discrete Representation of Input Function with Respect to Input Changed

Working with the changes in the input provides a simple recursive formula

$$u_{j+1} = u_j + \Delta u_{j+1} \qquad (6)$$

Substituting for $u_j = u_{j-1} + \Delta u_j$ and continuing the process, equation (6) becomes

$$u_{j+1} = u_o + \sum_{i=1}^{j+1} \Delta u_i \qquad (7)$$

or, assuming u(x) can be represented by n piecewise continuous values, then

$$u = u_o + 1_L \Delta u \qquad (8)$$

where 1_L is an nxn matrix having the following form

$$1_L = \begin{bmatrix} 1 & & & & \\ 1 & 1 & & & \\ 1 & 1 & 1 & & \\ \vdots & & & & \\ 1 & 1 & \cdots & & 1 \end{bmatrix}$$

The application of piecewise constant functions to control problems is not new. It was first applied informally at the Shell Development Company in the late 1960's by introducing the following matrix:

$$A = \begin{bmatrix} a_1 & & & \\ a_2 & a_1 & & \\ a_3 & a_2 & a_1 & \cdots \\ a_4 & a_3 & a_2 & \\ a_5 & a_4 & a_3 & \\ \vdots & \vdots & \vdots & \end{bmatrix} \qquad (9)$$

where the entries a_i $\left[a = \frac{\partial y}{\partial \Delta u}\right]$ are the numerical values of the unit step response imposed on a linear dynamic system. Shell called this matrix the "Dynamic Matrix" and since this technology was applied to a linear control problem they called their algorithm "Dynamic Matrix

Control". Note that in the above matrix the jth column represents the jth input change i.e., $\Delta u_j = u_j - u_{j-1}$ and the ith row is the ith future value of the output.

Application to Nonlinear Problems

Piecewise constant functions can be used to solve general variational optimization problems subject to differential equations. Suppose we have the following general optimization problem.

$$\underset{(y,u)}{\text{Minimize}} \; \psi(x_f) = \int_0^{x_f} \phi(y,u) \, dx \qquad (10a)$$

subject to

$$D \frac{dy}{dx} = g(y,u) \qquad (10b)$$

$$y_{min} \leq y \leq y_{max} \qquad (10c)$$

$$u_{min} \leq u \leq u_{max} \qquad (10d)$$

$$y_o = \text{known}$$

where D is a square matrix of size y and may be singular for general treatment. Without loss of generality, all the inequalities are represented by simple bounds, equation (10c-d).

Problem (10) can be written equivalently as

$$\underset{(y,u)}{\text{Minimize}} \; \psi(x_f) \qquad (11a)$$

subject to

$$\frac{d\psi}{dx} = \phi(y,u) \tag{11b}$$

$$D\frac{dy}{dx} = g(y,u) \tag{11c}$$

$$y_{min} \leq y \leq y_{max} \tag{11d}$$

$$u_{min} \leq u \leq u_{max} \tag{11e}$$

$$\psi(0) = 0 \tag{11f}$$

$$y(0) = \text{given} \tag{11g}$$

Let the input function u(x) be represented by a simple function according to equation (5) with n distinct values in interval $[0,x_f]$. Using these input values equation 11b and 11c can be integrated exactly (continuously) for dependent variables (y) as the function of x. The optimization variables (the independent variables) are those discrete finite values of u(x) that must be adjusted by the optimization algorithm to minimize the objective function.

To proceed with the optimization process some vital information is needed and must be supplied to the optimization algorithm, such as the derivative of the objective function and the constraints with respect to the independent variables i.e., $\frac{\partial \psi}{\partial \Delta u}$ and $\frac{\partial y}{\partial \Delta u}$.

Rigorous Computation of Derivatives

In this section a rigorous treatment of derivative computation, which is clearly very time-consuming, will be given. Later, a new approach through the use of the classical sensitivity equations will be presented to overcome the computational burden associated with these derivative calculations.

Straightforward approximation of $\frac{\partial \psi}{\partial \Delta u}$ and $\frac{\partial y}{\partial \Delta u}$ can be given by

$$\frac{\partial \psi_i}{\partial \Delta u_j} = \frac{\psi_i(u_1, u_2, \ldots, u_j + \varepsilon, \ldots, u_n) - \psi_i(u_1, \ldots, u_n)}{\varepsilon} \quad (12a)$$

$$\frac{\partial y_i}{\partial \Delta u_j} = \frac{y_i(u_1, u_2, \ldots, u_j + \varepsilon, \ldots, u_n) - y_i(u_1, \ldots, u_n)}{\varepsilon} \quad (12b)$$

where

$$u_j = u_o + \sum_{\alpha=1}^{j} \Delta u_\alpha$$

For the sake of brevity, only the computation of the matrix $\frac{\partial y}{\partial \Delta u}$ will be given; $\frac{\partial \psi}{\partial \Delta u}$ is computed the same way.

The matrix $\frac{\partial y}{\partial \Delta u}$ which is the sensitivity matrix can be computed rigorously as

$$\frac{\partial y}{\partial \Delta u} = \begin{bmatrix} [y_1(u+\varepsilon e_1) - y_1(u)]/\varepsilon & 0 & \\ [y_2(u+\varepsilon e_1) - y_2(u)]/\varepsilon & [y_2(u+\varepsilon e_2) - y_2(u)]/\varepsilon & \\ \vdots & \vdots & \cdots \\ [y_M(u+\varepsilon e_1) - y_M(u)]/\varepsilon & [y_M(u+\varepsilon e_2) - y_M(u)]/\varepsilon & \end{bmatrix} \quad (13)$$

where e_i is the basis vector with 1 in its ith entry.

Note that every column must be generated by integrating the nonlinear model equation (11c) by perturbing the particular variable associated with that column. This, of course, is a time-consuming process which will be dealt with later.

The final Jacobian matrix has the following form:

$$\frac{\partial y}{\partial \Delta u} = \begin{bmatrix} a_1 & 0 & 0 & 0 & \\ a_2 & b_1 & 0 & 0 & \\ a_3 & b_2 & c_1 & 0 & \\ a_4 & b_3 & c_2 & d_1 & \\ a_5 & b_4 & c_3 & d_2 & \cdots \\ a_6 & b_5 & c_4 & d_3 & \\ \vdots & \vdots & \vdots & \vdots & \\ a_M & b_M & c_M & d_M & \end{bmatrix} \qquad (14)$$

which has the same structure as the DMC matrix except that the columns of the above matrix are no longer obtained from the values of the first column as in the linear case [compare with (9)].

Generation of the derivatives by perturbation is computationally expensive. Each column of the Jacobian matrix is generated by integrating the nonlinear model equation while perturbing the variable associated with that column. The same may be done to compute $\partial \psi(x_f)/\partial \Delta u$; this is the last row of the matrix $\partial \psi / \partial \Delta u$.

Before closing this section it should be pointed out that problem (11) was formed from problem (10) by introducing a new variable and a new differential equation. In practice this reformulation is usually not needed. For sampled data systems under digital control measurements are available only at discrete points and the integral equation given by (10a) is represented in terms of the summation over discrete values of ϕ. The discrete values of ϕ are computed at discrete values of independent and dependent variables. The values of independent variables are given as a set of piecewise discrete points and the dependent variables at discrete points evolve from the integration of the model equation (10b).

Efficient Computation of the Differential Matrix

The work of this section is based on the work of Morshedi, Lin and Luecke (1986). This work was designed to obtain a simple and yet efficient solution to the most expensive part of the UDMC algorithm.

It was pointed out that computation of this matrix by means of perturbation is not acceptable for on-line implementation of the UDMC algorithm on the digital computer. In this section a method will be presented that is tailored for repetitive use in the non-linear optimization subject to differential equations when the input is given by simple functions. It differs considerably from other methods and uses the solution of the piecewise continuous linear differential equations as interpolating functions. These functions can be obtained at lower computational expense during the optimization sequence because the results of the previous iteration can be used as initial estimation for the next.

Suppose the model can be described by the following non-linear ordinary differential equation:

$$\frac{dy}{dx} = g(y,u) \qquad (14)$$

$y(0) = y_0$
u = independent variable

To simplify the discussion, the first part of this work is restricted to single dimensional systems. The extension to multiple-input-multiple-output systems is straightforward and will be given later.

The independent variable in (14) is a simple function and mathematically can be described by equations (5), (6) and (7). The input variable, u, can thus be completely characterized as a vector where each element, u_i, represents the value of the input u for the ith interval.

Based on this finite-dimensional vector of inputs it is required to compute $\partial y/\partial u = \left[\partial y_i/\partial u_j\right]$. This task is accomplished by using the sensitivity equations. Equation (14) may be differentiated with respect to u:

$$\frac{\partial}{\partial u} \frac{dy}{dx} = \frac{\partial g}{\partial y} \frac{\partial y}{\partial u} + \frac{\partial g}{\partial u} \qquad (15)$$

The order of differentiation of the left hand side term may be reversed if u is continuous. The resulting differential equation is then the classical sensitivity equation. Of course, in UDMC formulation u is not continuous; it is a series of step functions and the reversal is not legitimate. However, because of the nature of the input being piecewise continuous the classical equation from the sensitivity can be used in the vector form. The key to the following derivation is to rewrite equation (14) in terms of the vector u instead of single variable:

$$u = (u_1, u_2, \ldots, u_n)^T \qquad (16)$$

Then,

$$\frac{dy}{dx} = g(y, u_1, u_2, \ldots, u_n) \qquad (17)$$

If the derivation is restricted to a single interval, j, the input during this interval u_j is continuous and it is permitted to reverse the order of differentiation:

$$\frac{d}{dx}\left(\frac{\partial y}{\partial u_j}\right) = \frac{\partial g}{\partial y}\frac{\partial y}{\partial u_j} + \frac{\partial g}{\partial u_j} \quad (18)$$

For the discrete input vector of length n there are n columns of the Jacobian to be computed. To generate the full Jacobian matrix, (n+1) ordinary differential equations (14) plus (18), (which is the same number of equations as for perturbation approach), must be integrated simultaneously. However, if the structure of these equations is exploited, the Jacobian matrix can be computed even more efficiently.

Although equation (17) is nonlinear, it does not contain any dependent variable from equation (18). Hence, it can be integrated alone without any interference from the sensitivity equation.

With $y(x)$ known $\frac{\partial g}{\partial y}$ and $\frac{\partial g}{\partial u}$ [see equation (18)] can be computed; hence, equation (18) can be treated as a linear differential equation having time-varying coefficients.

Application of the mean value theorem to equation (18) for computation of $\frac{\partial g}{\partial y}$ and $\frac{\partial g}{\partial u}$ leads to analytical solution for the sensitivity equation as follows:

$$\frac{\partial y_i}{\partial u_j} = \frac{\overline{\frac{\partial g}{\partial u}}}{\overline{\frac{\partial g}{\partial y}}}\left(1-e^{\overline{\frac{\partial g}{\partial y}}\Delta x}\right) + \frac{\partial y_{i-1}}{\partial u_j}e^{\overline{\frac{\partial g}{\partial y}}\Delta x} \quad (19)$$

where

$$\frac{\partial y_i}{\partial u_j} = \frac{\partial y}{\partial u_j}\bigg|_{\text{at } x = x_i = i\,\Delta x}$$

Δx = sample interval

$\dfrac{\partial y_{i-1}}{\partial u_j} = \dfrac{\partial y}{\partial u_j}$ at the beginning of the interval (an initial condition)

$\overline{\dfrac{\partial g}{\partial y}} = \dfrac{\partial g}{\partial y}$ = evaluation with an "average" value of y from the interval

$(i-1)\, \Delta x < x < i\, \Delta x$

$\overline{\dfrac{\partial g}{\partial u}} = \dfrac{\partial g}{\partial u}$ = evaluated with the "average" y.

Use of equation (14) with equation (19) improves the computational efficiency greatly in obtaining the Jacobian matrix. Only one nonlinear differential equation will be solved using a numerical procedure, such as adjustable interval Runge-Kutta method. The sensitivities are then computed analytically using the algebraic equation (19). The equations are used one interval at a time. The final value from one interval becomes the initial value for the next. The average value for y is updated for each interval. The value of the input variable u remains constant through each interval.

Further computational efficiency can be obtained since the same average values for $\dfrac{\partial g}{\partial y}$ and $\dfrac{\partial g}{\partial u}$ are used regardless of which u_j is being evaluated. Only the initial condition of each interval, $\partial y_{i-1}/\partial u_j$, changes as a function of u_j. For the first interval, $\partial y_0/\partial u_1$, $\partial y_1/\partial u_2$, etc., the initial conditions, are all zero. Thereafter, the initial condition for a given interval is the result of the previous interval.

Computation of Other Columns

After calculation of $\partial y_i/\partial u_1$, the value of the sensitivity for that interval in the second column $\partial y_i/\partial u_2$ is computed, again using equation (19). Note that only the value of the initial condition is changed in the implementation of equation (19).

The value of the sensitivity for a given interval in the jth column can be shown to differ from the value in the previous column only by the exponentially decayed value of the difference in the initial conditions for that interval. Thus it is not necessary for the nonlinear differential equation to be solved again, the new solution can be obtained simply from those of the first column.

Denote $S_{ij} = \partial y_i/\partial u_j$, then it follows from equation (19) that:

$$S_{i,j+1} = S_{i,j} + (S_{i-1,j+1} - S_{i-1,j})e^{\overline{\frac{\partial g}{\partial y}} \Delta x} \tag{20}$$

Since average values for $\frac{\partial f}{\partial y}$ are known, the entire Jacobian matrix will be generated using parameters calculated for the first column. Numerical integrations and the approximations procedure are required only for the generation of the first column. For more comprehensive treatment together with the complete algorithm and error analysis of the above discussion see Morshedi, Lin and Luecke (1986).

Extension to Multivariable Systems

The extension of this algorithm to multiple-input-multiple-output processes is straightforward. The principal problem is that the linear sensitivity equation (19) must be solved analytically which means the eigenvalues and eigenvectors are required. However, one set of eigenvalues apply to all columns of the Jacobian matrix and to all input of a multiple-input process.

The mathematical development proceeds as previously described except that the input variables and the response variables are interpreted as vectors, and the transition functions are matrices. Thus equation (14) are the Kth order differential system; y is a k-vector; u is a q-vector; g is a k-vector set of functions.

In sensitivity equation (19) $\frac{\partial y}{\partial u}$ is a qxk - matrix, $\frac{\partial g}{\partial y}$ is a kxk - matrix; $\frac{\partial g}{\partial u}$ is a qxk - matrix. In the analytical solution e.g., equation (19), $\partial y_{i-1}/\partial u_j$ is a qxk - matrix, and

$$e^{\frac{\partial g}{\partial y} \Delta x} \equiv P \, e^{\Lambda \Delta x} \, P^{-1} \tag{21}$$

$$-\frac{\overline{\frac{\partial g}{\partial y}}}{\frac{\partial g}{\partial u}} (1 - e^{\frac{\partial g}{\partial y} \Delta x}) = -PEP^{-1} \frac{\partial g}{\partial u} \tag{22}$$

where

$$e^{\Lambda \Delta x} = \begin{bmatrix} e^{\lambda_1 \Delta x} & & & \\ & e^{\lambda_2 \Delta x} & & \\ & & \ddots & \\ & & & e^{\lambda_k \Delta x} \end{bmatrix}$$

λ_i = the ith eigenvalue of $\frac{\partial g}{\partial y}$

P = the kxk - matrix of eigenvectors of $\frac{\partial g}{\partial y}$

$$E = \begin{bmatrix} \frac{1}{d_1}(1-e^{\lambda_1 \Delta x}) & & & \\ & \frac{1}{d_2}(1-e^{\lambda_2 \Delta x}) & & \\ & & \ddots & \\ & & & \frac{1}{d_k}(1-e^{\lambda_k \Delta x}) \end{bmatrix}$$

The number of eigenvalues and eigenvectors is equal to the number of simultaneous equation (14), or the number of dependent variables. The number of inputs does not affect the number of eigenvalue or eigenvectors. The greatest computational problem for this method is the calculation of eigenvalues and eigenvector at each subinterval for systems containing a large number of dependent variables. However an efficient and reliable iterative method can be designed to exploit the structure of the problem. Work is underway to obtain a complete row of the Jacobian matrix during the integation of the model equations while taking the sparsity structure as well as the problem stiffness (if needed) into account. The vital information such as availability of a good initial guess for the value of the Jacobian matrix from one iteration to the next and the knowledge that each column may be generated from the same linear differential equation with a different initial condition can be used to obtain the complete Jacobian matrix at once. Note that the fundamental information for integration of the sensitivity equation by numerical means comes from numerical integration of the model equation. In future work a comprehensive report on an efficient algorithm to deal with these issues will be presented.

The Complete UDMC Algorithm

The UDMC algorithm invokes a nonlinear optimization procedure. UDMC itself is independent of the optimization procedure chosen. The optimization variables are the Δu's as given by figures 1 and 2. The optimum Δu will completely characterize the input as a function of variable x. To compute the optimum input in addition to gradient information (i.e., $\partial\psi(x_f)/\partial\Delta u$; $\frac{\partial y}{\partial \Delta u}$) the information with regard to simple bounds must also be supplied. For simplicity all the inequalities are made into equalities by adding appropriate slack variables. The equalities are then added to the model equation as a set of algebraic equations. In equation (11c) the matrix D may be a diagonal matrix of one and zero elements. The nonzero elements of D describe the process model and the zeros are the constraint equations. The problem may be formulated differently based on the structure of the process models and constraint equations; different formulations do not affect the algorithm.

Note that the computation of the objective function derivative and the Jacobian matrix are obtained through the solution of the model equations. The final optimization problem can now be stated as follows:

$$\underset{\Delta u}{\text{Minimize}} \quad \psi_f(\Delta u) \tag{23a}$$

subject to

$$y_{min} \le y(\Delta u) \le y_{max} \tag{23b}$$

$$u_{min} \le u(\Delta u) \le u_{max} \tag{23c}$$

The notation $\psi_f(\Delta u)$ means $\psi(x_f)$ is a function of the vector Δu. Equation 23(c) can be replaced simply in terms of Δu using the linear equation (8).

Algorithm I

Step 0) Initialization : Chose Δu^o, set $\ell = 0$ and define ℓ_{max}.

Step 1) With known u_o, Δu [e.g., (7)] and $y(0)$, $\psi(0)$, compute

    a) $\dfrac{\partial \psi(x_f)}{\partial \Delta u}$ = the gradient of objective.

    b) $\dfrac{\partial y}{\partial \Delta u}$ = the Jacobian matrix.

Step 2) Call optimization algorithm to compute the next iterate $\Delta u^{\ell+1}$.

Step 3) Check for the convergence : if converged go to Step 5; otherwise go to Step 1.

Step 4) if $(\ell \geq \ell_{max})$ go to 5; otherwise go to Step 1.

Step 5) Stop.

On-line Implementation of UDMC

At this point an algorithm for nonlinear multivariable control problem will be given. Here the objective function is a special form of the one given by problem (11). The objective function is the sum of error squared between the output and the setpoint. More precisely, it is desired to solve the following optimization problem:

$$\text{Minimize } \psi(t_f) = \int_0^{t_f} (y - y_{sp})^T (y - y_{sp}) \, dt \quad (24a)$$
$$(y,u)$$

subject to

$$D \frac{dy}{dt} = g(y,u) \quad (24b)$$

$$y_{min} \leq y \leq y_{max} \quad (24c)$$

$$u_{min} \leq u \leq u_{max} \quad (24d)$$
$$\{u(0), y(0),\} = \text{known}$$

Using simple functions to characterize u, problem (24) can be equivalently formulated as:

$$\text{Min } \psi = \sum_{i=1}^{M} 1/2 \, (y_i - y_{sp})^2 \, w_i \quad (25a)$$

subject to

$$y_{min} \leq y \leq y_{min} \quad (25b)$$

$$u_{min} \leq u \leq u_{max} \quad (25c)$$

where y_i is obtained from the solution of the model equations (24b); w_i is an appropriate weighting factor and M, the upper limit of summation, represents the final time.

Algorithm II

Step 0)   Pick Δu^o.

Step 1)   Evaluate $\frac{\partial y}{\partial \Delta u}$, y, r = y - y_{sp},

$$\psi = 1/2 \sum_i r_i^2 = 1/2\, r^T r.$$

Step 2)   Compute $\frac{\partial \psi}{\partial \Delta u} = \left(\frac{\partial y}{\partial \Delta u}\right)^T r$ = gradient of the objective function with respect to Δu.

Step 3)   Call optimization algorithm to obtain new iterate Δu.

Step 4)   Check for convergence : if converged go to 5; otherwise go to 1.

Step 5)   Implement the first calculated value of each independent variable (i.e., 1st time interval value of each independent variable).

Step 6)   Update procedure:

   a) Set the initial value of Δu for next optimization problem.

   b) Integrate model over current interval to obtain initial condition for integration in the next interval (for more see below).

Step 7)   Correct for modeling error/feedback.

Step 8)   Check for stopping criterion.
          If converged go to 9;
          otherwise go to 1.

Step 9)   Go to Step 10 or select new setpoint and go to
          Step 0.

Step 10) Stop.

Some Implementation Considerations
Update Procedure and Feedback

Clearly Steps (6) and (7) in algorithm II are the important part of a good and robust control algorithm; in this section these two steps are expanded to clarify some important points.

For each input, once the n optimal future control moves have been determined, the first move in each manipulated input is implemented. It is now necessary to solve the optimization problem again at the next time interval for another set of n future moves.

To enhance the convergence a good initial estimate for the solution is required by an optimization algorithm. A good estimate for the solution can be obtained by shifting the vector of control moves for each input down by one (since the first move has already been implemented) and setting the nth value to zero or to the (n-1)th values of their corresponding previously calculated inputs.

The next item performed in the update procedure is to predict where the output will be at the end of the present time interval. This is necessary for two reasons i) the calculated values of the controlled outputs at the end of the present time interval will serve as the initial conditions for the integration in the next controller execution. The predicted values of the outputs must be used instead of the measured values of each interval to provide the model equations with a consistent set of initial conditions. Any difference between the predicted and measured values of the controlled outputs will be corrected by the feedback correction mechanism; ii) the predicted output values will provide vital information for the feedback correction mechanism as to how much correction is necessary.

The prediction is performed by integrating the model over the present time interval, reflecting the input moves just implemented. At the end of the first time interval, the values of the controlled output are measured and a term, δ, is calculated for each control output:

$$\delta = y_m - y_p \qquad (25)$$

where y_m and y_p are the measured and the predicted values of the output, respectively. This computed value of δ must now be reflected in the objective function in order to compute the new set of future moves. To reflect δ into the objective function, the residuals $r = y_i - y_{sp}$ may be redefined as

$$r = (y_i + \delta - y_{sp}) \qquad (26)$$

CONCLUSION

Universal Dynamic Matrix control is a powerful method for optimization problems where the underlying process models possess differential equations. The formulation of the UDMC is based on Shell's previously developed algorithm DMC, which was designed for linear control problems.

UDMC provides a solution of a difficult two-point nonlinear boundary value problem which is obtained from application of the calculus of variations to optimization subject to differential equations.

It has been shown that the DMC formulation can be extended by using piecewise constant functions to replace the input functions to solve nonlinear dynamic optimization problems. The use of simple functions transforms the continuous input function into a piecewise continuous function on each process input. Representing the input as such removes the need to revert to the Euler-Lagrange equations and hence alleviates the need for the solution of a tough two-point nonlinear boundary value problem. The original optimization problem can be solved directly using any available nonlinear programming algorithm.

Two complete algorithms, one for the general optimization subject to differential equations and another for online solution of nonlinear constrained control problems, have been presented.

Although the two algorithms given in this work do not depend on a particular optimization technique, excellent results have been obtained by coupling UDMC with the successive quadratic programming (SQP) method of optimization.

REFERENCES

Abadie, J., and J. Carpentier (1969). Generalization of the Wolfe Reduced Gradient Method to the Case of Nonlinear Constraints. In R. Fletcher (Ed.), Optimization, Academic Press, New York, pp. 37-47.

Bard, Y. (1972). Nonlinear Parameter Estimation. Academic Press, New York.

Chamberlain, R. M., M. J. D. Powell, C. Lemarechal, and H. C. Pedersen (1982). The Watchdog Technique for Forcing Convergence in Algorithm for Constrained Optimization. Mathematical Programming Study 16, 1-17.

Courant, R., and D. Hilbert (1953). The Calculus of Variations. Method of Mathematical Physics, Vol. I. Interscience publishers Inc., pp. 164-274.

Cutler, C. R., and B. L. Ramaker (1979). Dynamic Matrix Control. Paper No. 516, AICHE 86th National Mtg.

Cutler, C. R., J. J. Haydel, and A. M. Morshedi (1983). An Industrial Perspective on Advanced Control. AIChE Annual Meeting, Washington, D. C.

Fletcher, R. (1980). Practical Method of Optimization, Vol. I, Wiley, New York.

Fletcher, R. (1981). Practical Method of Optimization, Vol. II, Wiley, New York.

Fiacco, A. V., and G. P. McCormick (1968). Nonlinear Programming : Sequential Unconstrained Minimization Techniques, Wiley, New York.

Garcia, C. E., and A. M. Morshedi (1984). Quadratic Solution of Dynamic Matrix Control (QDMC). Proceedings of the Canadian Conference on Industrial Computer Systems.

Garcia, C. E., and A. M., Morshedi (1985). Solution of the Dynamic Matrix Control Proglem via Quadratic Programming (QDMC). Submitted for publication to Chemical Engineering Communications.

Hasdorff, B. A. (1976). Gradient Optimization and Nonlinear Control, Wiley-Interscioulem, New York.

Keller, H. B. (1968). Numerical Method for Two-Point Boundary Value Problems, Blaisdell.

Kubicek, M., and V. Hlavacek, Numerical Solution of Nonlinear Boundary Value Problems with Applications, Prentice-Hall, New Jersey.

Morshedi, A. M., C. R. Cutler, T. J. Fitzpatrick, and T. A. Skrovanek (1983). A Method of Controlling a Process by Combining The Quadratic/Linear Programming Method of Optimization with Dynamic Matrix Control, U. S. Patent pending.

Morshedi, A. M., C. R. Cutler (1984). Universal Dynamic Matrix Control. Technical Report, Shell Oil Company, Houston.

Morshedi, A. M., C. R. Cutler, and T. A. Skrovanek (1985). Optimal Solution of Dynamic Matrix Control with Linear Programming Technique (LDMC), American Control Conference Proceedings, pp. 199-208.

Morshedi, A. M., C. R. Cutler, and T. A. Skrovanek (1985). Optimal Solution of Dynamic Matrix Control with Quadratic Programming Technique (QDMC). <u>ISA Conference Proceedings</u>.

Morshedi, A. M., H. Y. Lin, and R. H. Luecke (1986). Rapid Computation of the Jacobian Matrix for Optimization of Dynamic Processes, to be published in <u>Computers and Chemical Engineers</u>.

Powell, M. J. D. (1977a). A Fast Algorithm for Nonlinearly Constrained Optimization Calculations. Presented at 1977 <u>Dundee Conference on Numerical Analysis</u>.

Powell, M. J. D. (1977b). Convergence of Variable Metric Methods for Nonlinearly Constrained Optimization Calculations. Presented at <u>Nonlinear Programming 3 Symposium</u>, Madison, Wisconsin.

Ray, W. H. (1981). <u>Advanced Process Control</u>. McGraw-Hill, New York.

Royden, H. L. (1963). The Lebesgue Integral. <u>Real Analysis</u>, The Macmillan Company, New York.

Sargent, R. W. H., and G. R. Sullivan (1978). In J. Stoer (Ed.), <u>Optimization Technique</u>, Proceedings of the 8th IFIP Conference on Optimization Techniques, Würzburg, Sept. 5-7, Part 2, Springer-Verlag, Berlin, pp. 158-168.

Schittcowski, K. (1980). Nonlinear Programming Codes. Lecture notes in <u>Economics and Mathematical Systems 183</u>, Springer-Verlag, Berlin.

Tapia, R. A. (1980). On the Role of Slack Variables in Quasi-Newton Method for Constrained Optimization. In L. C. W. Dixon and G. P. Szego (Eds.), <u>Numerical Optimization of Dynamic Systems</u>, North-Holland Publishing Company, New York, pp. 235-246.

THE INTEGRATION OF PLANNING, SCHEDULING, AND PROCESS CONTROL

Leon S. Lasdon
University of Texas, Austin, TX 78712

Thomas E. Baker
Chesapeake Decision Sciences, Inc., Berkeley Hts., NJ 07922

Abstract

In varying degrees, all companies in the process industry have suffered from the gap between the logistics planning function and process control. Optimal operations selected by linear programming planning models often turn out to be unattainable in practice. The detailed and timely information on process capabilities, which can be gathered at the process control level, is seldom transmitted to the rest of the logistics system.

The first part of this paper proposes an architecture for the planning, scheduling, and control hierarchy. It also attempts to demonstrate how the various tools and applications relate to each other and how they can be developed so as to improve the effectiveness of the overall logistics system via the integration of the planning, scheduling, and control functions. Such an integration can be realized, not by building larger LP models, but by developing a set of compatible tools which can be used at all levels of the planning, scheduling, and control hierarchy. The last two parts present primal and dual methods for coordinating the optimization of several process units, and discuss three leading nonlinear programming algorithms for process unit optimization.

keywords: Production planning and scheduling, higher level control, supervisory control, nonlinear optimization

1. The Integration Framework

Throughout the oil industry there has been a tremendous gap between the Linear Programming (LP) models used at the refinery planning level and the oil movements and process control functions carried out at the daily operations level. This gap renders most logistics systems, i.e., the company's supply, refining, distribution, and marketing organizations, less than responsive to the frequent disturbances with which they must contend.

The reasons for the gap between planning and process control are largely technological. The LP models used at the planning level are approximate representations of refinery capabilities which, in varying degrees, involve averages of daily operations over longer periods of time. These averaged representations make it difficult to translate LP solutions into actual practice; oil movements and operations scheduling functions are concerned with inventory feasibility on a daily or hourly basis. The averaged LP representations also produce product costs (or dual values) which are generally unusable at the process control level where decisions are made on a real-time basis.

In practice, the operations scheduling function ignores the signals passed from the planning function as soon as a crisis occurs. Since few analytical tools are used to assess the effects of operations scheduling decisions, crisis operation generally becomes the norm.

At the process control level, the gap is equally apparent. Operating targets taken from the planning level are often unrealistic (or not economical) because the current actual process yields and the day-to-day inventory constraints differ from the average representations found in the planning model. If the corporation decides to give process control more latitude by providing marginal values (short term economic incentives) for optimization at the supervisory control level, it is soon realized that the marginal values produced by the planning model are inadequate for the purpose.

Thus, if we are to close the gap between logistics planning and process control, we need to think in terms of new formulations at the planning level, new tools at the scheduling level, significant coordination capabilities at the higher level control level, multi-purpose process models, and a data base which links all functions in the logistics system. This report proposes that "integration" will be accomplished, not by building larger models, but by developing a set of complementary tools which can be used at all levels in the hierarchy.

The point of view taken in this paper is to define the best set of tools and applications possible with the technology which will be available in the next ten years; the scope of the design has not been limited by considerations of what is currently in place, e.g., MPSX or WHIZARD. However, the reader will note that much of what is currently in place forms the basis for future developments. It is also comforting to note that, while projected technological advances will make the implementation easier, the complete, integrated set of tools and applications could be successfully implemented with today's technology.

Planning/Scheduling/Control System Functions

The planning/scheduling/control system functions described in this paper may be somewhat less detailed than the applications areas found in any given company, e.g., strategic planning, tactical planning, operations planning, coordination, etc. In any case, we'll use this short section to define our terms. Although this paper is not geared to the industry's current practice, the reader will note that the system functions described below correspond to organizational structures already in place.

Strategic and Tactical Planning

The Strategic and Tactical Planning Functions deal with establishing and implementing the organization's long range future plans. Some of the decisions made at this level include capital budgeting, facility expansion, market development, competitive strategies, etc. Obviously, there are many direct connections between the Strategic and Tactical Planning Functions and the Logistics Operations Functions (Planning/Scheduling/Control) described below.

System Operations Planning

The Systems Operations Planning Function establishes multi-plant(multi-refinery) logistics system tactics dealing with uncertainties in crude buying, product demand, and plant capabilities. The time horizon (total planning horizon) usually associated with this level is three months to one year. Some of the functions performed at this level include producing plans and budgets, evaluating crude contracts, allocating crudes, allocating products, evaluating sales contracts, establishing crude and product values, establishing interim crude and product inventory (or production) targets.

Plant Operations Planning

The Plant (Refinery) Operations Planning Function establishes short term operating tactics for a single refinery or plant. The time horizon associated with this level is usually one to three months. Some of the functions performed at this level include allocating streams, determining unit operations and blends, and establishing inventory targets which are compatible with the overall system plan.

A model at the Plant Operations Planning level would generally be a one or two time period refinery LP model with detailed nonlinear representations of yields, qualities, pools, and blends. Inventory targets from the end of the first time period in the System Operations Planning model would serve as boundary conditions for the last time period of the Plant Operations Planning model.

Scheduling

The Scheduling Function establishes feasible short term operations within a single refinery with a time horizon of four to six weeks. This level schedules crudes, units, flows, and blends. It controls inventories to meet liftings and other major oil movements.

Relationships between operations' run lengths and throughputs, not represented in the LP models at the Systems and Plant Operations Planning levels, are developed in detail at the Scheduling level. Projected infeasibilities are fed back to the Plant Operations Planning level for replanning.

Higher Level Control

The regulatory process control is the first (lowest) level of control. Second level control can be defined as the single unit optimization implemented via supervisory process models. The Higher Level Control Function monitors the actions taken at the second level and coordinates the optimization across several units and operations. The time horizon associated with this level is one day to one week. This level establishes operating conditions and provides data on operations and process status.

The planning, scheduling, and control functions will be presented in the following as elements of a Logistics Integration Framework. As mentioned before, the Strategic and Tactical Planning functions are considered to be separate from the logistics system under study in this report.

Integration Framework Overview

The four major logistics operation system functions -- system Operations Planning, Plant Operations Planning, Scheduling, and Higher

Level Control--are shown in Figure 1 as part of an integrated framework

                              System
                              Operations
                              Planning
                              Function

                                                              Multiple
Plants
....
                                                              Single
Plant

                              Plant
                              Operations
                              Planning
                              Function

         Data Base   Process
         Management   Modeling

                  Scheduling   Higher
                  Function   Level
                                       Control

                              Process ....Process
                              Control   Control

            Integration Framework Overview -- Figure 1

    Figure 1 represents logistics system functions, not components. Logistics system components, e.g., a Model Management System or a Data Base Manager, are implied in Figure 1.

The Integration Framework shown in Figure 1 makes it clear that the system is hierarchical in its basic design. Planning/scheduling/control functions have occurred traditionally in segmented, decomposed forms. The projected changes in technology, e.g. local computing capabilities and improved communications, will probably cause these functions to remain disjoint. Thus, future progress in the integration of planning, scheduling, and control functions will probably be effected through the improved transfer of information and the development of multi-purpose tools which can be used across several functions rather than through the development of larger, more comprehensive models.

Thus the underlying design principle of the Logistics Integration Framework is:

- Accept decomposition by function as something which is natural.
- Allow business planning models, scheduling and control functions, to continue to focus on individual functions or subproblems.
- Provide for a hierarchy of models and submodels from large-scale, system-wide models to small, unit-specific models.
- Develop tools--process models, a model management system--which can be used at any functional level in the system.
- Establish a framework for passing information from one functional level to the next.
- Turn our attention to the difficulties of improving the quality of the information in our framework.

Information Flow

We now consider how the Integration Framework lends itself to the transmission of information during the logistics planning process. Figure 2, shown below, illustrates the flow of information which takes place as part of the implementation of the system operations plan.

These planning signals could be considered as the <u>forward flow</u> of information in the integrated system.

Forward Flow - Planning Process

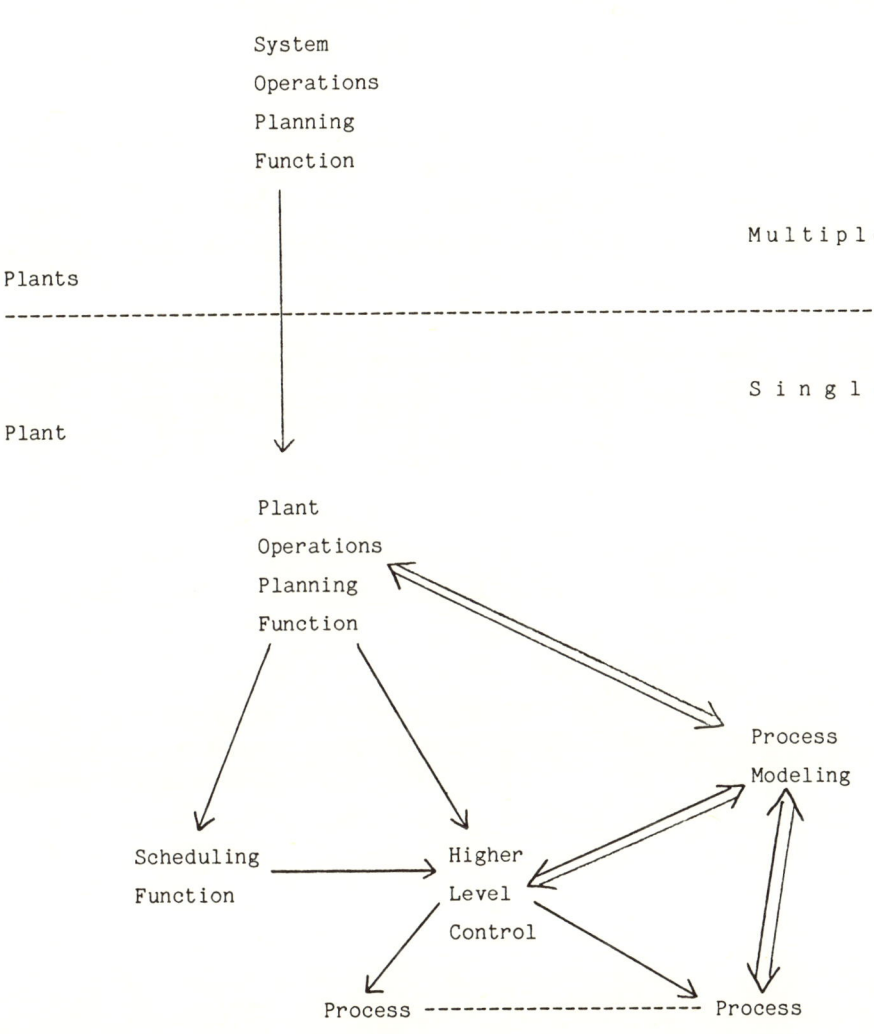

LEGEND:

———> passing targets and values
   (production and inventory targets and limits)
   (costs and cost limits)

<=====> iterative updating of nonlinear relationships

Information Flow for the Planning Process -- Figure 2

As noted in the legend of Figure 2, single-lined arrows represent the flow of information (production targets, inventory targets, product values and prices) which occurs as part of the planning/scheduling/control regulatory process. The double-lined arrows represent the iterative use of process models to update yields and gradients for the nonlinear representations found in Plant Operations Planning, Higher Level Control, and Process Control.

A model at the Plant (Refinery) Operations Planning level might be considered as a more detailed representation of a subset of the model at the System Operations Planning level. The model at the plant level encompasses only one plant and, in general, has a shorter time horizon than the system model. Thus, the flow of information from the system model to a refinery model should contain targets or prices for transfers between refineries, allocation of crudes and demands to refineries, and inventory or production targets at the end of the refinery model time horizon.

The Plant Operations Planning model should contain detailed nonlinear representations of process yields and blends. In our modeling hierarchy, the plant model is the last opportunity to coordinate operating conditions, stream allocations, and blends across the whole refinery. Thus, the plant model output defines optimal operations and blends for the Scheduling and Higher Level Control Functions. To the extent that the refinery model can produce realistic short term marginal values, these costs and associated cost limits are passed to the Higher Level Control Function.

The operations and blends from the refinery model are further refined at the Scheduling level, principally to ensure day-to-day operation feasibility. These refined operations and blends are passed as targets to the Higher Level Control Function.

The Higher Level Control Function further adjusts the marginal values passed from the refinery model to relate the product costs to short term operating conditions. To be useful guides for short-term operations, the marginal values produced by the refinery model (on a month's time horizon) are adjusted to reflect the current operation as well as future operations on a day-to-day or hour-to-hour basis. When realistic short term product costs are provided to the optimization models at the supervisory process control level, then the supervisory process control level can be given more latitude to operate away from predetermined higher level targets.

Backward Flow - Data Capture

Figure 3, below, illustrates the flow of information involved in data capture, model and system updating, i.e., feedback or a <u>backward flow</u> of information.

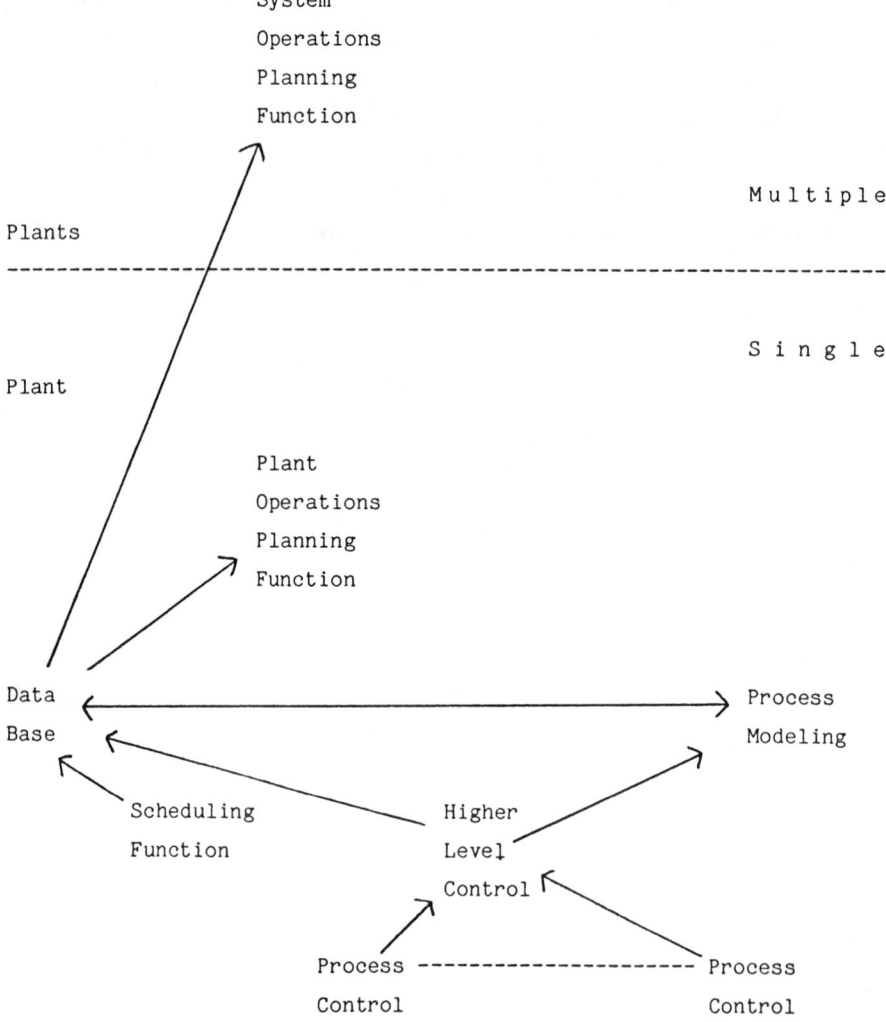

Fig. 3 Information Flow for Data Capture, Modeling Update

The key elements in this backward flow of information are the Data Base and the Higher Level Control Function.

The Higher Level Control Function gathers information at the process level in order to update process yields, stream qualities, processing capabilities, inventory levels and inventory qualities. This information, in conjunction with the Process Models, is used to update the plant's Data Base. The Data Base, in turn, forms the basis for the plant information system and serves as the source for most of the information used in the higher level functions of Systems Operations Planning and Plant Operations Planning.

Great care must be taken to organize the data in terms of its time horizon and range of applicability. For example, when should an observed change in process yield be used to update the Plant Operations Planning model? When should it be used to update the System Operations Planning model?

Our discussion of information flow deals with two types of information:

- Production and Inventory Targets and Limits (primal)
- Costs and Cost Limits (dual)

These two types of information parallel the math programming concepts of primal and dual information. In math programming terminology, production and inventory (physical) information would be considered as primal information; price and cost (economic) information would be considered as dual information. So far, we have not differentiated which type of information would be passed from level to level. However, we might state a general rule:

If we are concerned primarily with optimality in another function or level, it is better to pass dual information; if we are concerned with feasibility, it is better to pass primal information.

In most logistics and refining problems, we have sufficient flexibility such that feasibility is not of major concern except at the scheduling level. However, optimality is of concern--a fact which places a premium on realistic dual information.

Integration Framework Components

We now turn to the components upon which the framework is built. The functional specifications of the Logistics Integration Framework, discussed in previous sections, place stringent requirements on the components which make up the framework. Some of the major requirements will be discussed below.

Model Management System Requirements

The Model Management System (MMS) is one of the key components in the Logistic Integration Framework presented above; it would be impossible to implement the Integration Framework without a flexible, powerful MMS. For the purposes of this report, a MMS includes the functions of matrix generation, model optimization, solution analysis, solution reporting, data/model/solution cataloging, and data/model revision.

Since the System Operations Planning model and the individual Plant Operations Planning models should be generated from the same base data, it is clear that the model generation facility in the Integration System must be extremely flexible and independent of specialized data structures. The same set of base data might be used to generate a detailed, nonlinear, variable-time-period formulation at the Plant Operations Planning level or a less detailed, linear, fixed-time-period formulation at the System Operations Planning level. This is possible, but not easy.

Data Base Requirements

In most industry LP practice, the Data Base Manager and the data base itself are often considered as part of the Model Management System. However, in this report we will separate the two to underline the importance of data base management to all Integration Framework components. In the current context, the Data Base Manager supports system functions involving model management, reporting, scheduling, and higher level process control.

A primary requirement of the data base structure of the Integration Framework is that any given set of data is updated and maintained in only one place, i.e., each set of data has a single source point. All manipulations of a set of data proceed from a single source point. The collection of data which, for example, describes a refinery might be considered the <u>base data</u> for the refinery.

Designing a data base structure which will allow each refinery to model its own specific problems and yet allow a consistent representation of the refineries at the systems operations planning level is perhaps the most difficult task in the development of the Integration Framework.

Process Model Requirements

Process Models or Process Simulators predict the output of process units (product yields, qualities, performance measures, capacity consumption) as functions of input variables (feedstocks, feedstock properties, operating variables). Process Models are used to generate simplified process representations in the System and Plant Operations Planning Models, to explore alternative operating modes in the daily operation of the process units, and to provide process performance information at the supervisory process control level.

Figure 2 illustrates the universal role played by the Process Models in the Integration Framework. Process Models can be used in an iterative sense during the optimization process to provide nonlinear yield and gradient information at the Plant Operations Planning level, the Higher Level Control level, and in the supervisory control models at the Process Control level. The Process Models must be closely linked to the Data Base structures so that they can be used to generate the linearized representations of nonlinear processes found in model formulations at the System Operations Planning level.

Optimization Requirements

By far, the most widely used optimization technique in logistics optimization is Linear Programming (LP). Useful extensions to LP techniques include Successive Linear Programming (SLP) and Mixed-Integer Programming (MIP). SLP is a technique for including nonlinear relationships (nonlinear yields, nonlinear blending, quality pooling functions) in mostly-linear models. MIP is a technique for handling discrete functions (yes or no, run with minimum throughput or at zero) in linear models.

Outside of the traditional LP applications areas, nonlinear programming techniques have proven to be successful in the supervisory control area. Some combinatorial algorithms (for sequencing, routing, and loading) have had some success in the area of scheduling applications.

The Logistics Integration Framework will require a powerful LP optimizer at the System Operations Planning level, capable of solving large multi-time-period multi-refinery models. Comprehensive Successive Linear Programming (SLP) capabilities will certainly be required at the Plant Operations Planning level and, to some degree, at the System Operations Planning level. Some minor combinatorial algorithms will be required at the Scheduling level, and some practical techniques for data reconciliation and process unit coordination will be required at the Higher Level Control Function. All of these optimization requirements are within the current state of the art.

Interactive Interface Requirements

Users must be able to interact with system components at various levels throughout the Framework. For the purposes of this paper all user interfaces are considered to be Interactive Interfaces. In addition, the term Interactive Interfaces is used to encompass those aspects of Decision Support Systems and of Expert Systems which are applicable in the planning/scheduling/control areas.

Interactive Interfaces need to be full-screen, friendly, easy to learn and to use. Where required, color graphics can be used to reduce complex relationships and large amounts of data to recognizable patterns. Results, as well as user inputs, must be presented to the user in easily recognizable forms. The current state of the art in interactive color graphics is far beyond current practice in the oil industry (with some notable exceptions, i.e. control room applications).

2. The Higher Level Control Function

Process unit optimization problems are coupled by the fact that outputs of one unit are inputs to others, often with a storage medium in between. Hence, proper values and/or targets must be applied to interconnecting stream flows and qualities so that optimization of each unit will lead to an overall refinery optimum. This section deals with the issues of what coordination parameters should be used, and how their values can be found. Starting from the definition of a refinerywide optimization problem, we suggest a hierarchical system for coordination, and discuss primal and dual algorithms which find optimal values for the coordinating parameters. These parameters are targets, prices, or a mixture of both. Coordination using the refinery planning and scheduling models is also described. Finally, recommendations are made for future work.

Coordinating System Structure

Any decomposition for optimization must start with a clear statement of the problem to be decomposed. In the real time case, the answer is superficially obvious - maximize overall refinery profit subject to operating, supply and demand constraints. Since intermediate and final product streams are often stored prior to use, this must be a multi-period problem. Only in this way can current inventories, and future prices, supplies, and demands affect current decisions.

We propose a 2 time period model, illustrated in Figure 4. Period 1 has duration just long enough to complete the units' current operations, and should be about equal to the horizon of current process unit optimization problems. The second period extends to the point where targets are available from the refinery planning model. This latter model has two periods - the first is short (2-4 days, say)

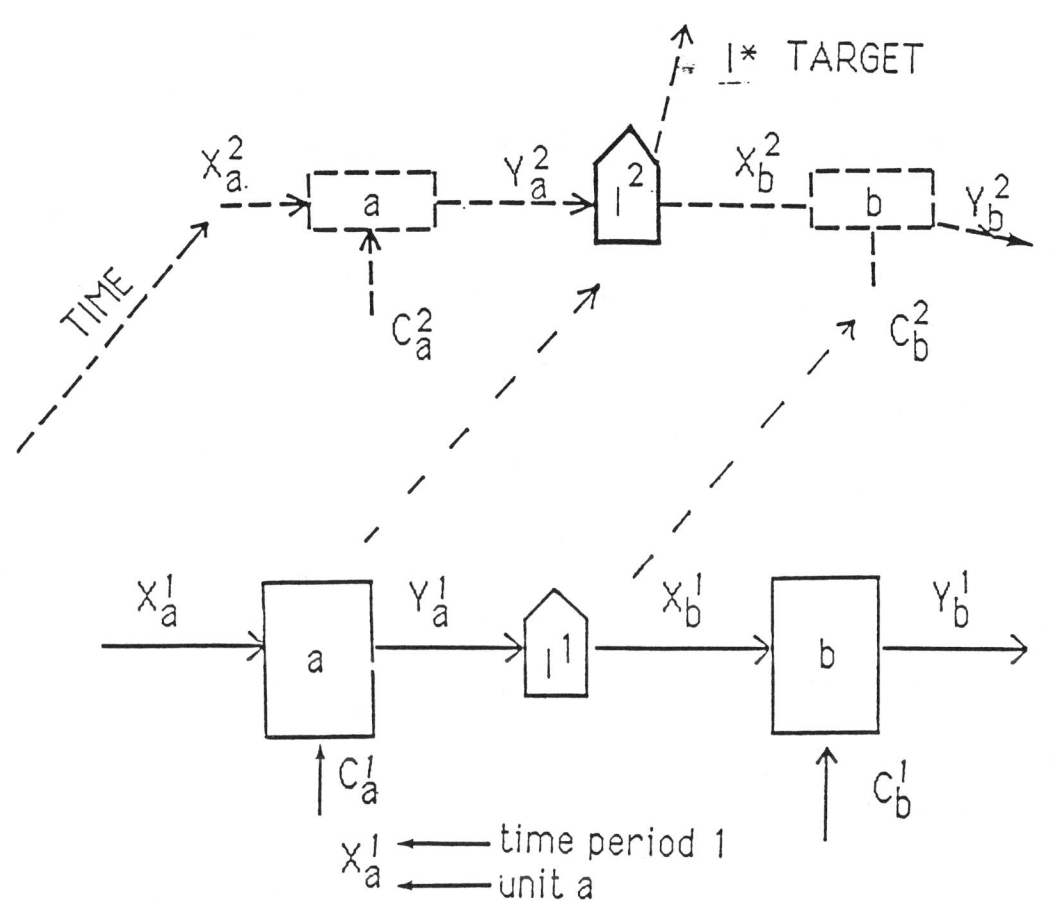

Two Time Period Formulation

Figure 4

to include the next set of operations in the refinery, and the total horizon is one month. These relationships are shown in Figure 5.

A Pricing (Dual) Coordination Algorithm

To make all this more concrete, let's define the objective and constraints of each process unit in each period respectively as z_a^i, z_b^i, and p_a^i, p_b^i, i=1,2. Then the overall problem is:

$$\text{maximize} \quad z = z_a^1 + z_a^2 + z_b^1 + z_b^2$$

subject to all operating constraints of both process units in both periods. In addition, as shown in Figure 4, we have a final inventory target I*, and a lower limit, LL, on the inventory I^1. For simplicity, no upper limit is shown -- it would be handled the same way. Using the symbols in figure 4, these constraints are

$$I' = I^0 + Y_a^1 - X_b^1 \geq LL \qquad (\lambda_1) \quad (1)$$

$$I^2 = I^1 + Y_a^2 - X_b^2 = I^* \qquad (\lambda_2) \quad (2)$$

The quantities λ_i shown next to these constraints are Lagrange multipliers, which may be thought of as prices which must be paid for violating them. The price λ_1 is nonnegative, and has an optimal value of zero if the optimal solution is strictly within the corresponding limit.

Each process unit's operating constraints and objective involve only its own inputs, outputs, and control variables. Only the constraints (1) - (2) involve variables of more than one process unit, so these couple the problem. Price-directed decomposition decomposes

Planning, Scheduling and Process Control 599

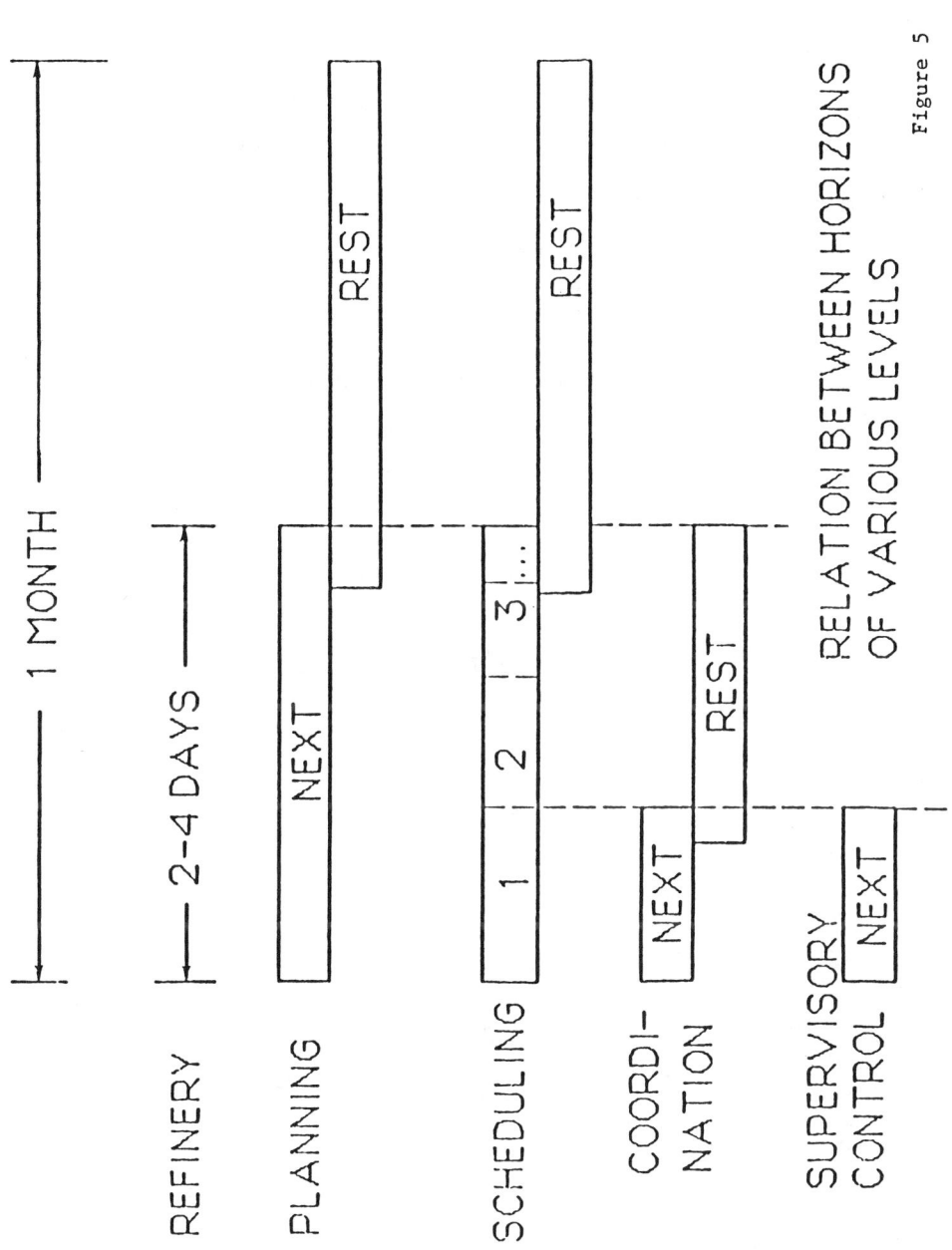

Figure 5. RELATION BETWEEN HORIZONS OF VARIOUS LEVELS

the problem by relaxing these coupling constraints, and costing out any violation in the objective using the prices λ_i. The constraint violations are:

$$e_1 = LL - Y_a^1 + X_b^1 - I^0 \qquad (3)$$

$$e_2 = I^0 + Y_a^1 + Y_a^2 - X_b^1 - X_b^2 - I^* \qquad (4)$$

Multiplying these errors by their prices, λ_i, and subtracting this penalty cost from the true system objective yields a relaxed problem:

$$\text{maximize} \quad L = z_a^1 + z_a^2 + z_b^1 + z_b^2 - \sum_{i=1}^{2} \lambda_i e_i \qquad (5)$$

subject to $p_a^1, p_a^2, p_b^1, p_b^2$. Now there are no coupling constraints, since they have been moved to the objective by the relaxation. The Lagrangian function, L, can be rearranged as follows:

unit a, period 1 \qquad\qquad unit b, period 1

$$L = z_a^1 + (\lambda_1 - \lambda_2) Y_a^1 \qquad + z_b^1 - (\lambda_1 - \lambda_2) X_b^1$$

both units, period 2

$$+ z_a^2 + z_b^2 - \lambda_2(Y_a^2 - X_b^2 - I^*)$$

constant

$$+ (\lambda_1 - \lambda_2) I_0$$

jence, maximizing L leads to three subprograms.

Higher Level Subproblem (P_{HL})

(1) $\max \; z_a^2 + z_b^2 - \lambda_2(Y_a^2 - X_b^2 - I^*)$

subject to $\quad p_a^2, p_b^2$

Supervisory Control Problems with Additional Cost Terms

Unit a \qquad\qquad\qquad Unit b

(2) $\max \; z_a^1 + (\lambda_1 - \lambda_2) Y_a^1 \qquad \max \; z_b^2 - (\lambda_1 - \lambda_2) X_b^1$

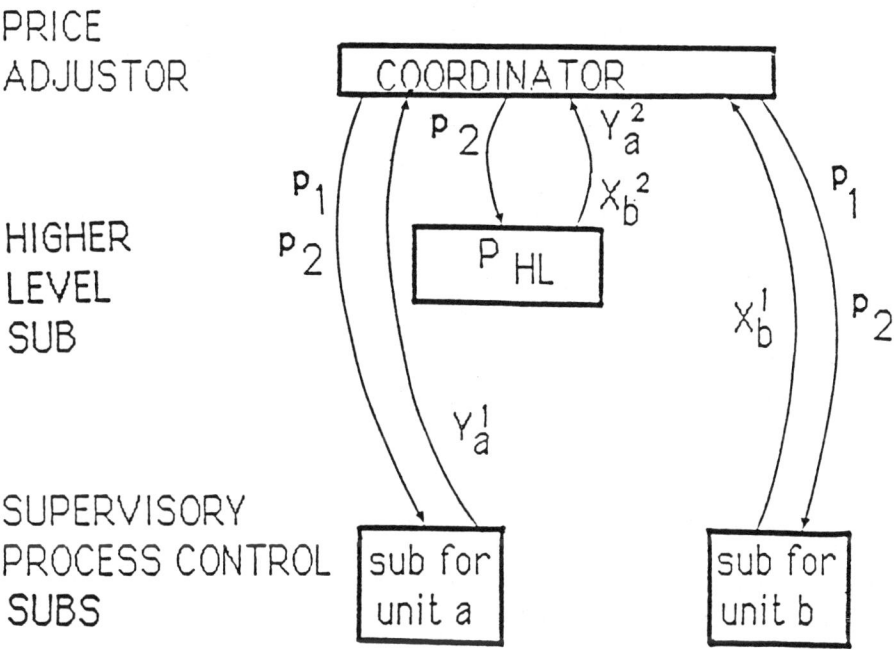

Three Level Hierarchy

subject to p_a^1 subject to p_b^1

Note that the period 1 subproblems credit unit a with a revenue of $(\lambda_1 - \lambda_2)Y_a^1$ for its production Y_a^1 and charges unit b an amount $(\lambda_1 - \lambda_2)X_b^1$ for its use of X_b^1.

Lagrangian duality theory (see [2]) shows how to compute the λ's. If all subproblems are solved, the error terms e_i in (3)-(4), evaluated at the optimal solution of the subproblems, are the gradient of a dual objective. This objective is just the maximum value of L in (5). If there are prices λ_1^*, λ_2^* which cause the subproblem solutions to be optimal overall, then these λ^*'s minimize this dual objective function. Hence, a simple steepest descent method applied to the dual will bring us closer to optimality. The adjustment of λ_i has the form

$$\lambda_i^{new} = \lambda_i^{old} + \alpha e_i \qquad (6)$$

where α is a step size. Hence, the overall system structure is as shown in Figure 6. Optimal values for process streams obtained by solving the subproblems are sent to the coordinator, which adjusts the prices according to (6). the new prices are sent back to the subproblems, and the process is repeated until all constraint violations are acceptably small.

Properties of Pricing Decomposition

Several aspects of this proposed approach need more study. A fundamental problem is that optimal prices λ_i^* may not exist, i.e., there may be no prices which cause subproblem solutions to be optimal in the refinerywide problem. This is mitigated by the fact that the dual objective is an upper bound on the overall maximum. If the optimal controls of each subproblem can be used to obtain a feasible solution to this problem, this yields a lower bound. Experience in other applications [Fisher, 1981] indicates that these bounds may be acceptably close. This experience also shows that only a few adjustments of the λ's may be needed to get the bounds close enough.

Other difficulties include the need to solve the higher level problem and the problem of different plant units having different time horizons in their subproblems. Solution of prototype problems, and eventual application to real problems, are needed to validate this methodology.

A Targeting (Primal) Coordination Algorithm

Coordination can also be achieved by specifying targets for interconnecting variables, and adjusting these targets to achieve overall optimality. In our example, we need targets for the unit a outputs y_a^i and for the unit b inputs and outputs X_b^i, Y_b^i. Let an overbar denote these targets, e.g., $\bar{Y}_a^i$ is the target for Y_a^i, etc. Then the subproblems are:

Higher Level Subproblem

(1) max $\qquad z_a^2 + z_b^2$

subject to $\qquad p_a^2, p_b^2$

Supervisory Control Subproblems with Target Constraints

(2) <u>Unit a</u> (3) <u>Unit b</u>

max z_a^1 $\qquad$ max z_b^1

$$\text{subject to } p_a^1 \text{ and } Y_a^1 = \bar{X}_a^1 \qquad \text{subject to } p_b^1 \text{ and } X_b^1 = \bar{X}_b^1$$

The inventory I^1 is determined by

$$LL \leq I' = I^0 + \bar{Y}_a^1 - \bar{X}_b^1 \qquad (7)$$

and the higher level target I^2 is enforced through the constraint

$$I^* = I^2 = I^1 + \bar{Y}_a^1 - \bar{X}_b^1 \qquad (8)$$

Relations (7) and (8) are constraints which are included in the problem which determines optimal targets. The objective of this problem is simply the sum of all subproblem maxima, which is refinerywide profit (as a function of the target values). It can be shown (see Lasdon, 1970) that the gradient of this objective with respect to a given target, say Y_a^1, is the optimal Lagrange multiplier corresponding to the constraint $Y_a^1 = \bar{Y}_a^1$. Hence, optimal multiplier values are sent from the subproblems to the coordinator, which uses them to determine new targets, applying a steepest ascent procedure. The targets are sent back to the subproblems which are re-solved. This is just the reverse of the information flow in pricing decomposition.

Properties of Primal Decomposition

Decomposition by targets has the advantage that optimal target values (unlike optimal price values) always exist. In addition, all subproblem solutions are feasible for the overall problem, which is not true for pricing decomposition. However, it is possible that a given set of targets cannot be met by some unit, i.e., subproblems can be infeasible. Fixing all intermediate stream inflows and outflows may overconstrain a unit. Fixed targets could be replaced by ranges, but an algorithm for varying these ranges needs to be developed. Targets or ranges could be applied to some streams and prices to others. Again, an algorithm for simultaneous variation of such a mixture has not been developed yet.

Treatment of Inventories

If there is no tank (or only a spill tank) between units a and b in Figure 4, the inventories I^i are all zero, as is the lower limit LL. This case of direct connection of process units is considered in [Lasdon, 1970.] Then the interconnection error term for pricing decomposition is simply

$$e_1 = X_b^1 - Y_a^1$$

If there is a tank between the units, but one is fairly certain that I^1 will not violate a limit, then the price λ_1 can be set to zero. However, the final inventory target constraint (2) still couples all units in all periods, requiring use of the price λ_2. If there is no target, the inventory I^1 decouples the two process units, since each can feed the tank or draw from it independently. Then both unit optimizations are independent, and there is no need for coordination.

Tactical Issues

The schemes discussed here need not be applied as frequently as the process units are reoptimized. The same prices or targets could be used for several single unit optimizations. In addition, not all process units need be involved in optimization. Some control inputs can be fixed, or chosen by schemes not including formal optimization. Prices for the input or output streams of such units play at most an accounting role, but higher level targets might be met by adjustment of unit operating conditions.

The second period subproblem, P_{HL}, need not use as accurate process unit models as the two first level subproblems. P_{HL} is only there to account for inventory targets and overall limits, and to bridge the time gap between the refinery planning and supervisory control models. It could use aggregated linear process models, in which case it could be solved quickly as an LP.

Coordination Using the Refinery Planning and Scheduling Models

It is possible to use the refinery planning or scheduling models for coordination of process unit optimization. They could provide either targets or prices for interconnecting streams. The short first period of the planning model, plus its inclusion of nonlinearities, should insure that these values are better suited for coordination than values from the current single period LP refinery models.

This may be the best approach for some refineries. It avoids the need for any additional coordination models. However, there is less frequent communication on a 2-way basis with the actual process unit models than with the schemes proposed here. In addition, the planning and scheduling models include less unit detail, and are not updated as frequently. Hence, tradeoffs between simplicity and solution quality need to be explored.

3. Nonlinear Optimization Algorithms for Supervisory Control

Many important algorithms for solving nonlinear programs have been developed or refined in the last 5 years. Not all can (or should) be considered here. The following classes of methods will be discussed, on the assumption that they are the most promising thus

far, and that they are the most suitable for solving supervisory control problems:
- (1) Generalized Reduced Gradient (GRG) methods
- (2) Successive Linear Programming (SLP)
- (3) Successive Quadratic Programming (SQP) methods

Descriptions of each of these classes must be brief, but more detailed information is available in [Powell, 1982] and [Powell, 1978].

Descriptions of the Algorithms
GRG Methods

By introducing appropriate slack variables any inequality constraints in the problem can be converted to equalities. Hence the nonlinear optimization problem can be written as

minimize $f(x)$
subject to
$$g(x) = 0 \qquad (1)$$
$$\ell \leq x \leq u \qquad (2)$$

where x contains both the original variables of the problem and the slacks, and g is the vector of all constraint functions.

There are many possible GRG algorithms. Their underlying concepts are described in [Lasdon, Waren, Jain, and Ratner, 1978]. This section briefly describes the version currently implemented in the code GRG2 [Lasdon and Waren, 1985] and [Lasdon and Waren, 1978].

GRG algorithms use the m equality constraints (1) to solve for m of the variables, called the basic variables (as in LP), in terms of the remaining n-m nonbasic variables. Denoting the basic and nonbasic variables as x_1 and x_{nb} respectively, the constraint equations become

$$g(x_1, x_{nb}) = 0 \qquad (3)$$

The Jacobian matrix of g may be similarly partitioned as

$$\partial g/\partial x = \left(\partial g/\partial x_1, \; \partial g/\partial x_{nb}\right) = \left(B_1, \; B_{nb}\right) \qquad (4)$$

where, for simplicity, the variables are assumed renumbered so the basics are the first m components of x.

Let $\bar{x}$ be the current feasible point (i.e., $\bar{x}$ satisfies constraints (1) and (2)). Then the specific variables to be chosen as basic must be selected so that B_1, evaluated at $\bar{x}$, is nonsingular. In this case the constraints (3) can be solved (at least conceptually) for x_1 in terms of x_{nb} to yield the basics as a function of the nonbasics, $x_1(x_{nb})$. This representation is valid for all x_{nb} sufficiently near $\bar{x}_{nb}$. The objective function is then reduced to a function of x_{nb} only,

$$f(x_1(x_{nb}), x_{nb}) = F(x_{nb}) \qquad (5)$$

and the original problem (1)-(2) (at least in the neighborhood of $\bar{x}$) is transformed to a simpler <u>reduced problem</u>

minimize $F(x_{nb})$

subject to the bounds on x_{nb}. The function F is called the <u>reduced objective</u> and its gradient ∇F the <u>reduced gradient</u>.

GRG algorithms solve the original problem (1)-(2) by solving (perhaps only partially) a sequence of reduced problems, which are usually solved by a gradient method. At a given iteration, with nonbasic variables $\bar{x}_{nb}$, the reduced gradient is computed as follows:

(a) Solve $\quad B_1^t \Pi = \partial f / \partial x_1 \quad$ (6)

for the simplex multiplier vector Π;

multiplier vector Π;

(b) Compute the reduced gradient

$$\partial F / \partial x_{nb} = \partial f / \partial x_{nb} - \Pi^t B_{nb} \quad (7)$$

Note that all partial derivatives are evaluated at the current point $\bar{x}$.

Following [Murtaugh and Saunders, 1978] the nonbasic variables are further partitioned into s superbasic variables, x_2, which are strictly between their bounds and n-m-s remaining nonbasic variables, x_3, which are at one of their bounds. In GRG2, the reduced gradient with respect to the nonbasic variables, $\partial F/\partial x_3$, is used only to determine if one of these variables should be released from a bound to join the superbasic set. The reduced gradient with respect to the current superbasics, $\partial F/\partial x_2$, is used to form a search direction, $\bar{d}$. Both conjugate gradient and variable metric methods can be used to determine $\bar{d}$. Then a one dimensional search is initiated, whose goal is to solve the problem

minimize $F(\bar{x}_{nb} + \alpha \bar{d})$

$\alpha \geq 0$

In the above, $\bar{d}$ has been extended to include zero components for the nonbasics at bounds. This minimization is done only approximately, and is accomplished by choosing a sequence of positive values $\{\alpha_1, \alpha_2, \ldots\}$ for α. For each value α_i, $F(\bar{x}_{nb} + \alpha_i \bar{d})$, must be evaluated. By (5) this is equal to

$$f(x_1(\bar{x}_{nb} + \alpha_i \bar{d}), \bar{x}_{nb} + \alpha_i \bar{d})$$

so the basic variables $x_1(\bar{x}_{nb} + \alpha_i \bar{d})$ must be determined. These satisfy the system of equations

$$g(x_1, \bar{x}_{nb} + \alpha_i \bar{d}) = 0$$

where $\bar{x}_{nb}$, α_i, and $\bar{d}$ are known and x_1 is to be found. If x_1 appears nonlinearly in any constraint then this system must be solved by an iterative procedure. Usually, a variant of Newton's method is used.

In the case of nonlinear constraints, the one dimensional search can terminate in three different ways. First, Newton's method may not converge. If this occurs on the first step, α_1 is reduced and we try again. Otherwise, the search is terminated. Second, if the Newton method converges, some basic variables may be in violation of thei bounds. Then GRG2 determines a new value such that at least one such variable is at its bound and all others are within their bounds. If, at this new point, the objective is less than at all previous points, the one dimensional search is terminated. A new set of basic variables is determined and solution of a new reduced problem begins. Finally, the search may continue until an objective value is found which is larger than the previous value. Then a quadratic is fit to the three α_i values bracketing the minimum. F is evaluated at the minimum of this quadratic, and the search terminates with the lowest F value found. The reduced problem remains the same.

An important feature of this algorithm is its attempt to return to the constraint surface at each step in the one dimensional search. This differs from earlier strategies suggested by Abadie [1969] and by Luenberger [1973], which involve linear searches on the tangent plane to the constraint surface prior to returning to that surface. GRG2 returns to the surface each time because it is simpler and because it is felt that it leads to a more reliable algorithm. Computational experience (see [Schittkowski, 1980]) shows that, if properly implemented, this strategy can be developed into an algorithm that is both reliable and efficient.

SLP Algorithms

Successive Linear Programming (SLP) methods linearize any nonlinear objective and constraint functions about some base point $\bar{x}$, i.e., g_i is replaced by $g_i(\bar{x}) + \nabla g_i(\bar{x}) \Delta x$. Upper and lower "step bounds" $\bar{s}$ are imposed on Δx to insure that the errors in this approximation are limited. The resulting LP with variables Δx is solved, and the point $\bar{x} + \Delta x$ is tested. If it is deemed improved, it is accepted and the process is repeated. If not, the step bounds are reduced and the LP is resolved. Successive points generated by this procedure need not be feasible, even if the initial point is. However, the amount of infeasibility generally is reduced as the iterations proceed.

SLP methods have the major advantage that they can use existing efficient LP codes as "subroutines". The idea was first described by Griffith and Stewart [1961], and was later extended by Buzby [1974] and Boddington [1979]. Recently, Zhang, Kim, and Lasdon [1985] have investigated several SLP strategies, and have developed an algorithm that is quite efficient. This algorithm is based on exact penalty and trust region concepts. Convergence has been proved, and improved performance on some large plant operations planning problems is presented in [Baker and Lasdon, 1985]. As discussed in section one, such algorithms are being used more and more at the Plant Operations Planning level, to account for important nonlinearities such as pooling and nonlinear yields and blends.

Successive Quadratic Programming Algorithms

Successive Quadratic Programming Algorithms solve a sequence of quadratic programs (QP's), i.e., quadratic objective, linear constraints. As in SLP, the linear constraints are linearizations of the actual constraints about the current point. The quadratic objective used is $x^TQx + \nabla f^T x$, where Q is an approximation of the matrix of second partial derivatives of the Lagrangian function

$$L = f - \lambda g$$

where λ is a row vector of Lagrange multipliers. This can be justified by showing that the QP solution yields the point one would get by applying Newton's method to the necessary (Kuhn-Tucker) conditions for optimality of (1)-(2) [Powell, 1978], [Fletcher, 1981]. Usually, the QP solution is not accepted immediately as the new point; it is used to specify a search direction, and a one-dimensional search in that direction is performed.

The connection with Newton's method is most easily seen in a problem with only equality constraints [Powell, 1978]. There the necessary conditions for optimality are

$$\nabla f - \sum_i \lambda_i \nabla g_i = 0$$
$$g = 0$$

Newton's method linearizes this system, yielding the linear equations

$$\begin{pmatrix} G & -J \\ J^T & 0 \end{pmatrix} \begin{pmatrix} \Delta x \\ \Delta \lambda \end{pmatrix} = \begin{pmatrix} -\nabla L \\ g \end{pmatrix} \qquad (8)$$

In the above, G is the matrix of second partial derivatives of the Lagrangian $L = f - \lambda g$, J is the matrix of first partial derivatives of g, and all quantities are evaluated at some point $(\bar{x}, \bar{\lambda})$. It is easily shown that, if $(\Delta x, \Delta \lambda)$ satisfy (8), then Δx, $\bar{\lambda} + \Delta \lambda$ satisfy the necessary conditions for optimality of the quadratic program

minimize $\quad \nabla f^T y + 1/2 \, y^T G y \qquad (9)$

subject to the linearized constraints

$$J^T y + g = 0 \qquad (10)$$

The extension to inequality constraints is straightforward; linearize them too and include them in the Lagrangian function when computing G. Of course, G need not be positive definite, even at an optimal solution of NLP, so the QP (9)-(10) may not have a minimum. Fortunately, positive definite approzimations, Q, to G can be used, and then (9)-(10) will have an optimal solution if it is feasible. Such approximations to G can be obtained by a slight modification of the popular BFGS updating formula used in unconstrained minimization [Powell, 1982]. This formula requires only the gradient of the Lagrangian function, ∇L, so second derivatives of the problem functions need not be computed.

It can happen that the linearized constraints (10) are infeasible. If so, [Powell, 1978] proposes adding an additional feasibility variable, but this does not always remedy the situation. In [Fletcher, 1981], a modified form of the QP (9)-(10) is proposed, which is $\quad$ minimize $\quad\quad\quad\quad\quad\quad$ minimize $\nabla f^T y + 1/2$

$$\nabla f^T y + 1/2 \, y^T G y + w^T(p+n)$$

subject to $J^T y + g = p - n$

$$p \geq 0, \quad n \geq 0$$

In the above, p and n are vectors of additional "deviation" variables, which cause the QP to always be feasible. The nonnegative vector w penalizes these deviations in the objective. If $w_i \geq |\lambda_i^*|$, where λ_i^* is the optimal Lagrange multiplier for the constraint $g_i = 0$, then a suitable algorithm using this QP will converge to a stationary point of the L1 exact penalty function associated with the original problem (see [Fletcher, 1981]). Whatever QP is used to determine the search direction y, there are two basic strategies for determining the distance to be moved along y; trust region and line search. Trust region strategies usually use upper and lower limits, called step bounds, on the components of y, choose a trial point x + y, and make a test to see if this trial point is acceptable. If not, the step bounds are reduced. Complete details are provided in [Fletcher, 1981]. Line search strategies vary in choice of the line search objective function. Many authors use the exact penalty function

$$P(x,w) = f(x) + \sum_{i=1}^{m} w_i |g_i(x)|$$

(which is easily extended if inequality constraints are present.) This can sometimes cause slow convergences near the optimum due to the nondifferentiability of P. An algorithm which uses an augmented Langrangian line search objective is described in [Gill, et.al., 1984]. Computational comparisons of line search and trust region strategies, and of the 2 QP formulations described above, appear in [Fan, et. al., 1985]. However, much work remains to be done, and new variations are proposed frequently.

Much work has also been done by Chemical Engineers on SQP algorithms for flowsheet optimization. Here a major issue is whether the recycle streams are to be included in the algorithm as equality constraints (and hence become satisfied only as the optimum is approached-this is called the infeasible variant) or whether these streams are to be converged at each iteration of the SQP algorithm (the feasible variant). Discussion and comparison of these options appears in [Biegler and Hughes, 1985].

Comparisons

Table 1 below summarizes the relative merits of SLP, SQP, and GRG algorithms, focusing on their application to supervisory control problems.

Algorithm	Relative Advantages	Relative Disadvantages
SLP	1. Easy to implement 2. Widely used in practice 3. Rapid convergence when optimum is at a vertex 4. Can handle very large problems 5. Does not attempt to satisfy equalities at each iteration	1. May converge slowly on problems with non-vertex optimal 2. Will usually violate nonlinear constraints until convergence, often by large amounts
SQP	1. Usually requires fewest function and gradient evaluations of all three algorithms (by far)	1. Robustness in practice undetermined 2. Will usually violate nonlinear constraints until convergence,

| | 2. Does not attempt to satisfy equalities at each iteration | often by large amounts
3. Harder than SLP to implement – requires good QP solver |
|---|---|---|
| GRG | 1. Probably most robust of all three methods
2. Versatile-especially good for unconstrained or linearly constrained problems, but also works well for nonlinear constraints
3. Can utilize existing process simulators employing Newton's method | 1. Hardest to implement
2. Needs to satisfy equalities at each step of algorithm |

Table 1 - Comparison of Optimization Algorithms

One feature appears both as an advantage and disadvantage -- whether or not the algorithm can violate the nonlinear constraints of the problem by large amounts during the solution process. SLP and SQP usually will generate points yielding large violations. This can cause difficulties, especially in models with log or fractional power expressions. Then, negative arguments for these functions may be generated. These problems have been documented in [Sarma and Reklaitis, 1979], using a complex chemical process as an example. There, SLP and some exterior penalty algorithms failed, while a GRG code described in [Lasdon, Waren, Jain, and Ratner, 1978] succeeded, and was quite efficient. On the other hand, algorithms which do not attempt to satisfy the equalities at each cycle can be far faster than any which does, as pointed out in [Berna, Locke, and Westerberg, 1980]. The fact that such algorithms usually will satisfy any linear constraints at each iteration should ease the difficulties referred to above.

Clearly, all three algorithms have advantages which will dictate their use in certain situations. For large problems, SLP software is used most widely, since it is relatively easy to implement given a good LP system. An SLP code capable of solving large problems, implementing the PSLP algorithm in [Zhang, et.al., 1985], is described in that reference. The GRG code GRG2 [Lasdon and Waren, 1985] is widely used for small to medium size problems, especially in the process unit optimization area. SQP software for process optimization problems is being designed as part of a chemical process simulation and optimization system by Westerberg et. al -- see [Locke and Westerberg, 1982].

Conclusions

Throughout the refining industry there has been a tremendous gap between the LP models used at the refinery planning level and the oil movements and process control functions carried out at the daily operations level. The tools are currently available to close this gap in the form of an integrated system for planning, scheduling, and control. In fact, many oil companies are actively engaged in programs to bring about the integration of these functions within the next five to ten years.

While the models proposed in this paper do not necessarily cross organizational lines, the Logistics Integration Framework components, e.g., process models, model management system, data base, do cut across the current organizational structure. We are proposing the development of tools which can be used at various levels in the system. This requirement for consistency across applications in various refineries and corporate organizations will pose some difficulties for higher level mangement.

The Logistics Integration Framework has the adaptive capabilities of a feedback control system in that its primary purpose is to respond to disturbances at all functional levels. At the System Operations Planning level a disturbance might be a change in long term crude availability or product demand. At the Process Control level a disturbance might be a short term variation in process yield. Any hierarchy, designed to achieve these adaptive capabilities will necessarily appear to have an apparent repetition of various functions at different levels in the system. This repetition of functions and the emphasis on information feedback between the various levels of the system make the Integration Framework responsive to a wide variety of disturbances and possible future business environments.

References

1. Abadie, J. and J. Carpentier, "Generalization of the Wolfe Reduced Gradient Method to the Case of Nonlinear COnstraints," in Optimization, Academic Press, New York, pp. 37-47 (1969).

2. Baker, T.E. and L. Lasdon, "Successive Linear Programming at Exxon," Management Science, Vol. 31, No. 3, pp 264-274 (1985)

3. Berna, T.J., M.H. Locke and A.W. Westerberg, "A New Approach to Optimization of Chemical Processes," AICHE J., Vol 26 , p. 37 (1980).

4. Biegler, L.T., and R. Hughes, "Feasible Path Optimization with Sequential Modular Simulators," Computers and Chemical Engineering Vol. 9, No. 4, pp 379-394 (1985)

5. Bodington, C.E., "Nonlinear Programs for Product Blending," Joint National TIMS/ORSA meeting, New Orleans, April/May (1979).

6. Buzby, B.R., "Techniques and Experience Solving Really Big Nonlinear Programs," in Optimization Methods, R.W. Cottle and J. Krarup (eds.), English Universities Press, London (1974)

7. Fan, Y., S. Sarkar, and L. Lasdon, "Experiments with Successive Quadratic Programming Algorithms," Working Paper, Dept. of General Business, University of Texas at Austin, April, 1985 (To appear in JOTA)

8. Fisher, M.L., "Lagrangian Relaxation Method for Solving Integer Programming Problems," Management Science, Vol. 27, No. 1. pp 1-18 (1984)

9. Fletcher, R. Practical Methods of Optimization, Vol. 2, Constrained Optimization, John Wiley (1981)

10. Gill, P.E., W. Murray, M. Saunders, and M. Wright, "Users Guide for NPSOL (Version 2.1): A Fortran Package for Nonlinear Programming", Systems Optimization Laboratory, Dept. of Operations Research, Stanford Univ., Sept. 1984

11. Griffith, R.E. and R.A. Stewart, "A Nonlinear Programming Technique for the Optimization of Continuous Processing Systems," Management Science, Vol. 7, pp 379-392 (1961)

12. Lasdon, L.S., "Optimization Theory for Large Systems," Chapter 8, Macmillan (1970).

13. Lasdon, L.S. and A.D. Waren, "Generalized Reduced Gradient Software for Linearly and Nonlinearly Constrained Problems," in Design and Implementation of Optimization Software, H. Greenberg (ed.), Sijthoff and Nordhoff, Holland, 363 (1978).

14. Lasdon, L.S. and A.D. Waren, "GRG2 User's Guide," Department of General Business, School of Business Administration, University of Texas, Austin, TX (1985).

15. Lasdon, L.S., A.D. Waren, A. Jain, and M. Ratner, "Design and Testing of a Generalized Reduced Gradient Code for Nonlinear Programming," ACM Transactions on Mathematical Software, 4, 34 (1978).

16. Locke, M.H., and A.W. Westerberg, "Large Scale Applications of Successive Quadratic Programming Methodology," paper presented at AICHE meeting, Los Angeles, California, Nov. 16-17, 1982, in the Symposium on Large Scale Systems.

17. Luenberger, D., Introduction to Linear and Nonlinear Programming, Addiston-Wesley, 158-162 (1973).

18. Murtagh, B.A. and M.A. Saunders, "Large-scale Linearly Constrained Optimization," Math Programming, Vol. 14. pp. 41-72 (1978).

19. Powell, M.J.D. (ed.), "A Fast Algorithm for Nonlinearly Constrained Optimization Calculations," in Numerical Analysis, Proceedings of the Biennial Conference Held at Dundee, June 1977. (G.A. Watson, ed.), Lecture Notes in Mathematics, Vol. 630, Springer-Verlag, Berlin, Heidelberg, New York, 1978.

20. Powell, M.J.D., "Algorithms for Nonlinear Constraints that Use Lagrangian Functions," Math Programming, Vol. 14, No. 2, p. 224, 1978

21. Powell, M.J.D. (ed.), <u>Nonlinear Optimization, 1981</u>, Academic Press, 1982.

22. Sarma, P.V.L.N., and G.V. Reklaitis, "Optimization of a Complex Chemical Process Using an Equation Oriented Model," Mathematical Programming Study 20, 113-160 (1982)

23. Schittkowski, K., "Nonlinear Programming Codes--Information, Tests, Performance," Lecture Notes in Economics and Mathematical Systems, Vol. 183, Springer-Verlag, Berlin, Heidelberg, New York, (1980)

24. Zhang, J., N. Kim, and L. Lasdon, "An Improved Successive Linear Programming Algorithm", <u>Management Science</u>, Vol 31, No. 10, pp. 1312-1331 (1985)

CONTROL OF CHEMICAL REACTORS

W. Harmon Ray
University of Wisconsin, Madison, WI 53706

In their natural habitat, chemical reactors (as they are designed) represent some of the most diverse and exotic processes encountered by the process control engineer. They are usually highly nonlinear, have a dearth of composition measurements, and are prone to sensitivity and stability problems. Thus there is not a single control problem for chemical reactors, but several. For example, a good process control scheme might include the requirements

(1) to maintain stable and safe operation under normal conditions

(2) to guarantee safe operation in the face of major upsets (no runaway, rational alarm response, etc.)

(3) to produce a uniform, high quality product

(4) to maintain high productivity, low cost operation (economic optimization)

(5) To achieve efficient swings between product grades for continuous multiproduct reactors

(6) to produce batch to batch uniformity of product quality at high production rates in batch reactors.

The relative difficulty of meeting each of these requirements depends on the specific problem. Thus chemical reactor control is very discipline-oriented, because each discipline (eg; polymers, metallurgy, pharmaceuticals, petroleum refining, etc.) has its own special problems and particular control objectives. We often have a good understanding of these issues for traditional industries and older technologies. For example, we understand that the process control problems for a $100/lb specialty rubber centers on maximizing product quality and uniformity, while the control of catalytic reformers for upgrading the octane of gasoline fractions may focus on maximizing productivity and minimizing operating costs. However, for newer technologies, the difficult control problems and the control system objectives are not yet well understood. Thus the first two papers in this session will provide us with perspective in two relatively new areas of technology.

The first paper will survey some control problems which arise in the production of microelectronics materials. An increasing fraction of our chemical engineering graduates are being employed in the electronics industry, and this will be an important source of process control problems in the future. Thus we should begin to learn something about this industry.

In the second paper, we will see a review of control problems arising in biological reactors, and some of the future challenges to be faced in this area of technology. With the advances in microbiology suggesting new classes of products and even new industries, it will be important for process control engineers to understand some of the control issues which will arise.

The last two papers in this session will deal with process control system design strategies for chemical reactors. The first of these will discuss some new methods for dealing with complex nonlinear dynamics and potential reactor runaway. With our heightened awareness of process safety these days, new ideas such as these are needed to stimulate thought and work on control system design strategies to deal with these problems. The last paper of this session is a view from industry of the status of chemical reactor control in a large chemical company. From this survey, we shall see what techniques are (or are not) being used, and what are some of the difficult control problems one encounters frequently.

Taken together, these papers acquaint us with a broad range of issues which arise in chemical reactor control across many industries. It is hoped that this perspective will stimulate our thinking as we seek better ways to design control systems for chemical reactors.

CONTROL PROBLEMS IN MICROELECTRONIC PROCESSING

Klavs F. Jensen
University of Minnesota, Minneapolis, MN 55455

Abstract. Issues in chemical process control of microelectronics processing are reviewed with emphasis on problems in silicon integrated circuit (IC) manufacturing. Many of these control issues are similar to those of other chemical batch processing. A summary is given of general characteristics of IC fabrication and driving forces behind automation. Automation is discussed in terms of three areas: plant management, materials handling and unit operation control. The last area is of primary interest and common problems associated with difficult to measure process variables, complex interactions between variables, many parameters, nonlinearities, and inadequate process knowledge are illustrated with crystal growth, diffusion, and plasma etching examples. Current control schemes are based on single loop PID regulators whose setpoints are adjusted by microprocessors according to process recipes. Based on the discussion of control problems, it is proposed that significant process improvements can be realized through application of advanced control strategies combined with development of dynamic process models and sensors.

Keywords. Microelectronic processing; integrated circuits; automation; materials handling; unit operation control; crystal growth; dopant diffusion; plasma etching.

INTRODUCTION

The fabrication of microelectronic components involves long

sequences of complex chemical processes which can be divided into four general categories: bulk crystal growth, thin film deposition, doping and patterning. These are combined to produce three-dimensional microstructures on single crystal substrates such as the Metal Oxide Semiconductor Field Effect Transistor (MOSFET) shown in Fig. 1. Each substrate, commonly referred to as a wafer, serves as the basis for hundreds of integrated circuits each of which is formed from a hundred thousand or even more individual components (e.g. transistors, capacitors, and resistors). A typical process sequence can involve more than 30 individual steps (cf. Parrillo, 1983) and the ultimate measure of whether or not these have been done successfully is the performance of the final circuits. These circuits are highly sensitive to process variations and cannot be repaired if a particular chemical process step fails. Consequently, the yield of functional "chips" on a wafer is low for complex circuit structures.

The major portion of the microelectronic industry is based on silicon, but compound semiconductor (e.g. GaAs, GaAlAs, GaInAsP) based high speed and electro-optic technology is growing rapidly. The present review focuses on Si based process technology because of its predominance and further stage of development. The control tasks will be analogous for compound semiconductor processing; but also more difficult because of the complexity in handling the materials.

CHARACTERISTICS OF IC MANUFACTURING

Integrated circuit (IC) manufacture is composed of a large variety of batch processes (for an overview of these cf. Sze (1983) and Doane et al. (1982)). Single crystalline silicon is grown and sliced into wafers. These wafers are subsequently moved in batches among various deposition, patterning, oxidation, and doping equipment. The closest approximation to continuous processing is a so-called cassette-to-cassette system where each wafer is loaded and removed individually from the chemical process by a mechanical arrangement. Thus, the nearest chemical plant analogy would be a specialty chemical or pharmaceutical facility.

A fabrication line involves a large capital investment. The equipment is highly sophisticated and is placed in expensive, very clean environments. A new line may cost $10 million or more. Futhermore, the technology changes rapidly reducing equipment lifetime. As an illustration, during the past decade wafer size has gone from 75 mm to 150 mm requiring redesign of process internals and development of new equipment. This continuosly changing process technology also implies that a large engineering support and development staff is needed. Equipment operators must be skilled, but the line labor costs are usually small compared to the engineering cost.

The processing is done in class 10 or 100 clean rooms, i.e. the number of particles of diameter greater than 0.2 µm per cubic foot is less than 10 or 100, respectively. This is accomplished by cycling the clean room air through high efficiency filters. In addition, low particle content, deionized water must be available along with high purity H_2 and N_2. Most of the gases used are toxic so extensive safety and chemical treatment procedures are necessary. This in conjunction with the clean room procedures result in large operating expenses. Raw material costs are typically the lowest of the major production expenses. Figure 2 illustrates the relative costs, as estimated by Kaempf (1984).

AUTOMATION AREAS IN IC MANUFACTURING

Integrated circuit manufactures are showing considerable interest in automation with increasing circuit integration and complexity. This development is driven by the need to improve productivity and yields while maintaining quality and safety. The major automation problems can be grouped into three main areas:
   (1) plant management, including work in process (WIP) control
   (2) materials handling
   (3) unit operation control

The increased integration of circuits, made possible by reduced minimum feature sizes, has added to process complexity. For example, the fabrication of advanced microelectronic structures can involve two hundred

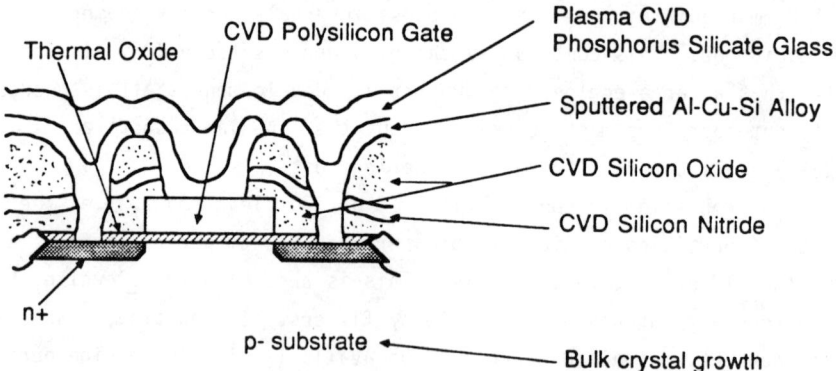

Fig. 1. Schematic of silicon MOSFET.

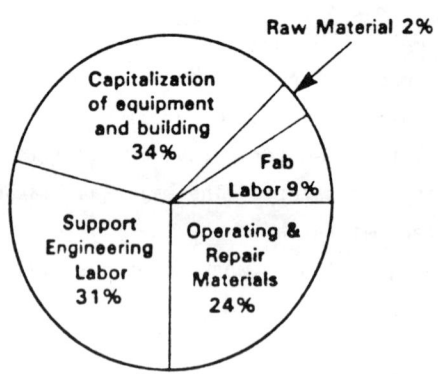

Fig. 2. Cost distribution in IC manufacturing (Kaempf, 1984).

or more process steps. It typically takes from two to six weeks to process a wafer of chips. Therefore, to improve productivity and process yield, it is necessary to have a supervisory system which (i) collects information on the the state of the system; (ii) gives status of work in progress; (iii) schedules work based on process priorities, product requirement, equipment readiness, and material availability; and (iv) controls product and raw material inventories.

The state of the system refers to whether or not deposition, etching, diffusion, etc., equipment is operating within product specifications. Because of the potentially hazardous chemicals, it also includes safety aspects such as monitoring ventilation, possible gas leaks, and gas scrubber performance. In addition, the clean room and utility functions have to be controlled. The work in progress (WIP) monitor provides information of where each wafer batch is in the line. A particular batch might be queuing for a diffusion step, be going through a lithographic sequence, or simply have been misplaced. In addition, the WIP monitor collects information on the availability of equipment, materials, and operators necessary to schedule the batches and optimize the line throughput, measured in terms of number of functional chips per unit time.

Materials handling concerns the physical movement of wafers through the fabrication line. To minimize contamination, in particular by small particles, the wafers are transported in closed cassettes between equipment and loaded automatically from cassettes into most equipment. Particle contamination is a major problem in VLSI technology and a large driving force behind automation. With minimum device features in the 1 µm range, particles greater than 0.1 µm will be "killer" defects. People are a significant source of particulates even with proper clean room garments a person can emit as many as 10^5 particles/minute (Harper and Bailey, 1984). This problem is the main reason for automation to remove line operators rather than a reduction in the minor labor cost.

The decrease in minimum feature sizes, responsible for the increased circuit integration, has come about through advances in process technology, in particular in the areas of lithography and pattern etching. This development has led to complex equipment for process "unit operations", i.e. crystal growth, thin film deposition, patterning, and doping. The operation of the equipment involves lengthy recipes that can only be executed by microprocessor control. Moreover, as the process limits are narrowed and the trend towards smaller feature sizes continues, the control of each step within the unit operations increases in importance.

CURRENT STATUS OF PROCESS CONTROL IN IC MANUFACTURING

Two recent conference papers by Atherton (1984) and Kaempf (1984) discuss issues and design in the implementation of process control systems for two major semiconductor manufacturers, Fairchild and Hewlett-Packard, respectively. Kaempf contribution gives a good overview of the issues mentioned in the previous section. Automation of small process lines dealing with wide variety of device structures has been described by Doyal et al. (1980) and Reid et al. (1984) with emphasis on the interface between the unit operation level and the supervisory computer system. Otherwise, published reports of automation in integrated circuit manufacture are scarce and concentrate on the area of plant integration and work-in-progress monitoring (Boehlert and Trybula, 1984; Harper et al., 1985). The monitoring and scheduling problems in this area are analogous to those involved in batch processing of specialty chemicals and it would seem that integrated circuit manufacture could benefit from experiences in the chemical industry and vice versa.

The materials handling problem is largely a mechanical problem involving wafer transport systems and control as well as design of robots for wafer handling. Thus, this problem is outside the scope of this review concerning chemical process control. Some of the issues in automated wafer handling are dicussed by Factor and Kaufman (1985), Harper and Bailey (1984), Parikh and Kaempf (1984), and Weiss (1984).

The equipment used in the unit operations is complex and increasingly microprocessor controlled to allow recipes to be performed automatically. However, advanced multivariable control schemes are rarely, if ever, invoked. The microprocessor adjusts setpoints according to some sequence of steps defined by the equipment manufacturer or the process operator. Flows, pressures, and temperatures are regulated independently by "off-the-shelf" PID regulators even though the control loops can interact strongly. For example, fluorocarbon concentration, substrate temperature, reactor pressure, and plasma power all influence silicon etch rates and uniformity, but they are typically controlled independently.

There are several reasons for this lack of advanced control applications. First, the rapidly changing process technology has involved quantum improvements in device processing rather than the more incremental changes one typically would expect from implementation of advanced process control. However, with the increasing size of wafers, decreasing minimum feature size and maturing of processing techonology, multivariable control schemes will become necessary to maintain profitability.

Second, the majority of equipment manufacturers are relatively small companies without the resources needed to develop control algorithms in the face of changing process technology. They have therefore chosen to use simple, third party control hardware.

Third, the fundamental understanding of many of the unit operations has been limited so it has been difficult to identify key operating variables. Recent experimental efforts and modelling studies have improved upon the situation, but more fundamental process investigations are needed. This will be further discussed in subsequent sections.

Fourth, it is difficult to measure whether or not a chemical process step has been done successfully since there are few on-line measurable electrical properties. For example, film thickness and grain structure of polycrystalline silicon can be measured after a deposition step. However, their effect on device performance might not show up until subsequent doping or patterning steps fail. Similarly, it is possible to measure etch rates on-line by laser interferometry, but the etch profiles must be checked by electron microscopy. Unexpected mask undercutting or undiscovered etch residues can result in subsequent contact and IC lifetime problems. Thus, most IC unit operations are run in open loop where corrective action is only taken after a problem has been discovered at a later stage or possibly even in the final circuit. These long measuring lags and feedback loops lead to wide swings in process yield and quality (Levy, 1984).

PROCESS EXAMPLES

Examples of potential process control applications are illustrated in the following subsections. These examples represent typical chemical process steps which could benefit from advanced control strategies similar to those used in other chemical plants.

Bulk Crystal Growth

Single crystalline silicon for ICs is produced from high purity polycrystalline silicon by Czochralski crystal growth. In this process the single crystal is pulled from a silicon melt as illustrated in Fig. 3. The growth consists of several steps (Kim et al., 1983; Pearce et al., 1983). First, a single crystal seed with the desired crystallographic orientation is brought in contact with the melt. Then by pulling the crystal up while controlling melt temperature, a short neck is formed to eliminate crystal dislocations. The seed and neck phases are followed by a fast outward growth under slow pull rates to reach the desired crystal diameter. After these initial steps, the bulk crystal is grown until the melt is nearly depleted. The doping concentration (if any) and the crystal diameter must be tightly controlled since this portion eventually will be sliced into single crystal wafers for IC manufacture. Finally, the crystal is "tailed-off" in a manner that minimizes the possibility of crystal structure loss.

Gains in process understanding, modelling, and control have led to large increases in crystal diameters during the past two decades. The standard size was 40 mm in 1965. 150 mm wafers are common today and 200 mm wafers are expected to dominate in the early 1990's (Larrabee, 1985). However, problems with dopant striations and crystal diameter variations remain (Fiegl, 1983; Kim et al., 1985). Some of these can be reduced by process modifications such as continuous replenishment of the melt (Fiegl, 1983) or growth under an applied magnetic field (Suzuki et al., 1981), but further improvements in crystal quality must include advances in process control.

A standard control approach in CZ growth uses a pyrometer to monitor the melt crystal interface. The crystal pull rate is manipulated to eliminate deviations in crystal diameter from its setpoint while the melt heater input is adjusted to maintain a constant pull rate. The loops are cascaded so that the output of the seed lift (PID) controller is the input to the melt temperature (PID) controller. In addition, the melt crucible is rotated to improve dopant uniformity across the crystal diameter. Modern CZ crystal growth equipment is microprocessor controlled so that the point changes can be programmed for transitions between seed, neck, bulk growth and tail-off stages. However, no advanced control schemes seem to have been implemented. Kim et al. (1985) have reported a reduction in crystal diameter variations with computer control of CZ growth using software implemented PID controllers, but no major changes in control strategy have been made.

Improvements in diameter and dopant control could result from implementing advanced multivariable control schemes within the existing sensor and final control element structure. However, even larger improvement could be achieved by incorporating detailed process models into the control approach. The heat transfer from the growing crystal to the surroundings, primarily by radiation, determines the crystal growth rate. Thus, by using a heat transfer model for crystal and melt in conjunction with spatial temperature measurements by pyrometer imaging, it should be possible to accurately control crystal diameter and pull rate. Derby et al. (1985) describes the necessary heat transfer model for the more difficult case of liquid encapsulated CZ growth,

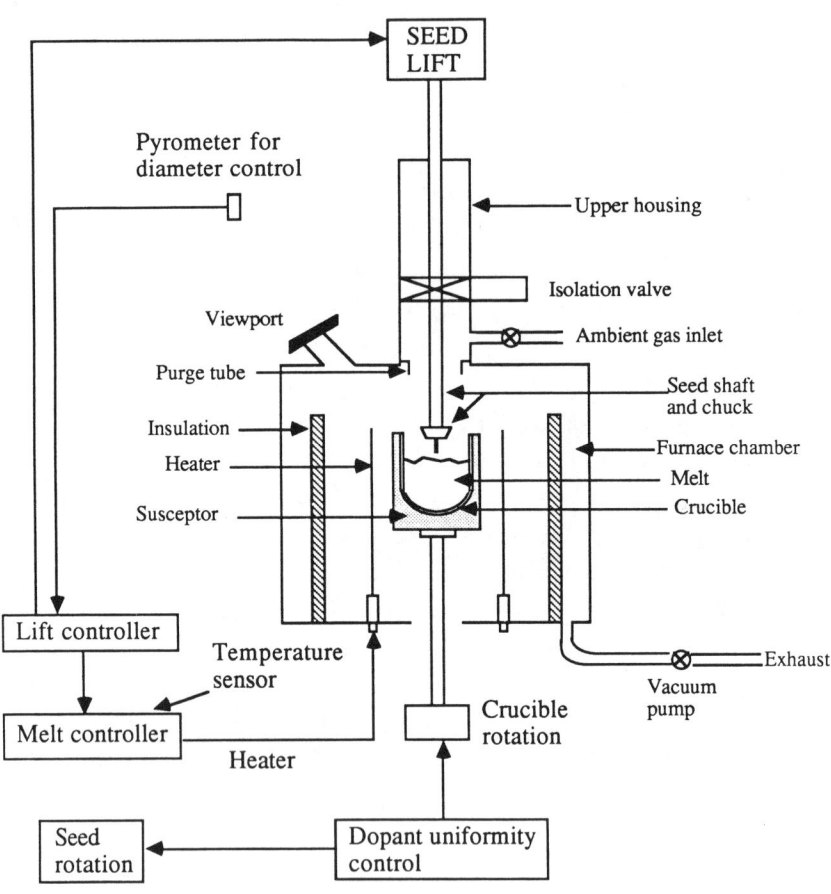

Fig. 3. Czochralski crystal growth (after Pearse, 1983).

which is used for gallium arsenide growth. Control of dopant incorporation will require a similar detailed model of the dopant distribution in the melt including effects of thermal convection and crucible rotation. This control approach is currently being explored (Brown, 1985).

Diffusion

Diffusion furnaces are used to heat wafers to high temperatures where dopant either external to the wafer or ion implanted into the substrate can diffuse into device structures and form electronically active regions. The furnace is tubular and wafers are supported concentrically and vertically in the tube with a spacing of approximately 5 mm. The tube is heated by a three or five zone resistance heating arrangement. The five zone configuration has been proposed to achieve a longer flat temperature zone and reduce end-losses in processing of 150 mm wafers (Maliakal et al., 1984). To minimize particulate contamination, the loading of wafers is automated on modern systems. The temperature in each zone is measured by one or two fixed thermocouples. It is controlled independently of the other zones by a PID controller whose setpoint can be ramped according to recipes in a central microprocessor.

Solid state diffusion is an activated process with activation energies in the order of 350 kJ/mole (Tsai, 1983). Therefore, to achieve reproducible, submicron diffusion depths, i.e. junction depths, it is crucial to control the furnace temperature within + 0.5 C at high temperatures, typically 1000°C. Moreover, as the minimum feature size continues to decrease the temperature limits will become accordingly tighter. However, the current control strategy already has problems and is not likely to be adequate for future diffusion processes. For example, temperature profiling along the length of the furnace tube has revealed temperature excursions significantly outside process limits (Maliakal et al., 1984). There are major problems in temperature monitoring. The thermocouples are usually located outside the quartz tube and the wafers may be cooled by axial radiation losses which cannot be detected by the thermocouples. Another key control problem is the disturbance caused by the loading of a cold set of wafers by gradually pushing them into the furnace. This involves the thermal capacitance of the wafers as well as natural convection driven flows. Thus, multivariable schemes for zone control including feed forward strategies for loading effects are issues that need to be addressed along with temperature sensor location decisions. Similar problems, e.g. tubular reactor control, have been considered extensively in conventional chemical process control.

The control issues associated with diffusion extend to other important IC manufacturing processes, specifically oxidation and low pressure chemical vapor deposition which are carried out in equipment similar to diffusion furnaces. Oxidation of silicon by oxygen is influenced by reaction and solid state diffusion (Deal, 1982). Therefore, temperature control is again critical. Low temperature chemical vapor deposition (LPCVD) is the dominant chemical deposition technique in IC fabrication where it is used to deposit thin films of polycrystalline silicon, silicon oxide and silicon nitride from gaseous reactant by chemical reactions (Jensen and Graves, 1983). The process is typically run under reaction controlled conditions. The chemical reactions have activation energies of 150 kJ/mole which implies that good temperature control is necessary to achieve uniform film thickness.

Plasma Etching

Pattern transfer from a lithographically defined mask of photoresist to an underlying thin film by etching is a key step in the fabrication of integrated microelectronic circuits. This etching process determines the size and shape of the features that eventually form the individual electronic components. Plasma etching allows control of the etch profile to produce vertical and sloped patterns depending on the application. This is not feasible with liquid etching which proceeds isotropically (Mucha and Hess, 1983). Thus, plasma etching is an essential component in any very large scale integration (VLSI) process sequence.

The plasmas that are used for etching are weakly ionized gases, sustained by a radio frequency electric field. The applied field, coupled with the relatively low rate of energy transfer between electrons and the much more massive neutral species, results in highly energetic electrons (10^4-10^5 K) and "cold" (300 K) ions and neutral species. The highly energetic electrons are capable of dissociating neutral species to form radicals which react with the film surface to form volatile etch products. For example, in the etching of silicon by fluorocarbons such as CF_4, highly reactive fluorine atoms are produced in the plasma region. These subsequently react with the exposed silicon surfaces to form volatile SiF_4 which is transported away to the flowing process gas.

The major advantage of plasma etching is the ability to obtain directional etching. This is a result of ion bombardment of the film combined with surface reactions of chemically active neutral species. The ion bombardment occurs because surfaces in contact with the plasma become more negative than the bulk plasma region, thus creating a field accelerating positive ions towards the surface. The ion energy can be manipulated by applying an external bias voltage to the electrode, but it depends also strongly on other plasma parameters, such as pressure, rf-frequency and power.

Control of plasma etching exemplifies process automation issues in microelectronics. The objective is to control submicron features, i.e. etch profile and etch rate, uniformly over a 150 mm wafer while using macroscopic inputs such as flow, pressure, reactant concentration, rf frequency and power. Figure 4 illustrates a typical plasma reactor, the desired microscopic features, and the various process inputs. The process is difficult to operate, is sensitive to small variations in process parameters, and only performs satisfactorily in a small region of a high dimension parameter space.

The objective, the etch profile, cannot be measured on-line but must be checked off-line by electron microscopy. In the case of trans-

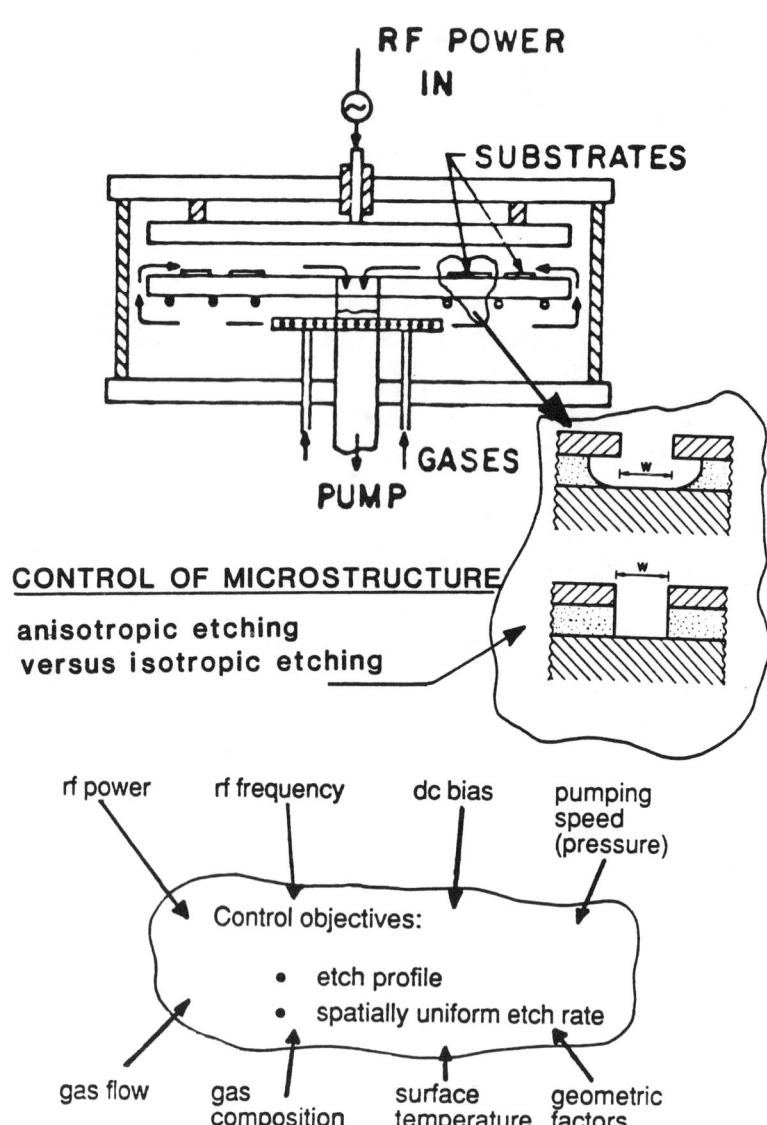

Fig. 4. Example of plasma etching system and major process variables.

parent films (in the visible or infrared), it is feasible to monitor local etch rates in situ by laser interferometry. To observe the etch uniformity, the laser must be scanned across the surface with associated signal interpretation problems. Alternatively, the optical emission from the discharge can be used to detect the etch end-point by emergence of emission from substrate etch products (Mucha and Hess, 1983, and references within this). These macroscopic observations clearly provide no information on the microstructure.

The control problem is compounded by the lack of a fundamental model which accurately describes the influence of process parameters on microstructure and etch rates. Such a model would not only have to include the standard transport and reaction equations for neutral species but also a description of the generation, transport, and reactions of charged species. The latter is crucial in predicting etch profiles and rates but the underlying physicochemical processes are not understood in sufficient detail (Dalvie et al., 1986; Graves and Jensen, 1986; Mucha and Hess, 1983).

The current approach to plasma reactor control appears to be independent single loops with standard PID regulators for reagent flows, pressure, substrate temperature and rf-power. The setpoints are scheduled by a microprocessor according to recipes that have been formulated by trial and error experimentation (Kaempf, 1984). When a recipe fails to produce the desired etch profile or uniformity as detected by offline microscopy, the process is retuned by an engineer. Thus, it would be advantageous to design model based schemes for closed loop control of etch performance. Although detailed process understanding is not yet available, it should be possible to improve plasma reactor performance by using multivariable control concepts in connection with simple models.

CONCLUSIONS

Fabrication of microelectronic devices involves long sequences of complex batch chemical processes. Raw material and fabrication line labor costs are low compared to capitalization, operation and engineering expenditures. The desire to maintain high yields, low cost and safety along with a simultaneous decrease in minimum feature size, growth in wafer size, and increase in circuit integration is driving the IC industry towards automation. The main areas of automation are plant management, materials handling, and unit operation control. Progress has primarily been made in plant management, in particular work in progress monitoring, and in mechanical aspects of materials handling.

Control of the individual chemical processes, such as crystal growth, diffusion and plasma etching, is mainly accomplished by single loops with PID regulators. Microprocessors are used extensively to schedule setpoints according to recipes formulated on the basis of offline trial and error experimentation. The chemical processes share the common difficulties of hard to measure process variables (e.g. microstructure), long feed back loops, nonlinear behavior, and complex interactions. The specific problems illustrated by examples from crystal growth, diffusion, and plasma etching occur in much the same manner in other IC unit operations, e.g. wafer pretreatment, chemical vapor deposition, resist coating and baking, and plasma deposition.

The needed advanced control approach is very similar to that used in other complex chemical applications such polymer processing. Considerable process improvements should be possible with existing control theory. Applications of estimation, identification, multivariable model based control are apparent from the above process examples. Advances in control will necessarily have to include fundamental process experiments and development of dynamic models since many of the unit operations are not yet completely understood.

ACKNOWLEDGEMENTS

Research leading up to this review was supported by the National Science Foundation (CBT-8551249).

REFERENCES

Atherton, R. W. (1984). The application of control theory to the automation of IC manufacturing - progress and problems. Proc. Am. Control Conf., San Diego, California, pp. 753-755.

Boehlert, B. G. and W. J. Trybula (1984). Successful factory automation. IEEE Trans. Components Hybrids Manuf. Technol., 7, 218-224.

Brown, R. A. (1985). Massachusetts Institute of Technol., personal communication.

Dalvie, M., K. F. Jensen and D. B. Graves (1986). Modelling of reactors for plasma processing. I. Silicon etching by CF_4 in a radial flow reactor. Proceedings Ninth Int. Symp. Chemical Reaction Engineering. Pergamon Press.

Deal, B. E. (1982). The thermal oxidation of silicon. In D. A. Doane, D. Fraser and D. Hess (Eds.), Semiconductor Technology, Electrochemical Society, Princeton, New Jersey, pp. 15-41.

Derby, J. J., R. A. Brown, F. T. Geyling, A. S. Jordan, G. A. Nikolapoulou (1985). Finite element analysis of a thermal-capillary model for liquid encapsulated Czochralski growth. J. Electrochem. Soc., 132, 470-482.

Doane, D. A., D. Fraser and D. Hess (1982). Semiconductor Technology. Electrochemical Society, Princeton, New Jersey.

Doyal, L. A., D. L. Weaver and C. W. Gwyn (1980). Computer control and data management in an LSI fabrication facility. IEEE Trans. Components Hybrids Manuf. Technol., 3, 339-344.

Factor, H. and K. Kaufman (1985). Wafer dicing - on the threshold of automation. Solid State Technol., 28, 7, 81-87.

Fiegl, G. (1983). Recent advances and future directions in CZ crystal growth technology. Solid State Technol., 26, 8, 121-131.

Graves, D. B., and K. F. Jensen (1986). A continuum model of dc and rf discharges. IEEE Trans. Plasma Science (April).

Harper, J. G. and L. G. Bailey (1984). Flexible materials handling - automation in wafer fabrication. Solid State Technol., 27, 7, 89-98.

Harper, J. G., J. B. Burkhardt, R. R. Nelson (1985). Automation, communication, and control in flexible wafer fabrication automation. Solid State Technol., 28, 1, 119-123.

Jensen, K. F., and D. B. Graves (1983). Modelling and analysis of low pressure CVD reactors. J. Electrochem. Soc., 130, 1950-1957.

Kaempf, U. (1984). Computer aided control of the integrated circuit manufacturing process. Proc. Am. Control Conf., San Diego, California, pp. 763-774.

Kim, K. M, A. Kran, P. Smetana and G. H. Schwuttke (1983). Computer simulation and controlled growth of large diameter Czochralski silicon crystals. J. Electrochem. Soc., 130, 1156-1160.

Kim, K. M, A. Kran, K. Riedling and P. Smetana (1985). Digital control of Czochralski silicon crystal growth. Solid State Technol., 28, 1, 165-168.

Larrabee, G. B. (1985). A challenge to chemical engineers - Microelectronics. Chemical Engineering, 92, 12, 51-59.

Levy, K., (1984). Productivity and process feedback. Solid State Technol., 27, 7, 103-106.

Maliakal, J. C., D. J. Fisher, Jr. and A. Waugh (1984). Trends in automated diffusion furnace systems for large wafers. Solid State Technol., 27, 12, 105-109.

Mucha, J. A. and D. W. Hess (1983). Plasma etching. In L. F. Thompson C. G. Wilson and M. J. Bowden (Eds.), Introduction to Microlithography, Am. Chem. Soc. Symp. Ser. 219, 215-286.

Parikh, M. and U. Kaempf (1984). SMIF - A technology for wafer cassette transfer in VLSI manufacturing. Solid State Technol., 27, 7, 111-115.

Parrillo, L. C. (1983). VLSI process integration. In S. M. Sze (Ed.), VLSI Technology. McGraw-Hill, New York, pp. 507-550.

Pearse C. W. (1983). Crystal growth and wafer preparation. In S. M. Sze (Ed.), VLSI Technology, McGraw-Hill, New-York, pp. 9-49.

Reid, B. K., J. D. Shott and J. D. Meindl (1984). Wafer fabrication and process automation research at Stanford University. Solid State Technol., 27, 7, 127-133.

Suzuki T., N. Isawa, Y, Okubo., K. Hoshi (1981). CZ silicon crystals grown in a transverse magnetic field. In H. R. Huff, R. J. Kriegler and Y. Takeishi (Eds.), Semiconductor Silicon/1981, Electrochemical Society, pp. 90-100.

Sze, S. M. (1983). VLSI Technology. McGraw-Hill, New-York.

Tsai, J. C. C. (1983). Diffusion. In S. M. Sze (Ed.), VLSI Technology, McGraw-Hill, New-York, pp. 169-217.

Weiss, M. (1984). Automated wafer processing using robots. Solid State Technol., 27, 7, 163-167.

COMPUTER CONTROL OF BIOREACTORS
PRESENT LIMITS AND CHALLENGES FOR THE FUTURE

Matthias Reuss

Technical University Berlin, D-1000 Berlin 65, West Germany

Abstract. This paper presents a discussion on the present state of bioreactor control. From the state of art in the off-line optimization of batch and fed-batch operations it may be concluded that a feedforward control of a biotechnological process is rather risky, not to say hazardous. It is therefore that additional feedback control is an essential requirement. The implementation of such feedback control is very often hampered by the absence of on-line sensors for the most important key variables. This is one of the reasons that development of various estimation schemes is of crucial importance for an effective control. Although progress in this area has led to improved on-line estimations of the bioreactor state, it is not likely that reliable identification in industrial processes with multiple substrates and products will be achieved by these methods. Special emphasis is therefore given to the description of a new automatic sampling device, which may help to move the analysers from the laboratory to the process. As in many chemical processes these process analysers may then become key inputs to more reliable computerized process control systems.

Keywords. bioreactor control; trajectory optimization; biotechnological processes; fed-batch fermentations; feedforward control; feedback control; state estimation; biofiltrator.

INTRODUCTION

In many respects the engineer concerned with the optimal control of a bioreactor faces far more difficult tasks than those involved in chemical or purely physical processing units. A bioreactor in its broadest sense is a complex biological system along with its physical environment. Of particular importance is the fact that the actual process we are interested in takes place inside individual living cells, which can be visualized as segregated biochemical micro - reactors interacting with their environment. What makes such a system extremely difficult to describe and hence hard to control is the expanding and self reproducing nature of these individual micro-reactors as well as their highly sophisticated intracellular control. In order to understand the overall dynamic behaviour of a bioreactor, obviously an essential prerequisite for its control, it is neccessary to closely examine the various interactions between the cells and their environment. And this, in fact, poses some of the most exciting and challenging problems for biochemical and control engineers alike.

The difficulties arise from the following peculiarities which are characteristic to the system:

a) The environment is a composition of all contributions to it and the degree of freedom in controlling this composition is very limited. The interplay of the various

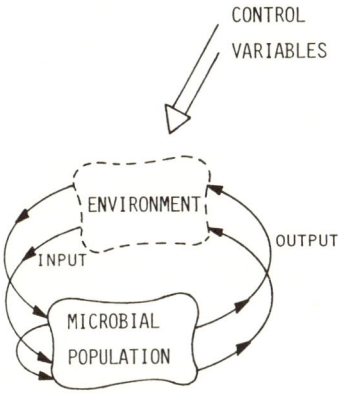

Fig. 1 Interconnections between the microbial population and environment

factors determining the composition of this environment (Fig. 1) is ordinarily quite complex because the system of microorganisms can exhibit different constitutive and inducible adaptabilities at the physiological level. The different regulatory mechanisms are characterized by relaxation phenomena which cover a broad range of time constants. Analysis of the dynamics of the various regulations may be furtherly complicated in the environment of an industrial fermentation process involving a multitude of substrates and products. Very often the "bottle-neck" control mechanism then changes during the process due to variations in the substrate composition and also due to the various products produced by the organisms. Furthermore, complex memory effects may lead to a dependency upon the history of the process.

b) The sensors which can be employed for the direct online measurements of the system state are very limited. As a matter of fact, most of the important bioreactor key parameters are impossible to measure on-line (in particular biomass concentration, growth rates, chemical composition of the environment, concentrations of intracellular key compounds). In the absence of sensors estimation schemes are needed, the reliability of which again depends on the complex behaviour of the microorganisms and may change during the entire process.

The limited knowledge about the overall behaviour of this system with its incredible levels of sophistication as well as the lack of reliable on-line sensors have not hampered, however, the attempts to tackle the problems of controlling the course of various processes by manipulating different environmental factors to force the population to function to our advantage.

This paper is not aimed at discussing in detail the enormous number of control strategies which has been suggested in the past. Recent reviews on various aspects of controlling bioreactors have provided a comprehensive treatment of the whole topic (Yamane and Shimizu, 1984; Parulekar and Lim, 1985).

It is therefore that the discussion will be focussed on identifying open problems and examining present limitations. Obviously, breakthroughs in tackling most of the present problems in the control of bioreactors can only be expected from further developments of reliable on-line analysis. Particular emphasis is therefore placed on the description of a new sampling device, recently developed in our laboratory, which presents a flexible avenue to on-line estimation of various broth constituents and thus also for improved process control.

OPEN LOOP FEEDFORWARD CONTROL – DETERMINATION OF OPTIMAL CONDITIONS

In spite of the large body of fundamental research devoted to the continuous operation of bioreactors the majority of fermentations is at present carried out under batch or fed-batch conditions. In this kind of operation the state of system changes dynamically and it is anticipated that the environmental factors need to be varied optimally. Several studies in recent years have examined the problem of such optimal operations with respect to various objective functions and control variables. The excellent reviews of Yamane and Shimizu (1984) and Parulekar and Lim (1985) illustrate that application of trajectory optimization has been rapidly developed, has made sufficient progress and seems to be an established technique in biochemical engineering these days. The most frequently used optimization strategies are based on the maximum minimum principle of Pontryagin. There are, however, also examples of using modifications of Green's theorem, dynamic programming techniques and simplex optimization strategies. The most common control variable is substrate feeding in order to optimize the productivity of the bioreactor. This paper is not aimed to repeat the state of this art, though a few examples have been summarized in Table 1 and Table 2.

Some important points, however, cannot be entirely ignored in order to draw some general conclusions and to find an orientation for further work. The first important

PROCESS	CONTROL VARIABLE	REFERENCE	
PENICILLIN	TEMPERATURE	CONSTANTINIDES ET AL.	(1970)
GLUCONIC ACID	PH, TEMPERATURE	RAI & CONSTANTINIDES	(1973)
YEAST	PH, TEMPERATURE	BOURDAND ET AL.	(1973)
PENICILLIN	SUBSTRATE FEEDING	FISHMAN & BIRYOKOV	(1973)
LYSIN	SUBSTRATE FEEDING	OHNO ET AL.	(1976)
AMINO ACIDS/ENZYMES	SUBSTRATE FEEDING	YAMANE ET AL.	(1977)
SCP/METHANOL	PH, TEMPERATURE	ALVAREZ & RICANO	(1979)
ERYTHROMYCIN	PH, TEMPERATURE	CHERUY & DURAND	(1979)
GLUCONIC ACID	PH, TEMPERATURE	NYESTE ET AL.	(1980)
ANTIBIOTICS	SUBSTRATE FEEDING	GUTHKE & KNORRE	(1981)
MIXED CULTURE / STEROID TRANSFORMATION	PH	YOSHIDA ET AL.	(1981)
ANTIBIOTICS	TEMPERATURE	BIRYOKOV	(1982)
GLUTAMIC ACID	SUBSTRATE FEEDING	KISHIMOTO	(1982)
GLUCONIC ACID	PH	REUSS ET AL.	(1984)
α - AMYLASE	SUBSTRATE FEEDING	STANISKIS ET AL.	(1984)
PENICILLIN	SUBSTRATE FEEDING	PARULEKAR & LIM	(1985)
BAKER'S YEAST	SUBSTRATE FEEDING	TAKAMATSU ET AL.	(1985)

TABLE 1   Application of trajectory optimization to biotechnological processes

OPTIMIZATION OF OPERATING MODE		
BATCH, SEMIBATCH, CONTINUOUS	OHNO ET AL.	(1978)
REPEATED FED-BATCH	WEIGAND ET AL.	(1979,1981)
CYCLIC BATCH	GUTHKE & KNORRE	(1982)
FED BATCH	SAN & STEPHANOPOULOS	(1984)

TABLE 2   Application of off-line optimization for different operating modes

point for discussion has been perfectly outlined by Aiba (1979) in a very compact statement, remembering that this kind of optimization will not satisfy the neccessary and sufficient conditions required for optimizing a problem in question, the solution pertains only the neccessary condition. The lack of sufficiency is caused by well known changes in microbial behaviour from run to run even if operating conditions are exactly the same. These uncertainties are likely to cause the actual transients to drift away from those predicted by the optimal policy. Thus, frequent recalculations of the optimal conditions are required. This in turn depends on the possibility to monitor the state of the bioreactor, and thus on the measurements of the key variables used as feedback information.

The second problem is related to the reliability of the mathematical models of the processes, which is obviously a fundamental prerequisite for any theoretical optimization study. The most rigorous simplification of reality in the kinetic description of the dynamical behaviour of batch and fed batch processes is the use of so called unstructured models. In these models the properties of the biomass are assumed sufficiently specified by the total amount of biomass present in the system. It is well known that the essential prerequisite for the validity of simple growth rate expressions, like that of Monod for example, according to which

$$dX/dt = \mu_{max} \, X \, S/(K_s + S) \quad (1)$$

is a quasi steady state condition between the cells and their environment. Any rapid change in the environment, for example in the substrate concentration, will cause various intracellular regulations. The dynamics of these regulations are usually quite complex and most often the Monod equation is not applicable to describe the dynamic behaviour of the microbes.

This kind of problems is particularly important for the trajectory optimization of fed batch processes because the control of substrate appears linearly in the Hamiltonian resulting in bang-bang control policies. Any abrupt changes in the feeding of substrate will probably not only drift the culture from the predicted optimal course it may also result in production of undesirable byproducts and other unforeseen changes in the intracellular metabolism. The situation is further complicated in those situations in which a feedback control based on an on-line estimation of the system's state is to be superimposed on the feedforward control. Due to the fact that the system's parameters required for the indirect estimation can drastically change in case of the mentioned variations in the cellular metabolism the system becomes hard to control and it may thus happen that the optimal feeding strategy totally fails and the fermentation altogether falls to the floor.

Further research in this area should therefore try to make more use of those structured models, in which a limited number of compositional aspects of the biomass are considered. Quite a number of promising examples concerning this subject have appeared in the literature in the last decades (Roels and Kossen, 1978; Roels, 1983).

A further important point for future work in this field of trajectory optimization is related to the incorporation of key transport phenomena into the mathematical modelling of the processes. It is indeed questionable if the results of an optimization of the feeding strategy for an aerobic process will be accepted in industry without taking into account the limitation of oxygen transfer capabilities in the production plants. With aid of a comprehensive mathematical model for the penicillin process (Bajpai and Reuss, 1980) it has been shown (Bajpai and Reuss, 1981, Reuss et al., 1980) that a seemingly optimal feeding strategy for substrate glucose may result in drastical losses in productivity during scale up if oxygen transfer capabilities of the plants are not properly taken into account.

ON-LINE MEASUREMENTS AND INDIRECT STATE ESTIMATIONS

From the foregoing discussion it becomes clear that realization of optimal feeding strategies without additio-

nal feedback control in case of biotechnological processes is rather questionable. It goes without saying that these feedback control requires on-line measurements or at least reliable estimates of the state variables of the process at any point in time. The problem with the measurement sensors is that the most important factors affecting the phenomena are usually out of read of conventional environment oriented instrumentation. The values to be measured have been subdivided in different ways:
(1) Effect: Physical - chemical - biological - molecular-biological parameters
(2) Localisation: extracellular - intracellular
(3) Kind of measurement: continuous - discontinuous.

Looking to the list of important parameters (Table 3) the remarkable gap between dreams and reality becomes obvious. Indeed, the applicability of the various

TABLE 3   Important measurements in biochemical reaction processes

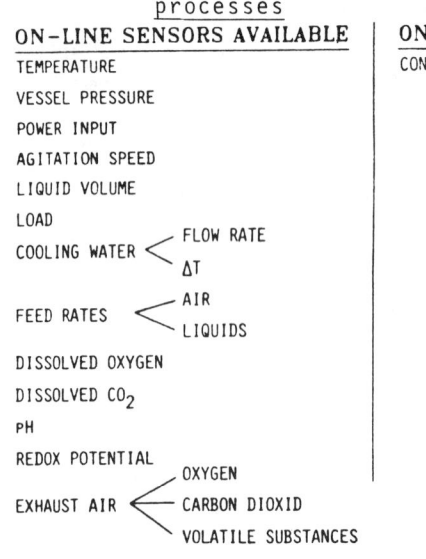
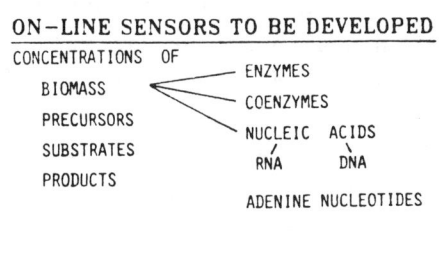

measurement systems suggested for key variables of the biological system (substrates, precursors, products, total biomass concentration and intracellular composition) is still restricted to very special applications in a laboratory environment. Recent reviews of instrumentation by Swartz and Cooney (1978), Tannen and Nyirit (1979), Mor (1981) and Flynn (1982) have provided a comprehensive treatment of the various sensors available and clearly demonstrate that there is an urgent need to develop new sensors for on-line measurements which can be applied in an industrial environment.

In the absence of sensors various estimation schemes are needed which allow estimation of variables which cannot be monitored on-line. It is interesting to realize that the estimation of the state of a bioreactor is an application of the biosensor principle at a large scale. The response of the biological system, which is the entire bioreactor, is monitored by on-line measurements of oxygen and carbon dioxide in the exhaust air (Fig. 2). Calculation of the oxygen consumption and CO_2 evolution rate as well as the respiration coefficient by an on-line computer aided balance around the reactor (Fig. 3) can be used to make a continuous estimation of actual biomass and substrate concentrations. In few examples the estimation of the product concentration is also possible. These techniques depend on the knowledge of the corresponding material balances and also on the stoichiometric coefficients therein.

The most simple approach is based on the following model for the component i (i = CO_2, O_2):

$$Q_i = (1/Y_{X/i}) \, dX/dt + m_i X \qquad (2)$$

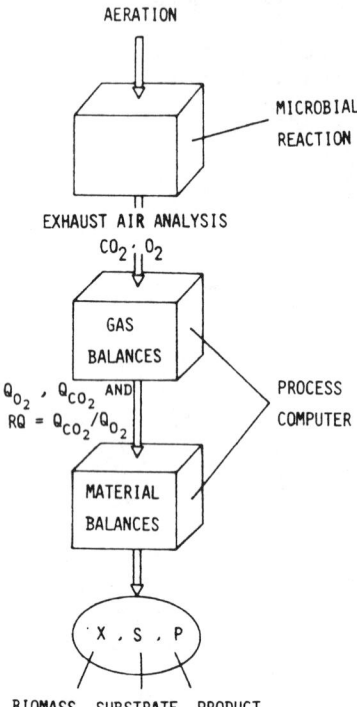

Fig. 2 State estimation from exhaust-air analysis

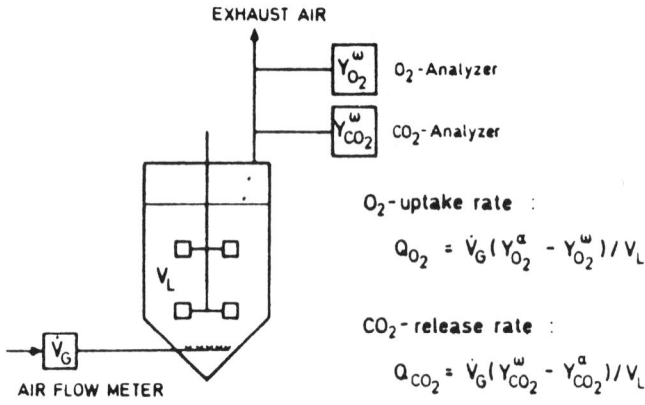

O_2-uptake rate:

$$Q_{O_2} = \dot{V}_G (Y_{O_2}^{\alpha} - Y_{O_2}^{\omega})/V_L$$

CO_2-release rate:

$$Q_{CO_2} = \dot{V}_G (Y_{CO_2}^{\omega} - Y_{CO_2}^{\alpha})/V_L$$

Fig. 3 Gas phase balance

Thus the uptake rate of O_2 or evolution rate for CO_2 is related to the sum of the individual terms for growth (dX/dt) and maintenance, which is proportional to the actual biomass concentration X. Fig. 4 presents an example of an

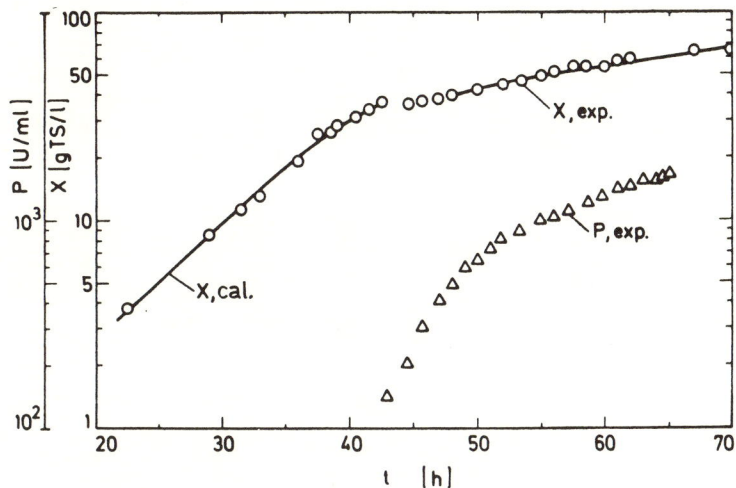

Fig. 4 Comparison between computed and measured biomass concentration during penicillin fermentation.

estimation of the biomass concentrations during cultivation of <u>Penicillium</u> <u>chrysogenum</u> for the production of penicillin. The results shown in this figure have been predicted from the CO_2 balance given by

$$Q_{CO_2} = (1/Y_{X/CO_2})\, dX/dt + m_{CO_2}\, X \tag{3}$$

The computer can solve the integrated equation:

$$X(t) = X(t=0) + Y_{X/CO_2} \int_0^t Q_{CO_2}\, dt - m_{CO_2} \int_0^t X\, dt \tag{4}$$

continuously by simple summation given the values of the

yield coefficient Y_{X/CO_2}, the maintenance constant m_{CO_2} and an initial value of the biomass concentration $X(t=0)$. Application of this technique has been described several times in the literature (Humphrey (1973), Reuss et al. (1976), Zabriskie et al. (1976)).

Before discussing the inherent problems and limits of this method a second example will be presented in order to illustrate the applicability for the additional estimation of substrate and product concentrations. This example is the production of gluconic acid with <u>Aspergillus niger</u> (Reuss et al. (1984)). In addition to the estimation of the biomass concentration based on the CO_2 balance, as shown before, product concentration can be estimated during this process because of the well known stoichiometry of the key reaction leading to the product, which is given by

$$C_6H_{12}O_6 + 1/2\ O_2 \longrightarrow C_6H_{12}O_7 \qquad (5)$$
Glucose   oxygen   gluconic acid

This reaction is catalyzed by the two enzymes glucose oxidase and catalase. Thus the balance equation for oxygen consumption takes the form

$$Q_{O_2} = (1/Y_{X/O})\ dX/dt + (1/Y_{P/O})\ dP/dt + m_{O_2}\ X \qquad (6)$$

It is further assumed that the CO_2-evolution in (3) requires an equimolar amount of oxygen. This assumption results in:

$$Y_{X/O} = Y_{X/CO_2} \quad \text{and} \quad m_{O_2} = m_{CO_2} \qquad (7)$$

Product concentration can be predicted by introducing (3) in (6) and integration:

$$P(t) = P(0) + Y_{P/O} \int_0^t (Q_{O_2} - Q_{CO_2}) \, dt \qquad (8)$$

with $Y_{P/O}$ = 2 mol gluconic acid/mol oxygen.
Substrate consumption can be related to growth, product formation and maintenance in the form:

$$dS/dt = -1/Y_{X/S} \, dX/dt - 1/Y_{P/S} \, dP/dt - m_S X \qquad (9)$$

and therefore actual substrate concentration can be computed from:

$$S(t) = S(t=0) - (X(t) - X(0))/Y_{X/S} - (P(t) - P(0))/Y_{P/S} - m_S \int_0^t X \, dt \qquad (10)$$

Figure 5 shows a comparison between on-line computed bio-

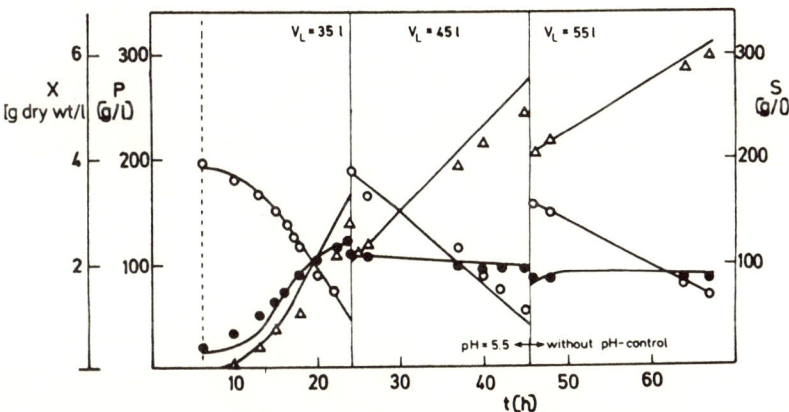

Fig. 5 Comparison between computed and measured biomass, glucose and gluconic acid concentration during gluconic acid production with Aspergillus niger.

mass, substrate and product concentrations and laboratory data. During this experiment fresh glucose was added twice and pH control stopped after the second addition. This procedure, as described in an industrial patent (Deutsches Patent) results in abrupt changes in the volume of the broth with corresponding dilutions of biomass and product concentrations.

Another important product which has been successfully predicted in a similar manner is citric acid (Kubicek, Zehentgruber and Röhr, 1979). According to the paper of Siebert and Hustede (1982) it seems that this technique is successfully applied during industrial large scale production.

The methods described so far are based on constant yield coefficients as well as constant maintenance parameters, an assumption which is not valid for a large number of fermentation processes. Also it is possible that the values of the coefficients change from run to run as illustrated by Nelligan and Calam (1983) in their experiments with Penicillium chrysogenum during penicillin fermentation. An example for this kind of problem, which has been predicted in our laboratory, is presented in Fig. 6. Shown in this figure are results from biomass estimations during penicillin production with a strain, which differed from that used in the experiment shown in Fig. 4. The estimated cell biomass systematically deviates from the laboratory dry weight measurements resulting in problems in the control of substrate feeding in this fed batch culture. The observation that no improvement is possible by readjustment of the parameters is in agreement with the results of Mou and Cooney (1983). These authors were the first repor-

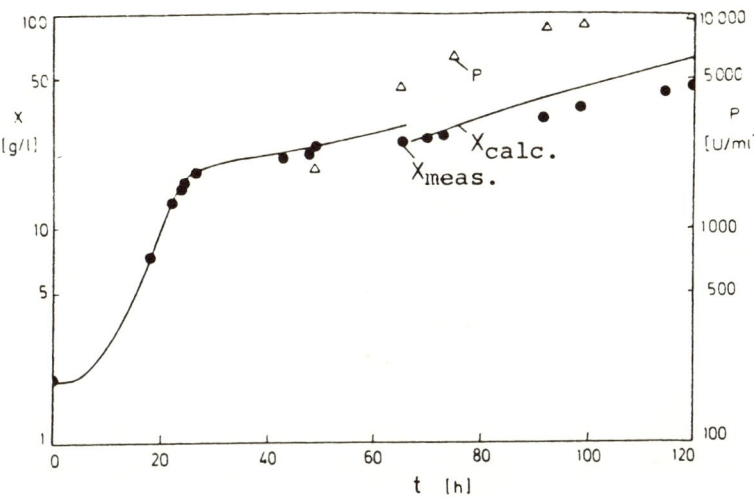

Fig. 6 Deviations between biomass estimations and measurements during penicillin fermentation.

ting about a pronounced time dependency of the maintenance requirement of Penicillium chrysogenum during fed-batch cultivation.

The assumption of constant yield coefficients has been bypassed by coupling the exhaust air analysis with simple elemental balances for carbon, hydrogen, oxygen and nitrogen according to the following reaction for the formation of biomass:

$$a\ C_xH_yO_z + b\ O_2 + c\ NH_3 \rightarrow C_\alpha H_\beta O_\gamma N_\delta + d\ H_2O + e\ CO_2 \qquad (11)$$

carbon energy   oxygen   ammonia   cell biomass   water   carbon
source   dioxide

If the compositions of the carbon and energy sources as well as the components of the cell biomass are known and assumed to be constant during the process there are five stoichiometric coefficients to be determined. Four equations are given from the elemental balances for C, H, O, and N. In addition an equation can be formulated from the direct measurements of oxygen and carbon dioxide in the exhaust air, which together with equation (11) delivers the fifth unknown in this estimation.

This method was proposed by Cooney, Wang and Wang (1977) and was successfully demonstrated for the biomass estimation in the production of baker's yeast (Wang, Cooney and Wang; 1977a, 1977b). In case of product formation other on-line measurements are needed for the estimation of the additional unknown stoichiometric coefficient. In certain cases the measurement of alkali/acid added to control the pH could serve for this purpose (Stephanopoulos and San, 1984).

Again this method only yields an estiamtion of the growth rate r_X, which is per definition:

$$r_X = dX/dt = \mu X \qquad (12)$$

For the purpose of control most often we need to have a reliable estimate of both, the specific growth rate μ and the biomass concentration X. The disadavantage of integration of equation (12), as described above, is the exact knowledge of the cell biomass $X(t=0)$. Stephanopoulos and San (1980) have discussed the problem involved in this integration in more detail and have shown that the variance of the estimated biomass concentration increases continuously with time. Thus:

$$Var(X) = Var(X(t=0)) + \sigma^2 t \qquad (13)$$

where σ^2 is the intensity of a Gaussian white noise with zero mean, $\xi(t)$, in the equation of growth rate:

$$dX/dt = (r_X)_{measured} + \xi(t)$$

in which $(r_X)_{measured}$ is the growth rate calculated from the measured exhaust air analysis data. This growth rate is related to the two state variables X and µ through

$$(r_X)_{measured} = \mu X + \eta(t) \tag{14}$$

where $\eta(t)$ again is a white noise. The estimates are then found by applying the extended Kalman filter technique. In order to allow also the estimates of time varying specific growth rates µ(t) the following equations have been added to the estimation scheme:

$$d\mu/dt = C + \eta_1(t) \tag{15}$$

and

$$dC/dt = -(F/V) C + \eta_1'(t) \tag{16}$$

where $\eta_1(t)$ and $\eta_1'(t)$ are white noises. For more details the reader is referred to the original papers (Stephanopoulos and San, 1982a, 1982b, 1984).
This approach was tested in a variety of cases. Simulations and experimental results indicate that the obtained on-line estimates are in very good agreement with the true values and the off-line measurements, respectively (San and Stephanopoules, 1984b).

Though this method represents a considerable and important contribution in the field of on-line estimations, a number of open questions is still remaining. These problems are related to those processes in which very complex media are used and products with unknown stoichiometry are produced. Also it has been clearly demonstrated (Mou, 1979) that during transient periods between fast and slow growth the stoichiometric relationships can change drastically. These effects are probably caused by unknown intracellular reactions related to the turnover of protein, nucleic acids and various storage materials. Also this approach will fail in processes where exhaust air analysis is neither correlated to the biomass nor to the product (Camus, Cheruy and Durand, 1982).

EXAMPLES FOR OPEN PROBLEMS IN THE COMPUTER CONTROL OF SUBSTRATE FEEDING

It is evident that a detailed discussion on all the different control strategies suggested in the past, including their very specific problems cannot be covered in this short review. Again the interested reader is referred to recent review papers (Yamane and Shimizu, 1984; Parulekar and Lim, 1985). Important contributions and many interesting examples, which have been worked out in recent years, can also be found in the Proceedings of the First IFAC Workshop on Modelling and Control of Biotechnical Processes (Ed.: Halme, 1982). Instead of repeating all the various aspects I would like to focus the attention to two illustrative examples on application of feedforward-feedback control in order to discuss some of the open problems and challenges for future work.

A massive amount of work has been devoted to the application of computer process control to the fed-batch baker's yeast production. (Aiba, Nagai and Nishisawa, 1976; Dairaku et al., 1981; Dairaku et al., 1983; Cooney and

Swartz, 1982; Pons et al., 1982; Wang et al., 1977 and 1979). The rational behind the different strategies is the maximization of biomass yield and productivity. This goal requires a control of the growth rate close to a critical value in order to prevent ethanol formation, which occurs in response to the so-called Crabtree effect. The fed-batch operation is then achieved by controlling the supply of the carbon and energy source molasses to the bioreactor. The open loop feedforward control, still applied in most of the industrial plants, is based upon a precalculated time depending feeding rate. For a variable reactor volume V the material balance equation for biomass X and substrate S can be written in the form:

$$d(XV)/dt = \mu X V \quad (17)$$

and

$$d(SV)/dt = -(1/Y_{X/S}) \, d(XV)/dt + F S_0 \quad (18)$$

Imposing the two conditions $dS/dt=0$ and μ = const. gives:

$$F = \mu (XV)/(Y_{X/S} S_0) \quad (19)$$

and for an open loop control:

$$F = \mu^{SP} X_0 V_0 \exp(\mu^{SP} t)/(Y_{X/S} S_0) \quad (20)$$

For realization of a closed loop strategy the required feed back information from the process can be either ethanol concentration in the fluid bulk or head space of the fermentor, measured with aid of an ethanol sensor or, alternatively, the respiration coefficient of the culture obtained from exhaust air analysis. Wang et al. (1979) suggested a strategy in which the feeding rate is calculated from a proportional controller according to

$$F = F_o (1 - K_p(Q_{CO_2} - RQ^{SP} Q_{O_2})) \quad (21)$$

where F_o is the feedforward part, calculated with equation (19) taking into account indirect estimations of actual biomass concentration (X) and growth rate μ (elemental balances together with exhaust air analysis). RQ^{SP} is the respiration coefficient in absence of ethanol formation used as the set point for the controller. Due to the fact that ethanol formation rate

$$Q_{ETOH} = Q_{CO_2} - RQ^{SP} Q_{O_2} \quad (22)$$

has been incorporated into the control strategy, it was possible to simultaneously exercise feedback control to avoid glucose overfeeding.

We applied a slightly different strategy for RQ-control, simular to the one suggested by Dairaku et al. (1983), who studied substrate feeding based upon constant ethanol concentration. In this concept control action is performed with aid of a PID controller

$$F = F_o + K(e(t) + \frac{1}{T_1} \int_0^t e(\tau) \, d\tau + T_D \frac{de(t)}{dt}) \quad (23)$$

with e(t) on-line calculated with (22). F_o is computed from (17), (18) and (19):

$$F_o = \mu^{SP} X V/(Y_{X/S} S_o) \quad (24)$$

with

$$XV = X_o V_o + Y_{X/S} \int_0^t F S_o \, dt \text{ and } \mu^{SP} = \text{fixed}$$
(25)

The integral in (25) is calculated from on-line weight measurements of the sugar feeding tank.

The point to be discussed here can be derived from the results presented in Fig. 7. Shown in this figure are

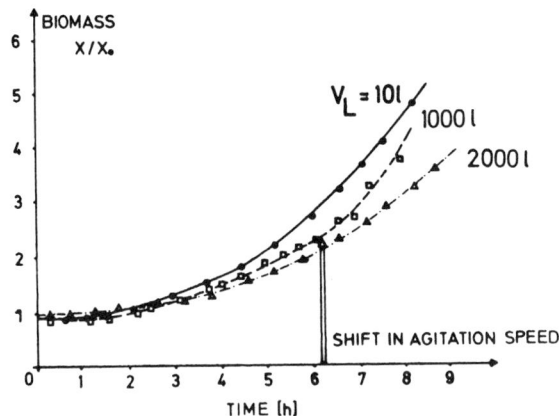

Fig. 7 Comparison of the growth of baker's yeast in computer controlled fermentations at different scales

growth curves obtained by applying the same control strategy in stirred vessels with different scales of operation. Using exactly the same setpoint ($RQ^{SP}=1.05$), productivity is obviously lowered with increasing scale. An explanation for this scale effect can be found from the differences in the intensity of mixing of the substrate fed into the reactor. The increase of growth rate after a shift in agitation speed in the 1000 l scale is an experimental proof for this hypothesis. With aid of a comprehensive mathematical model, taking into account the recirculation time distribution of the fluid as well as the kinetics of growth, substrate consumption and ethanol production it was possible to quantify these effects (Bajpai and Reuss, 1982, Reuss and Brammer, 1985). This example may serve as an example for the general problem of scale-up of control strategies for bioreactors, a field, which has not found sufficient attention in the past. From preliminary experiments and application of the mathematical model mentioned above for simulations of responses to step changes in the feeding rate we found that the process dynamics may totally change in different scale of operations. It is worthwhile to remember that empirical strategies for adjusting (or on-line tuning) the control parameters are not always applicable in biological processes, because the metabolism of the microorganisms may totally change in response to varying environmental conditions, even if the disturbances are small.

The reason that so much attention has been paid to the computer control of baker's yeast production is the very special metabolic regulation (Crabtree-effect), which results in a measurable by-product formation (ethanol). Thus, either ethanol concentration or respiration coefficient can be utilized directly as a feedback parameter. In the absence of such directly observable parameters any control of feeding depends on the reliability of the on-line estimation schemes, which has been discussed in the preceding paragraph. An important contribution in this direction has been made by Mou and Cooney (1983) for the control of different growth rates in the penicillin production. The feeding strategies suggested for this process are interesting for quite a lot of other industrial important products, for example antibiotics and enzymes. Many of these processes are divided into two phases: a growth phase and a production phase. Usually during the active growth period cell growth is to be controlled at a maximum, whereas during the production phase nutrient feeding must be reduced in order to avoid catabolite repression. Very often a precursor for product formation has also to be fed throughout this period. Overfeeding as well as underfeeding should be avoided because of the undesirable by-products, repression of product formation, oxygen limitation or irreversible damage due to degeneration of the biomass, respectively. Mou and Conney (1983) suggested a strategy in which biomass and growth rate during the active growth period are estimated from the carbon dioxide balance and throughout the carbon-limited production phase from carbon balancing. Sugar feeding in the two phases was controlled by feedforward-feedback strategies. For the transient period between the two phases an open loop control was suggested.

The problem with this kind of control strategies is that the underlying estimation scheme in the production phase (carbon balance) depends on the a priori knowledge of the specific production rates and the cell composition. In all those cases in which this information is not available or the productivities change with increasing process time and/or larger amounts of by-products are produced, the application of this technique may be questionable. Also it is useful to remember that very many feeding problems like for example precursor addition in penicillin fermentation can not be tackled by any indirect estimation from exhaust air measurements.

In trying to identify the limits and bottle necks in the present control of bioreactors, it is easy to see and there is also a general agreement that further progress in this field is not likely to be achieved without new concepts and developments for on-line measurements in order to increase the amount of information about the actual state of the system.

AN AUTOMATIC SAMPLING AND SEPARATION DEVICE FOR IMPROVED COUPLING OF ON-LINE ANALYSIS

As several times mentioned in this paper, the major bottleneck for application of computer control in fermentation is the lack of reliable, sterilizable and robust sensors for process key variables like biomass, substrate, precursor and product concentrations. Principally there are two ways to overcome these difficulties: a) design of new sensors like product or substrate specific electrodes for in situ application and b) automatic coupling of powerful process analyser systems like HPLC, process gas-chromatographs or auto-analysers. Undoubtedly, important progresses can be observed in the development of biosensors (Gronow, 1984). However, most of these sensors suggested in the past can not be sterilized. Also in many cases the problems of long time stability, reliability, accuracy and specifity have not been satisfactorily solved. As far as the biomass concentration is concerned development of a sensor will be a hopeless task if processes with mycelial organisms and complex media are envisaged.

The most important problem to be solved for using an automatic process analyser as an on-line measurement device is the design of the sampling system including a reliable on-line separation of biomass and solids from the sample. Most often this separation is performed with aid of membran systems, such as dialysis, ultrafiltration, cross-flow filtration installed in a bypass or directly introduced into the bioreactor (Fig. 8). Due to the well known problems of membran fouling and pore blocking in biological systems these separation techniques using non-renewable filter media may create process specific problems.

The approach developed in our laboratory (Lenz et al., 1985) is schematically illustrated in Fig. 9. The automatic sampling and separation device is a special designed filtration probe which operates in a cyclic batch mode. Liquid-solid separation is then performed with aid of conventional cake filtration. The advantages of this concept are the following: The media chosen for filtration can be optimized depending on the type of fermentation and the purpose of filtration. Choosing adequate filter media results in optimal filtrate yields for the filtration of bacteria, yeasts and mycelial organisms. Reducing the pore size of the filter band prolonges the life of HPLC-columns and precolumns. This optimization of the filter medium is one of the main differences between the here presented filtration probe and the one designed by Nestaas and Wang (1981) in which a multi - layered 40 mesh copper screen was used as filter medium. A further advantage of the filter is the concept of renewable filter medium, so that plugging and membran fouling are impossible. Due to the complete separation of the process and the filtration which are only coupled with the steam sterilized sample valve, a very high degree of security against contamination of the fermentation is guaranteed. Opposite to membrane separation techniques, where damage of the filter medium normally will cause the risk of process contamination such malfunctions

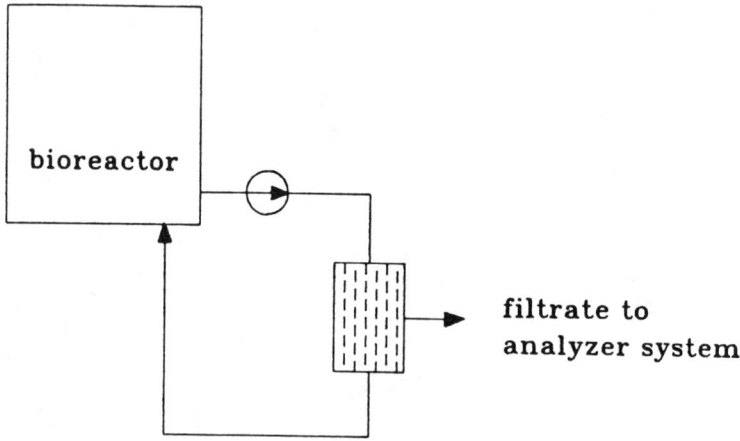

Fig. 8 Sampling via filtration in a by-pass

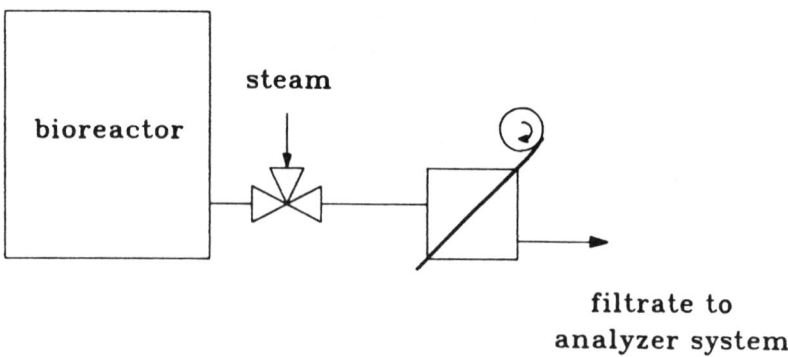

Fig. 9 Sampling via steam sterilizable valve and (batch-) cake filtration

in this apparatus will leave the fermentation unharmed.
The filter unit is aimed to fulfill two tasks:
a) Estimation of biomass concentrations from the analysis of cake filtration similar to the approach suggested by Neestas and Wang (1981) and Thomas et al. (1985) for penicillin fermentation and by Silvennoinen and Koivo (1982) for the Pekilo-Process.
b) Providing clarified liquid samples for the automatic transfer to various analyser-systems. As an example on-line HPLC-measurements will be presented.

Automatic Filtration

The equipment used for the automatic filtration is described in Fig. 10. The apparatus is computer controlled and works fully automatic. The entire cycle consists of the following actions and is shown in Fig. 11:
1) Sterilization of sampling valve 1 through three-way valve 3.
2) Transport of the filter band. Taring of the balance and closure of the filter unit.
3) Steam valve 7 closed, sampling valve 1 open for preliminary sample rejected through three-way valve 3.
4) Three-way valve switched for connecting sample valve with filter unit; sampling valve remains open (2.5 to 10 seconds depending on the viscosity of the broth).
5) Valve 3 in position 1; pressure line (valve 4) opened; ventilation valve closed.
6) Filtration with data acquisition and reduction with the aid of the computer.
7) Transfer of filtrate to HPLC and start of analysis.
8) Ventilation valve 6 opened; pressure valve 4 closed, filter body pneumatically lifted, cleaning with water and drying with air from the pressure line.

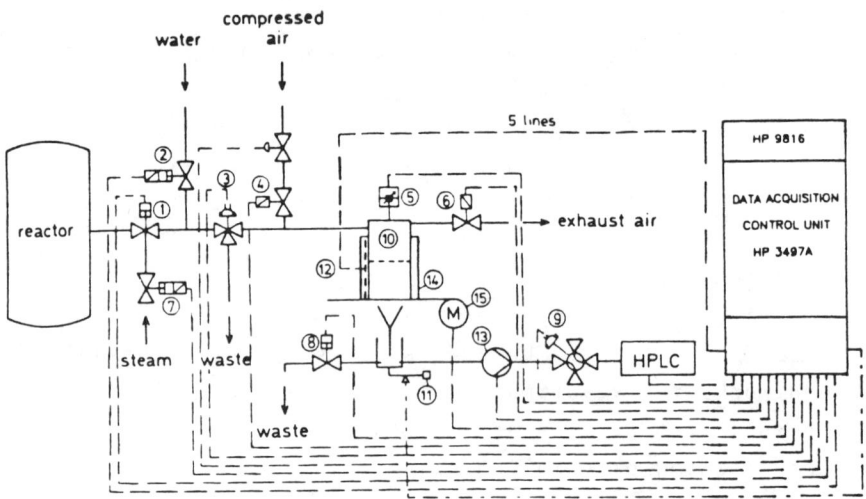

Fig. 10 Process computer coupled filtration unit (Bio-filtrator)

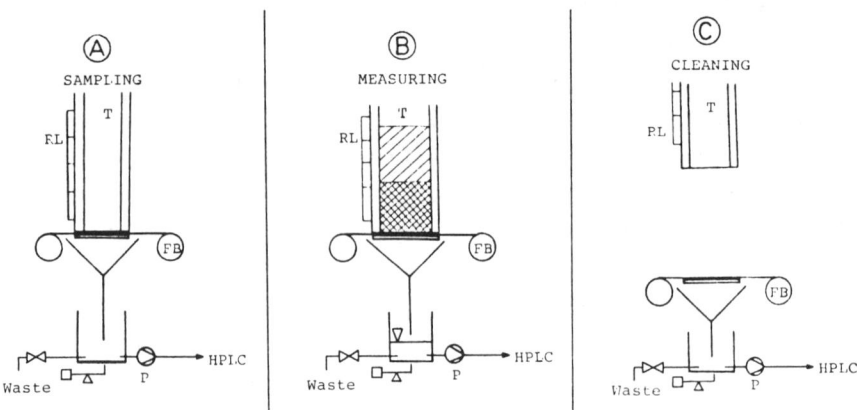

Fig. 11 Operation modes of the filtration unit

9) Sterilization of sampling line to valve 3.
10) Standby mode.

During the measurement phase B in Fig. 11 the computer reads the data from the reflection light barriers and the filtrate balance. All the data are stored and on the display a plot of filtration time/volume versus volume is presented (Fig. 12). Instead of using the signal of one single light barrier like Nestaas and Wang (1981) the filter tube is equipped with 5 reflection lights in different heights. This concept has some advantages compared to the measurement of the transmitted light at one point of the filter tube. The measurement of the reflected light is independent of the absorption of the fermentation fluids, which is a requirement for many industrial fermentation broths. The use of multiple signals for the detection of the height of the growing filter cake allows the use of regression methods for the elimination of individual measurement errors.

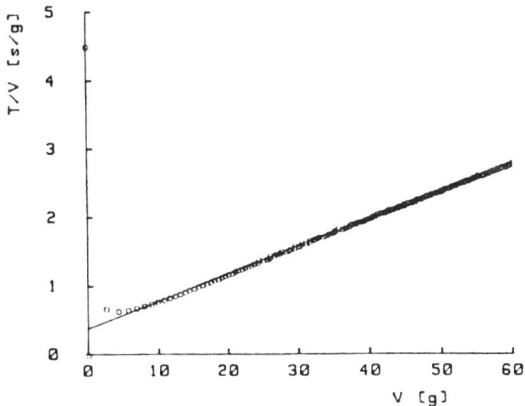

Fig. 12 Filtration characteristic for Penicillium chrysogenum (biomass concentration X = 22 kg dry weight/m^3)

After sampling of 150 to 500 data sets at intervalls between 1 to 2 seconds (depending on the filtration velocity) the computer starts the analysis of the filtration data.

Analysis of Filtration data

The analysis of filtration data is based on the Kozeny-Carman equation as presented by Nestaas and Wang (1981):

$$\tau = \frac{t_f}{V_c * V_f} = \frac{8K''\mu_F}{A^2 \Delta P} * \frac{g_h \bar{V}}{(g_h \bar{V} - 1)^3 d_h^2} \qquad (26)$$

with t_f=filtration time, V_c, V_f cake and filtrate volume, K''=Kozeny factor = 5.5 (cylinders), μ_F=dynamic viscosity, A=filter area, ΔP=filtration pressure, g_h=density of biomass (kgDW/m^3 hyphae), d_h=diameter of hyphae (4.5µm) and the specific cake volume

$$\bar{V} = \frac{1}{X} * \frac{V_c}{(V_c + V_f)} \qquad (27)$$

The only unknown in Eq. (26) is ($g_h \bar{V}$). Thus, using the measured data for t_f, V_c and V_f Eq. (26) can be iteratively solved for this unknown. Instead of using the raw data, cake and filtrate volumes are filtered with the aid of a linear regression. A typical example for this regression is shown in Fig. 13. The data for the cake volumes are taken from the reflection light signals as shown in Fig. 14. The vertical dashed lines are placed by the computer at the point of steepest ascent of the individual signals. The slope of the line in Fig. 13 m_1 divided by the slope of the filtration characteristics (m_2) in Fig.12 results in the reciprocal value of the specific filtration time.

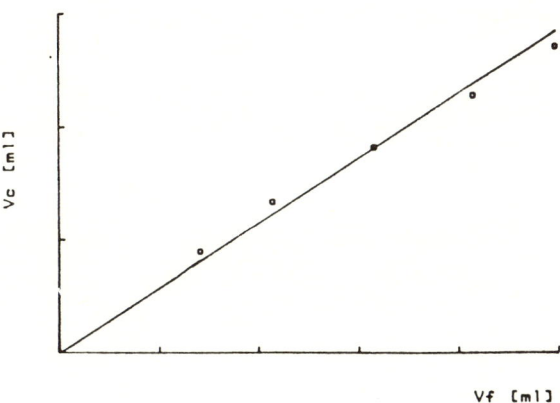

Fig. 13 Regression of filter cake vs. filtrate volume

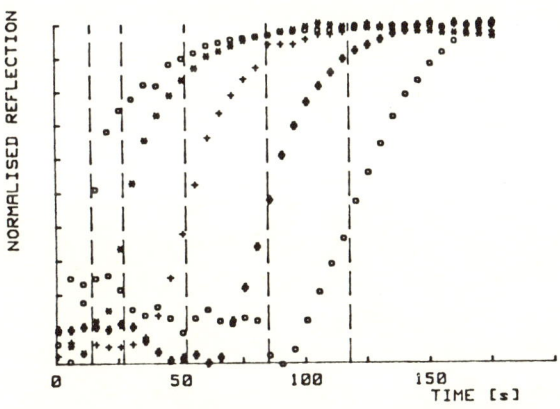

Fig. 14 Reflection light barriers data during penicillin fermentation

Thus

$$\tau = \frac{m_2}{m_1} \qquad (28)$$

Provided with the value of $(g_h \bar{V})$ either
a) hyphae density can be calculated with aid of external dry weight measurements, or
b) biomass concentration X can be estimated, if g_h is known.

From extensive experiences with computer coupled penicillin fermentations (using substrate feeding strategy based on carbon dioxide production) we found the hyphae density being constant for growth rates during production phase $\mu_{prod} \geqslant 0.1\ \mu_{max}$. This observation is in agreement with what has been reported by Nestaas (1980). This region of growth rate is reasonable for the production of Penicillin and we therefore decided to use filtration data for estimation of biomass concentrations.

Results. Fig. 15 and 16 present two examples in which the filtrate was automatically analyzed with aid of the on-line coupled HPLC. Fig. 15 shows the product concentrations during a fed-batch penicillin production. It must be emphasized that simultaneously with this penicillin concentration the HPLC is able to detect the concentrations of the precursor (phenylaceticacid) which is also fed to the reactor and a by-product of the process (o-hydroxy-phenylaceticacid). Highly reliable on-line measurements can thus be obtained which can subsequently be utilized for new control strategies. An interesting application in the foregoing example is the control of the feeding of the precursor. Fig. 16 shows an example for on-line glucose measurements with HPLC during a batch fermentation of Saccharomyces cerevisiae. The agreement between the on-line HPLC-data and the results from enzymatic analysis in the laboratory is satisfactory.

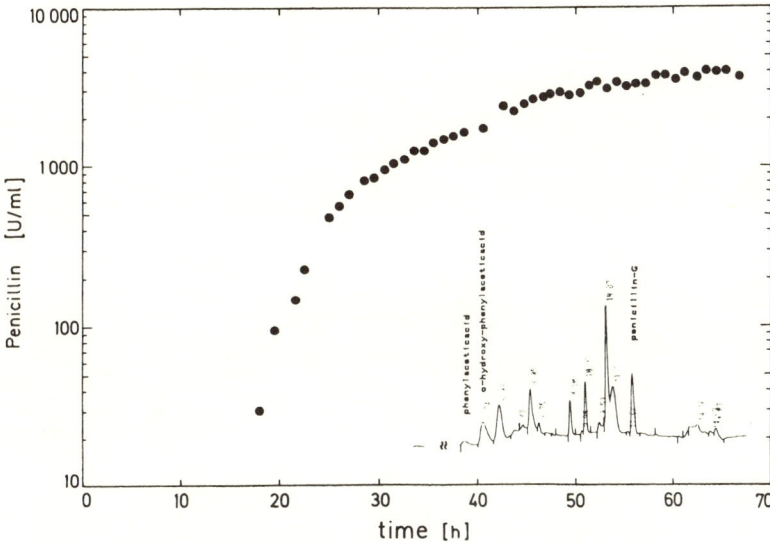

Fig. 15   On-line HPLC - measurements of penicillin

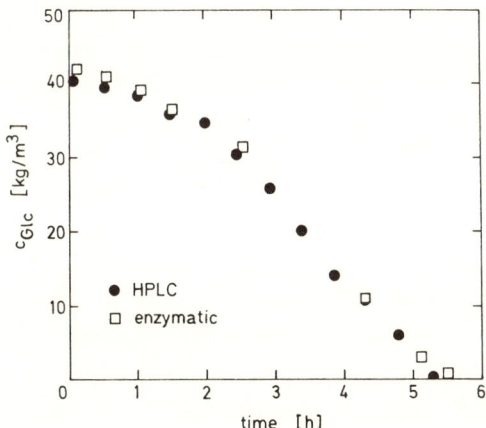

Fig. 16   On-line HPLC - measurements of glucose in a yeast fermentation

A typical result for calculated biomass concentrations from the quantitative analysis of the cake filtration is shown in Fig. 17. The open circles present on-line computed biomass concentrations, the closed symbols are dry weight measurements from the laboratory. For the purpose of control the estimations can be improved by applying various filtering methods.

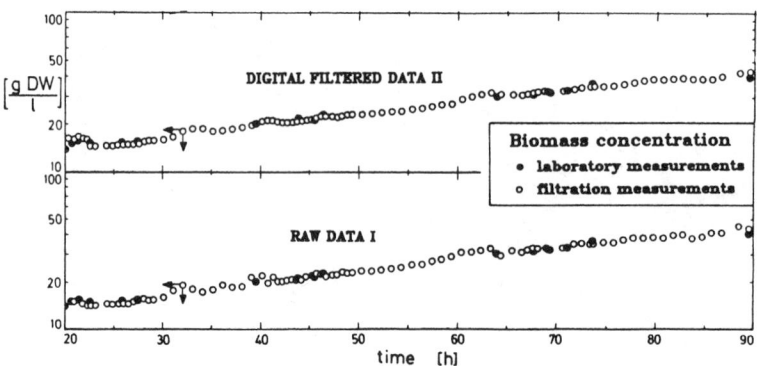

Fig. 17 Comparison between on-line estimation of biomass concentration (filtrator) and laboratory measurements.

In a first attempt to improve the feedforward-feedback control of glucose feeding in the penicillin process we utilized these cyclic informations about the actual biomass concentration to readapt the yield and maintenance coefficients in the simple Q_{CO_2}-estimator (equation (4)). Thus the new measurements are incorporated in the previous structure of fast state estimation using exhaust air-analysis but now increasing the accuracy and reliability of the estimation by taking into account the time variations of model parameter. Fig. 18 compares computed and measured biomass concentrations during a penicillin fermentation in which this new state estimation was successfully incorporated into the feedforward-feedback strategy for glucose feeding suggested by Mou and Cooney(1983).

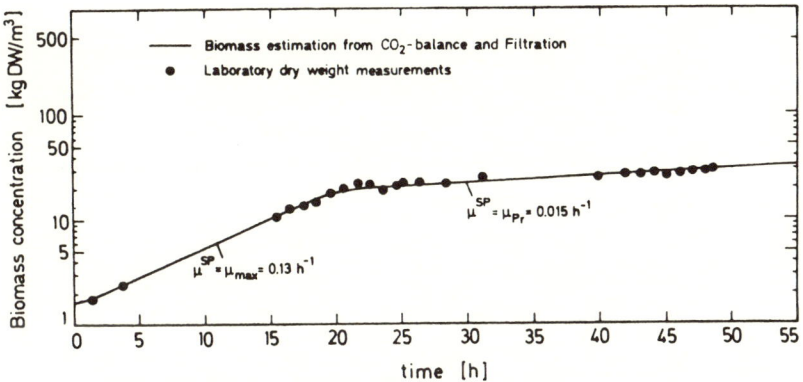

Fig. 18 Results of a computer controlled penicillin fermentation using on-line estimation for biomass from exhaust-air analysis and filtration.

It goes without saying that the impact of all these new measurements can be further improved by their incorporation into more sophisticated estimation algorithms, as for example in a way suggested by Stephanopoulos and San (1984). Work in this direction is presently underway.

CONCLUSIONS

In order to illustrate the present limits and challenges for future work this short survey has been primarily concerned with the state of the art of computer-aided bioreactor control from the point of view of applicability. The author is very much aware of the inadequacies of this brief treatment, particularly, the lack of detailed discussion of the many interesting contributions to the control of continuous cultures, the suggestions for application of adaptive control principles and on-line optimization. However, in attempting to apply these more sophisticated control ideas to the complex requirements of biotechnological processes one should always keep in mind that present on-line informations about the processes are still insufficient to allow a reliable on-line state estimation. Thus, it becomes clear that breakthroughs in this area very much depend on the improvements of on-line measurements. Also at present reliable models to describe transient states of biological systems do not exist, due to insufficient knowledge of the dynamics of the response of microorganisms. Work in this direction is expected to make important contributions to more sound application of sophisticated control strategies.

REFERENCES

Aiba, S., S. Nagai, and Y. Nishizawa (1976). Fed-batch culture of Saccharomyces cerevisiae: A perspective of computer control to enhance the productivity in baker's yeast cultivation. Biotechnol. Bioeng., 18, 1001-1016.

Aiba, S. (1979). Review of process control and optimization. Biotechnol. Bioeng. Symp. No. 9, 269-281.

Alvarez, J., and J. Ricano (1979). Modelling and optimal control of SCP-fermentation process. Biotechnol. Bioeng. Symp. No. 9, 149-154.

Bajpai, R.K., and M. Reuss (1980). A mechanistic model for penicillin production. J. Chem. Tech. Biotechnol., 30, 332-344.

Bajpai, R.K., and M. Reuss (1981). Evaluation of feeding strategies in carbon - regulated secondary metabolite production through mathematical modelling. Biotechnol. Bioeng., 23, 717-738.

Bajpai, R.K., and M. Reuss (1982). Coupling of mixing and microbial kinetics for evaluating the performance of bioreactors. Can. J. Chem. Eng., 60, 384-392.

Biryukov, V.V. (1982). Computer control and optimization of microbial metabolite production. In A. Halme (Ed.), Modelling and control of biotechnical processes, IFAC-Publication, Pergamon Press, pp. 135-144.

Bourdand, D., and C. Foulard (1973). Identification and optimization of batch culture fermentation processes. I. Europ. Conf. Process Control in Fermentation, Dijon (France), Proceedings.

Camus, J.-L., A. Cheruy, and A. Durand (1982). Estimation

and control of the lipids contents for an antibiotic production process. In A. Halme (Ed.), Modelling and control of biotechnical processes, IFAC-Publication, Pergamon Press, pp. 213-218.

Cheruy, A., and A. Durand (1979). Optimization of erythromycin biosynthesis by controlling pH and temperature: theoretical aspects and practical applications. Biotechnol. Bioeng. Symp. No. 9, 303-320.

Constantinides, A., J.L. Spencer, and E.L. Gaden (1970). Optimization of batch fermentation processes. I. Development of mathematical models for batch penicillin fermentation and II. Optimum temperature profiles for batch penicillin fermentation. Biotechnol. Bioeng., 12, 802-830 and 1081-1098.

Cooney, C.L., H. Wang, and D.I.C. Wang (1977). Computer-aided material balancing for predicting biological parameters. Biotechnol. Bioeng., 19, 55-66.

Cooney, C.L., and J.R. Swartz (1982). Application of computer control to yeast fermentation. In A. Halme (Ed.), Modelling and control of biotechnical processes, IFAC-Publications, Pergamon Press, pp. 243-251.

Dairaku, K., Y. Yamasaki, K. Kuki, S. Shioya, and T. Takamatsu (1981). Maximum production in a baker's yeast fed-batch culture by a tubing method. Biotechnol. Bioeng., 23, 2069-2081.

Dairaku, K., E. Izumoto, H. Morikawa, S. Shioya, and T. Takamatsu (1983). An advanced microcomputer coupled control system in a baker's yeast fed-batch culture using a tubing method. J. Ferment. Technol., 61, 189-196.

Deutsches Patent: DE 18 17 907.

Fishman, V.M., and V.V. Biryukow (1974). Kinetic model of secondary metabolite production and its use in computation of optimal conditions. Biotechnol. Bioeng. Symp., No. 4, 647-662.

Flynn, D.S. (1982). Instrumentation for fermentation processes. In A. Halme (Ed.), Modelling and control of biotechnical processes, IFAC-Publication, Pergamon Press, pp. 5-12.

Guthke, R., and W.A. Knorre (1981). Optimal substrate profile for antibiotic fermentations. Biotechnol Bioeng.,, 23, 2771-2777.

Guthke, R., and W.A. Knorre (1982). Efficiency of the cyclic batch antibiotic fermentation. Biotechnol. Bioeng., 24, 2129-2136.

Gronow, M. (1984). Biosensors. Trends in Biotechnol., 9. 336-339.

Halme, A. (Ed.) (1982). Modelling and control of biotechnical processes. IFAC-Publication, Pergamon Press.

Humphrey, A.E. (1977). The use of computers in fermentation systems. Process Biochem., March, 19-26.

Kishimoto, M., T. Sawano, T. Yoshida, and H. Taguchi (1982). Optimization of a fed-batch culture by statistical data analysis. In A. Halme (Ed.), Modelling and control of biotechnical processes. IFAC-Publication, Pergamon Press, pp. 161-168.

Kubicek, C.P., O. Zehentgruber, and M. Röhr (1979). An indirect method for studying the fine control of citric acid formation by Aspergillus niger. Biotechnol. Letters, 1, 47-52.

Lenz, R., C. Boelcke, U. Peckmann, and M. Reuss (1985) A

new automatic sampling device for the determination of filtration characteristics and coupling of HPLC to fermentors. Paper submitted to 1st IFAC Symp. on Modelling and control of Biotechnological Processes, The Netherlands, 11.-13. Dec.

Mor, J.R. (1982). A review of instrumental analysis in fermentation technology. Proc. Third International Conf. on computer application in fermentation technology, England, pp. 109-118.

Mou, D.G. (1979). Toward an optimal penicillin fermentation by monitoring and controlling growth through computer-aided mass balancing. Ph. D. Thesis, MIT, Cambridge, USA.

Mou, D.G., and C.L. Cooney (1983). Growth monitoring and control through computer-aided on-line mass balancing in a fed-batch penicillin fermentation. Biotechnol. Bioeng., 25, 225-255.

Neestas, E. (1980). Quantitative characterization of mycelial biomass and morphology using the filtration probe in the penicillin fermentation. Ph. D. Thesis, MIT, Cambridge, USA.

Neestas, E., and D.I.C. Wang (1981). A new sensor, the "filtration probe" for quantitative characterization of the penicillin fermentation. I. Mycelial morphology and culture activity. Biotechnol. Bioeng., 23, 2803-2813.

Neestas, E., D.I.C. Wang, H. Suzuki, and L.B. Evans (1981). A new sensor, the "filtration probe" for quantitative characterization of the penicillin fermentation. II. The monitoring of mycelial growth. Biotechnol. Bioeng., 23, 2815-2824.

Nelligan, I., and C.T. Calam (1983). Optimal control of penicillin production, using a mini-computer. <u>Biotechnology Letters</u>, <u>5</u>, 561-566.

Nyeste, L., B. Sevella, L. Szigeti, A. Szoke, and J. Hollo (1980). Modelling and off-line optimization of batch gluconic acid fermentation. <u>European J. Appl. Microbiol. Biotechnol.</u>, <u>10</u>, 87-94.

Ohno, H. E. Nakanishi, and T. Takamatsu (1976). Optimal control of semibatch fermentation. <u>Biotechnol. Bioeng.</u>, <u>18</u>, 847-864.

Ohno, H., E. Nakanishi, and T. Takamatsu (1978). Optimum operating mode for a class of fermentation. <u>Biotechnol. Bioeng.</u>, <u>20</u>, 625-636.

Parulekar, S.J., and H.C. Lim (1985). Modelling, optimization and control of semi-batch bioreactors. In A. Fiechter (Ed.), <u>Adv. Biochem. Eng./Biotechnol.</u>, <u>32</u>, 207-258.

Pons, M.N., J. Bordet, and J.M. Engasser (1982). A micro-mini computer hierarchical control system for a laboratory scale fermentor. In A. Halme (Ed.), <u>Modelling and control of biotechnical processes</u>, IFAC-Publication, Pergamon Press, pp. 259-265.

Rai, V.R., and A. Constantinides (1973). <u>AIChE Symp. Ser.</u>, <u>132</u>, 114.

Reuss, M., R.P. Jefferis III, and J. Lehmann (1976). Application of an on-line system of coupled computer to fermentation modelling. In R.P. Jefferis III (Ed.), Workshop Computer Application in Fermentation Technology 1976, Verlag Chemie, Weinheim, New York, pp. 107-124.

Reuss, M., R.K. Bajpai, R. Lenz, H. Niebelschütz, and A. Papalexiou (1980). Scale up strategies based on the interaction of transport and reaction. Keynote, 6. Int. Ferm. Symp., Ontario, Canada, July 20/25.

Reuss, M., S. Fröhlich, B. Kramer, K. Messerschmidt, and H. Niebelschütz (1984). Mathematical modelling of microbial kinetics and oxygen transfer for gluconic acid production with Aspergillus niger. Proc. 3. Eur. Congr. Biotechnol., Verlag Chemie, München, Vol 2, pp. 455/460.

Reuss, M., and U. Brammer (1985). Modelling and control of biotechnological processes. Paper submitted to 1st IFAC Symp. on Modelling and control of biotechnological processes, The Netherlands, 11.-13. Dec.

Roels, J.A., and N.W.F. Kossen (1978). On the modelling of microbial metabolism. In M.J. Bull (Ed.), Progr. Industr. Microbiol., 14, Elsevier, Amsterdam, Oxford, New York, pp. 95-203.

Roels, J.A. (1983). Energetics and kinetics in biotechnology, Elsevier Biomedical Press, Amsterdam, New York, Oxford.

San, K.-Y., and G. Stephanopoulos (1984a). A note on the optimality criteria for maximum biomass production in a fed-batch fermentor. Biotechnol. Bioeng., 26, 1261-1264.

San, K.-Y., and G. Stephanopoulos (1984b). Studies on on-line bioreactor identification. II. Numerical and experimental results. Biotechnol. Bioeng., 26, 1189-1197.

Siebert, D., and H. Hustede (1982). Citronensäure-Fermentation - biotechnologische Probleme und Möglichkeiten der Rechnersteuerung. Chem.-Ing.-Tech., 54, 659-669.

Silvennoinen, E., and H.N. Koivo (1982). Estimation of the biomass concentration in the Pekilo-process from the filtrate flow rate of the concentrator. In A. Halme (Ed.), Modelling and control of biotechnical processes, IFAC-Publication, Pergamon Press, pp. 219-258.

Staniskis, J., and D. Levisanskas (1984). An adaptive control algorithm for fed-batch culture. Biotechnol. Bioeng., 26, 419-425.

Stephanopoulos, G., and K.-Y. San (1980). State estimation for computer control of biochemical reactors. In M. Moo-Young (Ed.), Adv. Biotechnol. Vol I, Scientific and engineering principles, Pergamon Press, pp. 399-406.

Stephanopoulos, G., and K.-Y. San (1982a). On-line estimation of time-varying parameters. Application to biochemical reactors. In A. Halme (Ed.), Modelling and control of biotechnical processes, IFAC-Publications, Pergamon Press, pp. 195-199.

Stephanopoulos, G., and K.-Y. San (1982b). On-line estimation of the state of biochemical reactors. In J. Wei and C. Georgakis (Eds.), Chemical Reaction Engineering - Boston, ACS Symp. Ser. 196, pp. 155-163.

Stephanopoulos, G., and K.-Y. San (1984). Studies on on-line bioreactor identification. I. Theory. Biotechnol. Bioeng., 26, 1176-1188.

Swartz, J.R., and C.L. Cooney (1978). Instrumentation in computer - aided fermentation. Process Biochem., Febr., 331-374.

Takamatsu, T., S. Shioya, and H. Chikatani (1985). Comparison of simple population models in a baker's yeast fed-batch culture. Chem. Eng. Sci., 40, 499-507.

Tanner, L.P., and L.K. Nyiri (1979). Instrumentation of fermentation systems. Microbial Technology, 2, 331-374.

Thomas, D.C., V.K. Chittur, J.W. Cagney, and H.C. Lim (1985). On-line estimation of mycelial cell mass concentrations with a computer-interfaced filtration probe. Biotechnol. Bioeng., 27, 729-742.

Weigand, W.A., H.C. Lim, C.C. Creagan, and R.D. Mohler (1979). Optimization of a repeated fed-batch reactor for maximum cell productivity. Biotechnol. Bioeng. Symp. No. 9, 335-348.

Wang, H.Y., C.L. Cooney, and D.I.C. Wang (1977). Computer-aided baker's yeast fermentation. Biotechnol. Bioeng. 19, 67-86.

Wang, H.Y., C.L. Cooney, and D.I.C. Wang (1979). Computer control of baker's yeast production. Biotechnol. Bioeng., 21, 975-995.

Yamane, T., T. Kume, E. Sada, and T. Takamatsu (1977). A simple optimization technique for fed-batch culture. J. Ferm. Technol., 55, 587-598.

Yamane, T., and S. Shimizu (1984). Fed-batch techniques in microbial processes. In A. Fiechter (Ed.), Adv. Biochem. Eng./Biotechnol., 30, 147-194.

Yoshida, T., M. Sueki, H. Taguchi, S. Kulprecha, and N. Nilobol (1981). Modelling and optimization of steroid transformation in a mixed culture. European J. Appl. Microbiol. Biotechnol., 11, 81-88.

Zabriskie, D.W., W.B. Armiger, and A.E. Humphrey (1976).

Application of computers to the indirect measurement of biomass concentration and growth rate by component balancing. In R.P. Jefferis III (Ed.), Workshop Computer Application in Fermentation Technology 1976, Verlag Chemie, Weinheim, New York, pp. 59-72.

NOTATIONS

A	filter area
d_h	diameter of hyphae
F	sugar feeding rate
K"	Kozeny factor
K_s	Michaelis-Menten constant
m_i	maintenance coefficient of component i (kg i/(kg dry weight h))
m_1, m_2	slopes of regression lines
P	product concentration
ΔP	filtration pressure
r	growth rate (dX/dt)
Q_i	volumetric rate of substance i (mol i/m^3 h)
RQ	respiration coefficient (Q_{CO_2}/Q_{O_2})
S^x	substrate concentration
S_o	substrate concentration in the feed
t	process time
V	reaction volume
V_o	reaction volume at time zero
V_c	filter cake volume
V_f	filtrate volume
$\overline{V}$	specific cake volume
X	biomass concentration
$Y_{i/j}$	yield coefficient (kg i/kg j)

Greek symbols

η	white noise
μ_F	dynamic viscosity
μ	specific growth rate
μ_{max}	max. specific growth rate
μ^{SP}	set point for growth rate
g_h	hyphae density
σ^2	variance
ξ	white noise

SOME NEW APPROACHES FOR CONTROLLING COMPLEX PROCESSES IN CHEMICAL ENGINEERING

E. D. Gilles
Universitat Stuttgart, Stuttgart, West Germany

Abstract. Model based measuring techniques find a wide field of application in chemical engineering. In order to run chemical reactors or separation processes under well defined operation conditions, the knowledge of those variables which cannot be measured directly is often requisite. In these cases estimation techniques can be applied when process models are available. It is the aim to demonstrate, that these methods gain increasing significance in practical application, when complex processes in chemical industries are to be controlled. These methods are necessary to ensure safe operation of reactors, to monitor and control the product quality and to facilitate the starting up of process units with strong interaction. In solving these problems the development of reduced mathematical models for nonlinear systems is often an important task.

Keywords. Kalman filter, alarm system, model reduction, fixed-bed reactor, polymerization reactor, distillation column, distributed parameter systems.

INTRODUCTION

In monitoring and controlling processes of chemical engineering the knowledge of those quantities which are not directly measurable is often required or at least desirable. Such quantities can include the concentrations of the various components of the mixture, the activity of a catalyst or certain properties of the product such as characteristic numbers of the chain length distribution of a polymer. It is well known that the Kalman filter and the Luenberger observer (Kalman, 1960; Luenberger, 1966; Gelb, 1974; Brammer, Siffling, 1975; Zeitz, 1977) are powerful tools to estimate or reconstruct those quantities from process variables which can be easily measured such as temperature and pressure. A characteristic feature of the filter- and observer-techniques is that the estimation is based on an a priori knowledge of the process in form of a mathematical model. The general structure of such a model-based measuring technique, valid for both the Kalman filter and the Luenberger observer is shown in Fig.1.

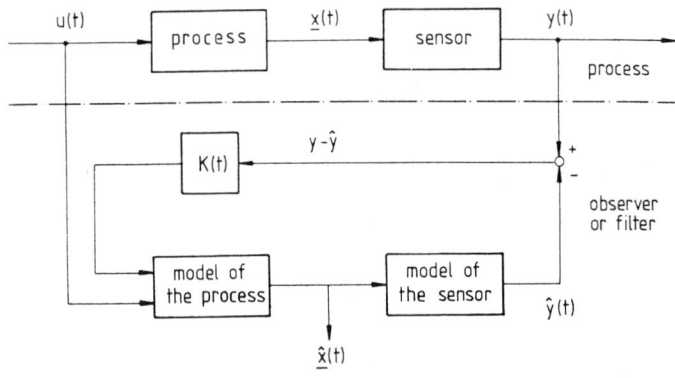

Fig. 1. Model-based measurement system

The application of these methods in chemical engineering is sometimes difficult. This is due to the complex nature of these processes, which complicates the formulation of a sufficiently accurate mathematical model. Considering for example chemical reactors, the behavior is often determined by the complex interaction of the chemical reaction and the physical transport phenomena. Such transport phenomena include convection, diffusion and heat conduction.

Because the success of the filter and observer-techniques primarily depends on the quality of the mathematical model, particular attention must be paid to the special problems of modelling the processes in chemical engineering. Regardless of these difficulties in modelling it can be stated that the application of model-based measuring techniques can decisively contribute to the solution of measuring problems. Furthermore, they give the basis to ensure safe operation of chemical reactors by an early detection of critical states. The information on the process made available by these techniques can also be used to improve monitoring and controlling of complex plants. This is particularly important during start-up and shutdown. In addition, strategies for flexible operation characterized by variable throughput can be facilitated.

It is the intention to illustrate these facts by means of some practical problems in chemical engineering. These examples which can be considered to be of technical interest, have in common that the applied monitoring and control techniques have been practically tested either on laboratory- or large-scale plants.

ESTIMATION OF TEMPERATURE AND CONCENTRATION PROFILES IN A FIXED-BED REACTOR

Chemical reactor operation in the industrial practice is characterized by several problems involving yield optimization, safe reactor operation and control. To tackle these problems the values of the state variables of the system, e.g. temperature and concentrations of the reactants and the reaction products, should be sufficiently known. As the fixed-bed reactor is a system with distributed parameters the state variables are functions of at least one spatial coordinate. Often the concentrations cannot be measured at all. It is also impossible to obtain temperature measurements at each location of the catalyst bed. It should also be stated, that even if product samples are available, they can be usually processed only at relatively long intervals, owing to the retention times of the measuring device. The estimation of the local temperature and concentration profiles knowing only temperature measurements at a few locations in the catalyst bed is therefore an important task. The theory of state estimation for linear systems with distributed parameters is well developed (Ray, Lainiotis, 1978). However, in the case of strongly nonlinear systems such as the fixed-bed reactor there are still many questions to be answered. Therefore it is requisite to test the efficiency of an estimation algorithm experimentally on a pilot-plant.

The laboratory type reactor used for the experiments is shown in Fig. 2. The reactions studied are the partical oxidation of methanol to formaldehyde and the consecutive oxidation of formaldehyde to carbon monoxide.

The air feed is supplied by an air cylinder and is controlled by a mass flowrate controller. Methanol is supplied in liquid phase; its flowrate is controlled by a feed pump. Both feed streams are mixed in a vaporizer. The methanol-air gas mixture is then conducted to the reactor entrance through a heated pipeline. The methanol concentration in the gas feed must be kept under the lower explosion limit for catalyst stability and safety reasons; therefore it is continuously measured by a detecting system. Before entering into the catalyst bed, the feed gas is heated up to the desired inlet temperature in a calming section filled with inert material.

The stainless steel reactor tube is packed with pellets of an iron oxide-molybdenum oxide industrial catalyst. The cooling jacket is divided in three independent sections. In this way different cooling temperature profiles can be achieved by means of three thermo-

State Estimation of a Fixed-Bed Reactor
Mathematical Model

Reaction:
$$CH_3OH + \tfrac{1}{2}O_2 \xrightarrow{r_1} HCHO + H_2O$$

$$HCHO + \tfrac{1}{2}O_2 \xrightarrow{r_2} CO + H_2O$$

Energy Balance:

$$\frac{\partial x_1}{\partial t} + \frac{\partial x_1}{\partial z} - \varepsilon\frac{\partial^2 x_1}{\partial z^2} + B(x_1 - x_{1k}) = Da_1 \cdot r_1 + Da_2 \cdot r_2$$

$$x_1(0,t) = x_{10}(t) \quad ; \quad \left.\frac{\partial x_1}{\partial z}\right|_{z=1} = 0$$

Material Balances:

$$\frac{\partial x_2}{\partial z} = -Da_3 \cdot r_1 \quad ; \quad x_2(0,t) = x_{20}(t) \quad ; \quad (CH_3OH)$$

$$\frac{\partial x_3}{\partial z} = Da_4 \cdot r_1 - Da_5 \cdot r_2 \quad ; \quad x_3(0,t) = x_{30}(t) \quad ; \quad (HCHO)$$

Reaction Rates:

$$r_1(x_1, x_2) = e^{-\frac{A_1}{x_1}} \cdot x_2^{n_2} \quad ; \quad r_2(x_1, x_3) = e^{-\frac{A_2}{x_1}} \cdot x_3^{n_3}$$

Fig. 2. Fixed-bed reactor plant

stats. The length of the catalyst bed is 1 m and its radius 13 mm. 65 thermocouples enable the measuring of temperature over the reactor length at 36 axial cross-sections and 3 radial positions. Axial and radial temperature profiles can be obtained continuously by means of a fast multiplex selecting system. For concentration measurements, capillary tubes are used to sample reaction gas at 27 positions in the packing. By means of a selecting valve, the gas samples are conducted to a gas chromatograph for quantitative composition analysis. The carbon monoxide concentration at the reactor exit is measured continuously by an infrared spectrophotometer.

Process control (feed units, thermostats, analytical instruments) as well as storage and processing of experimental data are conducted by means of a process computer in interactive mode.

The reactor model formulated in terms of normalized, dimensionless variables and parameters is a quasihomogeneous one-dimensional model comprising the following equations:

$$\frac{\partial x_1}{\partial t} + \frac{\partial x_1}{\partial z} - \varepsilon \cdot \frac{\partial^2 x_1}{\partial z^2} + B(x_1 - x_{1K}) = Da_1 r_1 + Da_2 r_2 \quad (1)$$

$$x_1(0,t) = x_{10}(t) \;;\; \left.\frac{\partial x_1}{\partial z}\right|_{z=1} = 0 \;;\; x_1(z,0) = x_1^*(z)$$

$$\frac{\partial x_2}{\partial z} = -Da_3 r_1 \quad x_2(0,t) = x_{20}(t) \quad (2)$$

$$\frac{\partial x_3}{\partial z} = Da_4 r_1 - Da_5 r_2 \quad x_3(0,t) = x_{30}(t) \quad (3)$$

$$r_1(x_1, x_2) = e^{-\frac{A_1}{x_1}} x_2^{n_2}$$

$$r_2(x_1, x_3) = e^{-\frac{A_2}{x_1}} x_3^{n_3} \quad (4)$$

The numerical values of the parameters of this model are given in (Canavas, 1984a). In Fig. 3 temperature profiles at various time instants are given for the starting-up process of the reactor.

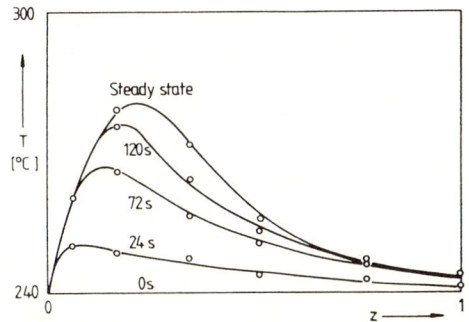

Temperature transients during the starting-up

Fig. 3. Temperature transients during the starting-up

These results and various other experiments have shown that the model is able to describe the reactor behavior with a sufficient accuracy.

Estimation of the steady-state profiles by smoothing

The efficiency of the model-based measuring techniques in reconstructing temperature and concentration profiles along the fixed-bed reactor was tested at first by considering the reactor at the steady state and by neglecting the contribution of the axial heat conduction to the heat balance (1). The latter assumption is justified by our previous experience with simulating the dynamic behavior and the steady state of the reactor. Under high (i.e. industrially significant) flowrates of the reaction gas, axial heat conduction affects the dynamic reactor behavior, but it has no significant influence on the final steady state.

Filter applications to fixed-bed reactors at the steady state reported by other authors (Kuruoglu, Ramirez, Clough, 1981, 1982; Canavas, 1984b, 1984c) consider a large amount of measured data.

This condition may be not satisfied in industrial implementations. The aim of this investigation was to obtain reliable temperature and concentration estimations, by using as little experimental data as possible. Since the data are to be processed in a non-real time scheme, the smoothing algorithm was employed (Gelb, 1974). The smoothed estimate of temperature and concentration at a certain axial position in the reactor is based on all the measurements available for the estimation along the reactor. A smoother can be thought of as a suitable combination of two filters. One of the filters, called a "forward filter", operates on all the data after position z and produces the estimate $\underline{\hat{x}}(z)$; the other filter, called a "backward filter" operates on all the data after position z and produces the estimate $\underline{\hat{x}}_b(z)$. Together these two filters utilize all the available information to yield the improved "smoothed" estimate. Since the system is nonlinear, the extended Kalman filter algorithm was used. The smoother was of the fixed-interval type. The filter parameters used are in (Canavas, 1984b).

The efficiency of the smoother was tested in the case of only one temperature measurement. The thermocouple considered was located in a region of high system sensitivity ($z = 0.12$). In Figs.4 & 5 the estimated profiles are compared with the additional experimental data, that were available but not used in the smoothing algorithm. By using the correct values of inlet concentration without any additional concentration measurement the smoother reconstructs successfully the temperature and concentration profiles over the whole reactor length, even if the inlet temperature value is unknown or incorrectly assumed.

If an error at the inlet concentration values is to be compensated, concentration measurements are indispensable. According to the previous considerations concerning the temperature measurements, it would be desirable to use experimental gas composition informa-

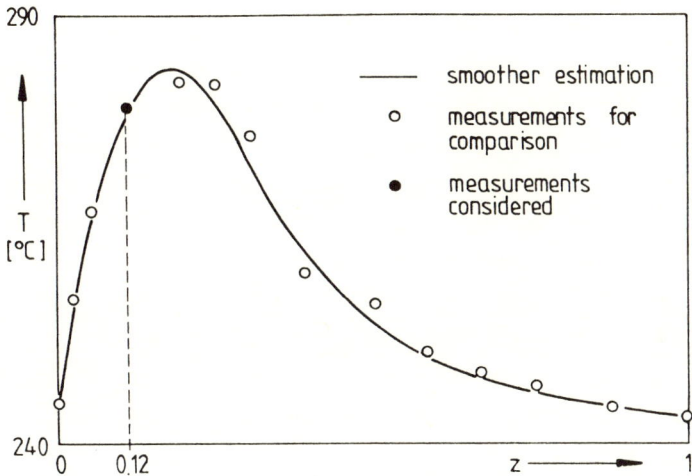

Fig. 4. Smoother estimation of the axial temperature profile

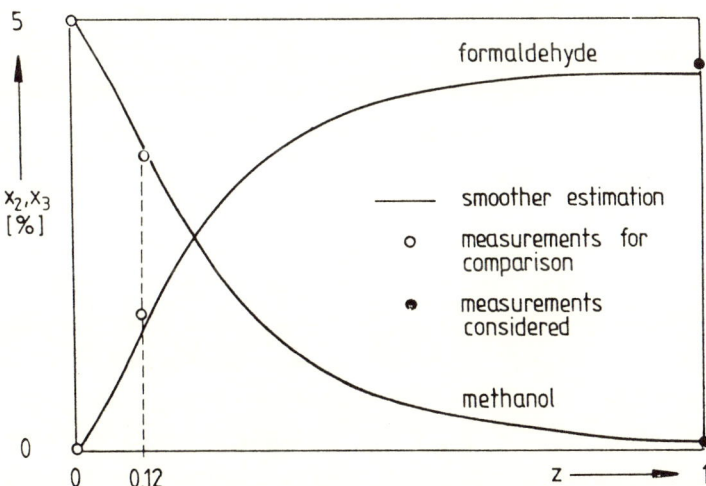

Fig. 5. Smoother estimation of the axial concentration profiles

tion obtained in a sensitive region of the reactor. However, sampling at intermediate points of the catalyst bed is difficult to realize in the industrial practice. Therefore a smoother was implemented by considering only directly available concentration measurements at the reactor outlet, additionally to the temperature measurement at $z = 0.12$. The profiles reconstructed by the smoother are independent on the inlet conditions and coincide with those shown in Figs. 4 & 5. Concentration measurements in the sensitive reactor region conducted for comparison but not considered in the smoothing agreed very well with the smoothed estimates.

Estimation of the dynamic behavior by filtering

The reported investigations concerning real time estimation of the dynamic behavior of fixed-bed reactors are mostly based on linearization assumptions (Alvarez, Stephanopoulos, 1982, Sørensen, Jørgensen, Clement, 1980). These assumptions however, reduce considerably the validity range of the estimators and increase the estimation sensitivity in respect to model and filter parameters. Therefore the efficiency of such estimators is diminished if reactor transients during starting-up or strong disturbances are to be evaluated. By using a heterogeneous reactor model Windes and Ray (1984) showed that a linearized filter with constant gains yields much poorer results than those obtained by means of an extended Kalman filter with time variable gains.

In our current research the nonlinear pseudo-homogeneous reactor model (1) - (4) is used in the implementation of a distributed parameter extended Kalman filter. By using second order spline functions, finite difference approximations for the spatial differential quotients in eqs. (1) - (3) are obtained. In this way one ordinary, nonlinear differential equation for the temperature is generated at each grid point, whereas concentrations, according eqs.(2) - (3), are assumed to be in a pseudo-steady state. The set of nonlinear ODE's obtained above is used as system model in an extended Kalman filter. The measurement vector consists of continuously available temperature measurements over the reactor length.

In Figs. 6 & 7 the performance of the dynamic filter during the starting-up is demonstrated in the case of both incorrect initial guess of the reactor state and incorrect assumption of the inlet temperature value. By considering the continuous measurements of a single thermocouple in the sensitive reactor region the temperature estimates converge rapidly toward the actual profiles (Fig.6). A second temperature measurement downstream of the hot spot enables a faster approximation of the actual temperature values through the filter in that reactor section (Fig.7). For an on-line implementation of the dynamic filter described above, the number of the model equations should be kept as low as possible, in order to minimize the computing effort. Our studies with different discretization densities showed that we could diminish the number of the grid points up to 4 internal and 2 boundary ones without causing a deterioration of the filter performance. More details about the design and the applications of the dynamic Kalman filter described above are given in (Canavas,1986).

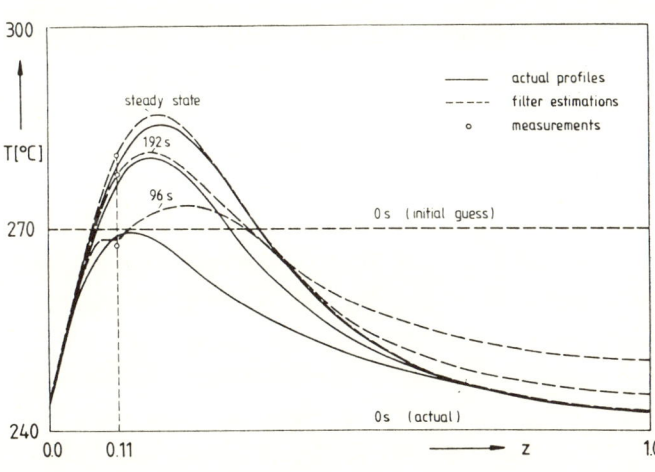

Fig. 6.

Dynamic filter estimation of the axial temperature profiles with a single thermocouple

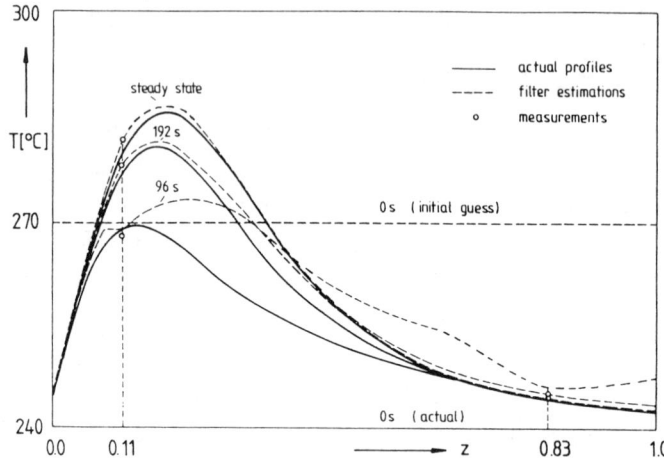

Fig. 7.
Dynamic filter estimation of the axial temperature profiles with two thermocouples

STATE ESTIMATION OF A POLYMERISATION REACTOR

The distribution of polymer molecules over their chain length determines important technological properties of the polymeric product. For this reason much effort has been undertaken in analytical techniques to measure the distribution function. The most important result in this development is gel permeation chromatography (GPC). This measuring technique has been successfully applied to problems in the laboratory. However, as the gel chromatography determination of the chain length distribution (CLD) requires sophisticated equipment and is corrupted by certain sampling lags, continuous supervision of plants by use of GPC measurements has not yet been implemented in industrial practice. Easily available and undelayed measurement variables are prefered in large-scale plants. Therefore model-based measuring techniques find a wide area of application in order to obtain, during plant operation, actual information on both conversions of reactants and product quality in polymerization reactors. In the last few years an increasing number of papers has been published on this topic (Jo, Bankoff, 1976; Kipporissides, MacGregor, Hamielec, 1981;

Adams, Schooley, 1969; MacGregor, Tidwell, 1980; Schuler, Zhang, 1985; Schuler, Papadopoulou, in press). However, the last two papers seem to be the first reports on experimental work, where the whole state vector of a polymerization reactor, also including the CLD of the polymeric product, is estimated during nonisothermal operation. Inspite of the good results of this work, some arguments exist against this procedure, so far presented, concentrated on the estimation of states determining the CLD, which are not observable and thus can only be obtained by simulation without any measurement update. From this point of view, the illustration of this estimation problem seems to be necessary.

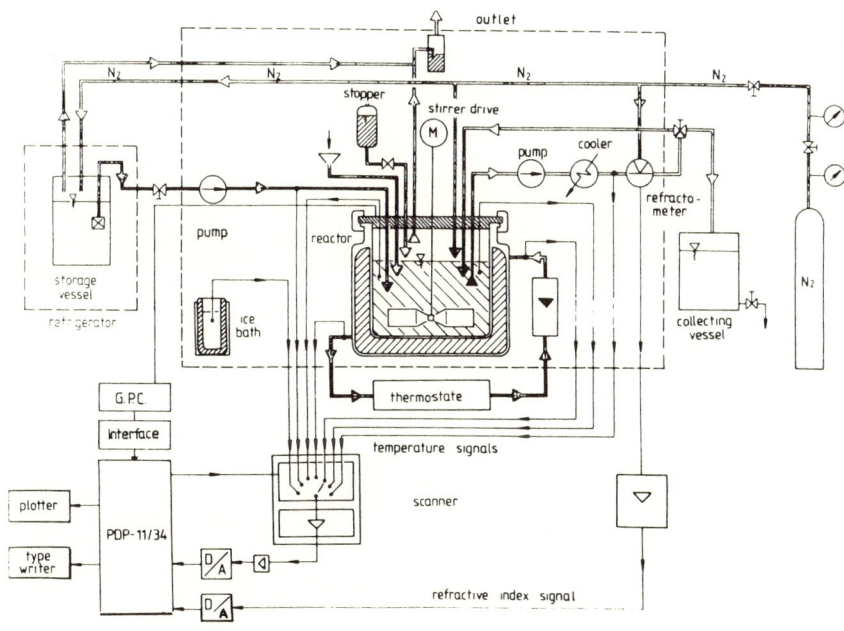

Fig. 8. Flow sheet of the laboratory plant

For experimental work the free radical solution polymerization of styrene initiated by AIBN was chosen, because the kinetics of this reaction are relatively wellknown and quantitatively describable. The flow diagram of the laboratory plant is shown in Fig. 8. As a reaction vessel, a stirred glass pot with 2 l capacity was used. The reaction mixture is covered by a nitrogene atmosphere. The coolant is pumped from a thermostate through the cooling jacket, Azo-bis-isobutyro-nitrile (AIBN) is used as an initiator, toluene as a solvent.

Thermocouples are installed to provide information on the reactor temperature, as well as the inlet- and outlet temperature of the cooling jacket. In order to supervise the conversion, the refractive index of the reaction mixture is measured with a process refractometer. All the measuring signals are fed to a process computer (PDP 11/34). A GPC is used for the analysis of samples of the reaction mixture at different points of time, in order to provide information about the CLD and the corresponding average weight lengths.

Mathematical model of the reactor

Under appropriate conditions, styrene polymerizes in a free radical mechanism, which has the following elementary steps (Ohlinger, 1955; Henrinci-Olivé, Olivé, 1969):

a) Dissociation of the initiator

$$I \xrightarrow{R_d} 2\, I^* + N_2 \uparrow \qquad (5)$$

b) Chain start by the initiator radical

$$I^* + M \xrightarrow{R_{pq}} P_1^* \qquad (6)$$

c) Chain start induced thermally

$$2\, M \xrightarrow{R_m} P_1^* \qquad (7)$$

d) Chain propagation

$$P_i^* + M \xrightarrow{R_p} P_{i+1}^* \qquad (8)$$

e) Chain transfer to monomer

$$P_i^* + M \xrightarrow{R_f} P_1^* + P_i \qquad (9)$$

f) Chain transfer to solvent

$$P_i^* + S \xrightarrow{R_s} P_1^* + P_i \qquad (10)$$

g) Chain termination by combination

$$P_i^* + P_j^* \xrightarrow{R_t} P_{i+j} \qquad (11)$$

The expressions for the various reaction rates in their simplest form are given in Schuler, Zhang, 1985. Some correction terms of these rates resulting from detailed modelling ideas are given in (Van Hook, Tobolsky, 1958, 1966; Henrinci-Olivé, Olivé, 1969; Kulkarni, Mashelkar, Doraiswany, 1980; Jainsinghani, Ray, 1977). The individual reaction rates appear on the right hand side of the balance equations for the species involved. Assuming quasistationary balances of radical species, one obtains the following equations for the mole numbers n of the reactants and products:

$$\frac{dn(M)}{dt} = -r(M)V = -(2fR_d + 2R_m + R_p + R_f)V \qquad (12)$$

$$\frac{dn(I)}{dt} = -r(I)V = -R_d V \qquad (13)$$

$$\frac{dn(S)}{dt} = -r(S)V = -R_s V \qquad (14)$$

$$\frac{dn(P_i)}{dt} = + r(P_i)V = [(R_f+R_s)(1-\alpha)\alpha^{(i-1)} +$$
$$+ \frac{1}{2} R_t(i-1)(1-\alpha)^2 \alpha^{(i-2)}]V \qquad (15)$$
$$i = 1,2,3...$$

The chain propagation probability α is given by

$$\alpha = R_p/(R_p+R_f+R_s+R_t) \qquad (16)$$

The equations (15) for the polymer chains P_i are the basis for the calculation of the chain length distribution. Their mean values can be determined from the moments

$$n(\lambda_j) = \sum_{i=1}^{\infty} i^j n(P_i); \quad j = 0,1,2,... \qquad (17)$$

The following differential equations are obtained for the first three moments:

$$\frac{dn(\lambda_0)}{dt} = r(\lambda_0)V = [(R_f+R_s) + \frac{1}{2}R_t \cdot 1)]V \qquad (18)$$

$$\frac{dn(\lambda_1)}{dt} = r(\lambda_1)V = [(R_f+R_s) + \frac{1}{2}R_t \cdot 2]V \cdot \frac{1}{1-\alpha} \qquad (19)$$

$$\frac{dn(\lambda_2)}{dt} = r(\lambda_2)V = [(R_f+R_s)(1+\alpha) + \frac{1}{2}R_t \cdot$$
$$\cdot 2(2+\alpha)]V \cdot \frac{1}{(1-\alpha)^2} \qquad (20)$$

From these moments, one can determine the number average chain length $\bar{m}_n$, the weight average chain length $\bar{m}_w$ and the polydispersity u:

$$\bar{m}_n = \frac{n(\lambda_1)}{n(\lambda_0)}; \quad \bar{m}_w = \frac{n(\lambda_2)}{n(\lambda_1)}; \quad u = \frac{\bar{m}_w}{\bar{m}_n} - 1 \qquad (21)$$

In order to achieve sufficient accuracy, one must solve a relatively large number of balance equations for chains P_i. In (Schuler, Zhang, 1985) 20 different chain lengths were selected so that the resulting CLD could be considered to be a smooth function of n.

The polymerization of styrene is a strongly exothermic reaction. The temperature of the reaction mixture can be obtained from the energy balance:

$$C \cdot \frac{dT}{dt} = k_w F_k (T_k - T) + (\rho c_p) \dot{V}_{Refr} (T_{Refr} - T) + (-h_R) r(M) \cdot V \quad (22)$$

The overall heat capacity C is the sum of that of the reaction mixture and the heat capacity C_p of the empty reactor

$$C = m c_p + C_p \quad (23)$$

The desired model equations which are able to describe the behavior of the polymerization reactor under nonisothermal conditions, are a special case of the general nonlinear state equation

$$\frac{d\underline{x}}{dt} = \underline{f}(\underline{x},\underline{u},t) + \underline{G}\ \underline{v} \quad \underline{x}(0) = \underline{x}_0 \quad (24)$$

The initial state $\underline{x}_0$ is given by the composition of the recipe $(n_0(\bar{M}), n_0(P_i), n_0(\lambda_j))$ and the starting temperature T_0. As to a more sophisticated model involving also secondary effects we refer to (Schuler, Zhang, 1985).

The signal flow diagram of the desired reactor model is shown in Fig. 9. It is composed of two subsystems in a serial structure. As a consequence of this structure, the equations for the temperature, the monomer and the initiator can be treated separately from the equations for the polymer products. The state variables of the first subsystem act as an input variables of the second subsystem, which describes the CLD and its characteristic parameters. No feedback exists from its output to the dynamics of the temperature and the conversion of the monomer.

Measurement

The actual values of the measured temperature T and the refractive index χ allow access to a part of the state vector. From a calibration with standardized mixtures of monomer, solvent and polymer, the empirical relation

$$\frac{[M]_0 - [M]}{[M]_0} = c_1 + c_2(\chi - \chi_B) \tag{25}$$

were found valid at constant solvent concentrations. Using a thermostated refractometer, one can determine from (25) the actual monomer concentration [M]. The measurement equations

$$y_1 = T + w_1 \tag{26}$$

$$y_2 = [M] + w_2 \tag{27}$$

are special cases of the general vectorial measurement equation

$$\underline{y} = \underline{h}(\underline{x}) + \underline{w} \tag{28}$$

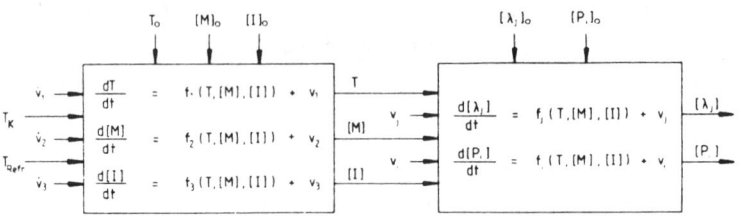

Fig. 9. Signal flow diagram of the process model describing a polymerization reactor

State estimation results

A Kalman filter of the standardized form (Gelb, 1974; Krebs, 1980; Gilles, 1979) was developed to determine a best estimate $\hat{x}$ of the actual state x from the measurements y corrupted by a white noise process w. Uncertainties in the process model $\{f\}$ were taken into account by the white noise v.

A necessary condition for the application of this filtering algorithm is the observability of the state vector from the available measurement variables. The analysis of local observability shows that the rank of the observability matrix is only three. From temperature and refractive index measurements one can only reconstruct the states of the first subsystem in Fig. 9. The other state variables, particularly those describing CLD of the polymer are not directly observable. The limited observability is a consequence of the system structure. Fig. 10 shows that the measurements are input variables of the second subsystem describing the CLD which has no feedback effect on the first subsystem. Because of this special structure, a filter of reduced order must be designed. The part of the state vector to be updated contains only the observable states. All other states, particularly those determining CLD can only be obtained by simulation of the corresponding balance equation without any measurement update. As the second subsystem is asymptotically stable, the influence of its unknown initial conditions decays exponentially with time.

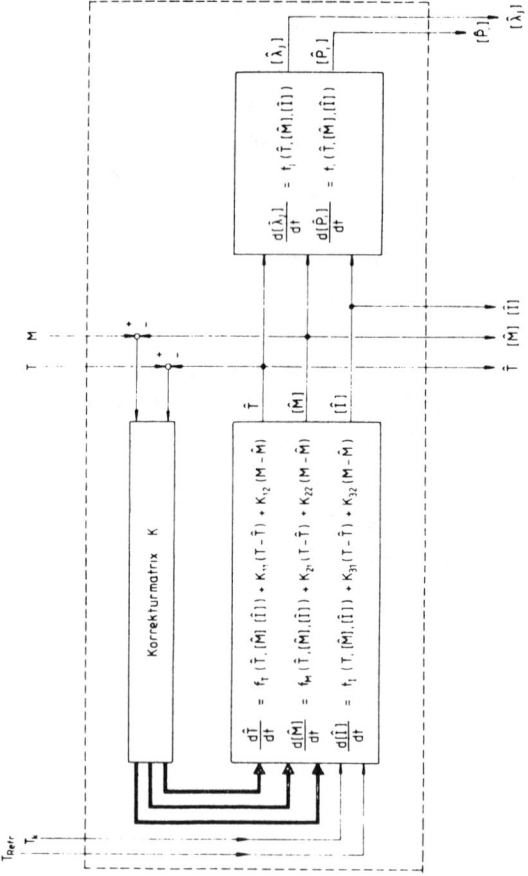

Fig. 10. Signal flow diagram of the special filter structure adapted to the polymerization reactor model

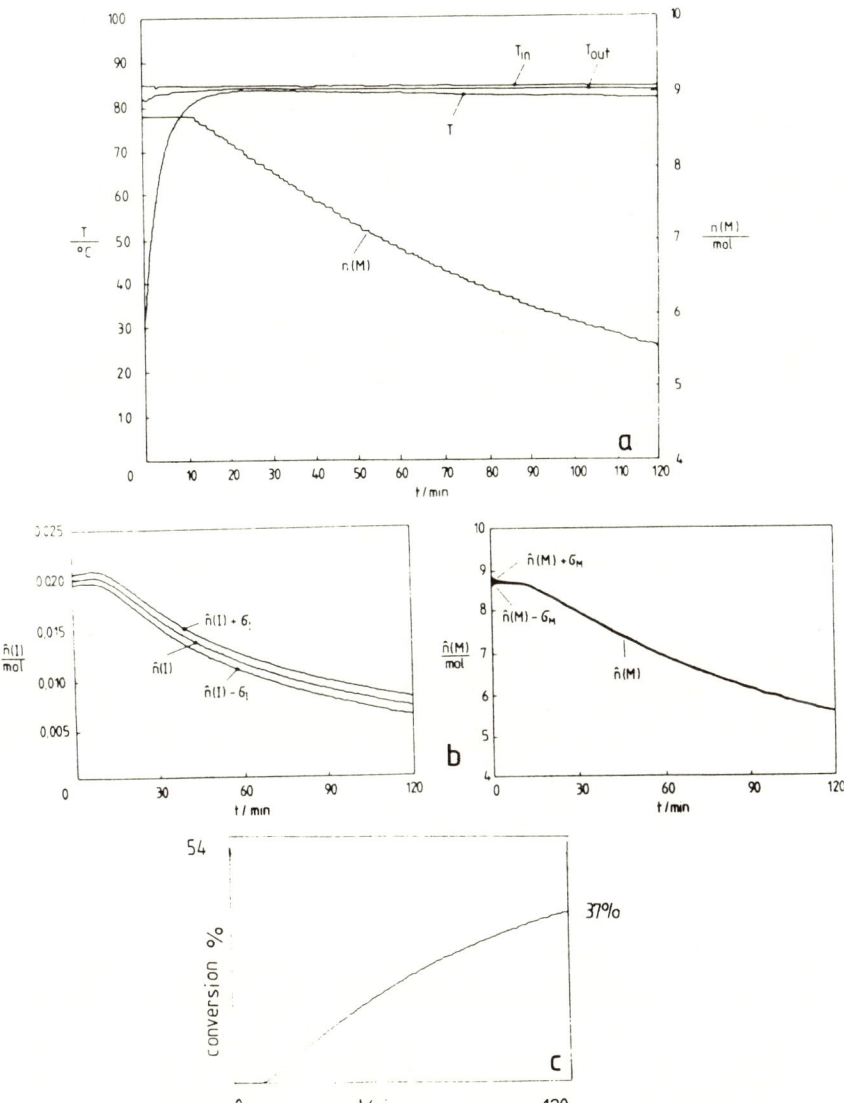

Fig.11. State estimation during heating up of the polymerization process
a) Measurement variables
b) estimates of the observable state variables
c) conversion

Fig. 11 shows some estimation results during heating up of the process to the reaction temperature of approximately 85°C. The measured variables T, T_K and [M] are given in Fig. 11a. Both jacket temperatures are higher than the reaction temperature. This fact is caused by the 25°C stream passing the thermostated refractometer. Fig. 11b shows the estimated values of the observable state variables. The estimate ñ(M) is practically identical to the corresponding measurement. The estimated mole number of the initiator decreases continuously because of the onset of reaction at higher temperatures. The 1σ-confidence intervals depicted additionally are very small for the measured states. The estimate of the initiator mole number is less accurate because this estimation is based mainly on model information.

The predictions of states of the second subsystem provide the desired information on product quality. Fig. 12a shows the estimated CLD at equidistant time intervals of 5 min. These estimates are obtained on the basis of the underlying model information of the process. Fig. 12b shows a sufficient agreement existing between the CLD-estimates and real GPC measurements. In spite of these good simulation results some weighty arguments exist against the procedure so far presented. There are various disturbances which might affect the CLD without having influence upon the measuring variables. The resulting errors in the estimation of the CLD are then not perceptible and cannot be corrected. These errors can be considerably reduced when the information on the CLD obtained from sporadic GPC-measurement is used in the estimation procedure. Since the analysis and its evaluation is not instantaneous the result of the GPC-measurement at time t_k will be obtained after a time delay τ_A, at $t'_k = t_k + \tau_A$.

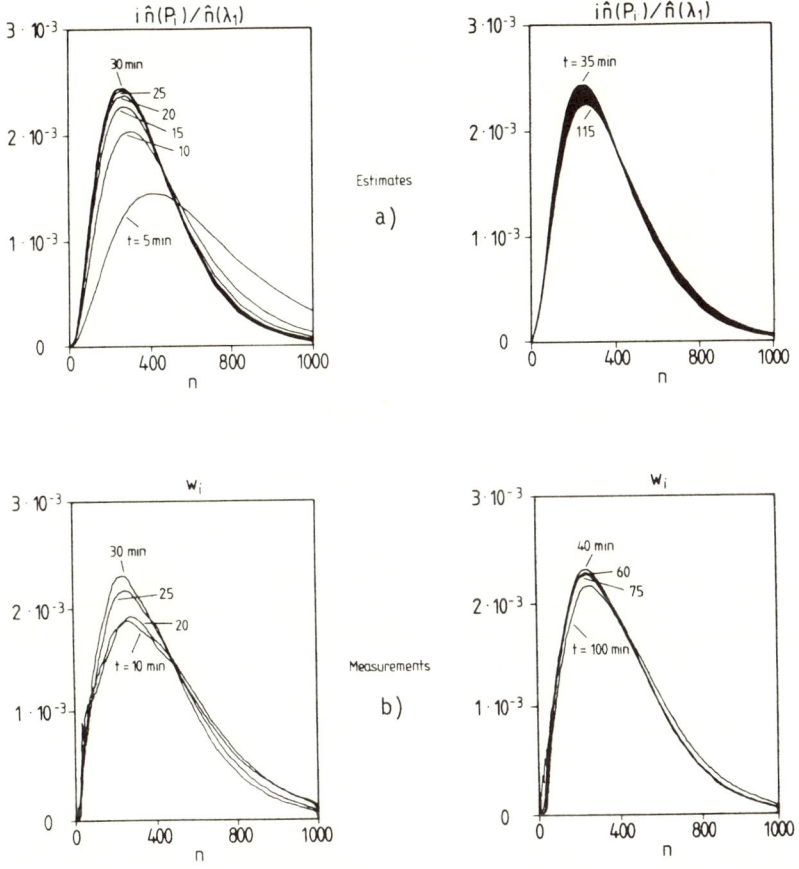

Fig. 12. a) Estimated chain length distributions
         b) Measurements of CLD with GPC

In the estimation of chains P_i for the calculation of chain length distribution, the measurements of the weight fractions W_i can be considered via a second estimator as shown in Fig. 13.

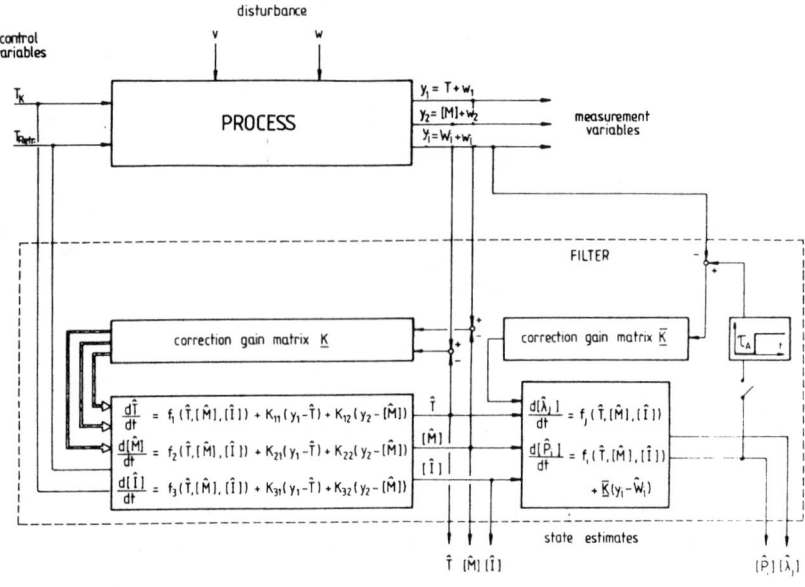

Fig. 13. Signal flow diagram of the special filter structure with the consideration of the GPC measurements

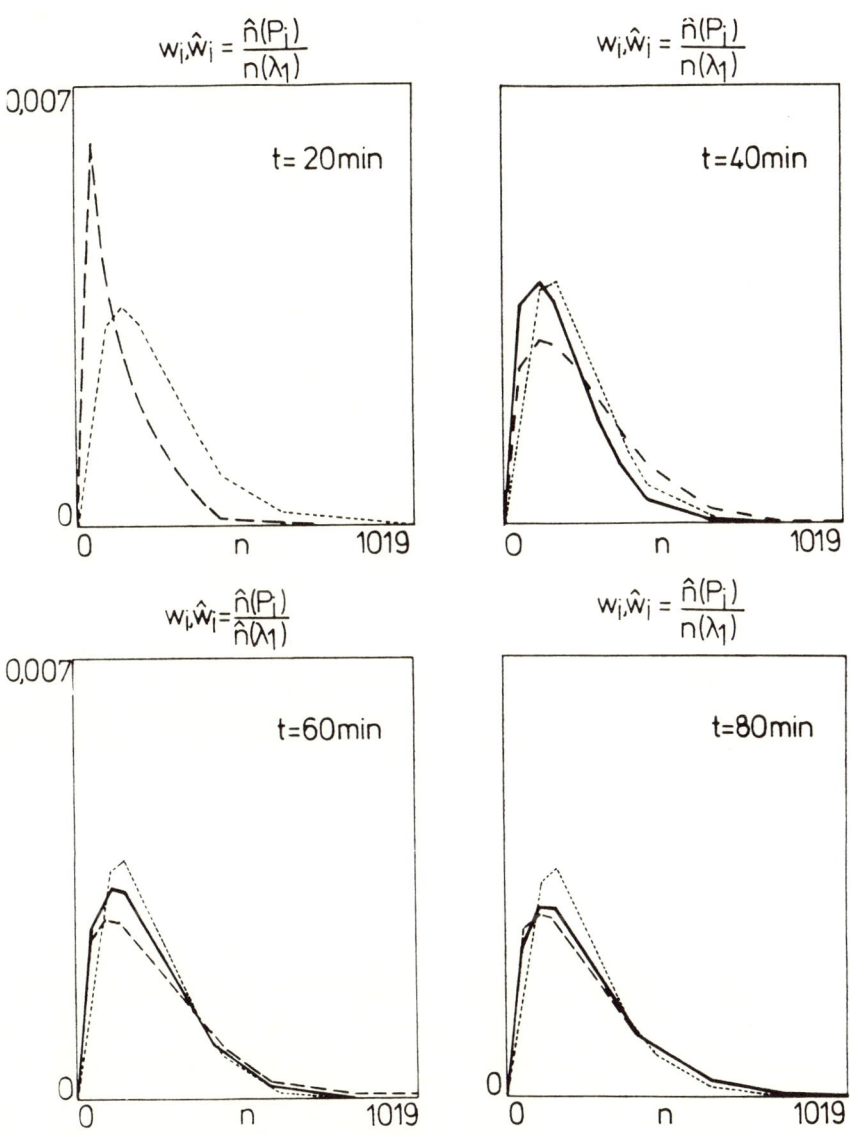

Fig. 14. Convergence of the CLD estimation considering GPC measurements
-------- Measurements of CLD with GPC
———— Estimated CLD considering GPC measurements
.... Simulated CLD

In Fig. 14 (Papadopoulou, S., Dissertation) one sees the convergence of the CLD when both estimators are working, in comparison with the results obtained by simulation. One can see that the consideration of the GPC measurement improves the estimation of the CLD. If there are modelling errors in the second subsystem, the estimation will help to correct the diverging estimation. The more accurate the reactor model is, the longer the time interval between two successive GPC analysis may be.

EARLY DETECTION OF HAZARDOUS STATES

Model-based measuring techniques find a wide field of application when hazardous states in chemical reactors are to be detected at an early stage, to ensure a safe operation. Here, these techniques are combined with methods exhibiting a close relation to pattern recognition. Kalman filters are used to estimate the state variables of the reactor. The classification of the process behavior is then based on explosion criteria or various model tests.

Criteria

As an introduction to these problems, a very simple batch reactor with a strongly exothermic reaction of first order is considered. In order to detect in such a system hazardous states various criteria can be applied. The question which of these criteria is the most appropriate one, can only be answered in the view of the specific conditions of such a process. The criterion, obviously privileged in industrial application, considers the state of the reactor as dangerous when both the first and the second derivative of the temperature are positive. When a simple P-mode temperature control is used the reactor behavior can be described by a two-dimensional model. Fig. 15a shows the developments of the temperature and the conversion over time for various values of the initial temperature. Corresponding to the above mentioned criterion the reactor exhibits a critical behavior when the initial temperature exceeds a certain value.

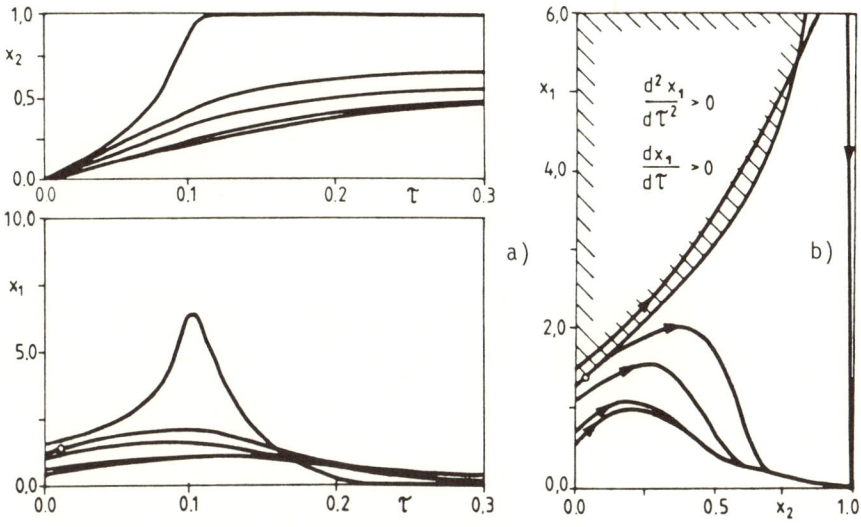

Fig. 15a,b. Reaction profiles with positive and negative inflections of time-dependent temperature function $x_1(\tau)$

In the phase-plane, Fig. 15b, the criterion defines the shaded area comprising all critical states. If all state variables are known the criterion can thus be evaluated in a very simple way. One must only examine whether the actual state of the reactor is within or outside the area defined as dangerous. In principle, this procedure is also applicable to higher-dimensional reaction systems, such as complex reaction mechanisms. Therefore, a process monitoring system should primarily make available the whole state vector by means of a model based measuring technique. If the state is estimated, even very complex criteria can be evaluated in a simple and systematic way to detect critical states. The structure of such a monitoring system is given in Fig. 16.

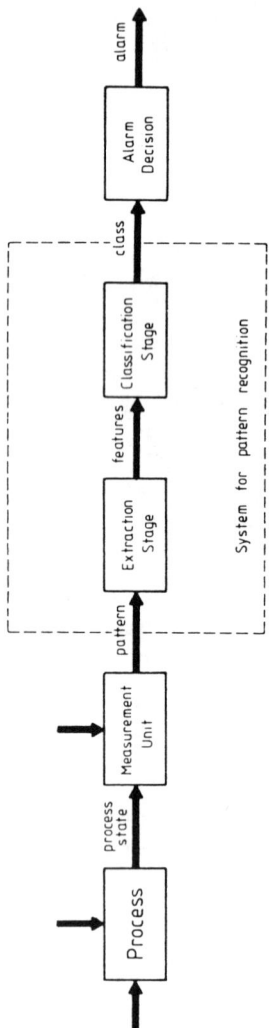

Fig. 16. Structure of pattern recognition

Sulfonation reaction

As an example of technical interest the sulfonation of an aromatic compound A_1 is considered:

$$A_1 + A_2 \xrightarrow{\tilde{r}_1} A_3 + (-\Delta h_R)_1 \qquad (29)$$

In processing industry, the heat of reaction is carefully removed, for it is known that the product A_3 may decompose in an undesired reaction step releasing a great amount of heat (Regenass, 1984; Gruber, 1974). From numerous heat-storage tests a formal kinetic is found to describe the reaction rate $\tilde{r}_1$

$$\tilde{r}_1 = k_{1o} \exp(-E_1/RT) A_1^\alpha \cdot A_2^\beta \qquad (30)$$

The time history of the temperature in an adiabatic vessel is shown in Fig. 17 for the first reaction step. The simulation of this experiment with (30) leads to a good result. A rapid increase in temperature at the beginning is due to the release of heat of mixing. Although the release of heat is not of minor magnitude, it is not considered in the mathematical model. Industrial sulfonation usually takes place in a semi- and not in a batch-reactor. Hence, it will be possible to approximate the reactor behavior by adding this heat to the heat of reaction.

If the mixture is left in an adiabatic vessel, a decomposition will take place. Fig. 18 shows the whole temperature time history of three different heat storage tests. All experiments are started with the same initial composition, but with different initial temperatures. Every run shows either a short or a rather long, almost isothermal phase. This is an indication for an autocatalytic reaction rate. Fitting of model parameters for the decomposition reaction

$$A_3 \xrightarrow{\tilde{r}_2} A_4 + (-\Delta h_R)_2 \qquad (31)$$

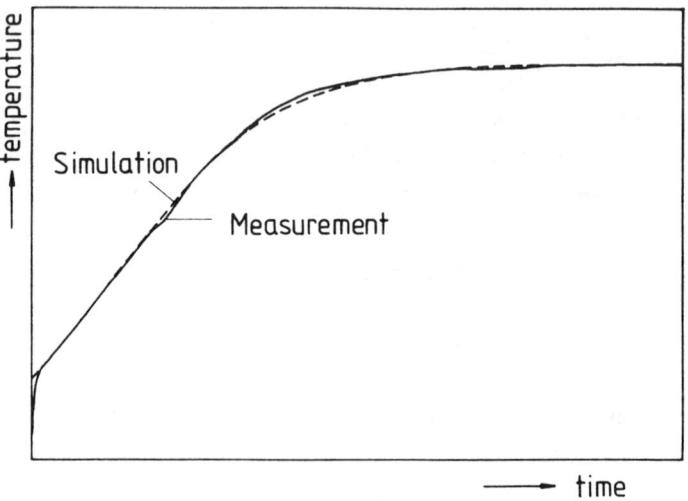

Fig. 17. Adiabatic heat storage test
First reaction step

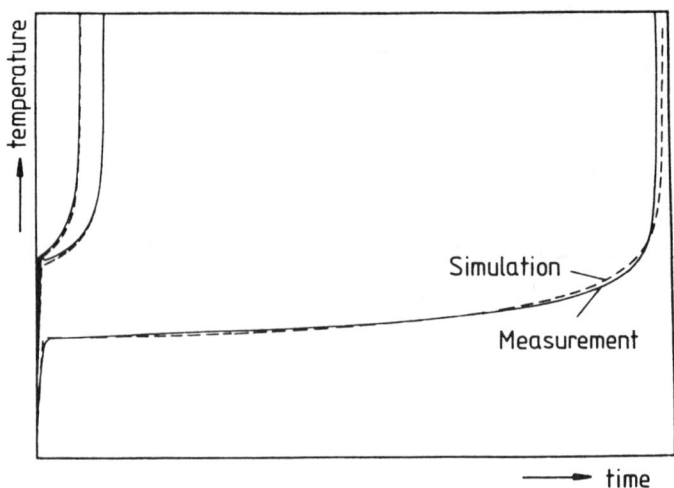

Fig. 18. Adiabatic heat storage test
decomposition reaction

leads to the following reaction rate expression

$$\tilde{r}_2 = k_{20} \exp(-E_2/RT) A_2^{\gamma} + k_{30} \exp(-E_3/RT) A_3 \cdot A_4 \qquad (32)$$

The first part is responsible for the starting of the decomposition, the second part indicates the autocatalytic character. With conventional cooling facilities, the very fast release of heat reaction $(-\Delta h_R)_2 \gg (-\Delta h_R)_1$ cannot be removed fast enough. A runaway of the reactor will be the result.

The mathematical model of an adiabatic batch reactor used for experiments consists of a balance equation for the enthalpy describing the change of temperature in time, and four balance equations for the components $A_1 \ldots A_4$.
Normalizing these equations with

$$x_i = \frac{u_i}{u_{i_o}} \qquad i = 1 \ldots 4$$

$$x_5 = \frac{T - T_u}{T_u} \cdot \frac{E_1}{RT_u} \qquad (33)$$

$$t = t_n \cdot \tau$$

leads to a set of nonlinear equations for the state variables $\underline{x}$:

$$\begin{aligned}
\dot{x}_1 &= -q_1 r_1 & x_1(o) &= x_{1o} \\
\dot{x}_2 &= -q_1 r_1 & x_2(o) &= x_{2o} \\
\dot{x}_3 &= q_1 r_1 - q_2 r_2 & x_3(o) &= x_{3o} \qquad (34) \\
\dot{x}_4 &= q_2 r_2 & x_4(o) &= x_{4o} \\
\dot{x}_5 &= q_3 r_1 + q_4 r_2 & x_5(o) &= x_{5o}
\end{aligned}$$

with

$$r_1 = \exp\left(\frac{x_5}{1+x_5/c_1}\right) x_1^{\alpha} x_2^{\beta}$$

$$r_2 = \exp\left(\frac{c_2}{c_1} \frac{x_5}{1+x_5/c_1}\right) x_2^{\gamma} + q_5 \exp\left(\frac{c_3}{c_1} \frac{x_5}{1+x_5/c_1}\right) x_3 x_4 \quad (35)$$

$$c_i = \frac{E_i}{RT_n}$$

The only measurement available is the signal y, which is the temperature x_5 disturbed by some noise w.

In order to detect in such a reactor type a hazardous state very early, the above mentioned criterion is applied, which considers the first and the second time derivative of the temperature. To avoid a repeated differentiation of the measurement, a Kalman filter is used for state estimation. The time derivatives of the temperature can then be obtained from the estimated state variables as following

$$\frac{d\hat{x}_5}{d\tau} = f_5(\underline{\hat{x}})$$

$$\frac{d^2\hat{x}_5}{d\tau^2} = f_6(\underline{\hat{x}}) \quad (36)$$

Keeping in mind that the main operations in filters are integrations giving signal smoothing, instead of differentiations giving signal roughing, better results are expected.

For the real experiment in Fig. 19 the alarm variable is indicated. The accelerated release of heat is detected at an early stage, earlier than process personnel could observe it. In Fig. 19 a starting temperature of the filter $\hat{x}_5(o)$ is chosen, that is different from the first measurement y. The unmodelled heat of mixing is the reason for this choise.

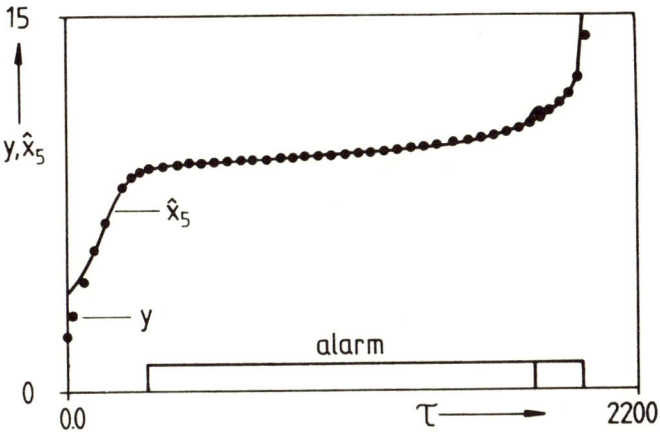

Fig. 19. Alarm creation for a heat storage test (experiment)

Prediction

When the whole state vector of the process is known at every time τ by state estimation, the mathematical model can be used to predict the future development of the reactor behavior by means of a fast simulation. The time of runaway is of particular interest.

Fig. 20 gives predictions for two experiments with the sulfonation reaction performed in an adiabatic vessel. Both experiments are started at 8.00 h. In one case the runaway takes place at 9.15 h in the other at 11.30 h. Although the system was disturbed by unmodelled mixing effects, the predictions of the remaining time until the runaway give satisfactory results at an early stage. The derivations from the exact predictions given by the broken lines in Fig.20 are due to wrong estimates of the filter. Bad mixing is responsible for the deviations at the beginning of the experiments.

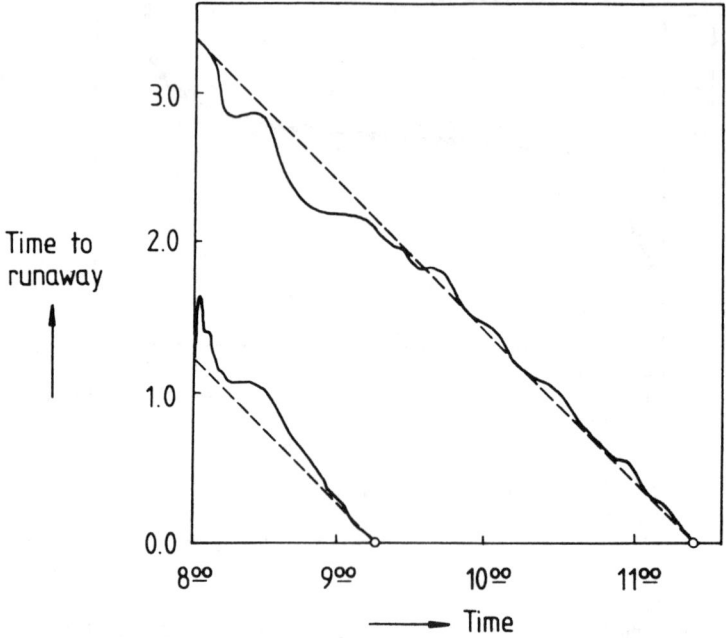

Fig. 20. Prediction of the time to runaway (experiments)

Predictions have always to be looked at with care, as modelling errors and disturbances of the process can lead to considerable errors in the simulation results.

Multiple filter method

In Gilles, Schuler (1982) it is shown, that the use of model based measuring techniques for process monitoring can even be expanded considerably beyond the procedure as described here. The knowledge of the complete state vector enables us to identify a certain mode of disturbance or to discriminate between different modes causing hazardous states.

The principal procedure can be explained by means of a sulfonation reactor (King, Gilles, 1984; King, 1985). Sulfonation of aromatic compounds is usually performed in a cooled semibatch-reactor. Some of the possible operating modes and adequate countermeasures are given in Table 1.

TABLE 1 Working modes of a sulfonation process

No	Mode	Countermeasures
1	normal operation	-
2	exothermic decomposition	→ increase cooling
3	penetration of cooling water	stop cooling, → stop reaction by other means
4	exothermic decomposition and then	→ increase cooling
	penetration of cooling water	→ stop cooling,...

If one compares modes 2 and 3, one sees that totally different countermeasures have to be taken into account. Penetrating water in the vessel will form with SO_3 new H_2SO_4 under almost instantaneous release of heat. In the most dangerous mode, 4, the emergency strategy is changed during operation. Increased cooling, as a result of a detected decomposition reaction, is replaced by a cut off of cooling.

This example demonstrates the importance of distinguishing between different faults. A possible warning system is shown in Fig. 21. For every mode a special filter is built that knows only the effects of this mode of operation. With every filter, the state variables are estimated. To discriminate between rival models, additional informations of the Kalman-Bucy-Filters are used. A filter not only estimates unmeasured state variables, but also gives some statistical information about the quality of the estimates. This statistical information is used to give a probabilistic statement about which filter describes reality best. The mode of the filter having the highest probability will therefore represent the present state of the process.

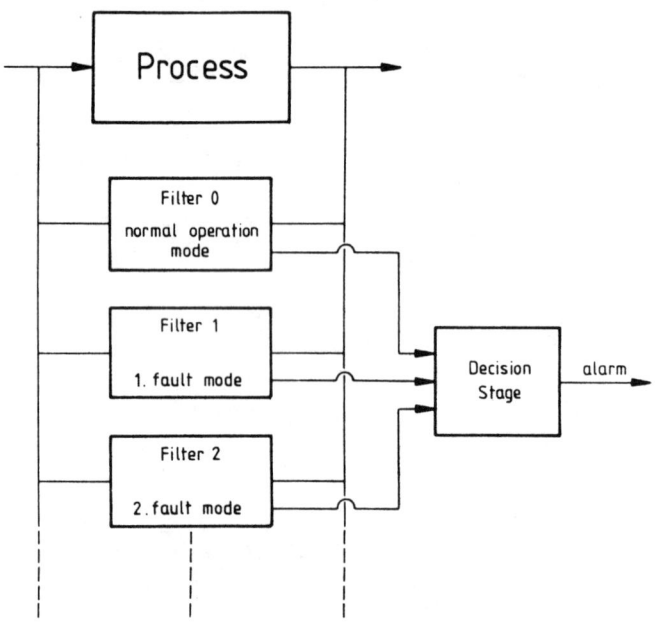

Fig. 21. Multiple filter method for fault discrimination

In the example of the sulfonation process three different model-based estimation schemes have to be built for

1) Normal operation mode
2) First fault mode: decomposition of product
3) Secondary fault mode:
   decomposition and water penetration.

Every filter tries to estimate its process state with its inherent model. Fig. 22 shows the temperature reading and the estimates of the different filters for a temperature-controlled semibatch reactor. In the considered example cooling is not strong enough to maintain the reactor at an isothermal behavior. The probability p_0 for normal operation mode will soon decrease.

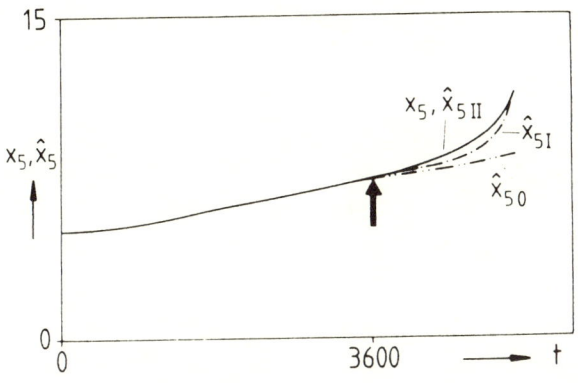

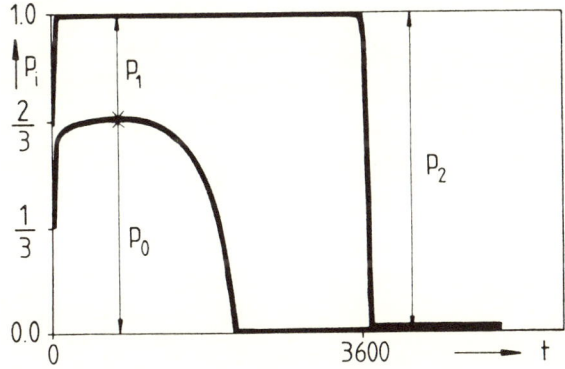

Fig. 22. Alarm creation in a semi-batch reactor with runaway

The probability p_1 of the decomposition mode increases to almost one. If, moreover, water penetrates at time t = 3600, the temperature rises even faster. Comparing the qualities of estimations show that only filter 3 can interprete the measurements well. The alarm for decomposition only is removed and replaced by an alarm for the simultaneous breaking-in of water and decomposition ($p_2 \approx 1$). If the cooling is not stopped, a runaway of the reactor will occur as shown in Fig. 22.

CONTROL OF A DISTILLATION PLANT

There are several reasons for which distillation columns are a worthwhile topic of research in the field of modelling, model-based measuring techniques and control. The dynamic behavior of this most used separation process is rather complex, due to the interaction of counter current twophase flow, thermodynamics and heat- and masstransfer. For operating such a plant it is therefore of considerable help if as much information as possible about the temperature and concentration profiles along the column is available. This is especially true during the startup procedure of the plant. Although practical experience over many years has shown the good operating characteristics for most of these units, serious problems occasionally arise, if a strong nonideality of the mixture causes sharp temperature profiles. In addition, the rising demand for saving energy and increasing product quality also requires the application of more up-to-date measuring techniques such as observers or filters, and modern control concepts. Considerable efforts are necessary to solve the various measuring and control problems, if two or more interacting columns are used to perform the desired separation.

Our investigations were initiated several years ago by serious control problems which arose in a large scale extractive distillation column in industry (Silberberger, 1979). In this column a binary

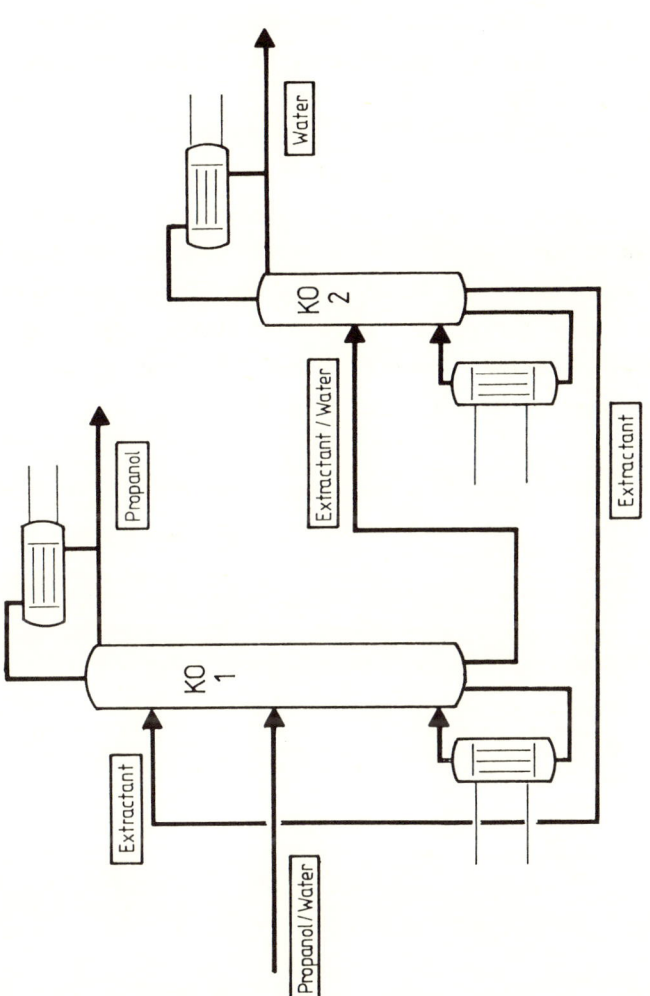

Fig. 23. Large scale extractive distillation plant

azeotrop isopropanol/water,was separated by using the extractant glycol. Under normal operating conditions the distillate has to be pure isopropanol, whereas the bottom product is a mixture of water and glycol only. The bottom product mixture is separated in a second column, so that the extractant can be fed back to the first one. The connection of the two columns, conventional for an extractive distillation is shown in Fig. 23.

While simulating the large scale plant, it was recognized that a binary mixture of water and extractant containing some 98 % molepercent water can be obtained by using a vapour side stream. According to Fig. 24 which shows the steady state profiles of the first column this vapour side stream must be removed at the stage where the maximum concentration of the intermediate boiling component water occurs. The flow sheet of such a plant is given in Fig. 25.

In order to investigate this new concept for extractive distillation in more detail, a pilot plant was constructed, of which the flow-sheet is given in Fig. 26.

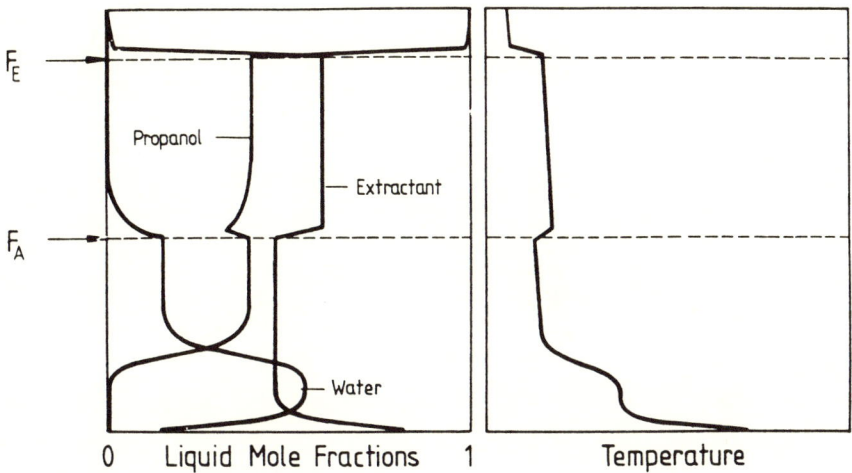

Fig. 24. Steady state profiles of the large scale extractive distillation

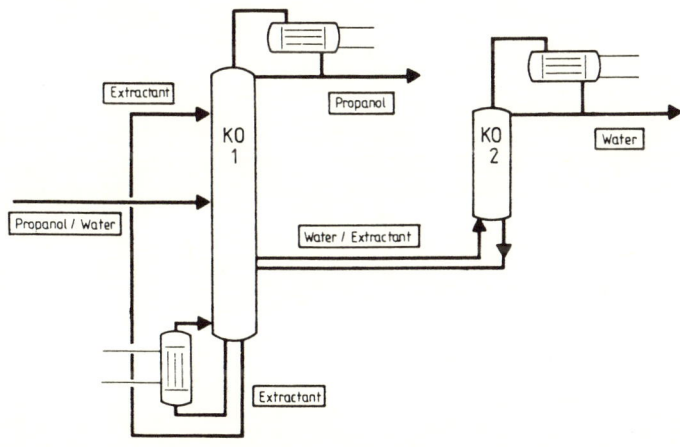

Fig. 25. Extractive distillation with vapour side stream

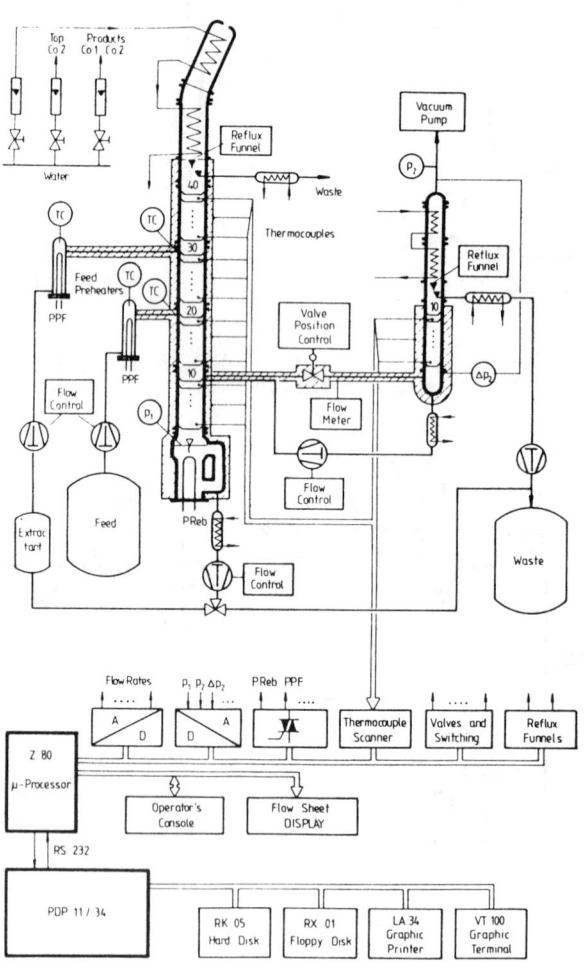

Fig. 26. Pilot plant for extractive distillation

This plant provides excellent facilities to test different concepts of model-based measuring and control techniques. The technical data of this plant can be found in Table 2. The two-column process was designed not only for extractive distillation but also for a multi-purpose use in research and education. Two and three component mixtures can be separated in continuous operating. By switching between two feed streams, step disturbances of feed concentration and feed temperature may be achieved. A liquid and a vapour side stream are installed to connect the two columns. In order to provide the full information about the temperature profile of the column, every tray is equipped with a thermo-couple and can be optionally equipped with a device to take liquid or vapour samples.

To study the extractive distillation the additional feed is used for the extractant. The bottom product of the first column consists of the pure extractant. A small amount of it is removed with the side stream into the second column and separated there from the water. Due to the high water concentration in the vapour feed, this column can be operated at a very low reflux ratio.

Detailed mathematical model

Digital simulation based on an appropriate mathematical model is the best tool for understanding the dynamic behavior of a column. We used a set of equations which model every stage as an heterogeneous two-phase system, consisting of the two totally mixed homogeneous phases, vapour and liquid, Fig. 27.

TABLE 2. Technical data of the laboratory plant

material	glass
trays	bubble cap trays one cap each, efficiency > 70 % main column 40 trays, 10 cm diameter side column 10 trays, 5 cm diameter
height	main column 7 m side column 2 m
reboiler	electrically heated 0,...,12 kW level control by overflow pipe
condensers	water cooled, reflux ratios adjustable by switching funnels, main column vented to atmosphere side column top pressure adjustable by a vacuum pump
feed preheaters	electrically heated 0,...,3 kW
pumps	adjustable piston pumps
heat jacket	12 sectors independently controlled
computers	Z 80 µ processor for operation, secondary control loops, supervision and data aquisition. PDP 11/34 process computer for control, display and data storage.

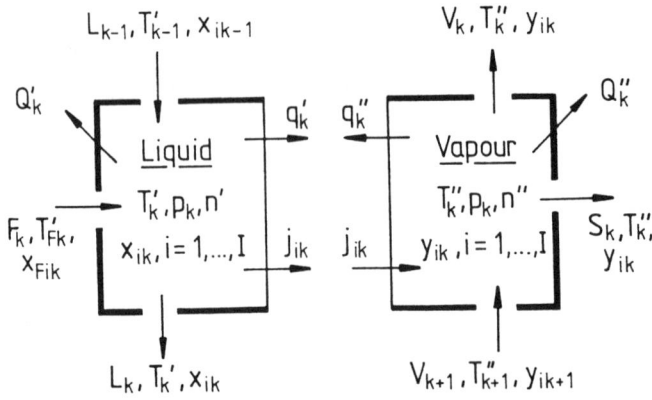

Fig. 27. Model system of stage k

Assuming constant molal holdup n', the balance equations of mass in the liquid phase on the k-th stage read:

$$n' \frac{dx_{ik}}{dt} = L_{k-1}x_{ik-1} - L_k x_{ik} - j_{ik} + F_k x_{Fik} ;$$

$$0 = L_{k-1} - L_k \sum_{i=1}^{I} j_{ik} + F_k \tag{37}$$

The molal holdup of the vapour phase is very small, n" << n', therefore:

$$0 = V_{k+1} y_{ik+1} - (V_k + S_k) y_{ik} + j_{ik} ; \quad i=1,\ldots I-1$$

$$0 = V_{k+1} - V_k - S_k + \sum_{i=1}^{I} j_{ik} . \tag{38}$$

The pressure drop Δp_k of a stage is mainly affected by the vapour flow V_k. Taking only this flow affect into consideration, the pressure of the k-th stage is approximately

$$p_k = p_0 + \xi \sum_{m=1}^{k} V_m^2 , \tag{39}$$

where p_0 denotes the top pressure.

Further simplifications are introduced to compute the mass transfer rates j_{ik} and the influence of energy on them. The heat transfer is assumed to be perfect and all mass transfer resistance is concentrated on the vapour phase side of the phase boundary. With these assumptions, we have $T_k'' = T_k' = T_k$, and the stage temperature may be computed from the liquid compositions. This so-called boiling point procedure also yields the equilibrium composition of the vapour y_{ik}, $i=1,\ldots,I-1$. The difference between y_{ik}^* and the bulk composition y_{ik} is considered to be the driving force of the mass transfer. As energy and mass transfer between the phases are connected by the energy balance equation of the phase boundary, there is an additional influence due to the energy impressed from the surroundings. Writing r_i for the heat of vapourization, one derives:

$$q_k'' + q_k' = \sum_{i=1}^{I} r_i j_{ik} \qquad (40)$$

The heat transfer rates q_k'', q_k' are computed from the quasi-steady state energy balance equations of the phases, and finally we get the expression

$$j_{ik} = \frac{r_I}{r_i} \alpha_I (y_{ik}^* - y_{ik}) + \frac{\Delta E_k}{IR_i} \; ; \; i=1,\ldots,I$$
$$\Delta E_k = L_{k-1} c_{pk-1}' \Delta T_{k-1} - V_{k+1} c_{pk+1}'' \Delta T_k + F_k c_{pFk}' \Delta T_{Fk} - Q_k. \qquad (41)$$

Here, the ΔT_s denote the temperature differences $\Delta T_j = T_j - T_{j+1}$. The loss of heat to the surroundings is summarized by $Q_k = Q_k' + Q_k''$.

In the condenser, the vapour is condensed totally, whereas the vapour V_1 leaving the reboiler is assumed to be of equilibrium composition $y_{i1} = y_{i1}^*$, $i=1,\ldots I$. Additional lag elements may be included to account for the heat flow resistance in the reboiler installations and the energy balance of the reboiler holdup.

The developed mathematical model is based on certain serious simplifications. However, measured data were matched at a sufficient accuracy by adjusting the parameters holdup n', the vapour flow resistance ξ and the mass transfer coefficient α_I (Silberberger, 1979).

Simulation Results

In the steady state, the distillate is high purity propanol, and the bottom product consists of the nearly pure extractant. The intermediate boiling component water is removed by the vapour side stream (98 mole % water). In the lower part of the column, below F_A, the changing compositions cause sharp increases of the tray temperatures, the so-called temperature fronts.

If a step disturbance reduces the feed rate F_A and - by means of a ratio control - F_E and the reflux R, the regions of mass transfer and the temperature fronts are forced to move upward in the column. The state variables of all other regions are only slightly altered, Fig. 28.

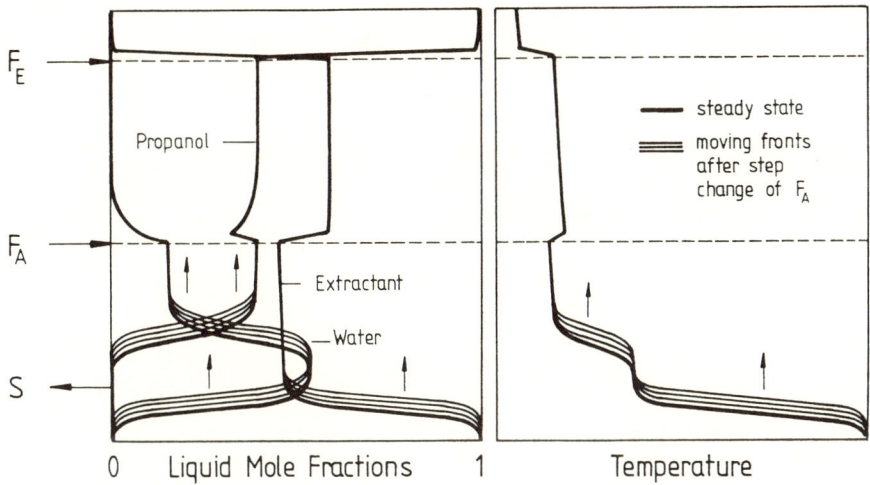

Fig. 28. Steady state and dynamics of the extractive distillation with a vapour side stream

Simplified Dynamic Model

The simulation results make the choice of appropriate state variables for a reduced order model apparent. The essential dynamic quality of the extractive column is the movement of the regions of high mass transfer isopropanol/water and extractant/water. Therefore, the loci of the temperature fronts are suitable state variables. To obtain the simplified model, the column is separated into two parts by cutting it off below the feed stage F_A. In the lower part, Fig. 29, two sections can be distinguished. The interphasial exchange of the isopropanol and the water takes place above the side stream S. The separation of the extractant from the water occurs below S. It is assumed that the regions of mass transfer remain completely within these sections. As a consequence, the compositions of all flows leaving the lower part of the column remain constant.

Writing x,y for the mole fractions of the liquid and of the vapour and using L,V for the flow rates of the two phases, the overall mass balance equation of isopropanol reads

$$\frac{d}{dt} n' \int_0^1 x_1(z)dz \triangleq -n' x_{10} \frac{dz_1}{dt} = L_0 x_{10} - V_0 y_{10} \tag{42}$$

and for the extractant we obtain

$$\frac{d}{dt} n' \int_0^1 x_2(z)dz = -n'(1-x_{20}) \frac{dz_2}{dt}$$
$$= L_0 x_{20} - L_i + V_1 - S y_{2S} - V_0 y_{20} \tag{43}$$

where n' denotes the liquid holdup of a stage.

The vapour phase dynamics and the hydrodynamics are considered to be in a quasi-steady-state. Hence, the following equations are valid:

$$L_0 - L_e + V_e - S - V_0 = 0$$
$$L_0 h_0' - L_1 h_1' + V_1 h_1'' - S h_S'' - V_0 h_0'' = 0 \tag{44}$$

where h', h" are the enthalpies of the liquid and the vapour. In these equations, the rates and concentrations of the entering flows L_0 and V_1, as well as the flow rate of the side stream S, are considered input variables. The vapour concentrations y_{10}, y_{20}, and y_{2S} are the parameters which were taken from the steady-state simulation results.

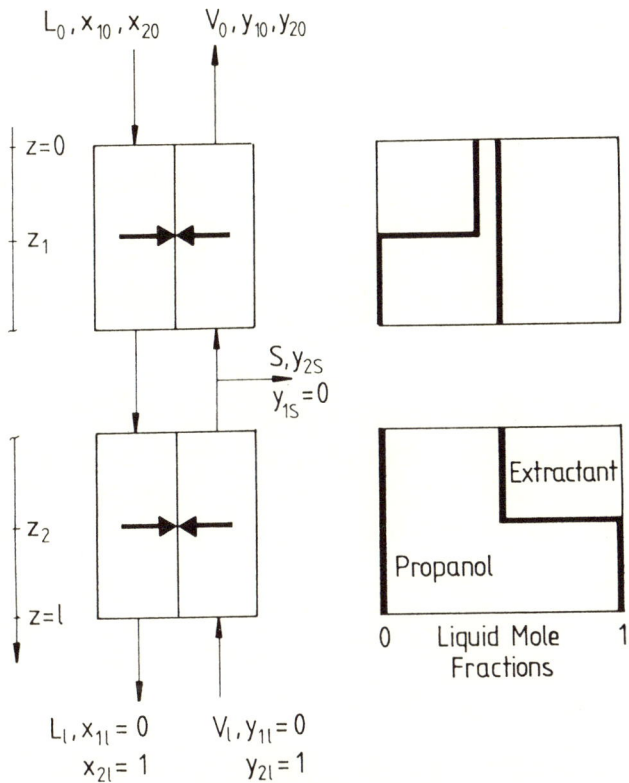

Fig. 29. Lower part of the column and approximated mole fractions profiles

As is shown in Fig. 28, the profiles in the column above F_A are not affected if the temperature fronts are moving. Hence, the influence of the feeds on the liquid flow entering the lower part is computed from the quasi-steady-state-balance equations of the upper part. The column is equipped with a ratio control, so that $R = f_1 F_A$ and $F_E = f_2 F_A$. Thus we obtain L_0, x_{10}, x_{20} in terms of F_A and x_{FA1}. To complete the simplified dynamic model, we may include the reboiler dynamics which may cause significant time lags between changes of the rate of the heating steam u_1 and the vapour flow V_1.

After introducing the state vector $\underline{x} = (\Delta V_1, \Delta z_1, \Delta z_2)^T$, the input vector $\underline{u} = (\Delta U_1, \Delta S)^T$, the disturbance vector $\underline{\xi} = (\Delta F_A, \Delta x_{FA1})$, and linearizing, the simplified model is represented in the block diagram Fig. 30 (Gilles, Retzbach, 1983).

If the movements of the fronts are measured by the average voltage of thermocouples installed at the desired loci of the fronts, the effective ranges of the measurements are limited due to saturation. Within the limits, the measurement variables $\underline{y} = (\Delta T_1, \Delta T_2)^T$ change nearly linearly with the deviations of the fronts. If the limits are exceeded, the output of the measuring device can be considered to be constant.

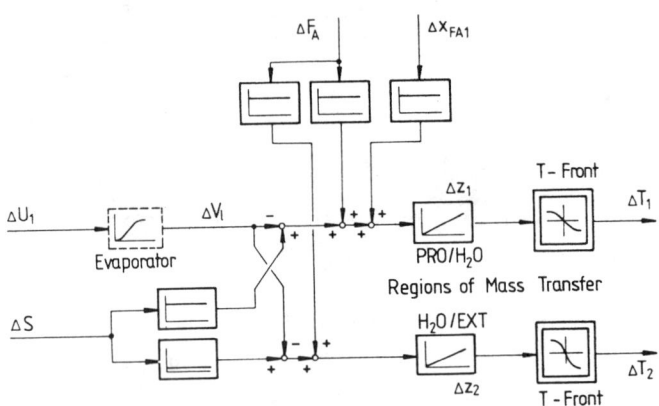

Fig. 30. Block diagram of the simplified dynamic model

Control system

The simplified dynamic model is used to design a state observer that estimates the state variables, the unknown feed composition and steady-state offsets of the feed flowrate. The estimation is based on the measurement of the two temperature deviations ΔT_1 and ΔT_2. Constant disturbances are assumed, so that the dynamic models of these influences are integrating elements.

The control loop is stabilized by state feedback. The estimated values of Δx_{FA1} and ΔF_A are used in a feed forward branch to compensate for the influence of the disturbances, Fig. 31.

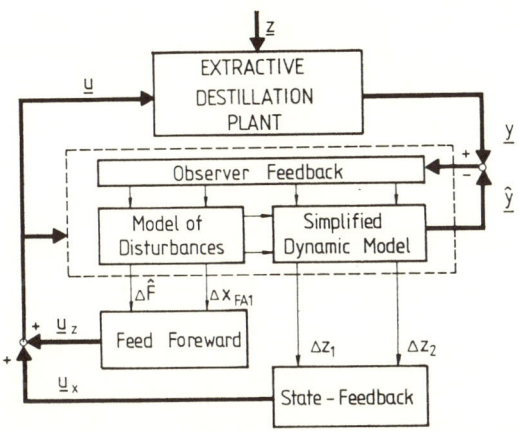

Fig. 31. Control System

If a step disturbance of the feed concentration occurs, the deviation of it is estimated with satisfactory accuracy, Fig. 32. The resulting deviations of the temperatures are insignificant. These simulation results are achieved by applying the control system to the stage-to-stage model of the distillation column. The control system was implemented in the pilot plant using a PDP 11/34 computer. The experimental results proved the high robustness and reliability of the control system during startup and operation (Retzbach, 1985).

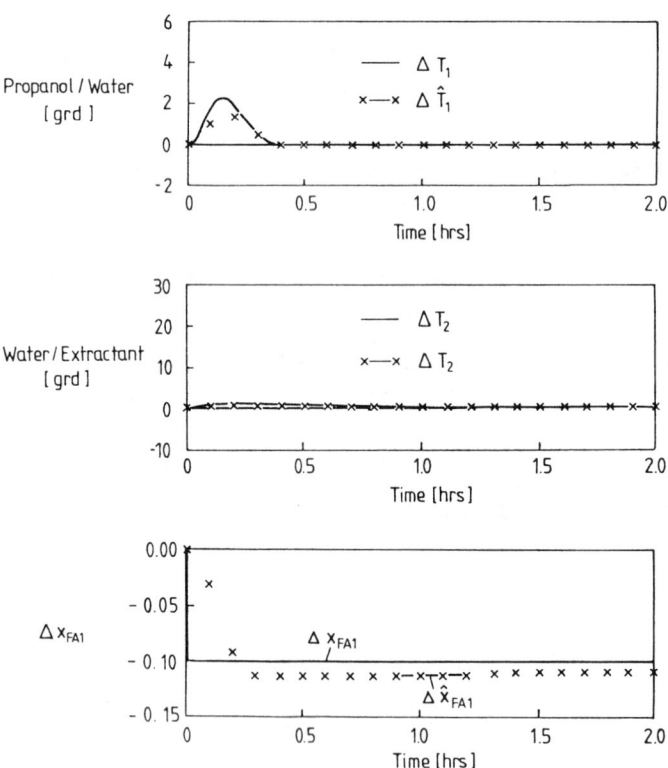

Fig. 32. Closed loop step response after a change of the feed composition

Loadchanges

For constant streams within the column sections the complete stage model may be written as

$$\underline{\dot{x}} = \underline{\underline{N}}(\underline{x},\underline{y}) * \underline{u} \tag{45}$$

The vectors $\underline{x}$ and $\underline{y}$ represent the liquid and vapor phase compositions. As shown in Fig. 33, the vector $\underline{u}$ represents the disturbances (feed flow-rate Z_A, feed composition x_{ZA}) and control variables (steam H, sidestream S, reflux R_1 and R_2, extractant flow rate Z_S). The steady-state equations are linear in u. Therefore the stationary composition profiles are unchanged, if all the input variables are multiplied by the same factor. Therefore, for changes in the feed flow-rate, all control variables will be adjusted by ratio control (Fig. 34).

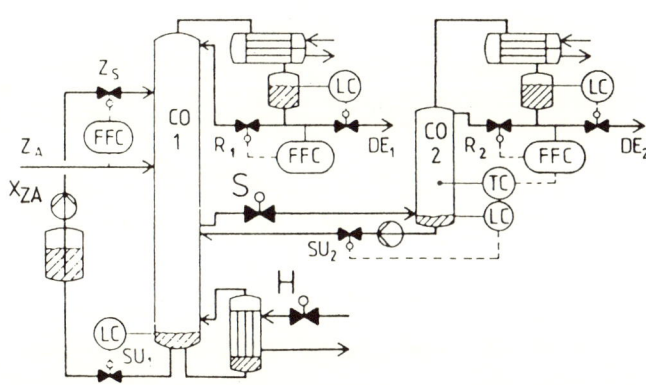

Fig. 33. Control variables and disturbances of the extractive distillation process

Start-up

The control strategy was tested with a laboratory distillation column, and a start-up procedure was developed, which produces almost no waste product. It is important for the product streams to always meet the specified product purities.

To start up, pure top-product is used as feed and the column is run with infinite reflux until the trays are filled (Fig. 35). Next glycol is fed to the column which forces the propanol out of the liquid phase. The glycol moves to the lower part of the column, whereas propanol is removed at the top (Fig. 36). As soon as there is pure glycol in the reboiler, it may be removed and the extractant loop closed. To stabilize the position of the mass transfer zone, that part of the observer which estimates the state variable s_2 and the disturbance ΔZ_A is activated. Also the control loop for the steam is closed. Next the feed is changed to the propanol/water mixture. The water removes the propanol from the liquid phase, but cannot reach the reboiler (Fig. 37). When the "water-bulge" is sufficiently large, the sidestream draw-off is opened, the estimations of the variables s_1 and Δx_{ZA} by the observer are activated, and the control-loops for the sidestream are closed (Retzbach, 1986).

CONCLUSIONS

In order to demonstrate the significance of modern measuring and control techniques for practical application in operating chemical plants four examples are considered in detail.

In the first example local temperature and concentration profiles of a fixed-bed reactor are estimated when temperature measurements at a few locations only are available. In a second example the estimation of the chain length distribution of the polymer in a polymerization reactor is discussed.

The problem of early detection of hazardous states in chemical reactors can be solved via state estimation. Finally, it is shown how reduced models are used to design an observer and a control system for two coupled distillation columns.

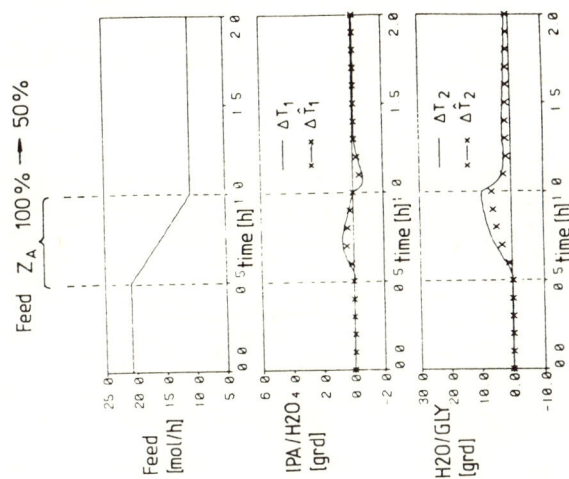

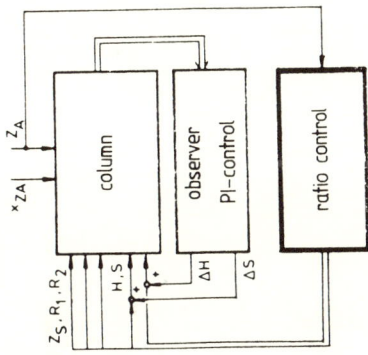

Fig. 34. Load change

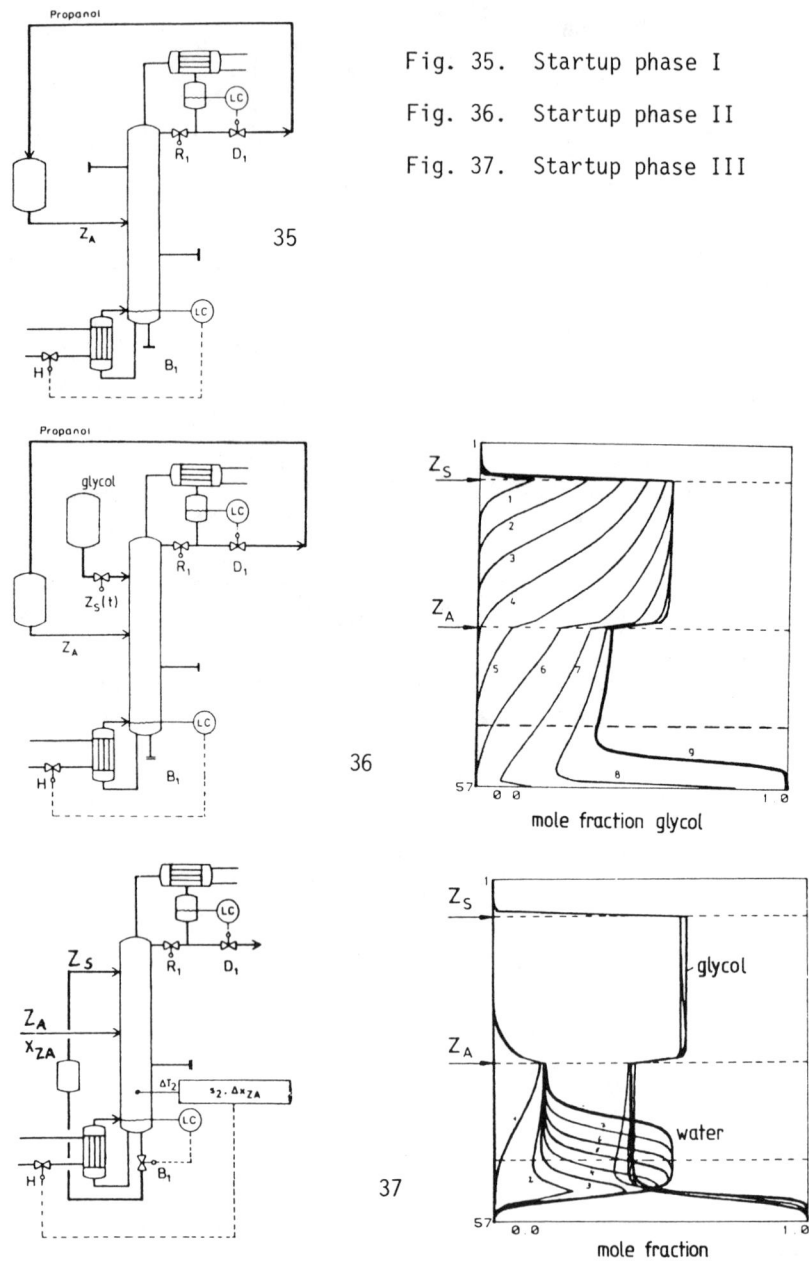

Fig. 35. Startup phase I
Fig. 36. Startup phase II
Fig. 37. Startup phase III

REFERENCES

Adams, P.C., Schooley, A.T. (1969). "Ada-Predictive Control for a Batch-Reaction. Instrumentation Technology, Jan., 57-62.

Alvarez, J., Stephanopoulos, G. (1982). An estimator for a class of non-linear distributed systems. Int. J. Control,36 (5), 787.

Brammer, W., Siffling, G. (1975). Kalman-Bucy-Filter. Deterministische Beobachtung und stochastische Filterung. R. Oldenbourg, München - Wien.

Canavas, C. (1984a). Modelling and Simulation of a fixed-bed reactor. Proceedings Int. AMSE Conf. "Modelling & Simulation", June 27-29, Athens, 3.4, 179-194.

Canavas, C. (1984b). On the Identification Problem of Fixed-Bed Reactor Parameters. Proceedings of the 3rd Mediterreanean Congress of Chemical Engineering, Nov. 19-21, 1984, Barcelona (Spain).

Canavas, C. (1984c). State and Parameter Estimation through Nonlinear Filtering. An Application to a Fixed-Bed Reactor. Proceedings of the TELECON '84 (IASTED International Conference), August 27-30, 1984, Halkidiki (Greece).

Canavas, C. (1986) (to be published). Parameter- und Zustandsschätzung an einem katalytischen Festbettreaktor. Ph.D. Dissertation, Universität Stuttgart.

Gelb, A. (Ed.) (1974). Applied Optimal Estimation, The MIT Press, Cambridge.

Gilles, E.D. (1979). Modellbildung, ein Hilfsmittel der Meßtechnik, Teil 1, Teil 2. Technisches Messen, 225-232, 271-274.

Gilles, E.D., Schuler, H. (1982). Early detection of hazardous states in chemical reactors. Ger. Chem. Eng., 5, 69-78.

Gilles, E.D., Retzbach, B. (1983). Reduced Models and Control of Distillation Columns with Sharp Temperature Profiles. IEEE Trans. on Autom. Control, AC-28, 628-630.

Grewer, T. (1974). Thermal stability of reaction mixtures. 1. Symposium on Loss Prevention and Safety Promotion in Process Industries. The Hague/Netherlands.

Henrinci-Olivé, G., Olivé, S. (1966). Über den Lösungsmitteleinfluß bei der Radikal-Polymerisation. Z. für Chem. Neue Folge, Bd.48, 35-50.

Henrinci-Olivé, G., Olivé, S. (1969). Polymerisation, Katalyse - Kinetik - Mechanismen, Verlag Chemie, Weinheim.

Jainsinghani, R., Ray, W.H. (1977). On the dynamic behavior of a class of homogeneous continuous stirred tank polymerization reactors. Chem. Eng. Sci.,32, 811-825.

Jo, J.H., Bankoff, S.G. (1976). Digital Monitoring and Estimation of Polymerization Reactors. AIChE Journal, 22, 361-368.

Kalman, R. E. (1960). J. Basic Eng., 82 D, 35-45.

King, R., Gilles, E.D. (1984). Dynamics and runaway of chemical reactors. Preprint of AIChE's Annual Meeting, San Francisco, 25-30 Nov. 1984.

King, R. (1985). Multiple Kalman filters for early detection of hazardous states. Proceedings of Industrial Process Modelling and Control, Hangzhou, 6-9 June 1985.

Kipperissides, C., MacGregor, J.E., Hamielec, A.E. (1981). Suboptimal Stochastic Control of a Continuous Latex Reactor. AIChE Journal, 27, 13-19.

Krebs, V. (1980). Nichtlineare Filterung. R. Oldenbourg, München.

Kulkani, M.G., Mashelkar, R.A., Doraiswamy, L.K. (1980). Solvent and viscosity effects in the decomposition of Azobisisobutyronitrile. Chem.Eng. Sci., 35, 823-830.

Kuruoglu, N.R., Ramirez, W.F., Clough, D.E. (1981). Distributed Parameter Estimation and Identification for Systems with Fast and Slow Dynamics. Chem. Eng. Sc., 36, 1357.

Kuruoglu, N.R., Ramirez, W.F., Clough, D.E. (1982). Steady-State Sequential Distributed Parameter Filtering. Theoretical Developments and Application to Packed-Bed Reactors. Preprints of IFAC 3rd Symposium on Control of Distributed Parameter Systems, 29.6.-2.7.1982, XI.13 - 17, Toulouse (France).

Luenberger, D.G. (1966). IEEE Trans. Autom. Control,11, 190-197.

MacGregor, J., Tidwell, P.W. (1980). Modelling and Control of Continuous Industrial Polymerization Reactors. Computer appl. to chem. eng. ACS, 251-268.

Ohlinger, H. (1955). Polystyrol, Springer, Berlin.

Papadopoulou, S.Ar. (to be published). Zustandsschätzung bei Polymerisationsreaktoren. Ph.D. Dissertation. Universität Stuttgart.

Ray, W.H.; Lainiotis, D.G. (Eds.) (1978). Distributed Parameter Systems: Identification, Estimation and Control. M. Dekker.

Regenass, W. (1984). The control of exothermic reactors. Symposium on Protection of exothermic reactors, Chester, U.K.

Retzbach, B. (1986) (to be published). Mathematische Modelle von Destillationskolonnenzur Synthese von Regelungskonzepten. Ph.D. Dissertation, Universität Stuttgart.

Schuler, H.; Zhang, S. (1985). Real Time Estimation of the Chain Length Distribution of a Polymerization Reactor. Chem. Eng. Sc., 40, 1891.

Schuler, H.; Papadopoulou, S. (in press). Real time estimation of the chain length distribution in a polymerization reactor. II. Comparison of estimated and measured distribution functions. Chem. Eng. Sc.

Silberberger, F. (1979). Modellbildung, Simulation und Regelung einer Extraktiv-Destillation. Ph.D. Dissertation, Universität Stuttgart.

Sørensen, J.P.; Jørgensen, S.B., Clement, K. (1980). Fixed-Bed Reactor Kalman-Filtering and Optimal Control. Chem. Eng. Sc., 35, 1223.

Van Hook, J.P.; Tobolsky, A.V. (1958). The thermal decomposition of 2,2'-Azo-bis-isobutyronitrile. J. Amer. Chem. Soc., 80, 779-782.

Windes, L.C.; Ray, W.H. (1984). Dynamic Estimation of Temperature and Concentration Profiles for Control of a Packed Bed Reactor. Preprint of AIChE's Annual Meeting, San Francisco, Nov. 25-30, 1984.

Zeitz, M. (1977). Nichtlineare Beobachter für chemische Reaktoren. Fortschrittsber. VDI-Z., Reihe 8 Nr. 27.

A REVIEW OF INDUSTRIAL REACTOR CONTROL: DIFFICULT PROBLEMS AND WORKABLE SOLUTIONS

Phillip D. Schnelle, Jr.
John R. Richards
E. I. du Pont de Nemours and Co., Inc., Wilmington, DE 19898

Abstract. This survey paper reviews the state of the art for industrial reactor control. It is intended to help define the current "level" of control as routinely practiced in industry. Industrial reactor control problems are difficult and varied. The solutions to these problems have to be robust and maintainable. It is hoped that this survey will help highlight some common difficult industrial problems and solutions.

The paper contains the results of an extensive literature survey, the review of 31 Du Pont reactor control applications, the results of interviews with 11 Du Pont process control professionals, and of course, the opinions and prejudice of the authors.

Keywords. Reactor Control; Control Survey; Modern Control; Traditional Control; Control Techniques; Control Theory.

1   INTRODUCTION AND BACKGROUND

The world of chemical process control and chemical reaction system design is in a constant state of change. New designs, strategies and techniques appear almost daily, making the practicing control engineer's job one of trading off optimum reactor operation versus the total cost of attaining that goal. What level of control is economically justified and what techniques stand the best chance of getting you there? These questions form the basis of this survey paper.

It is important for the reader to understand the structure of the paper. This understanding will help guide the reader through the many issues that will be discussed. The paper is broken into four major sections. The first section defines the reactor control problem and where it fits into the overall plant economics and summarizes reactor control issues. The second section covers the literature search and the survey of control professionals. This section contains a large amount of survey data that compares and contrasts reactor control techniques practiced in academia and industry. The third section reviews 31 industrial reaction systems to demonstrate where and how the various control techniques are utilized. The fourth section summarizes a logical approach to the reactor control problem as suggested by this survey.

Scope

The control of chemical reaction processes is a broad subject. The literature search indicates that there is considerable interest and research activity in this area. We will attempt to keep this paper generic in nature and give an unbiased evaluation of where industrial reactor control stands and some reasons why it is there.

What is the Industrial Reactor Control Problem?

This paper defines the reactor control problem as any aspect of reactor control from pressure, temperature, flow, and level (PTFL) measurement through control of final reactor product properties. PTFL control by itself, although difficult at times, is not sufficient in most cases to achieve the desired goals of reactor control.

There are many papers and texts which cover basic control (i.e. the PTFL problem). We will not dwell on this aspect as the papers are discussed in Section 2. The advanced control methods, where applicable, will typically fit at the higher levels of control. This level is integrally related to the PTFL control problem, but the difficulties of PTFL regulation will be considered only in reference to the property control question. This is particularly true of the temperature and pressure control considerations. Temperature, reaction rate, stability, molecular weight distribution, etc. are fundamentally related. There are many ways to address them as discussed, for example, by Ray (1985), and Hopkins (1979).

The industrial reactor control problem has several aspects which set it apart from the technical problems addressed by most researchers in the literature. These aspects are primarily ones of approach and necessity. Several of the references have touched on these issues for example Amrehn (1977), Foss (1973), MacGregor, et al. (1983), Luyben (1975), and Doss, et al. (1983). They usually revolve around the "keep it simple" approach. The following list will cover some problems the industrial control engineer must consider:

- The ECONOMICS of the reactor and reactor control system.

- The SAFETY and reliability of the proposed system.

- The PERFORMANCE of the control system (i.e. Control system integrity, measurement reliability, interlocks).

- The PRODUCT QUALITY constraint and allowances.

- The CAPACITY AND PRODUCTION COMMITMENT of the plant (operability).

- The MANPOWER required to design, install, and maintain the system.

- The TRAINING required to help operating personnel get the most out of the system.

- The EFFECT ON OVERALL PLANT CONTROL.

- The HARDWARE to be purchased or modified.

- The MAINTENANCE required for additional analyzers or new controls.

- The requirement for STEADY STATE OPTIMIZATION (off-line/on-line).

- The proper PLACE IN THE CONTROL HIERARCHY as discussed in, for example, MacGregor, et al. (1983).

How does one go about keeping a control system simple and address all of these critical questions? That is the industrial reactor control problem.

Reactor Control Economics

Today's process control economics can be viewed in Figure 1. This figure shows two example curves of economic incentive as a function of increasing control complexity.

The upper curve would represent a process where there is a large stake in getting the regulatory (PID) loops in automatic. The incentive slope then quickly flattens, indicating little additional stake for supervisory control or optimization. The lower curve represents a process where the incentive is almost linear with increasing control complexity.

Section 3, the survey of in-house reactors, indicates that the composite curve for chemical reactor control is somewhere in the middle of this region. However, there is considerable scatter amoung the reactors. There most certainly is a need to get the regulatory loops in automatic, and there typically is a stake for upper level property control and yield optimization. The reactor control problem clearly needs to be evaluated from an economic standpoint.

Where reactor control fits into overall plant economics. The chemical reactor often has a unique and complicated spot in plant economic considerations. Frequently the reactor is the most important piece of processing equipment from a dollar standpoint.

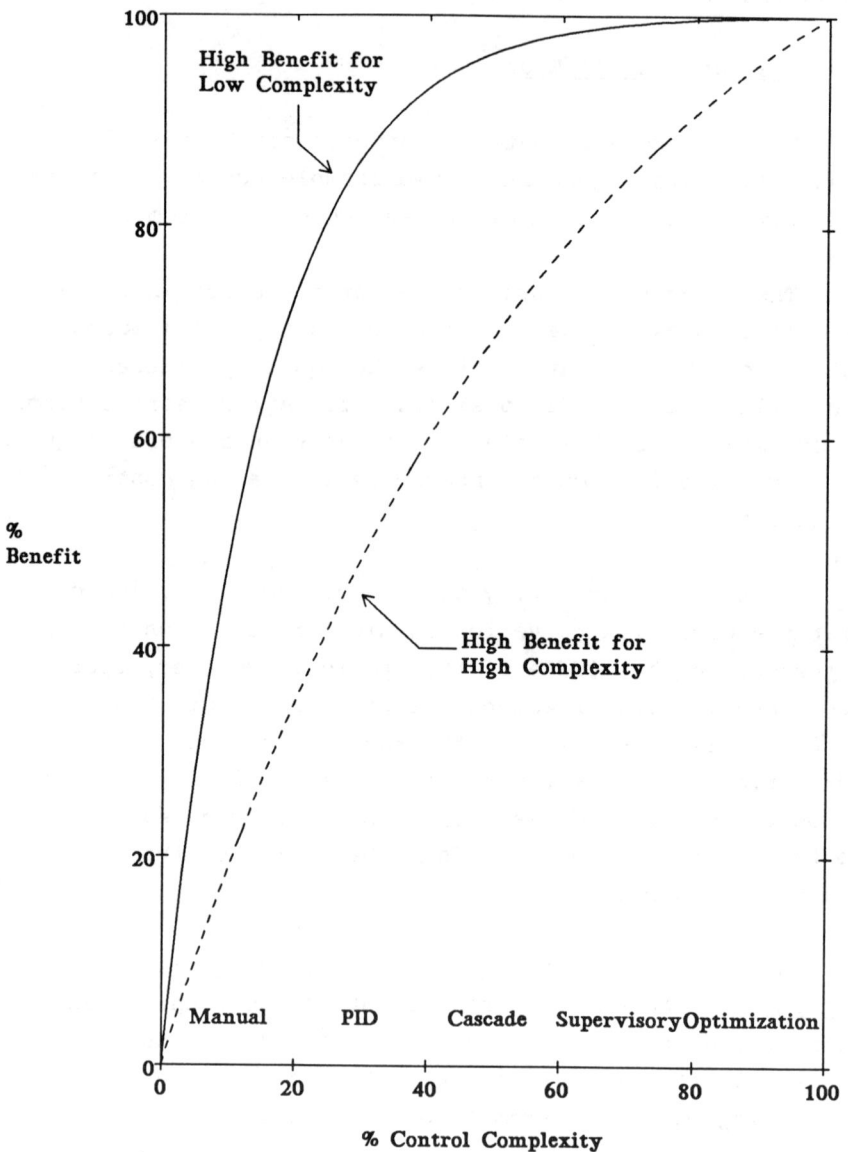

Fig. 1 Benefit vs. control complexity

Historically, the reactor control question was not one of economics but one of operability. Keeping the plant in an operable stable region was the primary focus of the process control system. In some situations today this is still the case and there is little economic incentive to do any more. As foreign competition has intensified and the process control systems have become more capable, the trend has shifted to automate our facilities more fully.

The question now becomes where to allocate the corporate effort for process control. Typically the answer is for production increases of sold out facilities, for improved quality to develop and maintain market share and for improved raw material yield for large volume businesses. Each of these three categories involves reactor control.

Consider the question of increasing production in a sold out market. Every pound squeezed out of the reactor can be sold for a profit. Experience indicates that increasing the yield is the single largest potential for improved control in existing facilities.

The yield problem may involve controlling the physical losses in raw material preparation, physical losses in reactor product handling (purge vent losses, etc.) and chemical losses to off specification product (not optimum reactor operating conditions).

Quality problems are also extremely important. The value of quality is often difficult to quantify and it may be the most important control problem of all.

Energy typically is not a major factor in the cost

benefit equation. The quality and yield economics often overshadow the cost of energy. Energy is, however, important from an operability standpoint. With plants that are highly integrated through heat balance or process recycle, reactor control becomes a much more complex problem. The control problem considerations then extend beyond the reactor area boundaries.

Manpower economics frequently surfaces as a potential cost saving area. Our studies have shown that improved control shifts manpower, it does not reduce it.

Summary of Reactor Control Issues

The above paragraphs have discussed the role that the reactor has in overall plant economics. As the business conditions dictate the need for improved control, Du Pont installations are evolving from the classic PTLF controls to more complex methods. The results of this migration effort have been varied. The following items briefly highlight our experience and some of the conclusions from data compiled in the following sections.

- There is a logical approach to the problem of reactor control design. This approach involves generating the best control system from an economic, maintenance and performance standpoint.

- Detailed analysis (i.e. simulation, system structure, and control concept) is frequently worthwhile. Straightforward implementation is essential for operability and maintenance. Analysis may show a simpler implementation could work better as in Longwell (1985).

- Advanced control methods do have applications in industry. Control problems involving nonminimum phase, drifting disturbances (with frequencies similar to process dynamic frequencies), noisy measurements, and process interactions are all too common. Industry needs ways to screen processes quickly and identify the problem areas where advanced methods can be useful.

- On-line (sampled) and in-line analyzers as well as intelligent use of lab data are becoming very important. Flow data reconciliation and Kalman filtering are being used with mixed success in the field. The technical issues are small compared to the maintenance issues.

- Feedforward control, for example ratio and cascade, are extremely important. Nontraditional methods of property control are also being used successfully.

- The use of optimization for yield improvement on large volume business is important. First pass quality control is critical for specialty and polymer systems.

The reactor control problem should be evaluated from an economic standpoint. Our experience has shown that good reactor control is critical for cost effective operation.

The following sections contain the data and results compiled for this study.

2 WHAT IS THE CURRENT STATE OF INDUSTRIAL REACTOR CONTROL?

Section 2 will review the literature as well as the opinions of individuals within Du Pont regarding reactor control system design.

Review of the literature

Introduction. This section will review the reactor control literature to evaluate where the state of the art is. This search, which contains 83 references, is not intended as an exhaustive search but as a fairly representative sample of the groups working in the field. An exhaustive search could easily yield several hundred references.

Our search, if biased at all, is biased towards the more complex problems. We would also like to apologize to any author whose paper may have been left out or erroneously categorized. As in any review paper, space does not permit a complete discussion of all the papers.

The literature review is summarized in Tables 1 to 5 (to be found in the Appendix) and Figures 2 to 7. These tables and figures try to categorize the papers with respect to whether it is theoretical or experimental, the type of reactor considered, whether traditional or nontraditional control techniques are discussed and the types of nontraditional analysis and synthesis techniques discussed. The tables can be used as a handy guide to find references on a particular topic. The figures show (in pie chart and bar chart form) the numbers and percentages of papers dealing with each topic. These figures will be useful in understanding the overview of the entire literature in this broad field.

Type of paper. Table 1 and Figure 2 characterize the types of papers in the review. The types are defined as university (67%), survey (17%), industrial (14%), or tutorial (2%). University or industrial papers denote the origin of the technical paper. Survey papers attempt to summarize the literature for a given time frame. Tutorials are similar to surveys, but are mainly instructive in nature.

Figure 2 shows that most of the papers are from universities. Very few papers are industrial, survey, or tutorial. We are pleased to see so many academic papers on this important topic. Coming from industry, it would be nice to see more industrial, survey, and tutorial papers. Companies, due to proprietary information protection, generally do not encourage publication for fear of revealing company secrets.

Theoretical or experimental. Table 2 and Figure 3 show whether the paper is theoretical (72%), experimental (12%), or both (16%). The number of theoretical papers is large compared to applications. It appears that the theory is driving the experiments. Another possibility is again, the proprietary information protection problem. Additionally, academic pilot scale equipment is generally expensive to set up and run. The most readable and applicable papers usually include both theory and experiment.

Type of reactor. Table 3 and Figure 4 show the types of reactors as continuous stirred tank reactor (CSTR) (46%), Fixed Bed (23%), Batch (13%), Tubular (8%), Semi-Batch (8%),

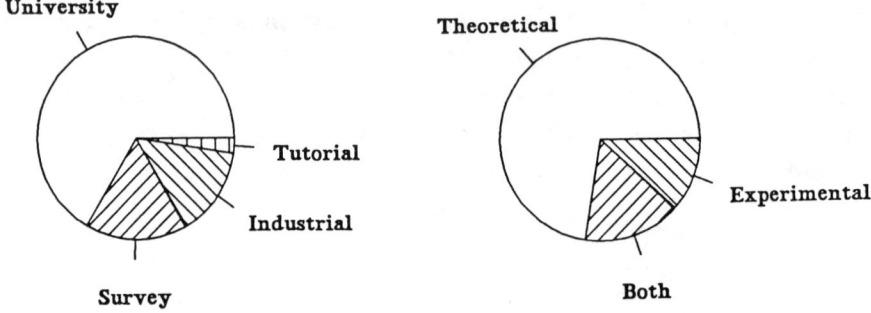

Fig. 2 Type of paper

Fig. 3 Type of literature

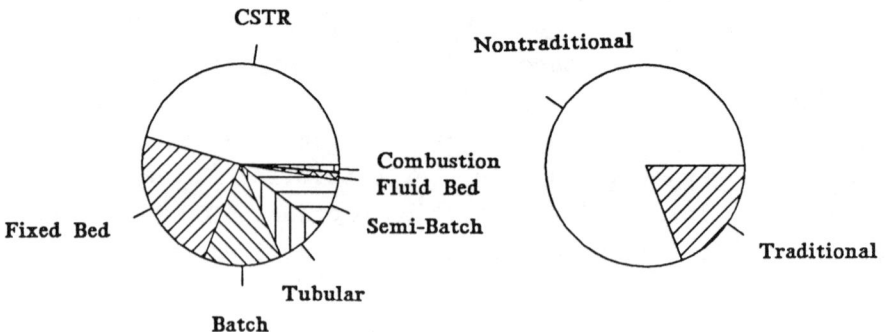

Fig. 4 Type of reactor

Fig. 5 Control techniques

Fluidized Bed (1%), or Combustion (1%). The definitions of the various reactors follow. A CSTR is a well mixed tank with inlet and outlet flows maintained during the reaction period. A semi-batch reactor has only inlet flows, and batch reactors have neither inlet or outlet flows after the initial charge. Fluidized bed reactors have gases passing a fluidized catalyst support. Tubular reactors consist of empty tubes while fixed beds are generally tubular with a catalyst bed. In combustion reactors the reactants are ignited upon entry.

The CSTR is a clear favorite for testing reactor control strategies. Fortunately, other reactors which are also industrially important have been studied in the literature.

Table 3 has a "General" category which is reserved for papers which, although relevant to the field, do not include a specific reactor type. This category was not included in Figure 4.

Traditional vs. nontraditional control techniques. Table 4 and Figure 5 show the split between those papers which discuss nontraditional (81%), and traditional (19%) techniques. The traditional techniques are those which have been routinely applied to reactor problems such as PID control of temperature. Nontraditional techniques control the reactor using unconventional tools such as linear quadratic gaussian (LQG). Many papers include both traditional control and nontraditional techniques such as Kwalik and Schork (1985).

The bulk of the literature surveyed is nontraditional. This is because most of the papers are from academia. By definition, academia is where the nontraditional research is being carried out. We feel that academia should invent and develop the nontraditional tools and hopefully make them "traditional". Most of the traditional papers were published before 1980 such as the pioneering work of Aris and Amundson (1958) and Nemanic, et al. (1959). However, this does not preclude interesting work continuing in the traditional area such as the paper by Cheung and Luyben (1979).

Types of nontraditional control techniques or issues. The papers involving nontraditional control are listed in Table 5 and Figures 6 and 7. Figure 6 shows the number of papers which used or discussed analysis techniques. Figure 7 shows the number of papers which used control design or synthesis techniques. Again, Table 5 shows the individual papers using each technique.

Figure 6 shows clearly that modeling and estimation (including state space modeling) are important techniques in nontraditional control analysis. Important issues are measurement considerations, system structure (what variable gets paired with what), maximum principle for servo problems and time delay considerations. Some of the other techniques and issues will gain importance such as singular value decomposition and flow data reconciliation.

Figure 7 indicates that many papers discuss control synthesis in terms of LQG, adaptive control, nonlinear model based control (using a nonlinear model in the controller), pole placement, optimal servo control, and steady state optimization. Other papers are beginning to include feedforward control, internal model control (IMC), and dynamic matrix control (DMC).

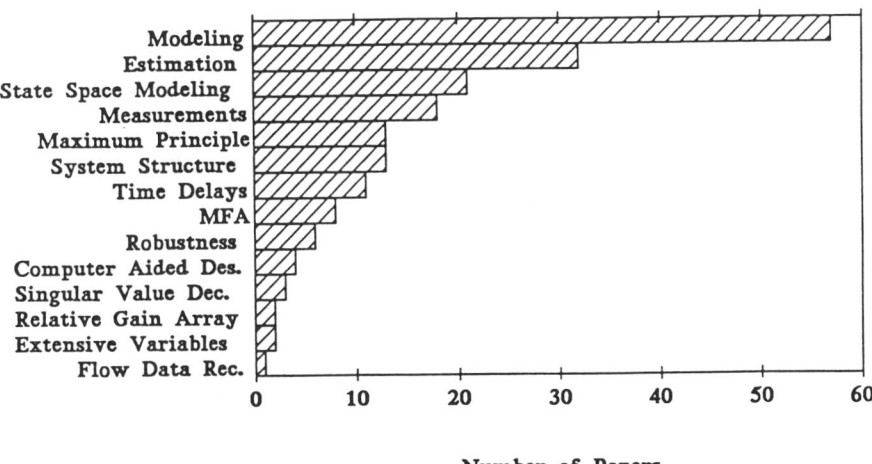

Fig. 6 Literature control analysis techniques or issues

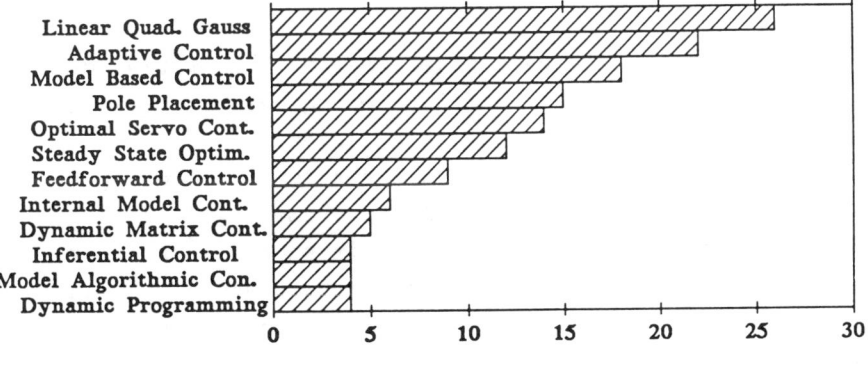

Fig. 7 Literature control synthesis techniques

The trend here is that the reactor is being looked at increasingly as a multivariable control problem. Also, modeling and a significant amount of analysis and model verification is being done from the beginning. Good examples of the current trend in academia are MacGregor, et al. (1983), Kwalik and Schork (1985), Penlidis, et al. (1985) and in industry Cardner (1984), and Ardell and Gumowski (1983). We feel this modeling approach for process understanding is worth the initial effort to obtain a good control design.

Review of In-house Survey

Introduction. Due to the shortage of industrial papers discussing matters relevant to the industrial reactor control professional, in-house interviews were conducted. The individuals were selected on the basis of their experience with installing actual control systems on reactors in Du Pont. The results of these interviews are shown in Figures 8 to 11. The interviews were divided into reactor control problems or issues and types of traditional or nontraditional solutions to these problems.

The individuals were asked to rank the problem or solution on a scale of 0 to 3, 3 being very supportive. The results for each question were averaged for all the individuals for that particular question. For example, for the problem of good analytical measurements, five individuals scored this a 3, and six scored it a 2 for an average score of 2.45. Therefore, these scores should be interpreted as a relative indicator of the importance of the problem.

Type of problems or issues. Figure 8 shows the type of reactor control problem or issue surveyed. The most important issues appear to be technical maintenance of the control system, general process understanding, yield, availability of good analytical measurements, reactor dynamics, and quality. Of less perceived importance are, simulation of the process, mass and energy integration, available hardware, and multivariable problems.

Interestingly, the two major items are not process control design issues but rather support issues (technical maintenance and general process understanding). Once an advanced control system is devised, it needs to be maintained in order to reap the economic benefit. Also, good process knowledge, as opposed to a "black box" approach, is important.

Yield and quality are important from a business viewpoint to decrease costs and retain or develop market share. However, they are often difficult to measure. Analytical measurements of reactor effluent properties are typically performed by gas chromatographs (GC's) or lab analysis of samples. This is a problem because they are either nonexistent, inaccurate, or have a significant time delay. We feel that much work can be done in developing and improving on-line measurement techniques both in academia and industry.

Available hardware seems adequate for most calculational needs, particularly if a host computer is installed.

Multivariable control does not rank as an important issue in this part of the survey. This result is contrary

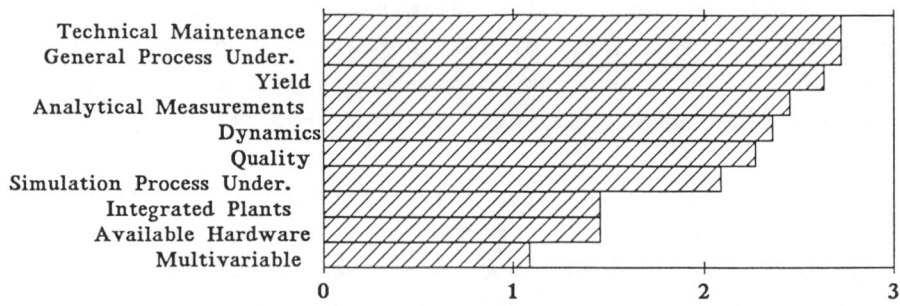

Fig. 8 Survey of problems vs. importance of problem

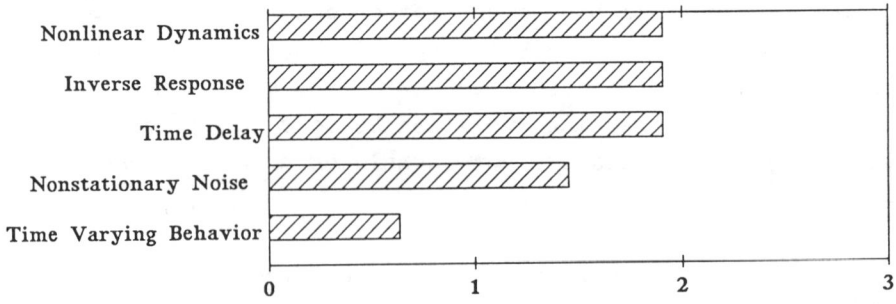

Fig. 9 Survey of dynamic problems vs. importance of problem

to the results of the literature survey. This may be because of the lack of analytical techniques in many cases. If the property cannot be measured, then it cannot be perceived as a control problem. Alternatively, many industrial multivariable problems are solved using conventional techniques.

Figure 9 shows a breakdown of the importance of various dynamic problems. The most important appear to be nonlinear dynamic behavior, inverse response, and time delays. Slow nonstationary noise and slow time varying behaviors are not perceived as important dynamic problems.

Types of traditional solutions. Figure 10 shows the relative importance of traditional solutions to reactor control problems. The most important traditional solution is getting good conventional measurements of temperature, pressure, level and flow. Also, feedforward and feedback control techniques are widely used. They are usually implemented in conventional control hardware. Overrides are sometimes used for critical situations.

The typical approach the designer takes for traditional types of problems is first to make conventional measurements. Then, the procedure is to design feedforward techniques to eliminate disturbances in feed flows, cooling water temperature, etc. Next, feedback control of reactor temperature, pressure and level is designed. Finally, constraints are handled using interlocks and overrides if necessary. Property control is usually accomplished by the operator changing the setpoints of this control system as property data from on-line instruments (such as GC's) or the lab becomes available.

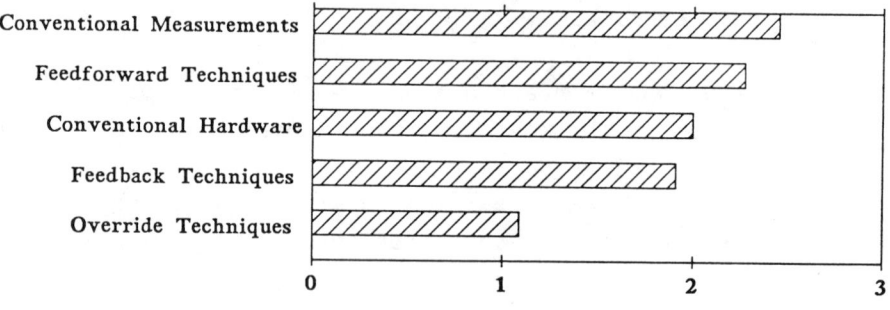

Fig. 10 Traditional solutions vs. importance of solution

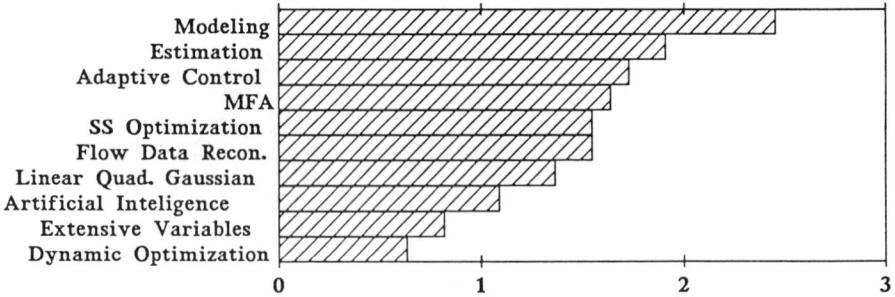

Fig. 11 Nontraditional solutions vs. importance of solution

Types of nontraditional solutions. If the above traditional solutions are not adequate to solve a difficult control problem, nontraditional solutions are applied. Figure 11 shows some of the nontraditional analysis and synthesis solutions. The most important tools are modeling and estimation followed by adaptive control, multivariable frequency analysis (MFA), and steady state optimization. Other techniques listed are lower on the relative importance scale such as flow data reconciliation and LQG. Due to the recent need for higher quality product, more of these techniques, such as LQG, will be utilized. We feel that some of the nontraditional solutions suggested, such as artificial intelligence, will be important in the future.

So, with a difficult problem, the designer will typically use modeling, estimation, MFA, adaptive control, and steady state optimization. We do not claim these are the only methods used. In fact, model based steady state control (which was not part of the survey) is also used to control fast reactors.

Differences Between Academia and Industry

Some differences do exist between academia and industry as shown by comparing the literature survey to the Du Pont control professionals survey. In new plants, for example, the traditional approach is to learn how to control a pilot plant, then design the equipment to get around the control problem. Unfortunately, this can lead to overdesign. Addition of large blending tanks to decrease product property variance is an example of this technique. This is not always the best policy, particularly if rapid product changes are desired.

Academia tends to emphasize the feedback control problem. This is because they use simulation of single pass, single unit reactors by themselves. These simulations tend to have continuous feedback of all variables with little or no time delay. Problems of recycle and interaction between units do not arise. Pathological types of disturbances as well as start-ups and shut downs are deliberately eliminated. Feedforward control is a tool heavily used in industry to solve some of these disturbance problems. Overrides and interlocks are used in start-up and shut down strategies. Nonlinear model based control strategies are also becoming important in industry.

3   SURVEY OF IN-HOUSE REACTORS

The industrial environment is the best place to separate the wheat from the chaff. The industrial application determines which control techniques work and which are too complex to maintain or too sensitive to keep in operation. If a control scheme has any merit (from an economic or improved operability standpoint), it will usually stay in service. If a control system makes the job of the operators or the mechanics more difficult for little or no perceived benefit, it will not be used.

In order to understand which types of reactor control schemes have been successful, a survey of installed reactor controls was taken. This survey will show why different reactor types, economics, dynamics, etc. dictate the method used to solve the control problem.

The following section is a review of 21 different industrial reactor control applications. Duplicate applications are not counted (only single "types") so this survey actually covers some 31 reactors. To be fair, if the survey is biased at all, it is biased in the direction of the more difficult reactor problems.

Reactor Characteristics

Let us start by reviewing some of the important properties and statistics concerning the reactors surveyed.

Reactor business type. Figure 12 shows the break down of large volume versus specialty products. Large volume implies a product made using large amounts of feedstocks and energy. Specialty implies a product made in relatively small quantities with a large profit margin. Most reactors in the survey (62%) produce product in large volume. This may currently be the largest sector. However, we feel the process control opportunities of the future will be in the specialty areas. As will be seen in the summary the large volume/specialty distinction has a major impact on the types of control required.

Reactor economic factors. Figure 13 shows the major economic incentive driving the control applications. The categories include:

Yield (61%). Maximizing a desired product from the minimum amount of feedstock.

Quality (25%). Controlling specific parameters of the product within some previously defined limits.

Energy (11%). The heat required or released by the reaction has an effect on cost or operability of

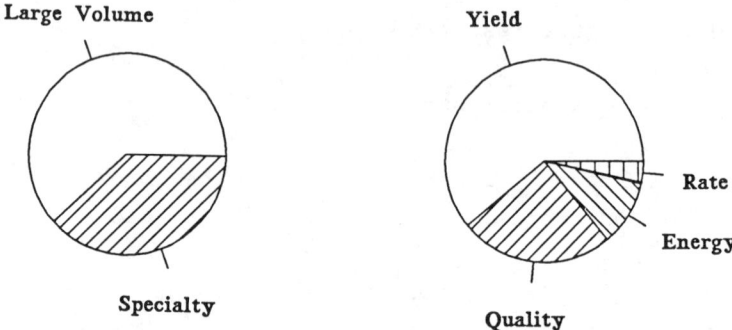

Fig. 12 Reactor business type

Fig. 13 Reactor economic factors

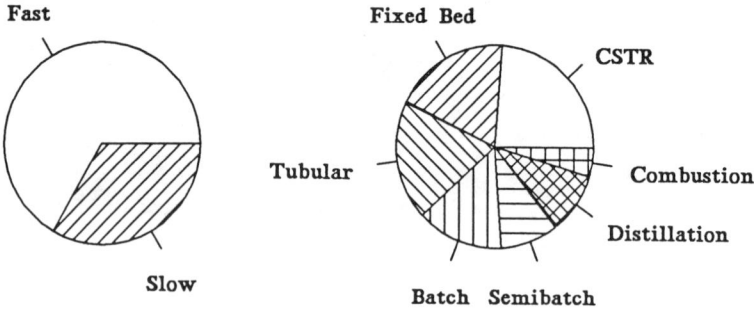

Fig. 14 Reactor dynamics

Fig. 15 Reactor types

reactor and other plant operations.

Rate (3%). Controlling rate to maximize production or match other area rates.

The figure shows that yield and quality are top considerations. Recent studies have shown that improved yield for sold out facilities is the largest single economic incentive. Quality is extremely important for market share but difficult to put a dollar figure on.

It is interesting to note that for reactors the cost of energy is typically not an important economic factor. It may have a significant impact on operability.

Reactor dynamics. Figure 14 indicates the percentage of the reactor systems tallied that have fast dynamics (67%). Fast dynamics means that there is high conversion in less than 10 minutes.

Fast reactor dynamics usually implies that steady state reactor property control is satisfactory. Manual control is typically not adequate for these processes because they are disturbed more frequently than operators can manage. Property control is attempted only on a slow basis (greater than 10 minutes) because, as mentioned previously, a feedback measurement of effluent properties is not often available faster than 1/5 of the dominant time constant. However, if a continuous feedback measurement is available, the decision to control dynamically or use steady state control depends on the frequency of the disturbances relative to the process time constant.

Process dynamics, frequency of property measurements,

and frequency of disturbances play a primary role in determining the type of control that can be used successfully.

Reactor types. Figure 15 gives a break down of the different reactor configurations that were in the survey. As the reader can observe, there is quite a variety. The definitions are covered in Section 2.

The figure shows that the most important reaction types are CSTR (24%), fixed bed (19%), tubular (19%), and batch (14%). The other categories were semi-batch (10%), distillation (10%), and combustion (4%). Reactor type plays a very important role in control system design. The literature review appears to have approximately the same equipment distribution as shown in Figure 4.

Reactor fluid phase. As shown in Figure 16, the majority of the reactions studied were liquid phase (62%). In liquid phase reactions, inventories typically determine reactor dynamics and thus control strategy. Reactors with large residence time need something better than steady state control. Gas phase reactions are common but almost always have small holdups and fast dynamics. Steady state control is often adequate for these reactors.

There are other types of reaction processes (e.g. gas/solid and even a gas/solid/liquid rotary kiln). They each have their own unique problems, but fortunately only constitute a small portion of our reactors.

Heating or cooling of reactors. Figure 17 shows a break down for the heating and cooling techniques used for the reactors in this survey. The majority of the reactors

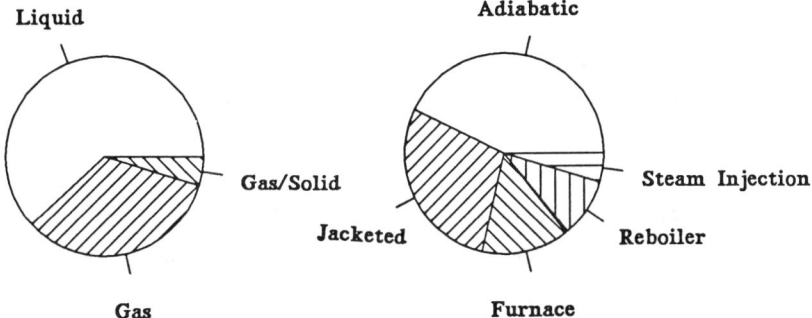

Fig. 16 Reactor fluid phase   Fig. 17 Reactor heating or Cooling

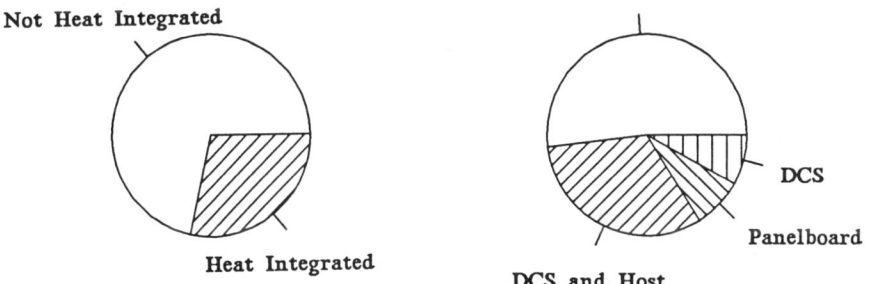

Fig. 18 Reactor heat integration   Fig. 19 Reactor control hardware

(43%) are adiabatic meaning little or no heat transfer through the reactor wall which makes reactor design simpler. Many types of reactors can fit into this class.

The next major category is the jacketed reactors. These are typically liquid phase reactions with slow dynamics. Dynamic temperature and product property control are frequently major concerns for this type of reactor.

The fired heater or furnace type reaction process presents another set of control problems and challenges. In addition to the normal reactor control concerns of temperature and composition, now the control engineer has to be aware of problems associated with combustible materials, whether feedstocks or fuels (i.e. burner codes, etc.).

Reactor heat integration. Figure 18 shows the percentage of non-heat integrated reactors (68%). The heat integrated reactor produces or consumes heat from other parts of the process. For this reactor the control problem becomes more complex. The boundaries of the problem have to go beyond the reaction area to consider the plant energy balance. Plant wide information systems and frequently optimization techniques are required to handle this situation.

Reactor control hardware. Figure 19 gives a breakdown of the types of control hardware.

Panelboard (8%). Analog, electronic, or pneumatic.

DCS (8%). Distributed control systems.

Panelboard/Host (52%), DCS/Host (32%). Panelboard or

DCS with a general purpose computer for data logging, accounting, calculations, supervisory control, etc.

The majority of the reactors surveyed have a panelboard/host or DCS/host configuration. As the DCS functionality improves, we may start to see more stand-alone DCS systems.

The current trend is to put modern control hardware on our reactors. We are not implying, however, that all reactors should have DCS or host systems. The decision is an economic one as was discussed previously.

Special reactor problems. There are frequently special considerations that play a very important role as part of chemical reactor control. These problems are often as important or more important than the economic considerations. Figure 20 shows the distribution and variety of problems found in our survey. Run away reactions (48%), toxic materials (26%), corrosion (13%), tube melt down (10%), and impurities (3%) all are important special problems.

Most of these special considerations are involved with safety. Many safety concerns affect the regulatory control level (measurements, valve speed, the need for feedforward, etc.) and basic reactor design.

Interlocks also play an important role in reactor control. They have their own classification and hierarchical structure; personnel safety (hardwired and/or redundant software), equipment protection (hardware depends on mechanical backup - rupture disc, etc.) and production/environmental protection. This is a very

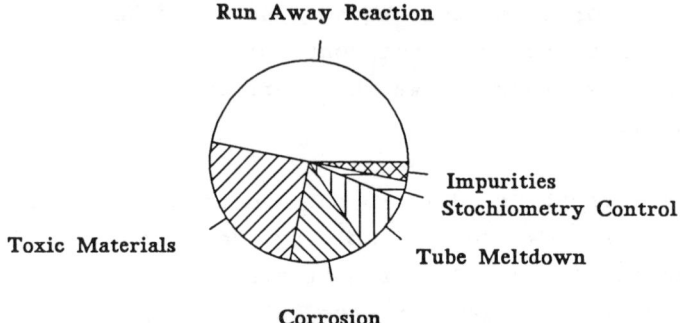

Fig. 20 Special reactor problems

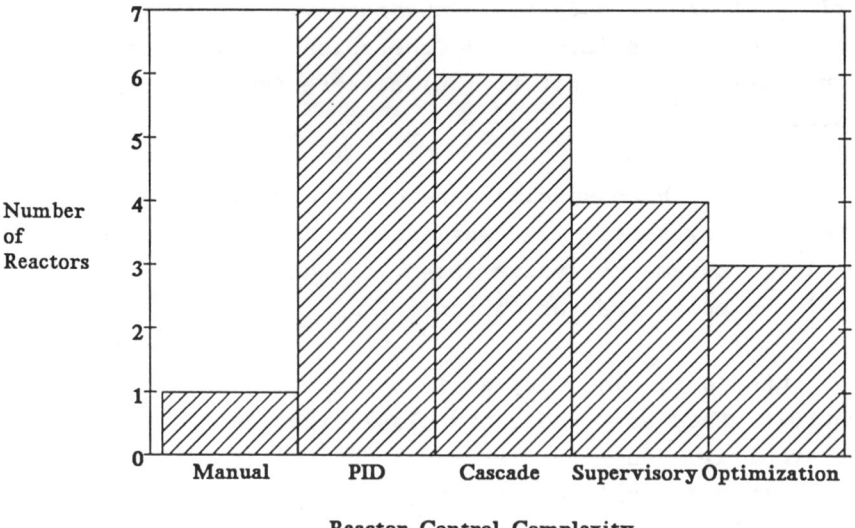

Fig. 21 Number of reactors vs. reactor control complexity

important subject but beyond the scope of this paper.

Level of Control Complexity

The reactor survey contained another important statistic. It assigned a level of control complexity to each reactor system. This complexity is shown as a function of frequency on Figure 21.

Figure 21 suggests that most of our plant reactor control schemes fall into the PID (33%) or cascade (29%) control level. The supervisory (19%) and optimization (14%) level are the next most frequent schemes. The manual type of operation exists in 5% of the reactors surveyed.

This data helps explain Figure 1, Benefits vs. control complexity. Figure 1 postulates a large incentive for getting regulatory loops in automatic. Figure 21 shows that the majority of our processes are PID level or above. The individual reactor results show examples that follow the upper and lower curves of Figure 1.

The upper curve of Figure 1 shows a declining economic incentive for upper level control. For the reactors which exhibit behavior like the upper curve, supervisory and optimization schemes cannot be justified economically.

The lower curve of Figure 1 shows that to get the full economic benefit, more complex levels of control are required. Figure 21 indicates that several of the reactors fall into this category.

In addition to economic factors it is important to point out that the more complex control system requires better measurements, technical support, and maintenance to reap the economic benefits. This suggests that manpower allocation problems may keep some of these advanced applications from being realized.

<u>Control complexity vs. reactor characteristics</u>. The reactor survey categories have been correlated against level of control complexity. Figure 22 shows the categories (or classifications) which have a 80% or higher correlation with a given level of complexity.

Starting from the top, OPTIMIZATION correlates with large volume businesses, yield, heat integrated, and host computer. This appears to be consistent with where one might think optimization could provide some economic benefit. Getting the most out of feedstocks (yield) and optimizing energy utilization from a heat integration standpoint is important. Obviously a host computer system is necessary for this function.

SUPERVISORY control correlates with non-heat integration, host computer, liquid phase, and run away reactions. Supervisory control appears to be important for liquid phase type reactions. This implies that it is important for reactors which have slower dynamics (i.e. liquid phase). These reactions are typically exothermic (i.e. run away) but usually do not contribute enough waste heat to be heat integrated. Again, a host computer is important for control.

CASCADE control correlates with yield and host computer. The correlation starts to break down at this level. The implication here is that cascade control is important for yield. The authors would argue that it is

Reactor Characteristics	Control Complexity				
	Manual	PID	Cascade	Supervisory	Optimization
Business Type					
Large Volume					X
Economic Importance					
Yield			X		X
Quality				X	
Heat Integration					
No		X		X	
Yes					X
Hardware					
Panel/DCS		X	X	X	X
Host Computer			X	X	X
Phase					
Liquid				X	
Special Problems					
Run Away				X	

Fig. 22 Control complexity vs. reactor characteristics

also important for quality. The host computer correlation suggests that our computers are being used to set panel or DCS controller setpoints. Conventional cascade control is done in stand-alone analog and DCS.

Conclusions of Reactor Survey

Several general observations can be made from the survey data.

- The many different types of reactions and reactor configurations lead naturally to different solutions. One control strategy cannot be replicated over and over as in other industries.

- Academia seems to be studying the right mix of reactor types for the chemical industry. More work could be done on combustion and distillation reactors.

- Reactor control schemes are driven by economics, maintenance concerns, and available technology.

4 REACTOR CONTROL TECHNIQUES

This section will summarize a logical approach to the reactor control problem as suggested by this survey. Figure 23 shows a general reactor control scheme. The reactor control design then becomes the problem of choosing the proper techniques that fit into the boxes of Figure 23. These must be chosen to do the best job from an economic, maintenance, and performance standpoint. The survey suggests that the boxes could contain the following techniques, listed in order of "popularity":

- **Measure**

Conventional Measurements (PTFL)
On-line Analyzers (GC, pH, IR, Mass spectrometer, Density, etc.)
Lab Analysis

- **Signal Condition**

Exponential Filtering
Flow Data Reconciliation
Kalman Filtering

- **Estimator**

Simple Calculations of Intermediate Variables
State/Output Observers
Kalman Filters
Complex Dynamic Model Calculations

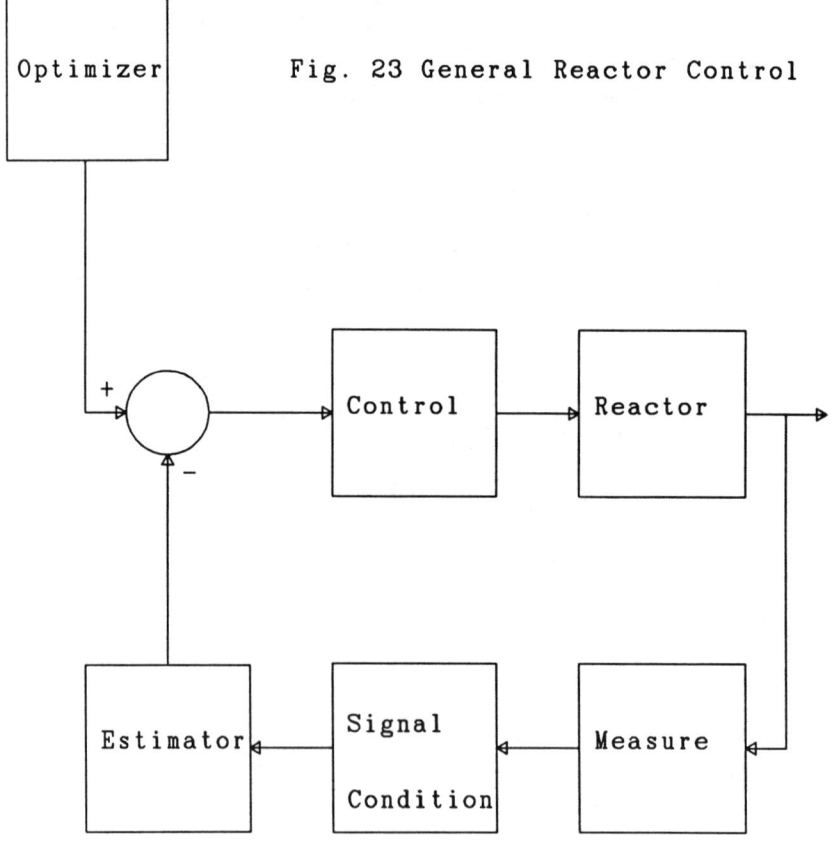

Fig. 23 General Reactor Control

- Control

Feedforward
Conventional Feedback (PID)
Overrides
Adaptive/Nonlinear Feedback
Nontraditional Feedback
    Steady State Control
    Multivariable Frequency Analysis
    Nonlinear Model Based Control
    Linear Quadratic Gaussian
    Pole Placement
    Internal Model Control
    Dynamic Matrix Control
    Inferential Control
    Model Algorithmic Control
    Optimal Servo Control
    Dynamic Programming
    Linear Programming

- Optimizer

Model Based Optimization
    Linear Programming
    Nonlinear Programming
Evolutionary Optimization (EVOP)

CONCLUSIONS

The reactor control problem should be evaluated from an economic standpoint. There is a need to get the regulatory loops in automatic. Often there is a stake for upper level property control and yield optimization.

Our experience has shown that reactor control is critical for cost effective operation.

The trend in the literature and industry is to view reactor control as a multivariable problem. Modeling and control system analysis are becoming standard reactor control design methods. We feel this modeling approach for process understanding is worth the additional effort to obtain good control.

Two major reactor control problems are technical maintenance and general process understanding. Once an advanced control system is devised, it needs to be maintained in order to reap the economic benefit.

Problems of recycle, interaction between units, pathological disturbances, as well as start-ups and shut downs are important industrial problems. Feedforward control is a tool heavily used to solve these disturbance problems. Overrides and interlocks are used in start-up and shut down strategies. Nonlinear model based control strategies are also becoming important in industry.

ACKNOWLEDGEMENTS

The authors would like to thank the following individuals for their help: D. V. Cardner, J. P. Congalidis, R. K. Cox, W. R. Ellingsen, T. J. Harris, D. Y. Ko, E. J. Longwell, R. L. Moore, M. C. Rominger, J. P. Shunta, and W. D. Smith, Jr.

REFERENCES

Ahlberg, D. T., and I. Cheyne (1979). Adaptive control of a polymerization reactor. <u>AIChE Sym. Ser.</u>, <u>159</u> (72), 221.

Ahn, Y. O., K. Y. Park, and T. K. Park (1983). Synthesis of vinyl chloride/acrylonitirile copolymer of constant composition. <u>PACHEC II</u>, 64.

Amrehn, H. (1977). Computer control in the polymerization industry. <u>Automatica</u>, <u>13</u>, 533.

Ardell, G. G., W. B. Field, and W. E. Crockett (1979). Practical features for control of multivariable systems in a real environment. <u>Ind. Process Control Proc. Workshop</u>, 106.

Ardell, G. G., and B. Gumowski (1983). Model prediction for reactor control. <u>Chemical Engineering Progress</u>, June, 77.

Aris, R., and N. R. Amundson (1958). An analysis of chemical reactor stability and control. <u>Chem. Eng. Sci.</u>, <u>7</u>, 121.

Arnold, K., A. F. Johnson, and J. Ramsey (1980). Adaptive control of the MWD in living anionic polymerization processes. <u>IFAC PRP 4</u>, 359.

Asbjornsen, O. A. (1984). Challenges in modern process control. <u>Comp. and Chem. Eng.</u>, <u>8</u>, 5, 275.

Bailey, J. E. (1972). On the theory of dynamic plug-flow processes: inlet control and distributed trajectory control. <u>Can. J. of Chem. Eng.</u>, <u>50</u>, February, 108.

Bonvin, D., R. G. Rinker, and D. A. Mellichamp (1983). On controlling an autothermal fixed-bed reactor at an unstable state-I. <u>Chem. Eng. Sci.</u>, <u>38</u>, No. 2, 233.

Bruns, D. D., and J. E. Bailey (1977). Nonlinear feedback control for operating a nonisothermal CSTR near an unstable steady state. <u>Chem. Eng. Sci.</u>, <u>32</u>, 257.

Cardner, D. V. (1984). Model inferential optimizing computer control of three series reactors. <u>ISA Conf.</u>, Houston.

Cheung, T. F., and W. L. Luyben (1979). P D control

improves reactor stability. <u>Hydrocarbon Proc.</u>, Sept., 215.

Clement, K., and S. B. Jorgensen (1981). Experimental investigation of a fixed bed chemical reactor control system designed by the direct Nyquist array method. <u>8th IFAC Conf.</u>, Kyoto, 2757.

Clough, D. E., P. M. Masterson, and S. R. Payne (1978). Computational problems in the determination of control policies for batch polymerization. <u>Proc. Summer Computer Simulation Conf.</u>, Los Angeles, 279.

Denn, M. M. (1972). Chemical reaction engineering. Optimization, control, and stability. <u>I and EC Annual Review</u>, 251.

Devia, N., and W. L. Luyben (1978). Reactors: Size versus stability. <u>Hydrocarbon Proc.</u>, June, 119.

Doss, J. E., T. W. Doub, J. J. Downs, and E. F. Vogel (1983). New directions for process control in the eighties. <u>AIChE Meeting</u>, Washington, D.C..

Edgar, T. F. (1980). Advanced control strategies for chemical processes: a review. <u>ACS Symp. Series</u>, <u>124</u>, 89.

Ellingsen, W. R. (1979). Implementation of advanced control systems. <u>AIChE Symposium Ser.</u>, <u>159</u>, 150.

Foss, A. S. (1973). Critique of chemical process control theory. <u>AIChE J.</u>, <u>19</u>, No. 2., 209.

Foss, A. S., J. M. Edmunds, and B. Kouvarltakis (1980). Multivariable control system for two-bed reactors by the characteristic locus method. <u>Ind. Eng. Chem. Fundam.</u>, <u>19</u>, 109.

Fox, J. M., W. J. Schmidt, Jr., and J. C. Kantor (1984). Comparison sensitivity and feedback control for optimized chemical reactors. <u>Proc. Amer. Control Conf., San Diego</u>, <u>3</u>, 1621.

Garcia, C. E. (1984). Quadratic/dynamic matrix control of nonlinear processes. An Application to a Batch Reaction Process. <u>Annual AIChE Meeting, San Franscisco.</u>

Garcia-Rubio, L. H., J. F. MacGregor, and A. E. Hamielec

(1982). Modeling and control of copolymerization reactors. ACS Symp. Series, 197, 87.

Georgakis, C., R. Aris, and N. R. Amundson (1977). Studies in the control of tubular reactors, parts I-III. Chem. Eng. Sci., 32, 1359.

Hallager, L. and S. B. Jorgensen (1981). Experimental investigation of self-tuning control of a gas phase fixed bed catalytic reactor with multiple inputs. 8th IFAC World Congress, Kyoto, Japan.

Hammarstrom, L. G. (1979). Control of chemical reactors in the subspace on reaction and control variants. Chem. Eng. Sci., 34, 891.

Harris, G. H., and L. Lapidus (1967). The identification of nonlinear systems. Ind. and Engg. Chem., 59, No. 6, 67.

Hashimoto, I., and T. Takamatsu (1981). New results and the status of computer-aided process control systems design in Japan. 2nd Eng. Foundation Conf., Sea Island, Georgia, 147.

Hopkins, B. (1979). Control of polymerization reactors. Proc. Ind. Process Control, Proc. Workshop, 1.

Jo, J. H., and S. G. Bankoff (1976). Digital monitoring and estimation of polymerization reactors. AICHE J., 22, No. 2, 361.

Johnson, A. R., B. Khaligh, and R. Ramsay (1982). Copolymerization reaction engineering. ACS Symp. Ser., 197, 117.

Jutan, A., J. P. Tremblay, J. F. MacGregor, and J. D. Wright (1977). Multivariable computer control of a butane hydrogenolysis reactor: parts I-III. AIChE J., 23, No. 5, 732.

Jutan, A., and A. Uppal (1984). Combined feedforward-feedback servo control scheme for an exothermic batch reactor. Ind. Eng. Chem. Process Des. Dev., 23, 597.

Kiparissides, C. and S. L. Shah (1980). Adaptive control of a polymer reactor. 4th IFAC PRP Conf., Ghent, 359.

Kumar, S., and J. H. Seinfeld (1978). Optimal location of measurements in tubular reactors. Chem. Eng. Sci., 33,

1507.

Kuruoglu, N., W. F. Ramirez, and D. E. Clough (1981). Distributed parameter estimation and identification for systems with fast and slow dynamics. Chem. Eng. Sci., 26, 1357.

Kwalik, K. M., and F. J. Schork (1985). Adaptive control of a continuous polymerization reactor. 4th Amer. Control Conf., 872.

Lapidus, L., and R. Luus (1967). The control of nonlinear systems. Parts I-III. AIChE J., 13, No. 1, 101.

Lappinga, A., and A. Foss (1984). Rapid set point attainment of reactor feed preheat system and coordination with reactor control. Proc. Amer. Control Conf., San Diego, 3, 1602.

Lee, K. S., and W. Lee (1983). Discrete-time multivariable adaptive control of a nonadiabatic fixed bed reactor. 3rd PACHEC, 147.

Leffew, K. W., and P. B. Deshpande (1981). A simulation study on the use of a dead-time compensation algorithm for closed-loop conversion control of continuous emulsion polymerization reactors. ACS Sym. Ser., 165, 533-565.

Lim, H. C., and R. J. Fang (1975). Optimal feedback control of a class of linear tubular process. AIChE J., 18, No. 2, 282.

Longwell, E. J. (1985). Temperature control of a cooling limited reactor; a complex control problem solved with a simple control system design. ISA Conference, Philadelphia, p. 9.

Luyben, W. L. (1974). Effect of reaction rate on the open loop stability of chemical reactors. AIChE J., 20, No. 1, 175.

Luyben, W. L. (1975). Batch reactor control. Instrum. Tech., 8, 27.

Lynch, E. B. and W. F. Ramirez (1975). Real-time optimal control of a stirred tank reactor using Kalman filtering for state estimation. AIChE J., 21, No. 4, 799.

MacGregor, J. F., A. Penlidis, and A. E. Hamielec (1983). Control of polymerization reactors. 5th IFAC PRP Conf., Antwerp.

MacGregor, J. F., and P. W. Tidwell (1980). Modelling and control of continuous industrial polymerization reactors. ACS Symp. Series, 124, 251.

Mandler, J. A., D. M. Strand, Khanna, R., and J. H. Seinfeld (1984). Control of a packed bed reactor with feed-effluent heat exchange. Proc. Amer. Control Conf., San Diego, 3, 1608.

Marini, L., and C. Georgakis (1984). Low-density polyethylene vessel reactors. Part II: A novel controller. AIChE J., 30, No. 3, 409.

McDermott, P. E., D. A. Mellichamp, and R. G. Rinker (1984). Multivariable self-tuning control of a tubular autothermal reactor. Proc. Amer. Control Conf., San Diego, 3, 1614.

McGreavy, C. (1983). On-line computer control system for chemical reaction processes. Comp. and Chem. Eng., 7, No. 4, 529.

Morari, M., Y. Arkun, and G. Stephanopoulos (1980). Studies in the synthesis of control structures for chemical processes. AIChE J., 26, No. 2, 220.

Mutharasan, R., and D. R. Coughanowr (1976). Sampled-data proportional control of a flow-forced tubular reactor. Ind. Eng. Chem., Proc. Des. Dev., 15, No. 1, 141.

Nemanic, D. J., J. W. Tierney, R. Aris, and N. R. Amundson (1959). An analysis of chemical reactor stability and control - IV. Chem. Eng. Sci., 11, 199.

Padmanabhan, L., and L. Lapidus (1977). Control of chemical reactors. In L. Lapidus, and N. R. Amundson (Ed.), Chemical reactor theory. Prentis-Hall, 814.

Parulekar, S. J., J. M. Modak, and H. C. Lim (1985). Optimal control of fed-batch bioreactors. 4th Amer. Control Conf., 849.

Penlidis, A., J. F. MacGregor, and A. E. Hamielec (1985). Continuous emulsion polymerization reactor control. 4th Amer. Control Conf., 878.

Ponnuswamy, S. R., S. L. Shah, and C. Kiparissides (1985). Optimal feedback control of a batch polymerization reactor. <u>4th Amer. Control Conf.</u>, 860.

Ray, W. H. (1972). Fixed-bed reactors; dynamics and control. <u>5th Chem. React. Eng. Eur. Symp., Amsterdam.</u>

Ray, W. H. (1981). New approaches to the dynamics of nonlinear systems with implications for process and control system design. <u>2nd Eng. Foundation Conf., Sea Island, Georgia</u>, 245.

Ray, W. H. (1983). Multivariable process control - a survey. <u>Comp. and Chem. Eng.</u>, <u>7</u>, No. 4, 367.

Ray, W. H. (1985). On-line monitoring, estimation, and control of polymerization reactors. <u>4th Amer. Control Conf.</u>, 842.

Rijnsdorp, J. E., and D. E. Seborg (1976). A survey of experimental applications of multivariable control to process control problems. <u>AIChE Symp. Series</u>, <u>159</u>, 112.

San, K., and G. Stephanopoulos (1983). Optimal control policy for substrate inhibited kinetics with enzyme deactivation in an isothermal CSTR. <u>AIChE J.</u>, <u>29</u>, No. 3, 417.

Schnelle, P. D., Jr., (1979). Continuous stable linear estimation of chemical process variable using deterministic interactive dynamic models. <u>ISA, Chicago.</u>

Schork, F. J., and Ray, W. H. (1981). On-line monitoring of emulsion polymerization reactor dynamics. <u>ACS Symposium Series</u>, <u>165</u>, 505.

Schuler, H. (1980). Estimation of states in a polymerization reactor. <u>4th IFAC PRP Conf., Ghent</u>, 369.

Seborg, D. E., S. L. Shah, and T. F. Edgar (1983). Adaptive control strategies for process control: a survey. <u>AIChE meeting, Washington.</u>

Seinfeld, J. H. (1969). Optimal control of a continuous stirred tank reactor with transportation lag. <u>Int. J. Control</u>, <u>10</u>, No. 1, 29.

Shunta, J. P. (1981). Control systems for complete plants - the industrial view. 2nd Eng. Foundation Conf., Sea Island, Georgia, 547.

Stephanopoulos, G. (1983). Synthesis of control systems for chemical plants - a challenge for creativity. Comp. Chem. Eng., 7, No. 4, 331.

Tatkar, V. K., R. E. Gilbert, and D. C. Timm (1985). Modern and classical control techniques applied to an anionic polystyrene reactor. 4th Amer. Control Conference, 855.

Timm, D. C., R. E. Gilbert, T. T. Ko, and M. R. Simmons (1982). Control of an isothermal polystyrene reactor. ACS Symp. Series, 197, 1.

Tsoukas, A., and M. Tirrell, (1982). Multiobjective dynamic optimization of semibatch copolymeriztion reactors. Chem. Eng. Sci., 37, 12, 1785.

Uppal, A., W. H. Ray, and A. B. Poore (1974). On the dynamic behavior of continuous stirred tank reactors. Chem. Eng. Sci., 29, 967.

Vakil, H. B., M. L. Michelsen, and A. S. Foss. Fixed-bed reactor control with state estimation. Ind. Eng. Chem. Fundam., 12, No. 3, 328.

Viswanadham, N., L. M. Patnaik, and I. G. Sarma (1979). Robust multivariable controllers for a tubular ammonia reactor. Trans. ASME, 101, Dec., 290.

Wei, J. (1978). The dynamics and control of coal gasification reactors. Proc. Jt. Autom. Control Conf., 2, 39.

Weiss, M. D. (1984). On line analyzers, key to effective chemical process control. Proc. Amer. Control Conf., San Diego, 1, 348.

Windes, L. C., W. H. Ray, and A. Cinar (1984). Dynamic estimation of temperature and concentration profiles for control of a packed bed reactor. Annual AIChE Meeting, San Francisco.

Yoo, Y. J., J. Hong, R. T. Hatch (1985). Sequential estimation of states and kinetic parameters and optimization of fermentation processes. 4th Amer. Control Conf., 866.

APPENDIX

TABLES 1 TO 5

TABLE 1 Type of Paper

University	University (Continued)	Industrial	Survey and Tutorial (*)
Ahn, et al. (1983)	Lim and Fang (1975)	Ahlberg and Cheyne (1979)	Asbjornsen (1984)
Aris and Amundson (1958)	Luyben (1974)	Amrehn (1977)	Denn (1972)
Arnold, et al. (1980)	Lynch and Ramirez (1975)	Ardell, et al. (1979)	Doss, et al. (1983)
Bailey (1972)	MacGregor and Tidwell (1980)	Ardell and Gumowski (1983)	Edgar (1980)
Bonvin, et al. (1983)	Mandler, et al. (1984)	Cardner (1984)	Foss (1973)
Bruns and Bailey (1977)	Marini and Georgakis (1984)	Ellingsen (1979)	Hashimoto & Takamatsu (1981)
Cheung and Luyben (1979)	Morari, et al. (1980)	Garcia (1984)	Luyben (1975)*
Clement and Jorgensen (1981)	Mutharasan & Coughanowr (1976)	Hopkins (1979)	MacGregor, et al. (1983)
Clough, et al. (1978)	McDermott, et al. (1984)	Jutan and Uppal (1984)	McGreavy (1983)
Devia and Luyben (1978)	Nemanic, et al. (1959)	Schnelle (1979)	Padmanabhan & Lapidus (1977)*
Foss, et al. (1980)	Parulekar, et al. (1985)	Shunta (1981)	Ray (1972)
Fox, et al. (1984)	Penlidis, et al. (1985)	Weiss (1984)	Ray (1983)
Garcia-Rubio, et al. (1982)	Ponnuswamy, et al. (1985)		Ray (1985)
Georgakis, et al. (1977)	Ray (1981)		Rijnsdorp and Seborg (1976)
Hallager and Jorgensen (1981)	San and Stephanopoulos (1983)		Seborg, et al. (1983)
Hammarstrom (1979)	Schork and Ray (1981)		Stephanopoulos (1983)
Harris and Lapidus (1967)	Schuler (1980)		
Jo and Bankoff (1976)	Seinfeld (1969)		
Johnson, et al. (1982)	Tatkar, et al. (1985)		
Jutan, et al. (1977)	Timm, et al. (1982)		
Kiparissides and Shah (1980)	Tsoukas and Tirrell (1982)		
Kumar and Seinfeld (1978)	Uppal, et al. (1974)		
Kuruoglu, et al. (1981)	Vakil, et al. (1973)		
Kwalik and Schork (1985)	Viswanadham (1979)		
Lapidus and Luus (1967)	Wei (1978)		
Lappinga and Foss (1984)	Windes, et al. (1984)		
Lee and Lee (1983)	Yoo, et al. (1985)		
Leffew and Deshpande (1981)			

COLUMN TOTALS = 28   27   12   16

GRAND TOTAL = 83

TABLE 2 Theoretical and/or Experimental

Theoretical	Theoretical (Continued)	Experimental	Both
Ardell, et al. (1979)	Lim and Fang (1975)	Ahlberg and Cheyne (1979)	Doss, et al. (1983)
Aris and Amundson (1958)	Luyben (1974)	Ahn, et al. (1983)	Hashimoto & Takamatsu (1981)
Arnold, et al. (1980)	Luyben (1975)	Amrehn (1977)	Jo and Bankoff (1976)
Asbjornsen (1984)	MacGregor and Tidwell (1980)	Ardell and Gumowski (1983)	Johnson, et al. (1982)
Bailey (1972)	Mandler, et al. (1984)	Cardner (1984)	Jutan, et al. (1977)
Bonvin, et al. (1983)	Marini and Georgakis (1984)	Garcia (1984)	Lynch and Ramirez (1975)
Bruns and Bailey (1977)	McGreavy (1983)	Hallager and Jorgensen (1981)	MacGregor, et al. (1983)
Cheung and Luyben (1979)	Morari, et al. (1980)	Lappinga and Foss (1984)	McDermott, et al. (1984)
Clement and Jorgensen (1981)	Mutharasan & Coughanowr (1976)	Schork and Ray (1981)	Penlidis, et al. (1985)
Clough, et al. (1978)	Nemanic, et al. (1959)	Weiss (1984)	Ray (1983)
Denn (1972)	Padmanabhan & Lapidus (1977)		Ray (1985)
Devia and Luyben (1978)	Parulekar, et al. (1985)		Rijnsdorp and Seborg (1976)
Edgar (1980)	Ponnuswamy, et al. (1985)		Windes, et al. (1984)
Ellingsen (1979)	Ray (1972)		
Foss (1973)	Ray (1981)		
Foss, et al. (1980)	San and Stephanopoulos (1983)		
Fox, et al. (1984)	Schnelle (1979)		
Garcia-Rubio, et al. (1982)	Schuler (1980)		
Georgakis, et al. (1977)	Seborg, et al. (1983)		
Hammarstrom (1979)	Seinfeld (1969)		
Harris and Lapidus (1967)	Shunta (1981)		
Hopkins (1979)	Stephanopoulos (1983)		
Jutan and Uppal (1984)	Tatkar, et al. (1985)		
Kiparissides and Shah (1980)	Timm, et al. (1982)		
Kumar and Seinfeld (1978)	Tsoukas and Tirrell (1982)		
Kuruoglu, et al. (1981)	Uppal, et al. (1974)		
Kwalik and Schork (1985)	Vakil, et al. (1973)		
Lapidus and Luus (1967)	Viswanadham (1979)		
Lee and Lee (1983)	Wei (1978)		
Leffew and Deshpande (1981)	Yoo, et al. (1985)		

COLUMN TOTALS = 30   30   10   13

GRAND TOTAL = 83

TABLE 3 Type of Reactor

CSTR	Tubular	Batch	Semi-Batch
Ahlberg and Cheyne (1979)	Arnold, et al. (1980)	Ahn, et al. (1983)	Garcia (1984)
Amrehn (1977)	Bailey (1972)	Amrehn (1977)	Johnson, et al. (1982)
Ardell and Gumowski (1983)	Georgakis, et al. (1977)	Clough, et al. (1978)	MacGregor, et al. (1983)
Aris and Amundson (1958)	MacGregor, et al. (1983)	Jutan and Uppal (1984)	Parulekar, et al. (1985)
Bruns and Bailey (1977)	Padmanabhan & Lapidus (1977)	Kiparissides and Shah (1980)	Tsoukas and Tirrell (1982)
Cheung and Luyben (1979)	Ray (1983)	Luyben (1975)	Yoo, et al. (1985)
Devia and Luyben (1978)		MacGregor, et al. (1983)	
Fox, et al. (1984)		Padmanabhan & Lapidus (1977)	
Hammarstrom (1979)		Ponnuswamy, et al. (1985)	
Harris and Lapidus (1967)		Ray (1985)	
Hopkins (1979)			
Jo and Bankoff (1976)			
Kwalik and Schork (1985)			
Lapidus and Luus (1967)			
Leffew and Deshpande (1981)			
Luyben (1974)			
Lynch and Ramirez (1975)			
MacGregor, et al. (1983)			
Marini and Georgakis (1984)			
Morari, et al. (1980)			
Nemanic, et al. (1959)			
Padmanabhan & Lapidus (1977)			
Penlidis, et al. (1985)			
Ray (1981)			
Ray (1983)			
Ray (1985)			
Rijnsdorp and Seborg (1976)			
San and Stephanopoulos (1983)			
Schnelle (1979)			
Schork and Ray (1981)			
Schuler (1980)			
Seinfeld (1969)			
Tatkar, et al. (1985)			
Timm, et al. (1982)			
Uppal, et al. (1974)			
COLUMN TOTALS = 35	6	10	6

GRAND TOTAL = 57

TABLE 3 Type of Reactor (Continued)

Fluidized Bed	Fixed Bed	Combustion	General
Ray (1985)	Bonvin, et al. (1983) Cardner (1984) Clement and Jorgensen (1981) Foss, et al. (1980) Hallager and Jorgensen (1981) Jutan, et al. (1977) Kumar and Seinfeld (1978) Kuruoglu, et al. (1981) Lapping and Foss (1984) Lee and Lee (1983) Mandler, et al. (1984) Mutharasan & Coughanowr (1976) McDermott, et al. (1984) Padmanabhan & Lapidus (1977) Ray (1972) Vakil, et al. (1973) Viswanadham (1979) Windes, et al. (1984)	Wei (1978)	Ardell, et al. (1979) Asbjornsen (1984) Denn (1972) Doss, et al. (1983) Edgar (1980) Eilingsen (1979) Foss (1973) Garcia-Rubio, et al. (1982) Hashimoto & Takamatsu (1981) MacGregor and Tidwell (1980) McGreavy (1983) Seborg, et at. (1983) Shunta (1981) Stephanopoulos (1983) Weiss (1984)
COLUMN TOTALS = 1	18	1	15

GRAND TOTAL = 35

TABLE 4 Traditional or Nontraditional Control Techniques

Traditional	Nontraditional	Nontraditional (Continued)
Amrehn (1977)	Ahlberg and Cheyne (1979)	Leffew and Deshpande (1981)
Ardell, et al. (1979)	Ahn, et al. (1983)	Lim and Fang (1975)
Aris and Amundson (1958)	Ardell and Gumowski (1983)	Lynch and Ramirez (1975)
Cheung and Luyben (1979)	Arnold, et al. (1980)	MacGregor, et al. (1983)
Devia and Luyben (1978)	Asbjornsen (1984)	MacGregor and Tidwell (1980)
Garcia-Rubio, et al. (1982)	Bailey (1972)	Marini and Georgakis (1984)
Hopkins (1979)	Bonvin, et al. (1983)	McGreavy (1983)
Jutan and Uppal (1984)	Bruns and Bailey (1977)	Morari, et al. (1980)
Kwalik and Schork (1985)	Cardner (1984)	McDermott, et al. (1984)
Luyben (1974)	Clement and Jorgensen (1981)	Padmanabhan & Lapidus (1977)
Luyben (1975)	Clough, et al. (1978)	Parulekar, et al. (1985)
Mandler, et al. (1984)	Denn (1972)	Penlidis, et al. (1985)
Mutharasan & Coughanowr (1976)	Doss, et al. (1983)	Ponnuswamy, et al. (1985)
Nemanic, et al. (1959)	Edgar (1980)	Ray (1972)
Schork and Ray (1981)	Ellingsen (1979)	Ray (1981)
Shunta (1981)	Foss (1973)	Ray (1983)
	Foss, et al. (1980)	Ray (1985)
	Fox, et al. (1984)	Rijnsdorp and Seborg (1976)
	Garcia (1984)	San and Stephanopoulos (1983)
	Georgakis, et al. (1977)	Schnelle (1979)
	Hallager and Jorgensen (1981)	Schuler (1980)
	Hammarstrom (1979)	Seborg, et at. (1983)
	Harris and Lapidus (1967)	Seinfeld (1969)
	Hashimoto & Takamatsu (1981)	Stephanopoulos (1983)
	Jo and Bankoff (1976)	Tatkar, et al. (1985)
	Johnson, et al. (1982)	Timm, et al. (1982)
	Jutan, et al. (1977)	Tsoukas and Tirrell (1982)
	Kiparissides and Shah (1980)	Uppal, et al. (1974)
	Kumar and Seinfeld (1978)	Vakil, et al. (1973)
	Kuruoglu, et al. (1981)	Viswanadham (1979)
	Lapidus and Luus (1967)	Wei (1978)
	Lappinga and Foss (1984)	Weiss (1984)
	Lee and Lee (1983)	Windes, et al. (1984)
		Yoo, et al. (1985)

COLUMN TOTALS = 16 33 34

GRAND TOTAL = 83

TABLE 5 Types of Nontraditional Techniques

Steady State Optimization	Modeling	Modeling (Continued)	Estimation
Cardner (1984)	Ahlberg and Cheyne (1979)	MacGregor, et al. (1983)	Ahlberg and Cheyne (1979)
Edgar (1980)	Ardell and Gumowski (1983)	MacGregor and Tidwell (1980)	Ardell and Gumowski (1983)
Ellingsen (1979)	Aris and Amundson (1958)	Marini and Georgakis (1984)	Asbjornsen (1984)
Fox, et al. (1984)	Arnold, et al. (1980)	McGreavy (1983)	Doss, et al. (1983)
Hashimoto & Takamatsu (1981)	Bailey (1972)	Morari, et al. (1980)	Edgar (1980)
MacGregor, et al. (1983)	Bonvin, et al. (1983)	Padmanabhan & Lapidus (1977)	Foss (1973)
Padmanabhan & Lapidus (1977)	Bruns and Bailey (1977)	Parulekar, et al. (1985)	Georgakis, et al. (1977)
Ray (1983)	Cardner (1984)	Penlidis, et al. (1985)	Harris and Lapidus (1967)
San and Stephanopoulos (1983)	Clough, et al. (1978)	Ponnuswamy, et al. (1985)	Jo and Bankoff (1976)
Stephanopoulos (1983)	Denn (1972)	Ray (1972)	Jutan, et al. (1977)
Tatkar, et al. (1985)	Doss, et al. (1983)	Ray (1981)	Kumar and Seinfeld (1978)
Timm, et al. (1982)	Edgar (1980)	Ray (1983)	Kuruoglu, et al. (1981)
	Ellingsen (1979)	Ray (1985)	Kwalik and Schork (1985)
	Foss (1973)	Rijnsdorp and Seborg (1976)	Lynch and Ramirez (1975)
	Georgakis, et al. (1977)	San and Stephanopoulos (1983)	MacGregor, et al. (1983)
	Hallager and Jorgensen (1981)	Schnelle (1979)	MacGregor and Tidwell (1980)
	Hammarstrom (1979)	Schuler (1980)	McGreavy (1983)
	Harris and Lapidus (1967)	Seborg, et al. (1983)	Morari, et al. (1980)
	Hashimoto & Takamatsu (1981)	Seinfeld (1969)	Padmanabhan & Lapidus (1977)
	Jo and Bankoff (1976)	Stephanopoulos (1983)	Penlidis, et al. (1985)
	Johnson, et al. (1982)	Tatkar, et al. (1985)	Ponnuswamy, et al. (1985)
	Jutan, et al. (1977)	Timm, et al. (1982)	Ray (1983)
	Kiparissides and Shah (1980)	Tsoukas and Tirrell (1982)	Ray (1985)
	Kumar and Seinfeld (1978)	Uppal, et al. (1974)	Rijnsdorp and Seborg (1976)
	Kuruoglu, et al. (1981)	Vakil, et al. (1973)	Schnelle (1979)
	Kwalik and Schork (1985)	Wei (1978)	Schuler (1980)
	Lee and Lee (1983)	Windes, et al. (1984)	Seborg, et al. (1983)
	Leffew and Deshpande (1981)	Yoo, et al. (1985)	Seinfeld (1969)
	Lynch and Ramirez (1975)		Tatkar, et al. (1985)
			Timm, et al. (1982)
			Windes, et al. (1984)
			Yoo, et al. (1985)
COLUMN TOTALS = 12	29	28	32

TABLE 5 Types of Nontraditional Techniques (Continued)

Feedforward Control	Model Based Control	Robustness	System Structure
Arnold, et al. (1980)	Ardell and Gumowski (1983)	Bonvin, et al. (1983)	Clement and Jorgensen (1981)
Asbjornsen (1984)	Arnold, et al. (1980)	Fox, et al. (1984)	Edgar (1980)
Cardner (1984)	Bonvin, et al. (1983)	Jutan, et al. (1977)	Foss (1973)
Johnson, et al. (1982)	Cardner (1984)	Lynch and Ramirez (1975)	Georgakis, et al. (1977)
Morari, et al. (1980)	Edgar (1980)	Morari, et al. (1980)	Jutan, et al. (1977)
Padmanabhan & Seborg (1977)	Georgakis, et al. (1977)	Stephanopoulos (1983)	Morari, et al. (1980)
Rijnsdorp and Seborg (1976)	Johnson, et al. (1982)		McDermott, et al. (1984)
Vakil, et al. (1973)	Kwalik and Schork (1985)		Ray (1981)
Viswanadham (1979)	Lapidus and Luus (1967)		Ray (1983)
	Lappinga and Foss (1984)		Stephanopoulos (1983)
	Lynch and Ramirez (1975)		Tatkar, et al. (1985)
	MacGregor, et al. (1983)		Timm, et al. (1982)
	Morari, et al. (1980)		Viswanadham (1979)
	Ponnuswamy, et al. (1985)		
	Seborg, et al. (1983)		
	Tatkar, et al. (1985)		
	Timm, et al. (1982)		
	Yoo, et al. (1985)		

Pole Placement Control	Measurements	Dynamic Matrix Control	Internal Model Control
Edgar (1980)	Ahn, et al. (1983)	Doss, et al. (1983)	Doss, et al. (1983)
Foss (1973)	Edgar (1980)	Edgar (1980)	Fox, et al. (1984)
Foss, et al. (1980)	Eillingsen (1979)	Garcia (1984)	Lappinga and Foss (1984)
Georgakis, et al. (1977)	Georgakis, et al. (1977)	Ray (1983)	Penlidis, et al. (1985)
Kwalik and Schork (1985)	Jo and Bankoff (1976)	Stephanopoulos (1983)	Ray (1983)
MacGregor, et al. (1983)	Johnson, et al. (1982)		Stephanopoulos (1983)
McDermott, et al. (1984)	Jutan, et al. (1977)		
Padmanabhan & Lapidus (1977)	Kuruoglu, et al. (1981)		
Ray (1983)	Lynch and Ramirez (1975)		
Schnelle (1979)	MacGregor, et al. (1983)		
Seborg, et al. (1983)	MacGregor and Tidwell (1980)		
Stephanopoulos (1983)	McGreavy (1983)		
Tatkar, et al. (1985)	Ray (1983)		
Timm, et al. (1982)	Ray (1985)		
Viswanadham (1979)	Schnelle (1979)		
	Schuler (1980)		
	Wei (1978)		
	Weiss (1984)		

TABLE 5 Types of Nontraditional Techniques (Continued)

Singular Value Decomposition	Pontryagin Maximum Principle	Inferential Control	State Space Models
Doss, et al. (1983) Ray (1983) Stephanopoulos (1983)	Clough, et al. (1978) Denn (1972) Kumar and Seinfeld (1978) Lapidus and Luus (1967) Lynch and Ramirez (1975) MacGregor, et al. (1983) Padmanabhan & Lapidus (1977) Parulekar, et al. (1985) Ponnuswamy, et al. (1985) San and Stephanopoulos (1983) Seinfeld (1969) Tsoukas and Tirrell (1982) Yoo, et al. (1985)	Asbjornsen (1984) Edgar (1980) Ray (1983) Stephanopoulos (1983)	Ahlberg and Cheyne (1979) Doss, et al. (1983) Edgar (1980) Fox, et al. (1984) Georgakis, et al. (1977) Jo and Bankoff (1976) Jutan, et al. (1977) Kumar and Seinfeld (1978) MacGregor, et al. (1983) Morari, et al. (1980) Padmanabhan & Lapidus (1977) Penlidis, et al. (1985) Ponnuswamy, et al. (1985) Ray (1983) Ray (1985) Rijnsdorp and Seborg (1976) Schuler (1980) Tatkar, et al. (1985) Timm, et al. (1982) Vakil, et al. (1973) Viswanadham (1979)

Linear Quadratic Gaussian	Flow Data Reconciliation	Multivariable Frequency Analy Extensive Variables	
Clement and Jorgensen (1981) Denn (1972) Doss, et al. (1983) Edgar (1980) Foss (1973) Hallager and Jorgensen (1981) Hammarstrom (1979) Jutan, et al. (1977) Lynch and Ramirez (1975) MacGregor, et al. (1983) McGreavy (1983) McDermott, et al. (1984) Ponnuswamy, et al. (1985) Ray (1983) Rijnsdorp and Seborg (1976) Stephanopoulos (1983) Vakil, et al. (1973)	Stephanopoulos (1983)	Clement and Jorgensen (1981) Doss, et al. (1983) Edgar (1980) Foss (1973) Foss, et al. (1980) Hashimoto & Takamatsu (1981) Ray (1983) Rijnsdorp and Seborg (1976)	Hammarstrom (1979) Marini and Georgakis (1984)

TABLE 5 Types of Nontraditional Techniques (Continued)

Model Algorithmic Control	Time Delays	Computer Aided Design	Relative Gain Array
Doss, et al. (1983)	Ardell and Gumowski (1983)	Doss, et al. (1983)	Edgar (1980)
Edgar (1980)	Cardner (1984)	Edgar (1980)	Stephanopoulos (1983)
Ray (1983)	Doss, et al. (1983)	Hashimoto & Takamatsu (1981)	
Seborg, et at. (1983)	Edgar (1980)	Ray (1983)	
	Kwalik and Schork (1985)		
	Leffew and Deshpande (1981)		
	MacGregor, et al. (1983)		
	Ponnuswamy, et al. (1985)		
	Ray (1983)		
	Ray (1985)		
	Seinfeld (1969)		

Adaptive Control	Dynamic Programming	Optimal Servo Control	
Ahlberg and Cheyne (1979)	Ahlberg and Cheyne (1979)	Bailey (1972)	
Asbjornsen (1984)	Denn (1972)	Bruns and Bailey (1977)	
Doss, et al. (1983)	Lapidus and Luus (1967)	Foss (1973)	
Edgar (1980)	Padmanabhan & Lapidus (1977)	Lapidus and Luus (1967)	
Ellingsen (1979)		Lappinga and Foss (1984)	
Hallager and Jorgensen (1981)		Lynch and Ramirez (1975)	
Kiparissides and Shah (1980)		MacGregor, et al. (1983)	
Kwalik and Schork (1985)		Padmanabhan & Lapidus (1977)	
Lee and Lee (1983)		Parulekar, et al. (1985)	
Leffew and Deshpande (1981)		Ponnuswamy, et al. (1985)	
MacGregor, et al. (1983)		San and Stephanopoulos (1983)	
MacGregor and Tidwell (1980)		Seinfeld (1969)	
Marini and Georgakis (1984)		Tsoukas and Tirrell (1982)	
McDermott, et al. (1984)		Yoo, et al. (1985)	
Padmanabhan & Lapidus (1977)			
Ponnuswamy, et al. (1985)			
Ray (1983)			
Ray (1985)			
Rijnsdorp and Seborg (1976)			
San and Stephanopoulos (1983)			
Seborg, et at. (1983)			
Stephanopoulos (1983)			

EXPERT SYSTEMS IN PROCESS CONTROL

George Stephanopoulos
Massachusetts Institute of Technology, Cambridge, MA 02139

Knowledge-based Expert Systems are creating a new "culture" among academicians and industrial practitioners. They offer promises, often over-sold and unduly exaggerated, open up new avenues in problem conceptualization and encourage the tackling of long-lasting "hard problems. But, a concise definition of what they are, a rational expectation of what they can do and a working facility of how to construct them, are still missing from the large majority of interested researchers. With this general background in mind, the Session on "Expert Systems in Process Control" was formulated to try to answer the the following: (a) What are some prototypical Expert Systems in Process Control? (b) What can they offer? (c) How are they developed?

Four presentations by five people gave a broad perspective on all three of the previous questions. Thus, the papers demonstrated:
1. Expert systems could be employed to enhance our abilities,
    (i) at the stage of a priori designs for control systems (Taylor, Shinskey, Umeda and Niida), or
    (ii) during on-line operations for the interpretation of process data, diagnosis, and control (Moore and Kramer).
2. How to achieve diversified objectives such as:
    (i) Encapsulate the knowledge of a recognized expert's expertise into a rule-based expert system (Shinskey).
    (ii) Organize diversified methodologies into an "intelligent assistant" for control system analysis and synthesis (Taylor).
    (iii) Create a software environment to support either a priori control system designs (Umeda and Niida), or on-line intelligent analysis of process data and control (Moore and Kramer).
3. The diversity of computers and computing environments that one could use to develop and run expert systems. For example:
    (i) Shinskey employed conventional means such as an IBM PC with

BASIC.

(ii) On the other hand, Umeda and Niida (SYMBOLISC 3670 with KEE), and Moore and Kramer (Lambda with PICON) employed special purpose LISP computers (SUMBOLICS, Lambda) and sophisticated development softward packages (KEE and PICON).

After the presentations, lively discussion emphasized certain important issues which are related to the nature of expert systems, their structure, what one might expect to see in the future, and the hardware and software requirements for their development and delivery. A free recollection of the discussion can be summarized by the following main topics.

A. The future of expert systems lies with the harvest of deep-knowledge which is based on the richness of chemical engineering science and control theoretical aspects. An unguided explosion of rule-based systems could lead to a very large number of disparate rules without reasonable coherence and counter-productive effects, e.g. poor maintainability.

B. Expert systems, for the time being, are being thought of as expert assistants or consultants, rather than expert solvers. This was found to be a strength rather than a weakness because it allows the interjection of a human's (control designer, process operator or engineer) accumulated knowledge about the specific problem at hand; knowledge which may not be available in the expert system.

C. The collection of the relevant knowledge was emphasized as a key issue. No specific guidelines of general utility were proposed, but specific suggestions directed towards the interviewing of expert designers or process operators were put forward.

D. The hybridization of expert systems, which can capture both quantitative and qualitative knowledge, was addressed and was concluded that LISP-based computers and software environments offer an excellent vehicle.

E. A serious criticism was raised concerning the lack of openly-available lists with the rules used by the expert systems. It was argued that given the fact that the essence and power of a rule-based expert system rests with its rules, a short list of published rules does not allow sufficient basis for evaluation of the system and learning by others. A counterargument was offered, pointing out that the purpose of the presentations was to describe the character of the expert systems rather than subject them to a scrutiny of their knowledge base.

F. The flexibility of the various computers and software environments was discussed in length. It was a general agreement that traditional languages such as BASIC, create inflexible expert systems requiring significant modifications in order to expand or modify their rule base, and adjust reasoning strategies. LISP environments it was argued, offer enormous flexibility, adaptability and continuous evolution.

G. It was suggested that future expert systems should include features such as: Automatic generation of rules; capabilities for learning; enhanced graphic interfaces, etc.

H. A series of questions was addressed to F.G. Shinskey. From the answers he revealed that: His expert system is in the field for testing; most of his rules could be found in his book on Control of Distillation Columns; some of his close associates had generated rules that he included in the system; he is planning to generate additional expert systems concerning the control of reactors and other unit operations.

I. J.H. Taylor indicated that the rules of their expert system tried to imitate the reasoning process followed by expert designers at G.E. He also emphasized that the expert system shell was developed internally.

J. K. Niida in his answers indicated that his efforts at CHIYODA were aimed at providing a prototype for testing purposes. Significant amount of work is still needed to be done especially in the areas of graphic interfaces and enrichment of the rule base with heuristics for the control structures of individual processing units.

K. Moore and Kramer indicated that the PICON shell offered to them a convenient environment for the development of rather transparent expert systems. Further discussion revolved around the use of deep knowledge (model-based rules) for fault diagnosis, which was agreed to be the proper direction for further developments.

EXPERT SYSTEMS FOR COMPUTER-AIDED CONTROL ENGINEERING

James H. Taylor
GE Corporate Research & Development, Schenectady, NY 12345

Abstract. The motivation, history, and process of creating an expert-system environment for computer-aided control engineering are outlined. The resulting expert system architecture is presented, and its generic aspects relative to engineering design are discussed. Developmental issues are also mentioned, including inference and numerical requirements and interfacing with conventional analysis and design software. The status of our project is summarized, and certain lessons regarding the difficulty of achieving certain ambitious goals are presented. Other projects in this area are also referenced.

Keywords. Control system design; CAD environments; expert systems; artificial intelligence; engineering design.

INTRODUCTION

Artificial intelligence in general and knowledge-based expert systems (KBESs) in particular have recently gained prominence in the general area of computer-aided problem solving. The earliest successes were in the area of diagnosis or troubleshooting (e.g., Mycin, Prospector, Internist, etc.). The application of this technology has been expanding with great rapidity in the last few years; refer, for example, to Hayes-Roth *et al* (1983), Coombs (1984), Michie (1982), IEEE (1984), or Davis and Lenat (1982) for a sampling of recent work. One of the more recent application areas of this technology is engineering design. The latter topic, particularly as it relates to control system analysis and design, is the focus of this presentation. The primary goals are to demonstrate what we have done in this area, to discuss what remains to be accomplished, to overview other related work, and to present a balanced, realistic view of the promise and cost of the KBES technology in this field.

HISTORY OF GE WORK IN CACE

The Control Technology Branch of GE Corporate Research & Development (**GE CRD**) has had a substantial interest and involvement in Computer-Aided Control Engineering (**CACE**) since about 1980. Some contributions are documented in Taylor (1982), which outlines some functional requirements of a CACE environment, Spang (1982, 1984), which describes the development and capabilities of our conventional environment called the **GE Federated System**, and Taylor and Frederick (1984), which documents a high-level conceptual definition for a KBES for CACE called **CACE-III**.

The General Electric experience and approach to CACE software development has been motivated by the need for an integrated, high-level environment for control engineering activities. When we started in 1980, there was nothing available that met our requirements. There was software, both public-domain and commercial, that could perform most of the functions that were required, but not in a single, convenient environment. Not having the resources to develop "the ideal environment" from scratch, we adopted the strategy of selecting packages with the best possible functional coverage, and "federating" the packages to create the desired capability and unity. The term "federated" signifies that the packages are still available for stand-alone use; generally, however, they are run under a supervisory program that manages the environment.

We developed our software in two major phases: In the first part of the effort, we purchased and/or developed software that could carry out the essential procedures of CACE with numerically robust algorithms and reasonably convenient and consistent user interfaces, and integrated them into a unified conventional system. This effort culminated in a software environment we call the *Federated System*, which is completely developed, documented, and delivered to a user group made up of controls engineers at about ten GE locations.

In the second part of our development, started in 1983 in a low-level exploratory manner, we focussed on making this software more accessible to less-than-expert users and relieving the user of much of the burden of using a sophisticated and complicated environment. We used the expert systems approach to accomplish this (Taylor and Frederick, 1984). This work was motivated by our own experience with the Federated System and other CACE software, and by the comments and suggestions of our user group.

The outcome of this research is an architecture, a detailed specification, and a prototype system called *CACE-III* which is, in our opinion, a *third-generation* CACE environment; hence its name. In contrast, first-generation tools are single-purpose software products (root locus routines, transfer function plotting programs, etc.) and often batch-oriented, and second-generation environments are functionally broader, interactive, conventional software; according to this classification, the Federated System is a second-generation environment for CACE. This third-generation project is still underway.

The GE Federated System

The Federated System was developed by GE CRD for internal use and for the GE Aerospace Business Group and Aircraft Engine Business Group. This sponsoring user group played a partnership role with GE CRD in defining the system. The basic approach that guided the development of the Federated System was as follows:

- define the required functionality,

- survey available CACE packages and choose a set of them that meets the functional requirements (to the extent possible), with numerically respectable algorithms and convenient user interfaces,

- extend packages or develop additional packages to meet remaining unmet needs,

- develop interface routines that can convert models and data into forms that are appropriate for each package,

- integrate the packages and routines into a unified, effective environment for CACE, and

- document and deliver the resulting software system to our user group.

The initial core package set (selected in 1981) included:

- CLADP, the Cambridge Linear Analysis and Design Program, which implements recent "British School" multivariable frequency-domain design methodologies (Edmunds, 1979),

- SIMNON, a nonlinear simulation package from Lund University (Sweden) that supports both its own modeling language and FORTRAN system models and allows the user to set up and execute simulations in a very flexible, command-driven environment (Elmqvist, 1977),

- IDPAC, a time-series analysis package from Lund University that performs statistical analysis (correlation and spectral) and parametric model identification (least squares and maximum likelihood) (Wieslander and Gustavsson, 1976), and

- SSDP, a modern state-space design package built at GE CRD using the I/O routines of CLADP and underlying numerical software from reputable sources.

More recently, MATLAB (Moler, 1982) was added to the Federated System. These core packages met most of the required functionality with the exception of strong nonlinear analysis and design capabilities and control system implementation.

This environment was integrated and extended to include appropriate model data-base transformations and overall functional compatibility (Taylor, 1982; Spang, 1982, 1984). In some cases, the data-base transformations were simply re-formatting linear system models obtained from mass storage, in other cases the interface was considerably more challenging. The most complicated transformation was from SIMNON nonlinear models to CLADP linear models, which required the installation of a new linearization routine. This capability also underscores the necessity of such interfaces: One of the most time-consuming and error-prone procedures in CACE is the manual generation and entry of linearized models.

Functional compatibility was achieved primarily through the use of interface routines. The most important feature in this sense is the provision for converting the compensator designed in CLADP into the SIMNON modeling language, so that the user can directly validate the design via simulation with the nonlinear plant model. Another feature incorporated for compatibility is equilibrium-finding, which is usually a necessary precondition for linearization.

Two additional capabilities were incorporated to deal with the analysis and design of nonlinear systems: a routine to generate describing function input/output models of nonlinear plants, and design routines to synthesize nonlinear compensators based on these models using a methodology developed at GE CRD (Taylor and Strobel, 1984). These enhancements to the Federated System are described in some detail in Taylor (1985b).

This approach to the design of a state-of-the-art conventional CACE environment has been highly successful. By making the best possible use of existing software, we have achieved broad and complete coverage of the CACE problem. Throughout this work, we have strived to make this environment as broad, powerful, practical, and "real-problem-oriented" as possible. The two most striking attributes of the Federated System are its breadth and its substantial capabilities to handle nonlinear problems (simulation, analysis, and design).

In the course of this activity, we had to consider a number of issues. In addition to functionality, primary considerations were software package selection, integration (e.g., "federating"), and data-base management (**DBM**). The solutions to the problems we encountered in these areas are detailed in Spang (1982, 1984). The following additional points should be considered:

- The trade-off between generic versus special-purpose environments may be an important consideration. The Federated System is a general-purpose environment; a narrowly-focussed environment would be quite different. There are two aspects to specificity: a special-purpose package might support just one approach (e.g., linear systems and linear quadratic regulator design), or it might support users from one discipline (e.g., chemical process control). The software selection would be influenced by both factors, and even the content and style of the user interaction could be "tailored" to suit the terminology and procedures generally used in a particular domain of application.

- The need for DBM should be assessed carefully. Few of our applications are multi-disciplinary, so the need for real DBM (e.g., keeping track of numerous models and analysis data files for a single project including the relations among the files) was not particularly stringent. We thus accepted the philosophy of the Federated System's core packages and limited DBM to file-naming conventions for files created by the various procedures. In many situations, a system that goes beyond simple file-naming conventions may be required.

- Finally, the CACE software scene has changed considerable in the last four years; it is quite possible that our software selection today would be different.

The impact of the first two issues has been discussed in Taylor (1985a), where some specific concepts in the areas of "tailoring" and DBM for integrated flight and engine control are outlined.

While much of the reaction to recently-developed CACE software (the Federated System and other packages) has been very positive, there have been were a substantial number of complaints related to "user friendliness". In the context of our work, these may be summarized as follows: It is difficult to remember all the details (commands and syntaxes, etc.) involved in using a comprehensive set of CACE packages, it is difficult to keep on top of the analysis and design process, database management is too rudimentary, it is not always clear what approach or even which package to use, it is difficult to deal with all the error conditions (error messages and corrective actions) and other aspects of using the software, etc. In short, one must be an expert, every-day user to take full advantage of the existing software, which can be a considerable frustration to less-than-expert or occasional users. These considerations prompted us to consider the use of KBES methods to improve the situation.

CACE-III

The basic thrust of our more recent research in CACE is the creation of a unified expert-aided environment for the full spectrum of control engineering activities, from high-fidelity modeling through control system design and validation (Taylor and Frederick, 1984). Two key issues are the complexity of control engineering problem solution and the broad spectrum of the user capabilities we wish to serve. The first point is clearly illustrated by listing the typical activities of a controls engineer: nonlinear modeling, simulation, equilibrium-finding, linearization, analysis of linear plant models, control system design (including trade-off analysis and tuning), and design validation. The latter issue we have summarized by saying that we are striving to create an environment "that can be a teacher to the novice, a partner to the more experienced controls engineer, and an assistant to the expert". In every case, we want to eliminate or reduce as much as possible the amount of unnecessary detail the user has to contend with, especially in the areas of software package usage (choosing the most appropriate algorithm, package and command selection, interpreting error messages and taking corrective action where possible, etc.) and keeping track of models and data (data-base management).

A fundamental conclusion of our research is that this goal can best be achieved by the use of the knowledge-based expert system (KBES) approach. This conclusion was reached by specifying the functional requirements of such an environment and showing how they can be met in a KBES, and contrasting such a system with existing conventional software environments. This comparison was made on the basis of our substantial experience in conventional CACE software development and integration as outlined above; our approach and findings are discussed in more detail in Taylor and Frederick (1984) and below.

In the discussion that follows, we present broad concepts and issues involved in creating an expert-aided environment for CACE. We accomplish this by outlining certain considerations regarding the use of the expert systems methodology, describing the development of CACE-III (a prototype KBES to demonstrate the capabilities and benefits of an expert-aided environment), illustrating the basic elements of a rule-based expert system, and discussing the lessons we have learned in the course of this project. The lessons specifically relate to the cost of developing "real" expert systems, the limitations of basing KBES development on "scenarios", and the difficulty of making software usage details invisible and supporting a variety of user capabilities.

EXPERT SYSTEMS CONSIDERATIONS

Expert or knowledge-based systems are software environments designed to aid in solving problems that require high levels of *expertise*, some degree of *inference* ("reasoning"), the use of *heuristics* (nonrigorous procedures or "rules of thumb"), and the systematic processing of *symbolic information*. Such problems are generally complicated and broad in scope, and are not amenable to clear-cut well-posed algorithmic solutions. Control system engineering is certainly such a task: The types of expertise that are required for CACE problem solving are: development and diagnosis of plant models, formulating a realistic controls design problem, selecting appropriate analysis and design methods, performing design tradeoffs, validating and documenting designs, and making effective use of conventional CACE software. Symbolic information to be processed in CACE includes descriptions of methodologies for CACE, the status of the current problem solution, the names and capabilities of CACE packages, their command sets, syntaxes, and error handling, data and model relations for DBM, etc.

Heuristics play a major role in a human expert's ability to formulate a well-posed CACE design problem, and reasoning capability is advantageous for directing and keeping track of the design process as it progresses. Also, the KBES approach provides a high-level, flexible, and supportive environment that can relieve the user of much of the low-level detail and drudgery involved in using a number of large software packages and in DBM. Further motivation for the use of the expert systems approach may exist for users in multi-disciplinary fields such as integrated flight and engine control: Few individuals can be expected to be expert in all aspects of a multidisciplinary problem, so expert-aiding is often essential to the production of meaningful analysis and designs. For more information on expert systems or for additional background reading in the area, refer, for example, to Hayes-Roth *et al* (1983), Coombs (1984), Michie (1982), IEEE (1984), or Davis and Lenat (1982).

The application of expert systems technology is often based on the idea of providing support for "less-than-expert users". This thought requires some caution, especially in our context. First, there are several factors involved in this phrase: In engineering design, such a user may not be aware of the latest theoretical developments in all fields involved in the problem to be solved, may not have had the experience required to synthesize theory into an effective analysis and design approach, and/or may not be a frequent practitioner of system design or user of the required software. Considering the breadth of knowledge required for CACE, especially in multidisciplinary fields, it is probably not reasonable to expect that many users of a major software environment for this activity will be experts in all aspects of the problem being solved. Providing support for "less-than-expert users" in this sense is desirable, reasonable, and feasible. On the other hand, it should be clearly understood that it is not appropriate to speak of developing an expert-aided CACE environment for use by personnel with little or no knowledge of controls: Too much understanding and common sense are required of the user for this to be practicable.

Another potential problem in the use of expert systems is the possibility of unrealistic expectations. We believe that it *is* realistic to use this methodology to raise the level of interaction for users that are basically competent and to support them in the ways outlined above, but it is *not* reasonable to expect that such a system will be *fool-proof* and able to provide the *optimal* solution to all problems. These factors make it imperative that such a system be designed to keep the user in a position of responsibility and authority. The latter issue - authority - requires that the system be flexible enough that the user will be supported to the extent needed and possible, without dominating the proceedings and forcing the user to accept undesired or meaningless solutions. It should be appreciated that this may be a major undertaking, and that there have to to be limits to the freedom of the user.

With these clarifications of the basic objective of an expert-aided system for CACE, we proceed to develop the idea further. This concept is based on the motivational issues, general functional requirements, and basic expert system ideas presented in Taylor and Frederick (1984).

CACE-III DEVELOPMENT

Our approach to conceiving and developing an expert-system CACE environment combines top-down and bottom-up CACE process modeling. By the first term, we mean working with experts in the fields of CACE to determine the conceptual and structural aspects of the problem and distilling that information into an expert system architecture. Bottom-up modeling referred to developing "scenarios" that show in detail how the expert system should serve and interact with the user (Taylor, Frederick, and James, 1984; Taylor, 1985a). A combination of these activities, in parallel with the development of corresponding real expert system architectures and rule bases, has proven to be extremely effective. Only top-down modeling is discussed in any detail in this presentation, so that the CACE-III architecture can be well explained and motivated. Certain generic aspects of this work are also pointed out.

Architecture Development

We used top-down modeling of the CACE methodology of a human expert as a way to develop detailed requirements and architectural concepts for CACE-III. The first outcome of this approach was the realization that a *complete, meaningful problem formulation* is a central issue in capturing the control system design process. This may be represented by a "list of facts" or, in artificial intelligence terminology, "frame". The information in this frame may be organized or partitioned into three components with the following informational content:

- *Plant model characterization* - ordinary or partial differential equation models, linear or nonlinear, stable or unstable, minimum or nonminimum phase, (un)controllable and/or (un)observable, et cetera.

- *Constraints* - architectural (e.g., centralized or distributed control), implementation (e.g., analog or digital), parametric (e.g., gains, data rate limits), et cetera.
- *Specifications* - time response, frequency response, performance indices, et cetera; sensitivity, disturbance rejection, robustness, et cetera.

This is illustrated in Fig. 1; such a *problem frame* is a major focal point for CACE.

Given a meaningful control design problem, we saw the human expert working in a parallel construct, which we call the *solution frame*. This is a list of facts that is developed as a "scratch pad" where the expert system keeps track of both the *data base* (models, analysis data) and the *process* (what has been done and what needs to be done, information required for decisions about the selection of design procedures and tradeoff analysis, and a log of the entire transaction). This is shown in Fig. 2.

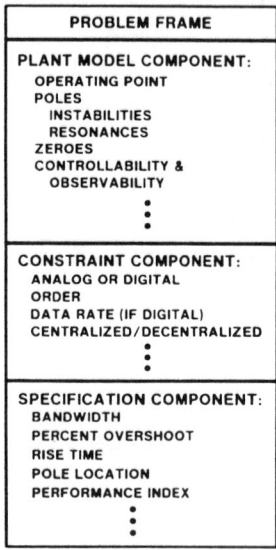

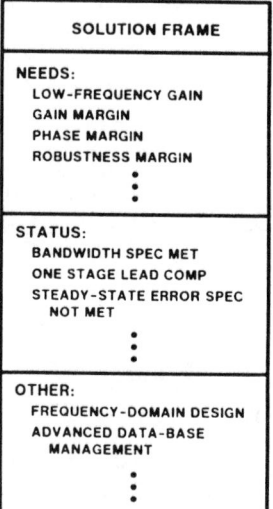

Figure 1. The CACE Problem Frame   Figure 2. The CACE-III Solution Frame

These structured lists of facts gave us a basis for developing *rule bases*. The process proceeds by asking: What are the *functions that must be implemented* in order to produce and manipulate these facts to solve the problem? This line of thought -

linking the activities of an expert with two key lists of facts (frames) and associated rule bases - gave rise to a functional structure of CACE-III that is depicted in Fig. 3. In particular, we have created a construct in which the rule base is partitioned into six functional parts, as shown, and a seventh supervisory rule base (not portrayed) which is discussed below. The functions of the rule bases in CACE-III may be summarized as follows:

- RB1 governs interactions among the *design engineer, plant models* (nonlinear and/or linear), and the *model component of the problem frame*. This rule base provides support in model development (including diagnostics relating to the physical process and suitability of models for control system design and numerical analysis), and sees to it that all required plant data are added to the knowledge base. Examples of the activities of RB1 are illustrated in scenarios in Taylor, Frederick, and James, 1984, and Taylor, 1985a.

- RB2 governs interactions between the *design engineer* and the *constraint and specification components of the problem frame*. Constraints are requested but are not mandatory; if supplied, these are also written into the list of facts that makes up the problem frame. These rules guide the user in entering design specifications and checks specifications for consistency, completeness, and achievability (realism). For example, consistency checks include being sure that the damping ratio dictated by percent overshoot and rise-time-to-settling-time ratio agree, etc., and realism tests include determining the specifications that can be achieved by a simple design approach (e.g., use of a standard PID design algorithm) and checking to see that the user's specifications are not excessively more demanding.

- RB3 and RB5 govern interactions between the *problem frame* and the *solution frame*. RB3 deals with specifications, constraints, and plant characteristics, and initializes the list of facts in the solution frame describing what needs to be done to achieve design goals. If design iteration or tradeoff analysis is called for, then RB5 supports the user in selecting the specifications to vary (relax or tighten) and modifies the problem frame appropriately, RB3 resets the solution frame accordingly, and the required set of designs is carried out by RB4. Finally, RB5 will present the results of the iteration or tradeoff analysis.

- RB4 governs interactions between the *solution frame* and the available *design procedures*. These rules decide what design approach(es) will best solve the problem (e.g., by matching approaches with specifications), executes the appropriate procedure(s) and algorithm(s) using conventional design software, and updates the solution frame to reflect the corresponding addition/change in the system. RB4 also performs a preliminary validation by checking that all specifications are met with the linear plant model and controller.

- RB6 governs the final control system *validation process* (which generally involves highly realistic simulation or emulation of the plant and controller), conversion from idealized controller design to *practical implementation*, and *documentation*. The last step involves archiving a record of the design process, including tradeoffs and information supporting all design decisions, and a record of the data base (model and data files, including information regarding assumptions and conditions for validity).

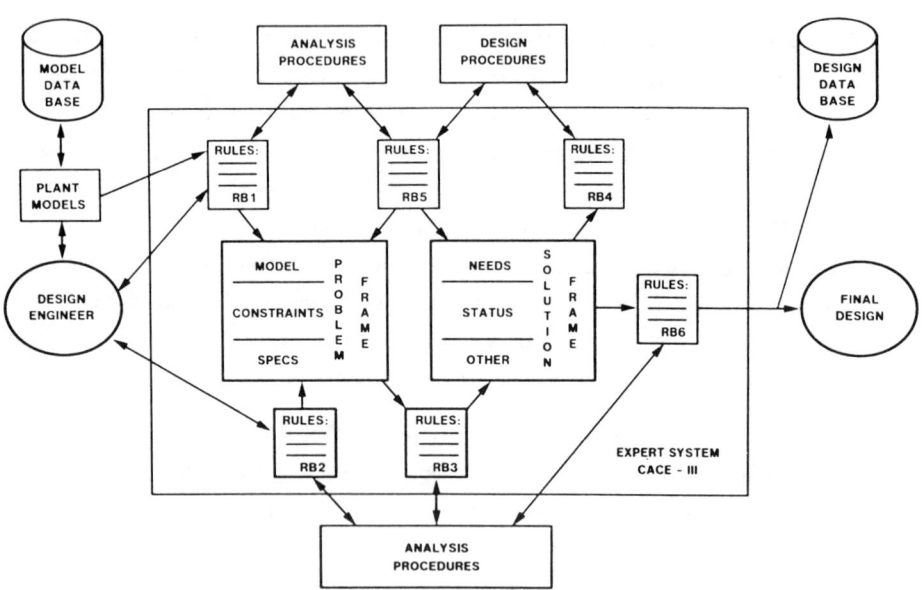

Figure 3. The Architecture of CACE-III

The goal of the first two rule bases is to have a well-formulated problem, thus ensuring a reasonable probability of success in the design phase. They were conceived as outlined above simply by considering the informational content of the problem frame and determining what functionality must be implemented in order to arrive at a meaningful problem. Observe that RB3, RB4 and RB5 may represent an iterative or dynamic "loop": In some cases the KBES must invoke these rule bases repeatedly until all specifications are satisfied, if possible. Again, these rule bases were established based on our model for how an expert controls engineer sets up and solves design problems. Finally, RB6 provides a more rigorous assurance that the user has a control system that will perform as required, with as little need as possible for additional engineering for implementation, and with the required engineering documentation.

We have found that the rule base partitioning depicted in Fig. 3 serves two important functions: First, it clarifies many conceptual aspects of the CACE process, and thus provides a basis for rule base development. In addition, this structure has proven to be useful for developing "meta-rules" (AI terminology for "rules about rules") that increase the efficiency of the expert system by partitioning the rule base and restricting the scope of the expert system to the rules needed for the task presently at hand. These meta-rules are contained in a seventh supervisory set of rules (James, Taylor, and Frederick, 1985) which monitors the status of the problem solution (as established by RB1-RB6) and determines when a rule base has completed its work and what rule base to invoke next.

The above rule bases cause the KBES to invoke conventional CACE software in performing all analysis and design functions. Expertise regarding the use of this software is "built-in", so that the user need not know command sets, syntaxes, and error messages. To the extent possible, the expert system knows how to interpret error messages and take corrective action. A protocol for coordinating the operation of the expert system and the CACE software is completely designed and implemented (James, Taylor, and Frederick, 1985). Finally, this expert system architecture can also support DBM, in the sense that the expert system has access to the relational information necessary for this function and can either perform the activity itself or supply the information to separate DBM software, as discussed in more detail below.

A Paradigm for CAE

The primary goal of our research was to provide a clear vision of an expert system for CACE that goes beyond the generality that "artificial intelligence promises to revolutionize engineering design". In retrospect, we believe that the CACE-III expert system architecture speaks to some generic aspects of engineering design, thus adding substance to this statement.

Several aspects of the development of CACE-III would seem to pertain to engineering design in general. First, the **motivation** for considering the creation of expert systems for engineering analysis and design in other fields is similar to that for CACE:

- The elements of engineering design are generally closely patterned on the tasks defined above for CACE: model, constrain, and specify to arrive at a well-posed problem; select and execute design approaches; perform trade-off studies if necessary; validate; document; implement.

- As a corollary to item 1, most engineering fields are inherently broad in scope, i.e., encompasses a wide variety of activities, and requires inference, judgement, and heuristics.

- We conjecture that in many areas engineering analysis and design software suffers from the same general deficiencies as we discussed previously - in particular, from being rather low-level, unsupportive, or otherwise not "user-friendly" to the less-than-expert or occasional user.

- Most areas of engineering are "dynamic", in the sense that new methods are continually being created and software is continually being developed to deal with the problem on a broader basis or higher level; the KBES methodology can accommodate evolution and transition new approaches to the user very effectively.

- There is substantial pressure to reduce the engineering time spent in analysis and design, to improve the quality of design, to aid the less-than-expert user, and to better document the design process.

Having summarized our reasons for believing that KBESs for engineering design will prove to be of substantial value in many disciplines, we will conclude by considering the **architecture** of such expert systems:

The structure portrayed in Fig. 3 was developed by a process that might be thought of as "modeling" the procedures of a human expert. For this reason, the similarities in terms of the activities required in engineering analysis and design (the first point above) and in the underlying thought processes of human experts gives us reason to believe that the expert system architecture depicted in Fig. 3 is strongly generic. In other words, it is our belief that the rule base structure of CACE-III can be used by expert system developers from other fields quite directly, merely by substituting the appropriate functions into RB1 through RB6.

AN EXPERT SYSTEM ENVIRONMENT

CACE-III was first developed using a GE CRD expert system environment called DELTA, a rule interpreter or "inference engine" created at GE CRD for diesel-electric locomotive trouble-shooting (Bonissone, 1983). This shell did not adequately support our needs, which included running and communicating with conventional software and performing numerical procedures. We then migrated to a more recently-developed and more powerful GE CRD expert system environment called **Delphi**.

While space does not permit a detailed discussion, we will outline some aspects of the operation of DELTA to the extent required to convey a basic understanding of rule-based expert systems. The goal is to illustrate the spirit of the expert system approach, not a working knowledge of DELTA or Delphi. This overview deals with considerations that are primarily generic to the use of rule-based expert systems, although the details differ from system to system.

Rules

The DELTA inference engine basically works with rules having the form:

IF [*Premise$_1$ Premise$_2$* ...]

THEN [*Conclusion$_1$ Conclusion$_2$* ...]

Premises may be thought of as defining a condition, situation, or state of a process; conclusions are actions to be taken if the premises of the rule are satisfied. Actions include adding facts to the knowledge base (list of facts), clearing facts from the knowledge base, or asking the user to determine what specific fact(s) to add to the knowledge base.

Each premise and conclusion in CACE-III has the form of a three-tuple which is true, false, or neutral (of unknown truth), i.e., each three-tuple may take on certainty values of +1, -1, and 0. A three-tuple has the form [OBJECT ATTRIBUTE VALUE]; for example, the fact [SPEC-CL-POLE MAX-REAL-PART UNASSIGNED] +1 denotes that no value has been specified for the maximum real part of the poles of the closed-loop control system being designed.

A *premise* may be that a three-tuple is true,

EQ [OBJECT ATTRIBUTE VALUE],

or false,

NE [OBJ ATTR VAL].

The *conclusions* of a rule may include any of the following actions:

1. DISPLAY, which indicates that the user should be notified about something. This conclusion must be followed by the message to be conveyed (in parentheses);

2. UDO, which indicates that the user should take some action. This conclusion must be followed by a statement of the action to be taken that will be displayed to the user (in parentheses);

3. WRITE [OBJ ATTR VAL], which indicates that a three-tuple should be written into the knowledge base;

4. CLR [OBJ ATTR VAL], which indicates that a three-tuple should be cleared from the knowledge base (certainty value set to 0);

5. ASK [OBJ ATTR VAL], which indicates that the user should be asked to determine the status of the three-tuple (true or false); or

6. MENU, which allows the user to make a selection from a list of options.

Observe that in the case of DISPLAY, UDU, ASK, and MENU each parenthetical expression is a message to the user; in other cases and in the premise portion of the rule, the phrases in parentheses are simply comments explaining the thinking of the rule base creator. These "comment lines" may serve as the basis for a "why" or "help" facility. Most of these ideas are illustrated in the two sample rules provided in Table 1, which are taken from the CACE-III rule base that governs the specification entry process.

Table 1. TWO RULES FROM CACE-III

```
RULE 106 ( Want to enter max Re part spec; already assigned )
IF:
  EQ [ SPEC-C-L-POLE MAX-REAL-PART REQUESTED ]
    ( User has asked to enter max-real-part )
  EQ [ MAX-REAL-PART VALUE + ]
    ( A value is currently assigned )
THEN:
  CLR [ SPEC-C-L-POLE MAX-REAL-PART REQUESTED ]
    ( Reset - clear the fact that triggered this rule )
  DISPLAY
    ( You want to enter a value for the maximum )
    ( real part of the poles, but a value has been )
    ( assigned previously. )
  ASK [ MAX-REAL-PART MODIFY VALUE ]
    ( Do you wish to replace the current value? {Y or N} ... ENTER: )
```

```
RULE 108 ( Current value of max Re part is to be replaced )
IF:
  EQ [ MAX-REAL-PART MODIFY VALUE ]
    ( User wants to modify the current value )
THEN:
  CLR [ MAX-REAL-PART MODIFY VALUE ]
    ( Reset - clear the fact that triggered this rule )
  CLR [ MAX-REAL-PART VALUE + ]
    ( Delete the old value )
  WRITE [ SPEC-CL-POLE MAX-REAL-PART UNASSIGNED ]
    ( Reset - pave the way for a new assignment )
  WRITE [ SPEC-CL-POLE MAX-REAL-PART REQUESTED ]
    ( Trigger the rule to request entry of new value )
```

Observe that a 'yes' response to the query in the conclusion of RULE 106 results in the three-tuple [MAX-REAL-PART MODIFY VALUE] being written to the list of facts as 'true'. This action satisfies the premise of RULE 108, thus triggering it. RULE 108, in turn, will activate the rule that asks the user for a value of SPEC-CL-POLE MAX-REAL-PART. Note that the '+' sign is a "wild card", denoting "any value".

Inference Mechanisms

DELTA and Delphi support inference in two modes: *fact driven* and *goal driven*. In the former case, the system systematically checks some or all the rules for premises that are known to be true so that the corresponding fact(s) can be written into the list of facts (knowledge base). In the *goal driven* mode, the system attempts to reach a specific conclusion (write a specific fact) by identifying a rule or rules that contain the desired goal in the conclusion and using its inference capabilities to prove that the premise(s) of the desired rule(s) are satisfied. The AI terminology for these two modes of operation are *forward* and *backward chaining*, respectively. An example of the first inference mode is filling in the plant characteristics in the problem frame, while attempting to achieve the goal "design specifications satisfied" exemplifies backward chaining.

The forward chaining inference mode operates by examining each rule and asking: Is $Premise_i$ known to be true? If the answer to this question is "yes" for every premise of the rule, then its conclusion is carried out, and the inference engine proceeds to consider the next rule. This process may be carried out exhaustively, or it may be limited either by features of the inference engine or by structuring the rules.

Backward chaining starts with a goal: write $Fact_k$ to the list of facts. First, the inference engine seeks a rule that can cause $Fact_k$ to be written, i.e., that has WRITE [$Fact_k$] among its conclusions. Then, a four-stage attempt to satisfy the premises $Premise_i$ of this rule is initiated, by asking:

1. Is $Premise_i$ known to be true (in light of the existing list of facts)?

2. Can $Premise_i$ be inferred from the current list of facts (using another rule or several rules)?

3. Can conventional analysis and design procedures be used to determine if $Premise_i$ is true?

4. Can the user determine if $Premise_i$ is true?

If the answer to any of the above questions is "yes", for each premise $Premise_i$ of the rule, then the goal has been achieved and $Fact_k$ is written into the knowledge base. In the case of an "if-and-only-if" rule, showing that a premise is false will terminate the backward chaining process. Otherwise, the backward chainer will continue to seek rules that can be satisfied so that $Fact_k$ can be written. The questions are always asked in the above order, so that running external procedures and asking the user are only done as last resorts.

Lists of Facts

The status of a session at any given time, or the outcome of a session when completed, is characterized by the list of facts that has been written in the course of the transaction. Such a list is illustrated in Table 2, which contains the list of facts that might exist in the CACE-III knowledge base at the end of a DIAGNOSE and SPECIFY session. In order to interpret Table 2, we provide the following brief guide to the shorthand notation in our list of facts: NL $\rightarrow$ nonlinear, FNAME $\rightarrow$ file name, DIAGN $\rightarrow$ diagnosis, L $\rightarrow$ linear, DIAG-DOMINANT $\rightarrow$ diagonally dominant, SS $\rightarrow$ steady-state, CHi $\rightarrow$ channel$_i$ of a multi-input multi-output plant. Observe that the underlying *data*, e.g., the numerical results of the nonlinear system diagnosis and the linearized model and its diagnosis, may be contained in ancillary files. The expert system often does not use this kind of data directly, but it must know where such information can be found so that it can be provided to external analysis and design procedures as required. The basis designation RBi refers to the rule base partitioning outlined above.

The preceding overview outlines the three major elements of a production rule-based expert system: rules, facts, and inference engine. Each component is in fact disarmingly simple. The power of this programming approach arises from this partitioning, which separates the knowledge into rules and facts, and deals with them in a dynamic fashion using a processor (inference engine) which may be simple or quite sophisticated. The capability of the processor is generally driven by the complexity of the problems to be solved; it is clearly beneficial to have the correct "match" to avoid the extremes of "overkill" (using a powerful inference engine to solve simple problems, usually at a high cost in terms of computational and response time) and "brute force" (having to develop an unnecessarily large rule base of simple-minded rules that may also take a long time to process). This division achieves a high degree of transparency, so that the rules can be kept lucid (assuming the expert system

Table 2. THE PROBLEM FRAME AFTER A DIAGNOSE AND SPECIFY SESSION

FACT			BASIS
OBJECT	ATTRIBUTE	VALUE	
PLANT	MODEL	NONLINEAR	User: RB1
PLANT-NL-MODEL	FNAME	EXOREACT	User: RB1
PLANT-NL-MODEL	TIME-TYPE	CONTINUOUS	Inferred: RB1
PLANT-NL-MODEL	STATE-TYPE	CONTINUOUS	Inferred: RB1
PLANT-NL-MODEL	ORDER	2	Inferred: RB1
PLANT-NL-MODEL	INPUTS	2	Inferred: RB1
PLANT-NL-MODEL	OUTPUTS	2	Inferred: RB1
PLANT-NL-MODEL	DIAGN-DATA-FNAME	EXORNDATA	Inferred: RB1
PLANT-NL-MODEL	NL-BEHAVIOR	MILD	Inferred: RB1
PLANT-L-MODEL	FNAME	EXOREACTL	Inferred: RB1
PLANT-L-MODEL	STABLE	NO	Inferred: RB1
PLANT-L-MODEL	CONTROLLABLE	YES	Inferred: RB1
PLANT-L-MODEL	OBSERVABLE	YES	Inferred: RB1
PLANT-L-MODEL	MINIMUM-PHASE	YES	Inferred: RB1
MODEL	DIAGNOSIS	DONE	Inferred: RB1
SENSOR	TIME-TYPE	CONTINUOUS	User: RB2
CONTROLLER	TIME-TYPE	CONTINUOUS	User: RB2
CONTROLLER	STRUCTURE	DIAG-DOMINANT	User: RB2
CONTROLLER	CHANNEL1-IN	U1	User: RB2
CONTROLLER	CHANNEL1-OUT	Y1	User: RB2
CONTROLLER	CHANNEL2-IN	U2	Inferred: RB2
CONTROLLER	CHANNEL2-OUT	Y2	Inferred: RB2
MAX-STEP-SS-ERR	CH1-VALUE	0.25	User: RB2
MAX-REAL-PART	CH1-VALUE	-1.4	User: RB2
MIN-DAMPING-RATIO	CH1-VALUE	1.0	User: RB2
MAX-STEP-SS-ERR	CH2-VALUE	0.5	User: RB2
MAX-REAL-PART	CH2-VALUE	-1.4	User: RB2
MIN-DAMPING-RATIO	CH2-VALUE	1.0	User: RB2
CONTINUOUS-SPEC	ENTRY	DONE	Inferred: RB2
CONTINUOUS-SPEC	ENTRY	REALISTIC	Inferred: RB2
CONTINUOUS-SPEC	ENTRY	COMPLETE	Inferred: RB2
CONTINUOUS-SPEC	ENTRY	CONSISTENT	Inferred: RB2
SPEC-SESSION	TERMINATION	NORMAL	Inferred: RB2

developer has a clearly-enunciated problem and solution formulation) and the inference mechanisms can be developed and validated separately.

Other Inference Engine Requirements

The above discussion provides a basic overview of selected expert system concepts, in particular, of rules, lists of facts (knowledge bases), and two modes of inference. There are other requirements for engineering design that go substantially beyond the needs for many other well-known expert system applications found in the general literature cited previously. Important additional capabilities are: the ability to run and communicate with external processes, the ability to perform numerical computations and tests, support for a partitioned rule base, non-monotonic reasoning, and the ability to be controlled or "steered" by the user.

In order for CACE-III to execute the analyses required for diagnosis and design, it is necessary for the inference engine to run external processes (conventional CACE software, e.g., the Federated System) that carry out tasks such as calculating the frequency response of the plant, determining gain margin and bandwidth, simulation, etc. CACE-III must initiate the process with the appropriate input parameters, and must be able to access the results of the analysis in order to update the list of facts. The DELTA inference engine we initially used in our research did not have this capability; converting to Delphi allowed us to perform this function.

The DELTA inference engine could not deal directly with numbers either, having only the ability to interpret and use literal strings, such as FOUR, POSITIVE, and BETWEEN-THREE-AND-FIVE. This restriction was not important in the original application (the diagnosis of diesel-electric locomotives; Bonissone, 1983), but in our case it required the use of numerous additional rules and menus for the entry of data in literal form. This limitation also does not exist in Delphi.

We found that CACE calls for a partitioning of the rule base into six parts plus a seventh supervisory rule base ("meta-rule base"). This was not supported by Delphi, so we extended the shell to accommodate this functionality (James, Taylor, and Frederick, 1985).

Non-monotonic reasoning is often required in engineering problem solving. This phrase designates a process in which the problem solution develops for a number of steps (procedures are carried out and facts accumulate), then it is necessary to backtrack and try again taking a different path. This happens whenever specifications cannot be met or a trade-off study is required, for example. In such instances, it is necessary to reset the facts by defining a previous point in the problem solution to be the new "current state" of the KBES and eliminating the facts developed after that point ("retracting belief"). It may be necessary to save the KBES state (in the case of design trade-off study), or the information might be partially or completely discarded (in the case of a dead-end). In any event, the retraction of belief may be difficult to manage in full generality, but can be implemented if it is permitted only in carefully-defined situations (e.g., design trade-off studies); refer to James, Bonissone, Frederick, and Taylor (1985).

The last requirement -- steerability -- will probably be crucial for the acceptance of any expert system for engineering design such as CACE-III by the engineering community. If a KBES is always in complete control, the engineer will become frustrated and/or bored. Frustration would be caused by having to accept the dictates of the software even when the user's experience contradicts them, while boredom would result from the tedium of being led by the hand through familiar parts of the task (e.g., specification entry).

We have identified the following mechanisms for allowing the designer to interact effectively with CACE-III in the design process:

1. The support provided by CACE-III is in the form of suggestions rather than constraints. If the user wants to ignore or not fully comply with the expert system's recommendations, then the software will not force compliance.

2. A 'why' facility can be invoked to determine the basis for a recommendation, so that the user can judge the advice before deciding whether or not to accept it.

3. A manual fact-entry capability can be added to allow the engineer to write facts directly into the knowledge base. For example, the goal of the specification entry portion of a transaction may be to write the following facts:

    MAX-STEP-SS-ERR   VALUE 0.005
    MAX-REAL-PART   VALUE -1.4
    MIN-DAMPING-RATIO VALUE 1.0

    Allowing a designer who does not require help to enter these facts directly will streamline the session considerably; the alternatives (e.g., by being led through a series of menus, Taylor, Frederick, and James, 1984) may be much less desirable to the experienced user. The use of a natural language interface would facilitate this type of user interaction greatly; we have not developed this idea at this point in our work.

4. The user can be permitted to invoke any underlying conventional CACE package and do whatever is desired.

These capabilities should help the user to maintain control of the design process in an effective way. The use of a 'why' facility is well known and exists in DELTA and Delphi; the other ideas (especially items 3 and 4) are specifically targeted for engineering users and may not be in commonplace usage. The last two features may also be difficult to implement; see LESSONS below.

The above comparisons of DELTA and Delphi were made primarily to contrast the requirements of an inference engine that is designed for pure symbolic manipulation tasks, such as diagnosis and maintenance, with those of an environment for CAD, where numerical processing and interactions with conventional analysis and design software are essential. The additional capabilities of Delphi are based on the original specification or definition of Delphi and were implemented using standard LISP functionalities.

CACE-III PROJECT STATUS

So far, we have developed a number of the necessary expert-system concepts, architectures, and real software implementations needed for an expert-aided environment for CACE:

- a *definition* for CACE-III, including a specific, detailed outline of the functional characteristics of the expert system and of the rule base,

- an *architecture* for CACE-III, including a partitioned rule-base structure that effectively implements all of the required activities (see Fig. 3) and an inference engine that supports it,

- *working rule bases* for several functions, including equilibrium finding and linearization, linear and nonlinear system model diagnosis, specification development, an automatic control system design procedure (lead-lag compensator design for a single-input/single-output plant, James, Frederick, and Taylor, 1985), and some aspects of validation (simulation with linear and nonlinear plant models).

- a proven means of interfacing the KBES with the required conventional modeling, simulation, analysis, and design software, and handling the interfaces among the various software packages (data and model transformations), and

- mechanisms providing the user with a high-level interface that provides support flexibly and without over-domination.

The CACE-III project we have described above is still in the exploratory or concept development phase. We have a working expert system that can perform major parts of the problem creditably, e.g., designing a controller using a conventional frequency-domain design package; this activity alone requires 70 rules which cause about 100 commands to be issued to the design package CLADP in the course of one design. To illustrate the capabilities of this system, CACE-III designed a high-order lead/lag compensator for a fifth-order plant G(s) with a pair of lightly damped poles; both the plant and compensator are portrayed in Fig. 4. The performance of the resulting closed-loop system is contrasted with that of a constant-gain compensated system in Fig. 5. This design exercise took about 15 minutes on a VAX 11/785, including specification development, design, and validation by simulation (James, Frederick, and Taylor, 1985).

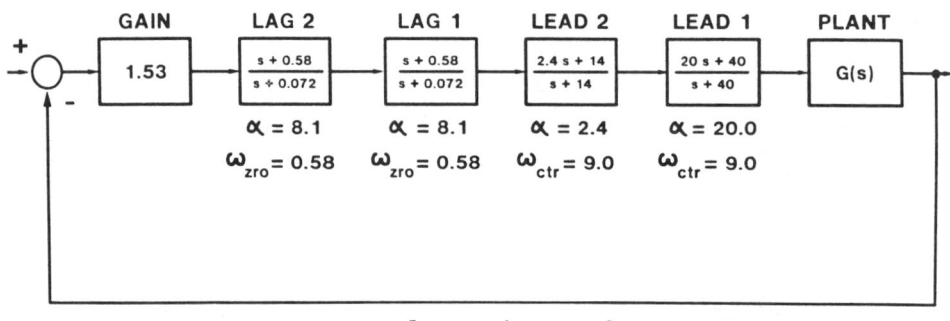

$$G(s) = 500(s+10)/s^5 + 33s^4 + 337s^3 + 1775s^2 + 4950s + 5000$$
POLES: $s = -2, -3 \pm j4, -5, -20$

Figure 4. Block Diagram of the Compensated System

In this effort, we have shown how expert systems may be used to provide meaningful solutions to the problems identified above. However, there is unquestionably considerable detail that needs to be filled in and implemented before we have a complete "real system".

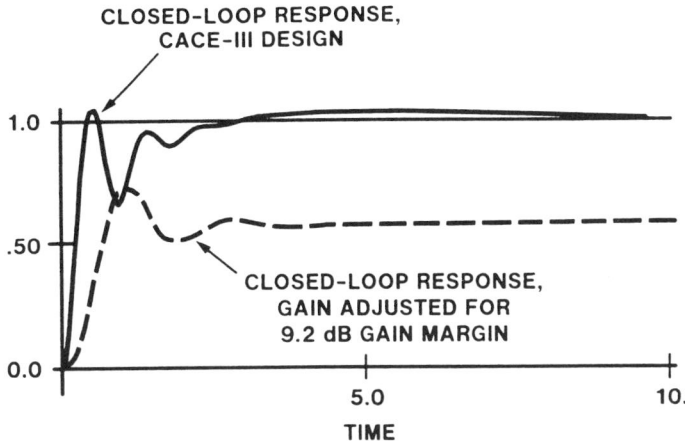

Figure 5. Closed Loop Time Responses

LESSONS AND UNFINISHED BUSINESS

There are many ways in which CACE-III is incomplete. We have come to realize that, in some senses, our present KBES represents only the tip of the iceberg. The following are several areas in which our present treatment is incomplete or totally lacking.

Support for Different User Capabilities

We have discussed our desire to support both the novice ("leading the user by the hand", if necessary) and the expert (without domination and boredom). To date, our primary focus has been on the support of the novice. Some ideas for providing an environment that would be more acceptable to the expert have been generated, including allowing the direct entry of facts (e.g., performance specifications) and allowing the user to invoke software to carry out other more arcane procedures. Neither of these ideas are easy to implement in a satisfactory way.

The direct entry of facts is simple, in the sense that providing the user the ability to input [MIN-DAMPING-RATIO VALUE 1.0] +1 is clearly feasible. The problem is that this, in effect, creates a new, very narrow command language that again requires careful attention to syntactic details. For example, [MINIMUM-DAMPING-RATIO VALUE 1.0] +1 would not be a useful input unless the KBES can perform the pattern matching with enough intelligence to recognize that MIN and MINIMUM are synonymous. At worst, the KBES would accept the erroneous fact (which would be ignored, because of the pattern mis-match *vis-a-vis* the rule base); at best the KBES would issue an error message and the user would have to find the correct formulation and enter it again. Neither course of events would be satisfactory.

The only worthwhile solution is to incorporate a natural language interface, so that the user could enter a statement such as **the minimum acceptable damping ratio value is 1.0** and have the input interpreted by the KBES appropriately. The size of this task is directly proportional to the generality of the interface. It is certainly feasible to implement this capability for a well-defined task such as specification entry using a rudimentary grammar, for example; to permit the user to command the KBES in greater generality would be a major undertaking.

Allowing the expert to carry out his own procedures in the middle of a design session and then let the KBES take over is also very difficult to accomplish in general. The expert system must be able to monitor the activity, interpret it, and update the list of facts accordingly as the user works; if this is not done, then the user cannot return to utilize the KBES capabilities based on the work done manually. Therefore, incorporating this capability is limited by our ability to foresee what actions a user might take so that these mechanisms could be built into the KBES, and by the amount of work required for implementation.

To be feasible, this option can be permitted only at well-defined points in the process, and only for well-defined activity. One way to implement this functionality is to incorporate some of the KBES procedural knowledge in the form of "macros" that the expert user can modify and execute from within the KBES; this will ensure that the KBES can understand the expert's activity so that the knowledge base is updated appropriately and the KBES can build on the results.

The above paragraphs outline our ideas for supporting the expert user. Based on the difficulty we see in implementing them with any generality, it seems clear that KBES environments are in fact easier to develop for the novice than for the expert user.

The Limitations of Scenarios

As mentioned above, the generation of scenarios (e.g., Taylor, Frederick, and James, 1984; Taylor, 1985a) played a large part in the development of CACE-III. The best way to acquire the knowledge that will be captured by any KBES is to carry out the given activity in full detail, with a realistic problem, using exactly the required tools. In our context, we selected a nonlinear plant model and first went through a "dry run" on paper, then developed rule bases to capture the process using software from the Federated System. At this point, we can carry out several of our initial scenarios in full.

There is one limitation or drawback to the use of scenarios: There is a tendency to think that a KBES that can carry out a scenario is a "real system". Often this is not true. In our case, there is often too much "branching", "groping", and "exploration of alternatives" involved in CACE to make a one-scenario KBES of general utility. In other words, an implementation that can only execute a single prescribed sequence of procedures would generally miss much of the judgment and selection of alternatives that has to be performed in the solution of real-life engineering design problems.

Making Software Usage Details Invisible

Eliminating much of the overhead in the use of conventional CACE software is clearly feasible, especially for the non-expert user who carries out procedures that have been anticipated and built into the system. The functions to be assumed by the KBES are invoking packages, executing algorithms, and error handling.

The knowledge of package capabilities and command sets and syntax can be incorporated readily into the KBES. Less-than-expert users would probably never have to issue a command in its native form; instead, the KBES would invoke packages and issue all commands based on the state of the problem solution and/or the wishes of the user. Implementing this capability may require a substantial amount of work, especially for a broad CACE environment such as the Federated System, but the effort is not conceptually demanding.

Exception handling is more difficult but still not a major problem, as long as the use of the software is limited to the activity foreseen by the KBES developers.

Again, there is a lot of attention to detail (starting with a catalog of possible error messages and recommended corrective action), but not much high-level reasoning involved.

There is an additional payoff to incorporating the knowledge of command sets, syntax, and exception handling into the KBES. If this knowledge is well organized, then the conventional software can be regarded as "interchangeable parts" in the sense that the architecture can be made highly modular, allowing the introduction of new packages that may extend the expert system's capabilities or improve its performance (through the use of numerically superior algorithms, for example).

Data-Base Management

The CACE data base has yet to be defined in complete detail; this step is a prerequisite to the rigorous implementation of DBM into a KBES such as CACE-III. The need for such a definition and preliminary ideas regarding its structure and contents have been presented by Maciejowski (1984) and LeVan (1984). Despite the incomplete definition of the data base, it is possible to consider the issue in general terms.

The fact that the CACE-III architecture is based on a conceptual model of CACE processes would seem to make some form of DBM a very natural and direct addition. We consider DBM functions in this context to include keeping track of all the files containing models (nonlinear and linear), the operating points and other conditions of validity of linearized models, analysis results, diagnostics, performance results, et cetera, along with the necessary relational information.

The CACE-III solution frame contains the list of facts in which the expert system keeps track of what has been done and what needs to be done, information required for decisions about the selection of analysis and design procedures and tradeoff analysis, and a log of the entire transaction; at this point in our work, we do not see a need for any DBM information that is not available in this knowledge base. The procedural and relational knowledge of the operation of the software is thus readily available so that the KBES can perform DBM activities itself (keeping track of the data base and writing DBM information into a master file) or invoke a conventional DBM system to carry it out. One use of such a master data-base management file is illustrated in Taylor (1985a). As mentioned previously in the context of software usage, DBM along these lines entails straightforward work that is not conceptually taxing. A rule base must be developed which "monitors" the activity of the KBES, and, whenever a procedure is performed which creates new information, ensures that the data are written to the appropriate file and/or the file is appropriately named and cataloged with the required relational information.

RELATED WORK

To the best of our knowledge, the first in-depth consideration of the application of KBES technology to CACE was by Taylor, Frederick, and MacFarlane (1983). That presentation was the genesis of our CACE-III project. Since that time, there have been a number of projects related to this concept:

- Gomez *et al* (1984) describes a prototype expert system for the treatment of stochastic control and nonlinear filtering problems. The program is written MACSYMA, LISP, and PROLOG, and accepts user input in natural language and symbolic form. It carries out the basic analysis of the user's problem in symbolic form and produces FORTRAN code implementing the controller or filter algorithm. PROLOG is used to check the well-posedness of the user's problem.

- LeVan (1984) considers the data representation requirements for CACE using frames and slots. The concept of frame is extended to provide slots for default/required information, procedures, nested frames, and ports for the input and output of CACE process "flows". The system modeling approach incorporated in this concept is the bond-graph methodology.

- Birdwell *et al* (1985) describes the overall concept of a KBES for CACE and a prototype system that implements these ideas for modern control and estimation problems (linear quadratic regulator design, Kalman-Bucy filters).

CONCLUSIONS

We believe that the research outlined in this paper has reached the point where the promise of KBES technology to solve many of the problems identified in existing conventional software can be seen to be real. However, the amount of effort involved in realizing such a higher-level environment in a useful form should not be underestimated, and the developer must be careful not to raise unrealistic expectations.

Research in this area is still in its infancy. The next few years should bring about a number of useful systems based on the ideas and work presented here and in the references.

Acknowledgements. Some of the material in the sections titled **EXPERT SYSTEM CONSIDERATIONS** and **AN EXPERT SYSTEM ENVIRONMENT** is taken quite directly from Taylor and Frederick (1984); the section **A Paradigm for CAE** is closely patterned on a section of GE internal report 84CRD127 by Taylor and Frederick. Special mention must be made of the work of John R. James, who, in the course of his PhD research at Rensselaer Polytechnic Institute, did much to bring CACE-III to reality.

REFERENCES

Birdwell, J. D., J. R. B. Cockett, R. Heller, R. W. Rochelle, A. J. Laub, M. Athans, L. Hatfield (1985). Expert systems techniques in a computer-based analysis and design environment. *Proc. Third IFAC Symposium on CAD in Control and Engineering Systems*, Lyngby, Denmark.

Coombs, M. J. (1984). *Developments in Expert Systems.* Academic Press, New York.

Davis, R. and D. B. Lenat (1982). *Knowledge-Based Systems in Artificial Intelligence.* McGraw-Hill, New York.

Edmunds, J. M. (1979). Cambridge linear analysis and design programs. *Proc. IFAC Symposium on Computer-Aided Design of Control Systems*, Zurich, Switzerland.

Elmqvist, H. (1977). SIMNON - An interactive simulation program for nonlinear systems. *Proc. Simulation '77*, Montreux, France.

Gomez, C., J. P. Quadrat, A. Sulem, G. L. Blankenship, P. Kumar, A. LaVigna, D C. MacEnany, K. Paul, and I. Yan (1984). An expert system for control and signal processing with automatic FORTRAN code generation. *Proc. 23rd Conf. on Decision and Control*, Las Vegas, NV.

Hayes-Roth, F., D. Waterman, and D. B. Lenat (1983). *Building Expert Systems.* Addison Wesley, Reading, MA.

IEEE (1984). *Workshop on Principles of Knowledge-Based Systems.* IEEE Computer Society, IEEE Cat. No. 84CH2104-8.

James, J. R., P. P. Bonissone, D. K. Frederick, J. H. Taylor (1985). A retrospective view of CACE-III: considerations in coordinating symbolic and numeric computations in a rule-based expert system. *Proc. Second Conf. on Artificial Intelligence Applications*, Miami Beach, Florida.

James, J. R., D. K. Frederick, and J. H. Taylor (1985). On the application of expert systems programming techniques to the design of lead/lag precompensators. *Proc. Control 85*, Cambridge, UK.

James, J. R., J. H. Taylor, and D. K. Frederick (1985). An expert system architecture for coping with complexity in computer-aided control engineering. *Proc. Third IFAC Symposium on CAD in Control and Engineering Systems*, Lyngby, Denmark.

LeVan, O. (1984). An expert system for control system design. *Proc. 18th Asilomar Conf. on Circuits, Systems, and Computers*, Asilomar, CA.

Maciejowski, J. M. (1984). Data structures for control system design. *Proc. EUROCON '84*, Brighton, UK.

Michie, D. (1982). *Introductory Readings in Expert Systems*. Gordon and Breach, London.

Moler, C. (1982). MATLAB users' guide. University of New Mexico Department of Computer Science Technical Report CS81-1 (Revised).

Spang, H. A. (1982). The federated computer-aided control design system. *Proc. Second IFAC Symposium: CAD of Multivariable Technological Systems*, Purdue, West Lafayette, IN, 121-129.

Spang, H. A. (1984). The federated computer-aided control design system. *Proc. of the IEEE*, **72**, 1724-1731.

Taylor, J. H. (1982). Environment and methods for computer-aided control systems design for nonlinear plants. *Proc. Second IFAC Symposium: CAD of Multivariable Technological Systems*, Purdue, West Lafayette, IN, 361-367.

Taylor, J. H., D. K. Frederick, and A. G. J. MacFarlane (1983). A second-generation (sic) software plan for computer-aided control system design. *Abstr. of IEEE Symposium on Computer-Aided Control System Design*, MIT, Cambridge, MA.

Taylor, J. H., D. K. Frederick, and J. R. James (1984). An expert system scenario for computer-aided control engineering. *Proc. American Control Conf.*, 120-128, San Diego, CA.

Taylor, J. H. and D. K. Frederick (1984). An expert system architecture for computer-aided control engineering. *Proc. of the IEEE*, **72**, 1795-1805.

Taylor, J. H. (1985a). An expert system for integrated aircraft/engine controls design. *Proc. National Aerospace and Electronics Conf. (NAECON)*, Dayton, O, 661-669.

Taylor, J. H. (1985b). Computer-aided control engineering environment for nonlinear system analysis and design. *Proc. Third IFAC Symposium on CAD in Control and*

Engineering Systems, Lyngby, Denmark.

Wieslander, J. and I. Gustavsson (1976). IDPAC - an efficient interactive identification program. *Proc. 4th IFAC Symposium on Identification and System Parameter Estimation, Tbilisi, USSR*.

EXPERT SYSTEMS IN ON-LINE PROCESS CONTROL

Robert L. Moore
Lisp Machine, Inc., Los Angeles, CA 90045
Mark A. Kramer
Massachusetts Institute of Technology, Cambridge, MA 02139

Abstract. Expert systems are an attractive vehicle for supervisory control of chemical processes, with applications to alarm handling, failure detection and diagnosis, optimizing control, and process management. The program PICON addresses the problems of applying expert systems in dynamic environments. The hardware and software issues involved in conducting inference processes with rapidly varying, dynamic data is the first focus of this paper. The second problem discussed involves construction of the knowledge base to carry out the task of process malfunction detection and diagnosis. Model-based reasoning schemes rather than heuristic-based methods are most powerful in this application. Recommendations for future research in this area are outlined.

Keywords. Expert systems; supervisory control; diagnosis; fault detection; process control.

INTRODUCTION

Modern distributed systems bring thousands of plant measurements to a central control room, where they are available for display to an operator. In general, the operator has only low-level information on the plant behavior from these measurements, including single point values, alarm status and trend plots on demand. An expert system, added to the distributed system, can apply some intelligence to this flow of information, to provide the operator with less, but more intelligent, information concerning plant conditions.

The most obvious application is during a major plant upset, where an operator must cope with hundreds of alarms in a limited time. In the case of a nuclear power plant, for example, the operator may have 800 alarms in the first two minutes of a major upset. An operator may be an expert, but in a short time it is difficult for any human operator to diagnose multiple problems involving such quantities of information.

An expert system could also be applied to detect measurement integrity or process integrity problems. An elementary example is to calculate a material balance around a process and to compare it with the measured flows. A discrepancy could indicate a measurement failure or a process failure, such as a leak. The application of such inference in a process control context was discussed by Fortin, Rooney and Bristol (1983) of Foxboro. They consider the alarm advisor application, and give a specific example of its proposed use. Their knowledge rules consist of deep knowledge (material balance models) as well as heuristics. We consider this synergistic use of both types of knowledge to be vital in process control applications.

The use of expert systems for control has been investigated by Astrom (1984). Expert control offers a compromise when accurate and tractable models for optimization are not available. This is the common case for industrial processes, and the current alternative is manual control, which may be subject to varying quality. Expert control offers the potential for providing consistent expert advice or direct control.

Expert systems for process startup and shutdown are a particular area of interest. These operations are done infrequently, but require considerable expertise, and errors can be very costly. One petrochemical company described an experience where an operator-mistake during a shutdown of a unit caused a shutdown of a large part of the refinery operations for more than a day.

The "teachable" nature of expert systems opens new opportunities for plant control approaches. Expert systems have the characteristic that the knowledge base can be augmented and modified. This allows adaptation to changes in plant condition and operating objectives - the equivalent of changes in operating instructions which is a common occurrence in plant operations. An expert system, once the knowledge is defined, could serve as a training vehicle for new operators. In many cases expertise is built-up over many years of operation, and is lost when an employee leaves or retires. The knowledge base of an expert system can be expected to become a corporate resource, available without disruption for training, simulation and other purposes as well as being used for advice and control for plant operations.

Real-time requirements force some special design considerations. The requirements of real-time expert systems have been defined in some detail by Sauers and Walsh (1983) of Martin-Marietta. They utilize a goal-directed production-rule scheme, with the capability of fetching selected rule sets into the environment at execution time. This allows a focus of system resources on current subtasks. They allow user-selected inference procedures as well as cost estimates of subgoal evaluation to control the search. They also improve the efficiency of pattern matching, via saving of pattern matching information between evaluation cycles.

Kawakami (1984) of Hitachi reports work done in Japan on diagnosis applied to large scale systems. He cites the following three problems limiting the success of the effort:

1. Insufficiency of speed and shortage of memory of the computers used for inference.
2. Lack of an effective knowledge acquisition method and of a proper knowledge description scheme.
3. Difficulty in detecting inconsistency in the knowledge base.

The work described in the present paper addresses these problems.

The discussion in this paper considers first the issues of expert system design for real-time process control applications. These issues have been addressed in the design of the Process Intelligent Control (PICON) expert system. Some issues of knowledge engineering for the problem of fault diagnosis are then addressed in some detail. Finally, projections on significant unsolved problems are presented, which will be addressed in the near future.

ISSUES OF REAL-TIME EXPERT SYSTEMS IN PROCESS CONTROL

Process control applications have a number of constraints and aspects which differ significantly from most expert systems applications. Among the more challenging aspects:

1. The large numbers of measurements in a process plant, sometimes exceeding 10,000 measurement points. Many of these measurements can change significantly within a few hours, minutes or even seconds. Furthermore, the plant state is not fully defined by the measurements at any one time, so that dynamic history of measurements may be important.
2. A large plant will require a large knowledge base. It is easy to project heuristic knowledge bases larger than any expert system yet implemented - knowledge bases of 10,000 to 100,000 rules are reasonable estimates. In a later section of this paper, we challenge the concept that plant knowledge can be adequately expressed by heuristics. In any case, process models are a convenient way to express a considerable amount of knowledge, and so the expert system must allow models and structure (deep knowledge) as well as heuristic representations.

3. In the plant emergency application in particular, the expert system must provide real-time advice or control action. Conventional expert system paradigms, which exhaustively search rule bases and fire rules, are inadequate for this task where tens of thousands of rules or more are involved.
4. In a large application the "knowledge engineering bottleneck" is a critical consideration. The process experts are typically in the plants, a long distance from centers of A.I. expertise. Unless extraordinary resources are placed on site, it is unrealistic to implement a knowledge base using complex AI programming methods. A natural language interface to the knowledge base is a practical necessity.
5. In a large knowledge base the issues of consistency and completeness of the knowledge base are serious. Since there is no known way to evaluate these issues in a general way, it is necessary to provide tools for the process engineer to use to test for the adequate behavior of knowledge and inference.

THE PICON PROJECT

These challenges were addressed in a project undertaken by LISP Machines, Inc. in cooperation with a group of process companies. The development project for PICON was started in 1983. A design which addresses the real-time process control issues was prototyped and tested in early 1985.

Interfacing with Real-Time Process Data

The Process Intelligent Control (PICON) package is designed to operate on a LISP machine interfaced with a conventional distributed control system, where as many as 20,000 measurement points may be accessed. This presents a major communication problem, which was solved by using the Lambda machine from LMI (Fig. 1). The Lambda provides two processors running in parallel: a LISP processor for

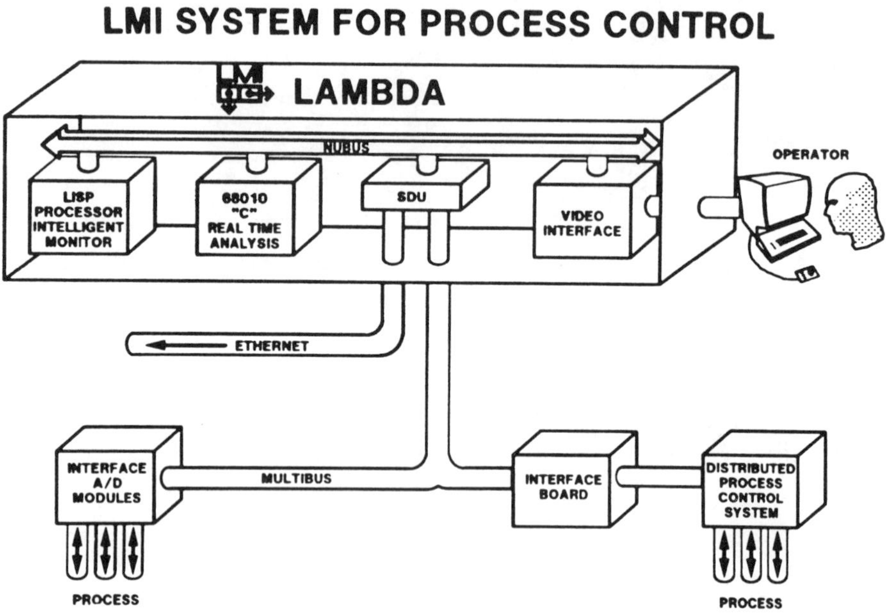

LMI SYSTEM FOR PROCESS CONTROL

Fig. 1. LMI system for process control.

the expert system; and a 68010 processor for real-time data access and certain low-level inference tasks. The 68010 uses an integral MULTIBUS to communicate with the distributed process control system. PICON requests data as required for inference, but does not force the distributed system to transmit all measurements and alarms on a fixed scan basis.

Real-Time Inference

The inference supported by PICON includes forward-chaining and backward-chaining. Within the context of alarm advising, there are requirements for both of these procedures. An expert process operator, during normal plant operation, scans key process information to monitor performance and detect problems which may not cause explicit alarms. The procedure which reflects this approach is a scanned forward-chaining inference, examining dynamic process conditions. The rules which determine possibly-significant-events are scanned, and condition-matching triggers a focus on the problem area.

In process control, the rules for possibly-significant-events are a combination of process measurements and process dynamic behavior. For example, a simple tank level may be viewed by an operator as:

IF tank-level is-greater-than 80%
OR (tank-level is-greater-than 60%
AND rate-of-increase tank-level is-greater-than 10% per-minute)
THEN tank-level is-too-high.

A more typical example would include multiple variables, with dynamic behavior more complex than rate of change.

An expert process operator, once alerted, will focus attention on the problem. This may involve invoking procedures for safety or other reasons, diagnostic rules, and it may involve assembling information to allow inference about the problem. Logic rules and procedures are used only when required for the diagnostic inference. PICON mimics the expert process operator in this regard: Logic rules and procedures are invoked specifically when they are required for diagnosis of a process problem, or as requested for a specific step in the inference.

Another use of this "focus" facility is to scan the plant in a background mode, focusing attention on parts of the plant to evaluate unit process performance and detect subtle problems, utilizing both the knowledge base of the the expert process operator and the expert process engineer. It is not practical to examine an entire plant continuously with this intensity, but the individual parts of the plant could be scanned in a background mode. This is equivalent to the way a process engineer would analyze plant performance during normal plant operation, looking at one plant area at a time.

It should be noted that the ability to focus not only emulates the way a human expert works, but also it avoids the problem

associated with overloading the distributed process system with
requests for information. While the expert system has latent
knowledge of how to access up to 20,000 measurement and alarm points
in the process environment, only those of interest to the expert
system at a given time need be accessed. This allows the expert
system to operate within the data transfer constraints of the
distributed system, which is generally a few hundred measurements
per second.

Organization of the Knowledge Base - Deep Knowledge

Plant knowledge is organized with a schematic at the top level.
Knowledge of the underlying plant structure and models ("deep
knowledge" in expert systems terminology) is entered into the
knowledge base by graphical construction and by specifying
attributes. Each item on the schematic has an associated frame of
knowledge, which includes such items as
- attributes of ports to other process equipment or
  instrumentation
- tag names of associated instrumentation
- parameters of models
- provision for dynamic model

Because the limiting task in developing an expert system is
knowledge engineering, we considered it important to provide
easy-to-use facilities for entering process engineer knowledge and
operator heuristics into the knowledge base. We developed two
knowledge capture tools: a schematic capture tool for process
knowledge, and a structured natural-language interface for heuristic
knowledge.

The schematic capture of process knowledge (Fig. 2) is an
interactive procedure. The knowledge engineer builds a schematic of

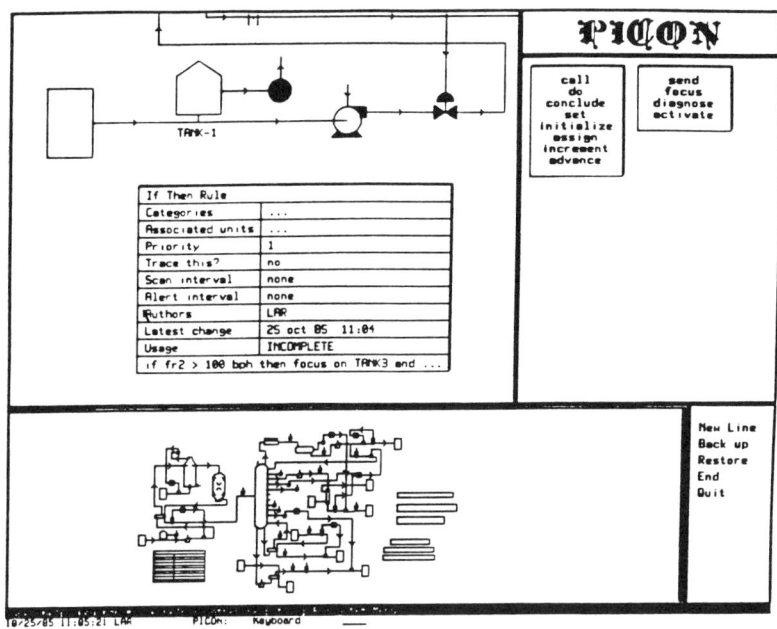

Fig. 2. Schematic capture of process knowledge.

the plant by selecting component objects. The inquiries then proceed, by pop-up windows, to request information required for the knowledge base, such as which of the various types of the component is being specified, specific parameter values, and a tag name which establishes access to the distributed system measurement in the case of sensor components. The schematic can be viewed in multiple windows, at a variety of scales, which allows conceptualization of large processes.

Organization of the Knowledge Base - Rule Frames

Heuristics are represented by frames of knowledge, which may have dynamic attributes. The heuristic frames include slots for:

  - A rule, which may include a conjunction of tests on

measurements, calculations such as material balance models and
dynamic model variables. The rule may result in conclusions
and activation of procedures.
- Attributes which relate to the rule, such as problem type and
  process unit.
- Attributes which are dynamic, such as certainty and data
  currency.
- Attributes relating to the use of the knowledge, such as
  scanning interval if the knowledge is to be periodically
  evaluated (see description of inference below), priority of
  knowledge, and currency requirements.
- A record of the authors and editors of knowledge, and the time
  of last edit.

Heuristics may be referenced in terms of attributes, so that the inference engine or a process engineer can access knowledge by referring to attributes such as process unit and problem type. Thus, heuristics can be invoked or edited by reference to specific types of knowledge, allowing the inference engine or a process engineer to focus attention on specific types of knowledge.

Inference Control

The inference engine has been specifically designed to gather data, arrive at conclusions, decide upon actions, and carry them out in an environment of continuously changing circumstances. It must be capable of changing its plans when circumstances change, and it must be able to operate as best it can with incomplete data. Furthermore, it must be able to operate in a situation in which not all information is up-to-date.

Measured or inferred values are time-stamped, so that it is possible to know how old the information is. Furthermore, every variable has what is called a currency interval. Information about states older than the currency interval will not be used, and the need to know the state when its latest value is no longer current will cause the variable to re-evaluate its state in the current circumstance. Different classes of variables may be given different currency intervals according to how volatile their state is.

The system provides backward chaining because a variable may request the values of other variables that it needs in order to determine its own state. This can generate a tree of backward references. Forward chaining occurs because variables also have the ability to excite other variables when they enter a certain state as specified by action lists in their heuristic frames.

The inference engine is not driven by logical unification as are several popular generic expert systems. It is a loosely procedural system which uses the concept of heuristic frames to provide structure and organization to the rule base. It makes use of IF...THEN... rules to arrive at conclusions, but (unlike unification driven systems) it does not use pattern matching to decide which rules to apply. Either the rules themselves have explicit suggestions as to what rules to apply when a given state is reached (and these can be provided by the expert), or it uses a database query facility to retrieve those rules pertaining to the current situation.

Timing in the Inference Engine.

Although we have described the inference engine as being loosely procedural, it is not procedural in the sense of a computer program that calls subroutines and then looses control until the subroutine which it calls has completed execution. When a heuristic frame has to evaluate its state, and it needs the states of other variables and these are not currently available, it excites those variables and terminates immediately. Just before terminating, it re-schedules itself to try again at some future time when its requisites may be available.

This brings us to an important piece of infrastructure of the inference engine: the event scheduler. This is capable of initiating specific activities at specific times in the future, and is the way that the inference engine reminds itself to do things. It is used in a variety of ways. For example, when a heuristic frame cannot determine its state, it reminds itself to retry after a short wait. Some activities schedule other activities rather than themselves.

Another piece of infrastructure is the cycle system which is used to initiate activities on a regular periodic basis, as distinct from the event scheduler which is used for scheduling single events. Those heuristic frames which perform periodic background checks (possibly-significant-event detectors) are examples of events that are scheduled cyclically.

The PICON system is designed so that several different process monitoring activities may be happening concurrently without interfering with each other. One might want to monitor a process in real time, simulate a possible future process, and run a test on recorded data from some time in the past all at once. The concept of time in PICON is abstracted, so that the system can run either in real time or simulated time.

Methods of Testing Knowledge

Several tools have been built to allow the process engineer to retrieve knowledge, to aid in the consideration of completeness and consistency. Using natural language menu choices, the engineer can retrieve knowledge by any of the attributes of the heuristic frames or schematic objects. For example, the engineer can request to see all knowledge relating to safety problems on a furnace. PICON promptly assembles all such knowledge for evaluation.

Simulation is another method of testing knowledge. To aid in this, simulation models can be constructed using the same natural-language facilities used in defining heuristic frames. Simulation frames can be connected to the process schematic and used for testing the knowledge base under various operating conditions.

Installation status

In May of 1985, the system was installed at the Port Arthur, Texas, facility of Texaco Chemicals. In August 1985, testing began at a pilot plant of Johnson Controls, and in September 1985 a system was installed at the Bayway, New Jersey facility of Exxon Chemicals. Several development groups are using PICON, including Oak Ridge National Laboratories, M.I.T. Chemical Engineering Department, Lockheed, Rockwell, Leeds & Northrup and Bell Laboratories.

KNOWLEDGE ENGINEERING FOR PROCESS MALFUNCTION DIAGNOSIS

One of the first tasks to be assigned to expert systems in process control is that of process malfunction detection and diagnosis. This task is a difficult one for process operators. Even well-trained operators may have difficulty diagnosing unanticipated failures, events of low probability, or incidents where multiple alarms are simultaneously activated. The appropriate experts may not be available to provide timely assistance to process operators. Expert systems can provide the critically-needed assistance for prompt detection and location of process faults.

There is no single prescription for construction of expert systems for process diagnosis. Several different designs have been successfully applied in the fields of medical, electronic, software, and process diagnosis. These approaches can be broadly classified into two categories. The shallow expert systems approach depends on pre-specified relationships between fault symptoms and malfunctions. The deep knowledge approach utilizes information on the design of the process in formulating the diagnosis. In the following, we discuss these approaches to building expert systems for process plant malfunction diagnosis emphasizing the merits and drawbacks of each approach.

Shallow Knowledge Approaches

The shallow expert systems approach tries to capture directly the relationships between irregularities in system behavior and system faults. The knowledge required in this type of expert system consists of empirical associations between the symptoms of malfunctions and the malfunctions themselves. As such, the approach can be viewed as a pattern recognition method. However, the expert system does not depend on a precalculated data base of measurement patterns, but instead utilizes production rules to mimic the deductive processes of the skilled operator. The shallow knowledge approach is typical in medical diagnosis programs, where underlying disease mechanisms are unknown or difficult to describe. The best-known example of a "shallow" diagnostic program is MYCIN (Shortliffe 1976).

The key task and frequent bottleneck in formulating an expert system of this type is knowledge acquisition. Expertise of sufficient breadth to cover a wide range of malfunctions must be encoded into the expert system. The difficulties are that:
- the knowledge requirements are unstructured and may be broad in scope
- there is no guarantee of diagnosability or completeness in the knowledge base
- the knowledge base is highly specific to the individual plant
- the number of rules may exceed practical limitations.

To reduce the complexity of building the expert system, it is desirable to decompose the problem into a number of small segments. This decomposition can be hierarchical (e.g. Davis et al. 1985) or according to unit operations. The focus facility of PICON provides a means of coordinating and controlling the execution of the modules, for any desired architecture.

A difficulty involved with modularization occurs when a single disturbance affects several units simultaneously. Unless special precautions are taken, the operator may receive conflicting diagnoses from the various modules activated in the course of the diagnosis.

To prevent this problem, <u>each module must contain sufficient knowledge to distinguish the effects of external disturbances from the effects of internal faults</u>. Without this knowledge, functioning units which are disturbed by fault propagation may be incorrectly diagnosed as faulty. Knowledge must be provided to deduce when the module is functioning <u>correctly</u>. Detailed models may be required for this determination; furthermore, unmeasured variables may make this determination difficult or impossible. The need to distinguish internal and external faults adds significantly to the complexity of the knowledge required for diagnosis.

Because of these drawbacks, the shallow knowledge approach is most appropriately applied to processes where model-based reasoning schemes cannot be applied, or to small operations where the scope of the knowledge required for diagnosis is limited, or to supplement model-based diagnosis schemes.

Deep Knowledge Approach

In the chemical engineering environment, much of the available knowledge is structured, and physical models based on thermodynamics, heat, mass and momentum transfer, kinetics, and process chemistry can be fruitfully incorporated into the diagnosis. The "deep knowledge" approach to diagnosis utilizes a model of the problem domain. Expert systems employing deep knowledge include ABEL, a program for diagnosis of electrolyte imbalances (Patil et al. 1981), and IDM, for diagnosis of electronic and mechanical devices (Fink 1985).

A deep knowledge expert system attempts to capture the underlying principles of a domain explicitly. In doing so, the need to anticipate every possible fault senario is eliminated. This results in the ability to handle a wider range of problem types and larger problem domains (Koton 1985). There are a number of model-based diagnostic methods which are a suitable philosophic basis for development of deep knowledge expert systems. In the following, we summarize three such techniques and offer criticisms of each.

<u>Casual Search</u>. Diagnosis can be described as the task of tracing process disturbances to their source. The fundamental premise of diagnostic techniques based on causality is that cause and effect linkages must connect the fault origin to the observed symptoms of the fault. The representation of causality employed most frequently in these techniques is the directed graph (digraph). The nodes of the digraph correspond to state variables, alarm conditions, or failure origins, and the edges (branches) represent the influences between the nodes. Positive or negative influences can be represented by the signs '+' or '-' given to the branches. The intensity of the effect, the time delay associated with the branch, and the probability of fault propagation along the branch may also be included in the digraph (Kokawa et al. 1983).

Methods of alarm management based on cause and effect analysis have been given by Andow & Lees (1975), Andow (1980), and Lees (1984). These methods group active alarms if they can be causally related to a single fault event. O'Shima and coworkers (Iri et al. 1979, Tsuge et al. 1985, Shiozaki et al. 1985) present a similar method which locates plant variables subject to external disturbances. This method attempts to locate nodes of the digraph which are causally connected to each observed disturbance. The method assumes that every process fault is directly associated with a deviation in the value of exactly one plant variable. The algorithm will not diagnose faults which simultaneously affect nominally independent variables, or affect causally-related variables in a manner inconsistent with the sign of the connecting branch. The presence of feedback loops degrades the diagnostic resolution of the method.

Despite these drawbacks, causal search is a suitable framework for a diagnostic expert system. Because causal search rarely yields a unique diagnosis, even when the process is well-instrumented (see Tsuge et al. 1985), this method is most powerful when combined with other techniques.

More work is needed in this area to improve the current algorithms to account for realistic factors such as noisy and time-varying data, and multiple faults. The following specific needs are perceived:
- Improvement of current algorithms to render the results insensitive to the selection of high and low thresholds for individual variables. Present algorithms can fail when a single variable expected to be high or low remains in the normal band, or when there is a spurious high or low reading.
- Improved qualitative models of the plant to treat conditional events. The digraph cannot account cases where two or more conditions must be present for a certain qualitative behavior to occur.
- Inclusion of time delay data to filter the fault possibilities on the basis of order of events.
- Extension of the analysis to more than one fault.

Some of these issues are discussed by Palowitch & Kramer (1985).

<u>Governing equations method</u>. A second class of model-based diagnostic techniques relies on redundancy between plant measurements. This redundancy can be explicit, as in the case where two or more sensors are applied to the same measurement, or calculated through the mathematical model of the plant. Several authors

have used this approach to detect sensor failures and mass balance violations (Mah et al. 1976, Romagnoli & Stephanopoulos 1981, Mah & Tamhane 1982, Crowe et al. 1983, Gertler & Singer 1985).

The governing equations method of Kramer & Palowitch (1985) applies not only to sensor failures and flow balances but also other classes of malfunctions and disturbances. This method involves associating each of the governing equations of the process (including material balances, energy balances, and modeling equations) with the fault conditions sufficient to cause violation of the equation. For example, suppose that $F1 - SP1 < \varepsilon$ represents the expected performance of a flow controller, where F1 is the measured flowrate and SP1 is the controller set point. If the left-hand side of this expression is significantly greater than zero, then one can infer: CONTROLLER-FAILS-HIGH or VALVE-STUCK-OPEN or F1-SENSOR-FAILS-HIGH. If it is significantly less than zero, then CONTROLLER-FAILS-LOW or VALVE-STUCK-CLOSED or F1-SENSOR-FAILS-LOW or CONTROLLER-SATURATION.

Fault diagnosis is accomplished by logical inference using the pattern of violated constraints. Let C_i^+, C_i^- and C_i^o represent positive and negative constraint violation and constraint satisfaction, respectively. Let F be the set of all possible faults, with members f. The set of faults which are sufficient to cause violation of the i-th constraint are defined as follows:

$$(\forall f)\ (f \rightarrow C_i^+) \in H_i^+ \qquad (1)$$
$$(\forall f)\ (f \rightarrow C_i^-) \in H_i^-$$

Let the condition of the plant be C^*, where $C_i^* = C_i^+$, C_i^-, or C_i^o depending on whether the i-th constraint is violated high or low, or satisfied. The set N_i^* of faults negated by the condition of the i-th constraint is:

$$\begin{aligned}(C_i^* = C_i^+) &\rightarrow (N_i^* = H_i^-) \\ (C_i^* = C_i^0) &\rightarrow (N_i^* = H_i^- \cup H_i^+) \\ (C_i^* = C_i^-) &\rightarrow (N_i^* = H_i^+)\end{aligned} \qquad (2)$$

Now let C_j be the j-th violated constraint. Let $H_j^* = H_i^+$ or H_i^- corresponding to the direction of constraint violation. Viable single-fault hypotheses are those which account for all violated constraints, and therefore lie in the intersection of the H_j^*. The set S_1 of viable single fault hypotheses is therefore:

$$S_1 = (H_1^* \cap H_2^* \cap \ldots H_{NV}^*) \sim (N_1^* \cup N_2^* \cup \ldots N_{NC}^*) \qquad (3)$$

If S_1 is empty, then multiple faults may be assumed. Formulae for resolution of multiple faults are given in Kramer & Palowitch (1985).

The topological procedures developed by Mah et al. (1976) and Romagnoli & Stephanopoulos (1981) provide alternate means of analysis of the structure of the governing equations which can be used to locate possible faults. However, Eq. (3) provides a more concise representation of the diagnosis than either topological method. Additionally, the set-based formulation is conducive to the application of fuzzy set theory (Zadeh & Fukanaka 1975) and similar non-Boolean reasoning techniques (Garvey et al. 1981, Barnett 1981). These techniques are needed in practice for formation of plausible diagnoses in the face of conflicting, noisy, and uncertain data. Kramer (1986) shows how Eq. (3) can be translated into a set of expert system rules using non-Boolean reasoning techniques.

<u>Hypothesis/test strategies.</u> Another deep-knowledge approach to diagnosis involves the process of hypothesis formulation and hypothesis testing. This approach mimics a typical human diagnostic technique of postulating a cause for an upset, determining the

symptoms of the postulated fault, and comparing the result to process observables.

A method of hypothesis formulation has been suggested by Fink (1985). This method involves three heuristics for narrowing the search for the location of a fault. The search locates units in the process whose inputs appear normal but whose outputs appear abnormal. While this method may work in the majority of cases, it will fail when inputs to the faulty unit are disturbed. This may be due to two causes: flow and pressure disturbances transmitted counter to the direction of material flow, and mass, energy, or control feedback around the faulty unit.

The method of causal search may be considered as an alternative method of hypothesis formulation. Under the assumptions that i) a single fault is present, ii) the primary effect of the fault is on one and only one variable, and iii) that the fault does not add new paths of causality to the process, the method of causal search as presented by O'Shima and coworkers is guaranteed in the absence of noise and transients to include in the set of possible causal origins the actual fault origin.

Hypothesis testing involves simulation of the effects of the postulated faults. Because of the real-time nature of the problem, numerical approaches are unsuitable. Qualitative simulation involves prediction of the direction of deviation of plant measured variables in response to faults. DeKleer & Brown (1984) and Kuipers (1984) have presented techniques for qualitative simulation of devices based on qualitative constraint equations which derive from the quantitative equations governing plant behavior. A constraint propagation method is used to determine the range of behaviors not disallowed by the qualitative constraints. Unlike numerical initial value problems, unique solutions cannot be guaranteed in qualitative simulations (Kuipers 1985). This is a result of the loss of information from the governing equations when they are converted to qualitative form.

FALCON (Chester et al. 1984) uses a qualitative representation of the plant similar to the signed digraph as the basis for qualitative simulation of processes. The difficulty with this form of qualitative model is that competing causal influences on a variable cannot be resolved without resort to additional information. FALCON uses heuristics and precedence rules to resolve these conflicts. Unfortunately, it can be proven that there is no single set of heuristics or rules that can resolve all such conflicts based solely on qualitative considerations. This is the result of the fundamental underspecification of the qualitative modeling problem.

Implementation

An expert system can use a more powerful representation of the plant than, for example, the signed digraph. A digraph in principle is limited to "or" logic cannot treat more complex logical conditions. A more general representation of plant structure is a set of rules describing the component types, the connectivity between components, the state of the component (referring to whether the component is on- or off-line, whether an input or output is measured, whether sensors are redundant, etc.), and behavior descriptions describing the relationship between inputs and outputs of the unit. Examples of the four types of rules are given by Yamada & Motoda (1983).

This rule-based representation of the plant is much more powerful and flexible than the signed digraph. Besides the ability to treat conditional events, the rule-based description lends itself naturally to the description of the plant as a set of unit operations. The behavior descriptions can include both qualitative and quantitative models. Back-up and safety systems which come on-line during the fault event can be treated with conditionals. The richer representations of plant structure and behavior available in the expert systems approach will facilitate more powerful diagnostics than are now possible by digraph-based techniques.

The knowledge required for diagnosis can be classified either as plant-independent or plant-specific. The plant-independent knowledge contains general information concerning diagnostic strategies, fault propagation, and models of unit operations. The plant-specific information includes plant topography, unit design specifications, and process chemistry. A powerful principle for design of an easily-implementable expert system is to separate the plant-independent knowledge from the plant-specific knowledge. This separation results in a transportable knowledge "core", to which plant-specific rules can be added subsequently. In the representation suggested above, the behavior rules are largely plant-independent. With these rules and a plant-independent inference engine to carry out the diagnosis, only the specific topography of the plant needs to be added to achieve a basic diagnostic capability. Construction of a portable modular diagnostic code in the style of FLOWTRAN or ASPEN is a goal which should be achievable within the next five years.

FUTURE OF EXPERT SYSTEMS IN PROCESS CONTROL AND OPERATIONS

Combination of Deep and Shallow Inference

In the foregoing, we argued the superiority of deep knowledge approaches to fault diagnosis. Nonetheless, certain processes will resist model-based approaches and will require heuristic description. Additionally, valuable shortcuts for diagnosis may be available in the form of heuristics. For maximum effectiveness, the expert system must combine these shallow inferences with the results of model-based analyses.

A straightforward means of combining inferences has been proposed by Kramer & Palowitch (1985). This approach, represented in Fig. 3, involves combination of evidence from diagnostic methods operating in parallel. Disregarding conflicting or uncertain data, the diagnosis lies in the intersection of the sets of possible faults suggested by each method. More sophisticated techniques (e.g. Garvey et al. 1981) can be applied where both positive and negative evidence exists for each possible fault. Fink (1985) has described an expert system with more intimate coupling between deep and shallow inference. The key to the next generation of expert systems for diagnosis will be the combination of shallow inferences of experienced personnel and the deep inferences derived from models of the process.

Design for Safety and Diagnosability

Research is needed for design strategies for improved fault recognition. The concern is for plant layout and sensor placement for optimum diagnosability, in new-plant design and for retrofit of existing processes. In retrofit, the goal is to suggest additional instrumentation to improve the diagnoses or to make certain unresolvable faults diagnosable. At the design state, the goal is to plan instrumentation and control of the plant in such a way that fault states will be readily apparent when the diagnostic methodology is applied. To achieve these goals, research is needed to address the following important questions:

- What are the minimum instrumentation requirements for detection of a fault, and what is the minimum instrumentation required for fault isolation, in the context of knowledge-based systems?
- What measurements can best serve to increase the certainty of the diagnosis beyond the minimum requirements?
- What additional measurements or criteria can help distinguish faults which may be easily confused?

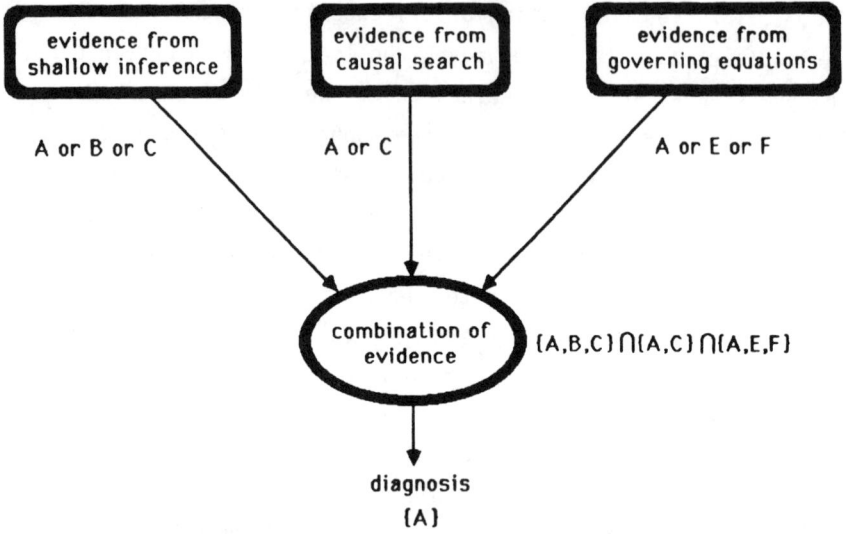

Fig. 3. Parallel approach to diagnosis

- How do failed-normal and latent sensor failures affect diagnosis?
- What is the optimum placement of redundant sensors?

Expert systems may also be able to assist in safety analysis of flowsheets, by simulating faults and the resulting effects on the plant, by simulating operating blunders, or by simply running safety checklists based on the types of chemicals, processing environments, and equipment used in the plant.

Learning Systems

At the present time expert systems have been found useful in a constrained context, where knowledge is strictly defined by rule-frames and models, and where inference proceeds within the limits of the defined knowledge. It is often noted that human experts display far broader capabilities, such as the ability to reason by analogy and to learn from observed behavior. We project that advances in expert systems will make inroads toward better performance in these areas, which will open new areas for application.

One of the simplest learning structures is to assume that the heuristics are correct except for parameters, such as certainty factors, to be adapted, e.g.

If tube-temperature rate-per-minute increase is-greater-than 10^0 and feedwater-temperature is-stable then conclude clogged-tube with certainty .6

This kind of learning, although beyond most expert systems capabilities today, is a simple parameter identification problem. However, the practicality of learning in this context is severely limited - it may only be useful in batch processes where repetitive data is available.

Deeper forms of learning involve inductive processes which are not well understood by cognitive scientists. It has been established that learning goes beyond observed experience - that humans know far more than they could have learned from experience. In linguistics, for example, it has been shown that a child knows more syntactic patterns than could ever have been overheard - establishing the existence of some deep knowledge not based on observation.

In summary, simple learning such as adapting parameters of models of the domain are a feasible near-term capability of expert systems. Truly creative learning - changing the fundamentals of the deep knowledge of the domain is analagous to scientific discovery and it is not a near-term prospect.

Testing Knowledge. One of the fundamental problems in expert systems is determination of the consistency of knowledge. In trivial cases, where heuristics directly conflict, an algorithmic determina- tion of inconsistency is possible. In the more general case, deep knowledge of process behavior and structure can be used to judge con- sistency. Deep knowledge will include process models, thermodynamics and physics laws and similar models of behavior. Heuristics will be balanced against deep knowledge and the knowledge engineer will be challenged in the event of conflicts. Deep knowledge will be used to evaluate observations and to detect discrepent behavior. We see the human knowledge engineer as the focus for resolving discrepencies, and for adapting heuristics and modifying deep knowledge structures.

Another fundamental problem is the need to incorporate expertise of multiple experts. For large problems, such as process plants, this is a necessity. Deep knowledge in the expert system could be used to evaluate the various chunks of knowledge for consistency with existing knowledge, and simulation based on deep knowledge could be used to test the integrated knowledge base.

The automatic generation of rules from deep plant knowledge is another promising area. The work on an automated reasoning system by Lusk and Stratton (1983) at the Argonne National Laboratory is an interesting example. In this application, the deep knowledge or physical knowledge of the plant is used to develop expert rules automatically. The potential use of the system is to allow nuclear plant operators to have expertise readily available, specifically to permit cooling to be directed to portions of the plant under various conditions of equipment failure. This is a simple example of an eventual necessity - the automated engineering of the knowledge base.

ACKNOWLEDGEMENTS

Robert L. Moore would like to acknowledge the technical contributions of LMI personnel Lowell Hawkinson, Carl Knickerbocker, Mike Levin, Mark David and Kumar Gidwani in the design of PICON. Also we wish to thank the personnel at Texaco and Exxon, who provided many valuable suggestions incorporated into the PICON design and who worked patiently with LMI during the in-plant evaluations. In particular we thank Pete Thompson and Warren Dalton of Texaco, and Phil Zaybekian and Brian Matthews of Exxon. Mark A. Kramer acknowledges the research efforts of B. Palowitch, Jr. and the support of A.D. Little Foundation and Shell Foundation.

REFERENCES

Andow, P.K. and F.P. Lees (1975). Process Computer Alarm Analysis: Outline of a Method Based on List Processing. Trans. Instn. Chem. Engrs, 53, 195.

Andow, P.K. (1980). Real-Time Analysis of Process Plant Alarms Using a Mini-Computer. Comp. Chem. Eng., 4, 143.

Astrom, K. J. and J.J. Anton (1984). Expert Control. IFAC World Congress, Budapest. (Forthcoming in Automatica).

Barnett, B.A. (1981). Computational Methods for a Mathematical Theory of Evidence. Proc. 7th IJCAI, Vancouver, BC.

Chester, D., D. Lamb and P. Dhurjati (1984). Rule-Based Computer Alarm Analysis in Chemical Process Plants. Proc. Micro-Delcon, 22.

Crowe, C.M., Y.A. Garcia Campos, and A. Hrymak (1983). Reconciliation of Process Flow Rates by Matrix Projection. AIChE J., 29, 881.

Davis, J.F., W.F. Punch III, S.K. Shum, and B.Chandrasekaran (1985). Application of Knowledge-Based Systems for the Diagnosis of Operating Problems. AIChE Annual Mtg., Chicago, IL.

DeKleer, J., and J.S. Brown (1984). A Qualitative Physics Based on Confluences. Artificial Intelligence, 24, 7.

Fink, P.K. (1985). Control and Integration of Diverse Knowledge in a Diagnostic Expert System. Proc. 9th IJCAI, Los Angeles, CA, 426.

Fortin, D.A., T.B. Rooney and E.H. Bristol (1983). Of Christmas Trees and Sweaty Palms. Proc. of Ninth Annual Advanced Control Conference, West Lafayette, Indiana, 49.

Garvey, T.D., J.D. Lowrance and M.A. Fischler (1981). An Inference Technique for Integrating Knowledge from Disparate Sources. Proc. 7th IJCAI, Vancouver, BC, 319.

Gertler, J., and D. Singer (1985). Augmented Models for Statistical Fault Isolation in Complex Dynamic Systems. Proc. ACC, Boston, MA, 317.

Iri, M., K. Aoki, E. O'Shima and H. Matsuyama (1979). An Algorithm for Diagnosis of System Failures in the Chemical Process. Comp. Chem. Eng., 3, 489.

Kawakami, J. (1984). Overview of Artificial Intelligence Applied to Control Technology in Japan and in Hitachi, Hitachi Research Laboratory, private communication, Japan.

Kokawa, M., S. Miyazaki, and S. Shingai (1983). Fault Location Using Digraph and Inverse Direction Search with Application. Automatica, 19, 729.

Koton, P.A. (1985). Empirical and Model-Based Reasoning in Expert Systems. Proc. 9th IJCAI, Los Angeles, CA, 297.

Kramer, M.A., and B.L. Palowitch Jr. (1985). Expert Systems and Knowledge-Based Approaches to Process Malfunction Diagnosis. AIChE Annual Meeting, Chicago, IL

Kramer, M.A. (1986). Malfunction Diagnosis Using Quantitative Models and non-Boolean Reasoning in Expert Systems. Submitted for publication.

Krigman, A. (April 1985). Expert Systems: Would you Trust One? INTECH, 9.

Kuipers, B. (1984). Commonsense Reasoning About Causality: Deriving Behavior from Structure. Artificial Intelligence, 24, 169.

Kuipers, B. (1985). The Limits of Qualitative Simulation. Proc. 9th IJCAI, Los Angeles, CA, 128.

Lees, F.P. (1984). Process Computer Alarm and Disturbance Analysis: Outline of Methods for Systematic Synthesis of the Fault Structure. Comp. Chem. Eng., 8, 91.

Lusk, E. and R. Stratton (1983). Automated Reasoning in Man-Machine Control Systems. Proc. Ninth Annual Advanced Control Conference, West Lafayette, IN, 41.

Mah, R.S.H., G.M. Stanley, and D.M. Downing (1976). Reconciliation and Rectification of Process Flow and Inventory Data. Ind. Eng. Chem. Proc. Des. Dev., 15, 175.

Mah, R.S.H., and A.C. Tamhane (1982). Detection of Gross Errors in Process Data. AIChE J., 28, 828.

Moore, R. L., L. B. Hawkinson and C. G. Knickerbocker, (1984). Expert Systems Applications in Industry, ISA/84, Houston, TX, 1081

Moore, R. L., L. B. Hawkinson, C. G. Knickerbocker, and L. M. Churchman, (1984). A Real-Time Expert System for Process Control, Proc. First Conference on Artificial Intelligence Applications (IEEE) Denver, CO, 529.

Palowitch, B.L. and M.A. Kramer (1985). The Application of a Knowledge-based Expert System to Chemical Plant Fault Diagnosis, Proc. ACC, Boston, MA, 646.

Patil, R.S., P. Szolovits and W.B. Schwartz (1981). Causal Understanding of Patient Illness in Medical Diagnosis. Proc. 7th IJCAI, Vancouver, BC.

Romagnoli, J.A., and G. Stephanopoulos (1981). Rectification of Process Measurement Data in the Presence of Gross Errors. Chem. Eng. Sci., 36, 1849.

Sakuguchi, T. and K. Matsumoto (1983). Development of a Knowledge Based System for Power System Restoration, IEEE Trans. on Power Apparatus and Systems, PAS-102, 320.

Sauers, R. and R. Walsh (1983). On the Requirements of Future Expert Systems, Proc. Eighth IJCAI, Karlsruhe, W. Germany, 110.

Shiozaki, J., H. Matsuyama, E. O'Shima, and M. Iri (1985). An Improved Algorithm for Diagnosis of System Failures in the Chemical Process. Proc. PSE '85, Cambridge, England, 133.

Yamada, N., and H. Motoda (1983). A Diagnosis Method of Dynamic System using the Knowledge on System Description. Proc. 8th IJCAI, Karlsruhe, W. Germany, 225.

Zadeh, L.A., and K. Fukanaska, ed. (1975). Fuzzy Sets and Their Applications to Cognitive Decision Processes. Academic Press, New York.

PROCESS CONTROL SYSTEM SYNTHESIS BY AN EXPERT SYSTEM

K. Niida
T. Umeda
Chiyoda Chemical Engineering and Construction Co.
Tsurumi, Yokohama, Japan

Abstract Presented is an approach to the synthesis of process control systems by the concept of expert systems. The purpose of the research is to obtain know-how on building commercially usable expert systems in synthesizing process control systems. Based on the experience gained by developing an experimental expert system for synthesizing distillation control systems, a system applied to chemical processes with various kinds of unit operations has been developed. The results of the experimental work and several requirements for developing commercially useful expert systems are presented.

Keywords Control system synthesis; chemical process; computer aided design; artificial intelligence; human expertise; heuristic knowledge; production system; frame model; CHIPS; KEE

INTRODUCTION

In the past decade, research on developing logical procedures for the synthesis of process control systems has been carried out by several authors (Nishida et al.,1979). Their efforts have contributed to deeper insights into process control system synthesis as well as to the development of synthesis procedures. However, practice in the synthesis of process control systems are still relies in large part on the engineer's intuitive approach. One of the major reasons why its computerization has not been accomplished is probably the restriction on presently available digital computer performance characterized by data processing.

Recently, much attention has been paid to developing expert systems involved in the area of applied artificial intelligence. Engineering activities are classified into those of synthesis and analysis, and computer applications related to the former class are in general much more difficult than those related to the latter by the conventional computer systems. It is expected that the expert systems will thus promote computer applications in the synthesis area. Up to the present time, there is no publication on expert systems for process control system synthesis except a few papers. PICON is a real-time expert system for advanced process control and has been introduced to aid plant operation. It is operated in a LISP machine directly connected to existing distributed control systems. A scenario of developing expert systems for Computer-Aided Control Engineering (CACE) is presented by Taylor et al. (1984). They intend to create third-generation man/machine environments for CACE. The expert systems to be developed on the basis of the scenario will provide a high-level design environment where many less-than-expert users can find and utilize appropriate CACE software without high-level knowledge of control theories or much experience in process control.

In this paper the authors present an approach by the expert system to the synthesis of process control systems. An experimental system was developed and various studies were carried out, using the system to obtain information for developing a practically useful system. For illustrative purposes, the synthesis of chemical process systems is taken up.

BRIEF REVIEW OF PROCESS CONTROL SYSTEM DESIGN

General Procedure

A general procedure of system design in process industries is summarized in Fig.1. It is assumed that the basic design of process systems has already been finished. In Steps 1 and 2, a process system is defined and its control objectives are specified. In Step 3, candidates of control loops in each unit are synthesized. In Step 4, the control performance of the total system is analyzed and coordination among control loops is carried out to eliminate interactions among them. In Steps 5, 6 and 7, more detailed analyses are carried out. In the present work, the first four items are discussed.

Control of Chemical Process Systems

In the operation of chemical process systems, there exist several kinds of operation modes such as start-up, shut-down, half-load and normal steady-state operations. The present work is concerned with control system synthesis in normal steady-state operations of chemical processes involving many kinds of units such as pumps, compressors, vessels, heat exchangers, furnaces, reactors and distillation columns. Pairing candidates between controlled and manipulated variables in the basic unit operations are shown in Fig.2. Controlled variables are indicated by symbols in circles and manipulated variables are indicated by symbols attached to the control valves. Control valves and streams represented by thin lines are optional. It is assumed that the flow rates, temperature and pressure of each feed stream to the units are predetermined except in the following case. If a pump or a compressor is directly connected to a tank or a header which has a large amount of storage capacity, one degree of freedom remains in the system and the feed flow rate can be controlled.

EXPERT SYSTEM

Expert systems consist of inference engines, data bases and knowledge bases. Human expertise is encoded and stored in the knowledge bases. The inference engines are software tools for building and utilizing expert systems. The data bases are stores of information or evidence provided directly by the users or derived by computers during inference. Programs for engineering calculations, inference controls, and user interfaces are stored in the data base.

In the present work, a commercially available inference engine, the KEE, has been used, along with ZETALISP, and inference is implemented on a Symbolics 3670. In the KEE both production rules and frame model are available. The production rules are used to represent heuristic knowledge. The frame models are effectively utilized for the implementation of data base and LISP program storage.

Knowledge Acquisition

Knowledge for process control system synthesis has been accumulated as facts or experience. A tremendous number of technical reports on process control have been published, and they include very good guiding rules for control system synthesis where the rules were derived from specific experience. Some knowledge in the form of "guides to process control" was found in the literature for the past three decades. Especially in the earlier stage of process control realization, experience was gained by practical use, and valuable reports were published in the literature. A technical report issued by Oil and Gas Journal - "How to Instrument the Fractionation Section Standard textbooks on process control (Buckley, 1964; Shinsky, 1982) also involve guides to process control system synthesis. Although some reports are available, for instance, in the area of distillation control, (Rademaker et al., 1975), they are specific and it is very difficult to accumulate generally useful information without continuous efforts over a long period. An important issue in developing expert systems is to extract heuristic rules from the experiential facts in the literature as well as from human experts. To overcome difficulties associated with acquiring useful knowledge, it is desirable to have process-oriented knowledge engineers who have good background knowledge of process design and control as well as computer science.

The above-mentioned knowledge in a part of so-called surface knowledge and it is often very hard to find precise reasoning on the validity of guiding principles. For analyzing these guiding principles, it is possible to make use of theoretical tools. Various methods for theoretical analysis of process control systems have been developed, and for given control structures these methods are useful to understand the system characteristics and to predict their control performance. It can be said that the surface knowledge becomes deeper if the validity is checked by theoretical analysis. If no appropriate theoretical tools are available, it is necessary to make the surface knowledge deeper by making careful observations of experts' decision-making in various events and extract their pattern of decision-making in the form of heuristic rules. It is advisable to use simulators to know control performance when a preset control system structure is given.

In the area of dillation system control, various concepts involving relative gains (Bristol, 1966; McAvoy et al., 1981; Takamatsu et al., 1982) are considered to be useful tools for the analysis and prediction of control performance. During the decision-making in control system synthesis, it is necessary to carry out computation the analysis by these methods. This feature is common in engineering decision-making and expert systems in practical use should involve the capability of engineering calculations.

Knowledge Representation

When the factual and experimental knowledge is in hand, the next step in developing expert systems is to encode the knowledge in the computers in the form of production rules. The production rules are the most common form of representation of the decision rules which are the IF-THEN rules for compiling the heuristic reasoning procedures, including the facts found in the literature. Several examples of production rules related to recycle streams are described as follows;

If there is a recycle gas compressor, then usually it does not require any process control in normal operation.

If there is a recycle gas loop, then the pressure of the recycle loop at the product separation vessel could be controlled by a make-up compressor or the flow rate of bleed gas stream.

If there is a bleed gas stream in normal operation in a recycle gas loop, then it is preferable to control the pressure of the loop at the product separation vessel by the flow rate of the bleed gas stream.

AN EXPERT SYSTEM FOR CHEMICAL PROCESS CONTROL SYSTEM SYNTHESIS

Inference Procedure

The procedure of basic inference in the expert system is shown in Fig. 3. In the figure, twelve context names are used in inference control. The context names are classified into the first four steps shown in Fig.1.

The first step is to define the problem to be solved. In the first context the user can specify the characteristics of the chemical process such as types of process and flow network, number of recycle loops, and number of bleed streams. In the second context, the user can specify the process flow network of the process. In the last context, the expert system checks the network to remove conflicts among the above specifications. The locations of units are automatically checked to determine whether they are on a recycle loop or not. The phases of the process streams are also estimated.

The second step is to determine controlled and manipulated variables to be synthesized. In the first context the user can specify the characteristics, disturbances, and requirements of control performance for each unit by selecting appropriate alternatives shown in the menus on the display. Contradictions, implications, and equivalences among the above inputs are found by using an in-house developed logic programming which is implemented on ZETALISP. The details of the technique will be described elsewhere. They are displayed immediately to the user. In addition, there are forward production rules used in finding recommendations for selecting controlled and manipulated variables. In the second context, the user can select some of the variables for each unit. Note that the degrees of freedom for process control will change by selecting some controlled and manipulated variables. Some of them in Fig.2 are shown as follows;

* compressor
    controlled variables   SYS.P, PD, PS
    manipulated variables   SB

* heat exchanger
    controlled variables   T
    manipulated variables   FB, FB*

In the last context, the expert system checks the above selected variables to find the conflicts between them. If any conflict is found, the inference control is given to the previous context. Some of the conflicts are removed automatically by using appropriate production rules. The degrees of freedom for the control system are checked.

The third step is to find and synthesize control loops. In the first content, the possible candidates of primary control loops are enumerated by taking the control characteristics of the process into consideration. In the second context, the primary control loops are synthesized by selecting some appropriate candidates. As an example, the strategies of synthesis of control system are given as follows:

(1) The resolution technique for conflicts between production rules is a position-ordered method.

(2) The production rules with the highest priority should be fired at first. For example, the higher priorities are assigned to dual-composition control and single-composition control.

(3) A control variable with only one available control loop should be selected.

(4) A control variable with a possibility to make two or more available control loops should be resolved later by using heuristic knowledge acquired in the domain of control system synthesis.

In the the last context, the secondary control loops are synthesized. If one of the specified requirements of control performance is not satisfied in the primary control loops, a secondary control loop can be synthesized. Otherwise, some indicators may be recommended.

In the last step, the result of synthesis is evaluated and displayed. In the first context, the expert system checks the above synthesized control loops to find conflicts among the loops and to find unused variables in the selected variables. In the second context, the result of synthesis is displayed with some explanation. In the last context, the user can interactively modify the result through the display.

<u>Data Base</u>

The data base structure of the chemical process control system is shown in Fig.4, which shows a standard graphical output in the KEE. The data base has a hierarchical structure implemented by using the frame model in the KEE. At the top of the structure, there is a general unit DB which has six kinds of units; LOOPS, EQUIPS, VARS, CONTROLS, PROGRAMS and B-NET. The unit B-NET has the role of defining process networks. If the user clicks B-NET on the display by using the mouse attached, a selection menu for the next equipment will emerge, and he can easily select one of them. Data in the equipment are automatically stored in a newly created unit for the equipment. The new unit is created under the unit EQUIPS, which has the role of gathering several kinds of equipment units such as DB-DIST and DB-PUMP. The relation between the new unit and its parent unit is expressed by dotted lines in the figure. All the slots used in these equipment units have already been defined in the unit EQUIPS with useful default values, and these slots can be made available in the lower units by using the inheritance technique in the KEE. Representative slots are given as follows;

CHARACTER	——————	Characteristics for processes and equipment
DISTURBANCE	——————	Expected disturbances
REQUIREMENT	——————	Requirement for control performance
NEXT-EQ	——————	Next equipment names
FROM-EQ	——————	Upstream equipment names
CV-SET	——————	Available controlled variables
MV-SET	——————	Available manipulated variables

The unit VARS is the role of storing controlled and manipulated variables which will be selected by the user. The unit LOOPS has the role of storing control loops. The unit CONTROLS has the role of storing information in inference control. The unit PROGRAMS has the role of storing LISP programs used in the expert system.

Knowledge Base

The knowledge base is summarized in Table 1. There exist 107 production rules in the knowledge base. A quarter of them are related to finding, synthesizing, and checking control loops. The rest are related to user input and its input help functions which are used in the first two steps shown in Fig.3. There is no rule related to inference control, which can be carried out by LISP programs stored in the data base. The object-oriented programming technique implemented in the KEE is used. A part of the knowledge base structure is shown in Fig.5. The knowledge base has a hierarchical structure. At the bottom of the structure, there exist production rules grouped into several classes such as 'B+INPUT-LOOPS', 'C+FIND-LOOPS', 'C+INPUT-CHECK' and 'D+FIND-LOOPS'. These classes are grouped into several superclasses such as 'KB-BLEED', 'KB-COMPRESSOR' and 'KB-DIST'. Each superclass has the role of gathering production rules related to each unit such as BLEEDER, COMPRESSOR and DISTILLATION. Production rules for each unit are stored in a different superclass so that it is easy to construct, maintain and execute its production rules. Relations between production rules and classes are also expressed in the dotted lines in the figure. Class 'C+FIND-LOOPS' has the role of gathering production rules related to finding possible control loops in the distillation. Class 'C+INPUT-CHECK' has the role of gathering production rules related to checking user input in the distillation. At the top of the structure, there exists a unit named 'KB' which has the role of gathering all the production rules in the knowledge base. Production rules related to both synthesis and checking of control loops are not shown in the figure. These rules are gathered in other superclasses under the unit 'KB'.

Some forward production rules in the superclass 'KB-COMPRESSOR' are shown in Fig.6. The rules 'C-F' and 'C-IC***RECYCLE' represent human expertise for finding possible control loops and checking the characteristics of compressors, respectively.

The first rule in the figure is simple. It has the role of finding a control loop whose controlled and manipulated variables are the feed flow rate F of a compressor. The rule consists of three parts: the rule name 'C-F', IF-part, and THEN-part. IF-part is described between symbol 'IF' and symbol 'DO'. THEN-part is described after the symbol 'DO'. IF-part consists of several clauses with logical combinations. THEN-part consists of LISP functions. When the contents of any clause in IF-part exist in the data base, the clause is evaluated to be true. If all clauses in a rule are true, the rule can be fired, and THEN-part of the rule will be evaluated. Any clause in IF-part consists basically of a unit name (an entity name), a slot name (an attribute name), and its slot value (its attribute value). In the first clause, 'GOAL' is a unit name, and 'EQ' is a slot name. '?EQ' is a variable which corresponds to its slot value in the data base. The information on inference control is stored in the special unit 'GOAL', and its slot 'EQ' has currently focused equipment in the inference. Therefore, the first clause means that the rule should be applied to the currently focused equipment, namely the compressor. The second clause is a logical combination of three clauses. The second clause means that there should be a selected controlled variable F of the compressor. The last clause in IF-part also means that there should be a selected manipulated variable F of the compressor. THEN-part has several LISP functions which generate a unit in the KEE for representing the flow rate control loop and store its slot values. The description of each function is too complicated for the present paper.

The second rule is the simplest, and has the role of both finding a recycle compressor in checking inputs and restricting the use of the following variables: flow rate of spill back 'SB', suction pressure 'PS', delivery pressure 'PD', and flow rate of feed 'F'

ILLUSTRATIVE EXAMPLE

A problem under consideration is to synthesize a set of conventional control loops on the basis of normal steady-state operation in a hypothetical chemical process whose structure is similar to that of desulfurization of kerosene by hydrogen. The feed of the chemical process is stored in a tank, and hydrogen comes from a header. The reactor effluent is separated to recycle gas and a liquid product which is purified in a distillation column. The top and bottom products of the column are stored in tanks. The process has a bleed gas stream in its recycle loop. The definition of the process network is shown in Fig.7, and its data base structure is already shown in Fig.4. Major specifications for the process and equipment are briefly shown as follows;

* the process
   ⟨C⟩   Hydrogen-desulfurization-type-flow-process
          With-gas-recycle
          With-gas-bleed

* the header
   ⟨D⟩   Pressure-may-be-varied

* the make-up compressor after the header
   ⟨C⟩   Make-up-compressor
   ⟨R⟩   Constant-flow
          Constant-suction-pressure

* the feed pump
   ⟨R⟩   Constant-flow

* the furnace
   ⟨R⟩   Constant-outlet-temperature

* the vessel
   ⟨C⟩   One-liquid-product
   ⟨R⟩   Constant-pressure

* the bleeder
   ⟨R⟩   Constant-pressure

* the heat exchanger after the vessel
   ⟨R⟩   Constant-outlet temperature

* the distillation
   ⟨C⟩   Normally-total-condenser
          Not-high-reflux-rate
          Not-high-product-purity
   ⟨R⟩   Constant-reflux-rate
          Constant-column-temperature

* the compressor in the recycle loop
   ⟨C⟩   Recycle-compressor

Where ⟨C⟩, ⟨D⟩, and ⟨R⟩ stand for characteristics, disturbances, and requirements of control performance, respectively. The expert system makes pairs between controlled and manipulated variables by taking the above specifications into account. As the result of inference by the KEE, a set of control loops for the chemical process was synthesized. They are shown in Figs.8, 9, and 10. Fig.8 is an intermediate result which shows fifteen candidates of primary control loops inferred after the context "FIND LOOPS". Consequently, three out of the fifteen candidates are deleted in the context "SYNTHESIZE PRIMARY LOOPS". By applying the synthesis strategies to the intermediate result, the following three control loops were discarded.

   control loop : SYS.P in VE (—— SB in COMPR
   control loop : P in DIST (—— H in DIST
   control loop : TS in DIST (—— B in DIST

  Fig.9 is a schematic diagram of the final result. The operating pressure P of the distillation column is normally controlled by the flow rate C of cooling medium and sometimes by DV, indicated as a secondary control loop. The detailed inferential process around the vessel is shown in Fig.10.

DISCUSSION

  In developing the experimental expert system, it was first necessary to define the basic structure of the data bases and knowledge bases, and the basic procedure of inference under the specified system boundary conditions and its objectives. In defining them, keywords such as unit, slot and value names (entity, attribute and value names) should be given precisely so that heuristic knowledge of the process control system can be correctly expressed. It is confirmed by several case studies that the basic structure of the system can be determined. For completion of the system, more specific expertise in many kinds of process systems should be acquired and examined for many chemical processes. The experience gained by developing the experimental expert system will be useful in developing the corresponding commercially usable system. In building it the following point should carefully be taken into consideration:

  The basic concept of conventional process control may be characterized by locality. This means that a controlled variable of an equipment unit may be effectively controlled by a valve for a manipulated variable which is available near the sensor for the controlled variable in their process network.

  In knowledge acquisition, it is important to obtain and enumerate possible characteristics of both chemical processes and their equipment in the form of symbolic data which can be handled in production rules. The knowledge engineer for the expert system should find and clarify the following relationships concerned with the enumerated characteristics.

- Relationships among the characteristics.
- Relationships between the characteristics of equipment and process networks around the equipment.
- Relationships between the characteristics and possible control loops.

If these relations can be obtained sufficiently the expert system can guide the users precisely and effectively by implementing the relations in the form of production rules and logic programmings.

Users have different levels of knowledge and experiences and it is desired for the expert system to be adopted to the user requirements. If they give the inputs on an alternative control system structure, the expert system has to report the reasons why the alternative structure is either acceptable or not. It is also useful to the users if the expert system can give information on the experience or reference.

CONCLUSION

An experimental expert system for synthesizing a chemical process control system are being successfully developed, and know-how on building commercially usable expert systems for synthesizing process control systems is being obtained. The extensive use of the presently developed system for the synthesis of process control loops of other unit operations remain for future work.

ACKNOWLEDGMENT

The authors wish to express their thanks to Chiyoda Chemical Engineering and Construction Company for supporting the study and for permitting its publication. They also express their appreciation to Mr. A. Kinoshita for acquiring human expertise on process control.

SYMBOLS

C	Flow rate of cooling medium
D	Flow rate of distillate
E	Flow rate of condensed liquid
F	Flow rate of a feed or a stream
H	Flow rate of heating medium
L	liquid level of a vessel
P	Column overhead pressure
R	Flow rate of reflux
T	Outlet temperature of an equipment unit
CV	Controlled variable
DV	Flow rate of vapour distillate
F1	Flow rate of stream 1 in an equipment unit
F2	Flow rate of stream 2 in an equipment unit
FB	Flow rate of a bypass stream in a heat exchanger
FO	Flow rate of a feed stream to a vessel
FV	Flow rate of a vapour stream from a vessel
LD	Liquid level of overhead drum
LB	Liquid level of column bottom
MV	Manipulated variable
TE	A temperature in enriching section

TS   A temperature in stripping section
TD   Distillate temperature
PD   Pressure Difference in column; Delivery pressure
PS   Suction pressure
SB   Flow rate of a spill back in a compressor
XD   A composition of distillate
XB   A composition of bottom
XF   A composition of feed
FB*   Flow rate of a bypass stream in a heat exchager in Fig.2
FI*   Flow rate of a feed stream in a heat exchanger in Fig.2
FO*   Flow rate of a outlet steam in a heat exchanger in Fig.2
SYS.P   A system pressure in a recycle loop
F.RATIO   Ratio of flow rates
L-TO-FLARE A line to a flare system

REFERENCE

Bristol, E. C. (1966). On a new measure of interaction for multi-variable process control. IEEE Transaction on Automatic Control, 133-134

Chin, T. G. (1979). Guide to distillation methods. Hydrocarbon Processing, 145-153

Hayes, R. F., D. A. Waterman and D. B. Lenart (1983). Building Expert Systems (Technowledge Ser. in Knowledge Engineering Series Vol. 1), Addison-Wesley

McAvoy, T. J. (1981). Connection between relative gain and control loop stability and design. AIChE J., vol.27, no.4, 613-619

Moore, R. L., L.B.Hawkinson, M.E. Levin, G.G. Knickerbocker (1985). Expert System, ACC, 885-888

Nishida, N., G. Stephanopoulos and A. W. Westerberg (1979). A review of process synthesis. AIChE J., vol.27, no.3, 321-351

Rademaker, O., J. E. Rijinsdorp and A. Maarleveld (1975). Dynamic and Control of Continuous Distillation Unit, Elsevier Scientific Pub. Co., New York

Reidel, J. C. (1954). How to instrument the fractionation section ..in field processing plant. Oil and Gas J., Technical Manual Report-Fractionation and Absorption, 2-17

Shinsky, F. G. (1982). PROCESS-CONTROL SYSTEMS, McGRAW-HILL Book Co., New York

Takamatsu et al. (1982). PSE-82, Kyoto

Taylor, J. H., D. K. Frederick and J. R. James (1984). An expert system scenario for computer-aided control engineering. Meeting of ACC, San Diego

Umeda, T, K. Niida (1983). Control of Distillation Systems (in Japanese), Chapter 7, Kagakukogaku no shinpo 17 -Jouryu Gijutsu, The Society of Chemical Engineering Japan

Umeda, T, K. Niida (1985). Process Control System Synthesis by expert system --- A progress report on Distillation Systems Control, IPMC, Hangzhou, China

Weiss, S. M., G. A. Kulikowski (1983). A Practical Guide to Design Expert System, Chapman and Hall, London

Table 1 Summary of knowledge base

NO. OF PRODUCTION ROLES IN THE KNOWLEDGE BASE	
FOR INPUT HELP	29
FOR INPUT CHECK	52
FOR FIND LOOPS, SYNTHESIS, AND CHECK	26
(TOTAL	107)

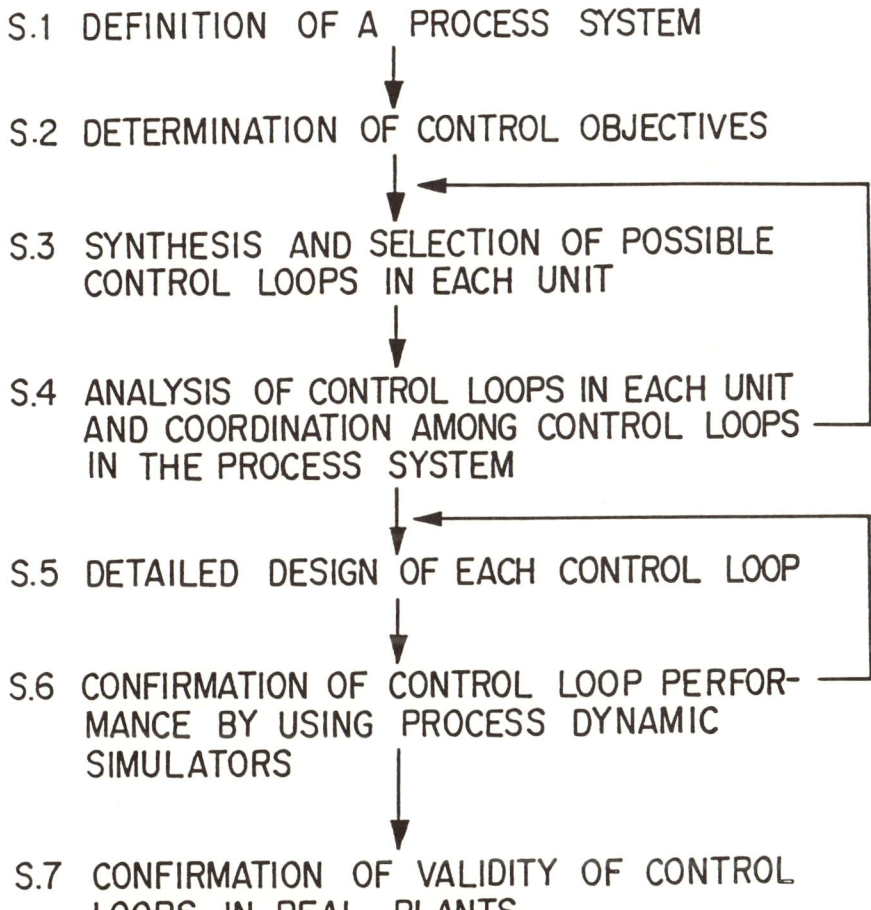

Fig.1 Procadure of control system design with major iterative loops

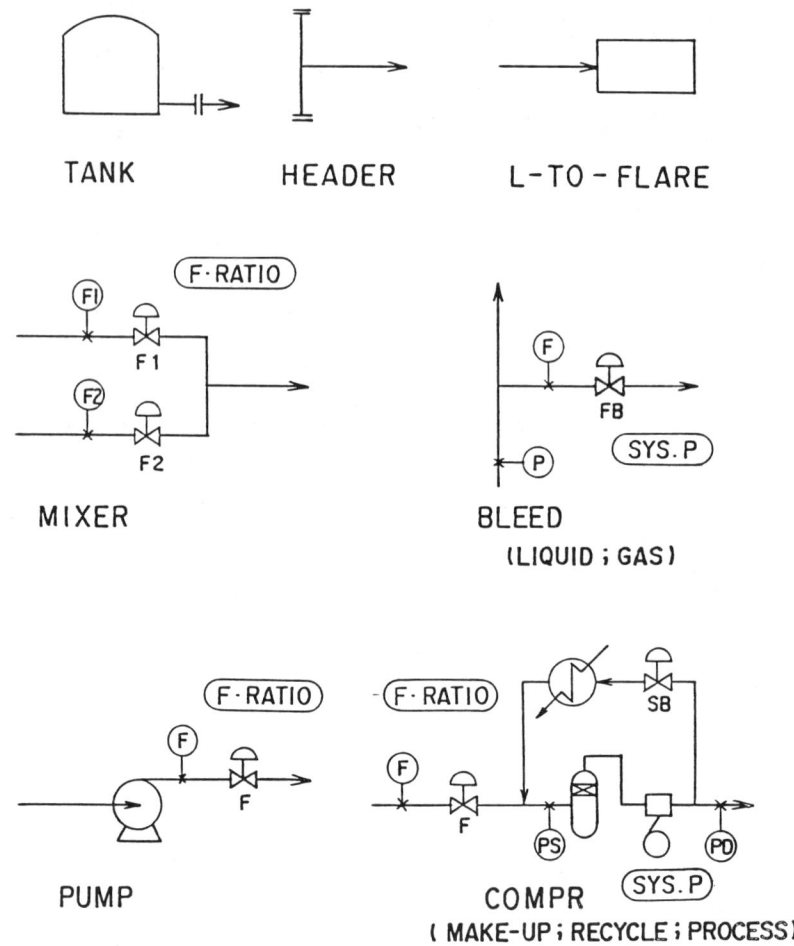

Fig. 2-A Pairing candidates between controlled and manipulated variables in each unit operation

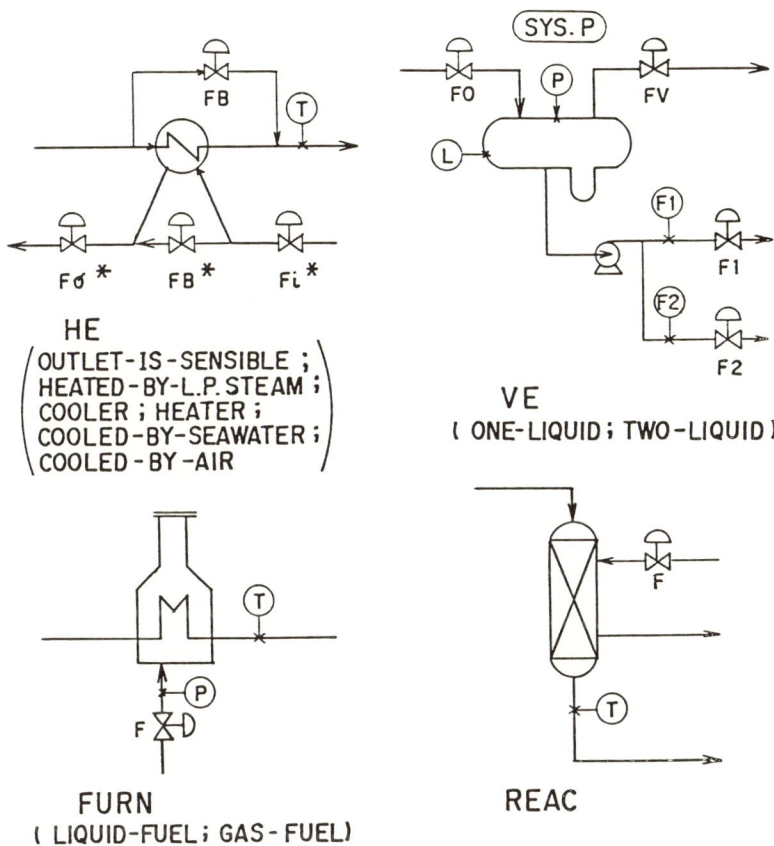

Fig. 2-B Pairing candidates between controlled and manipulated variables in each unit operation

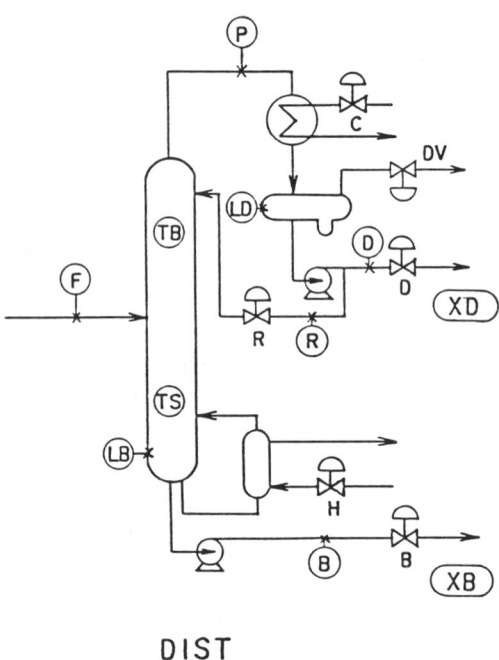

Fig. 2-C Pairing candidates between controlled and manipulated variables in each unit operation

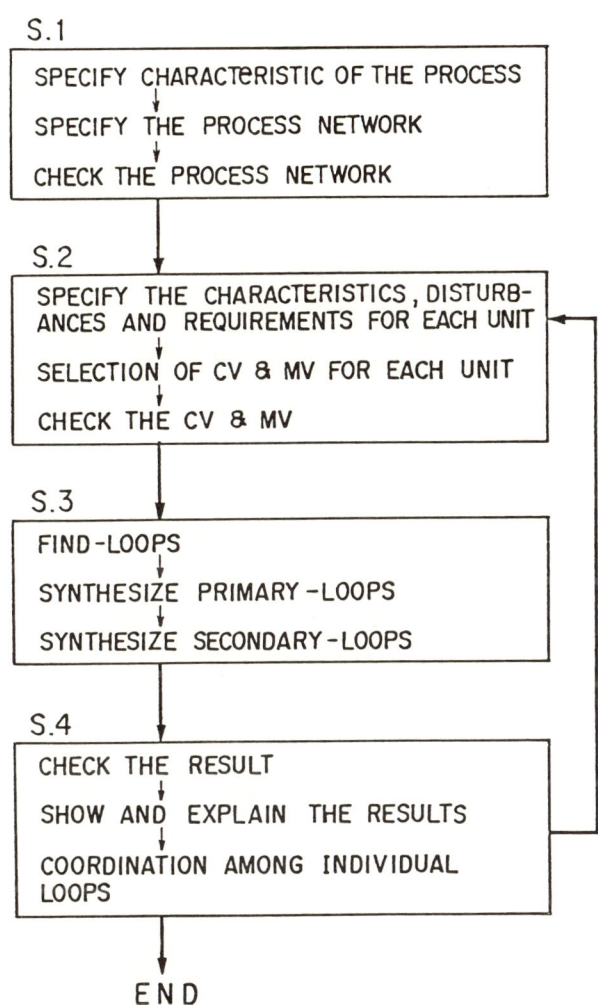

Fig. 3 Inferential procedure

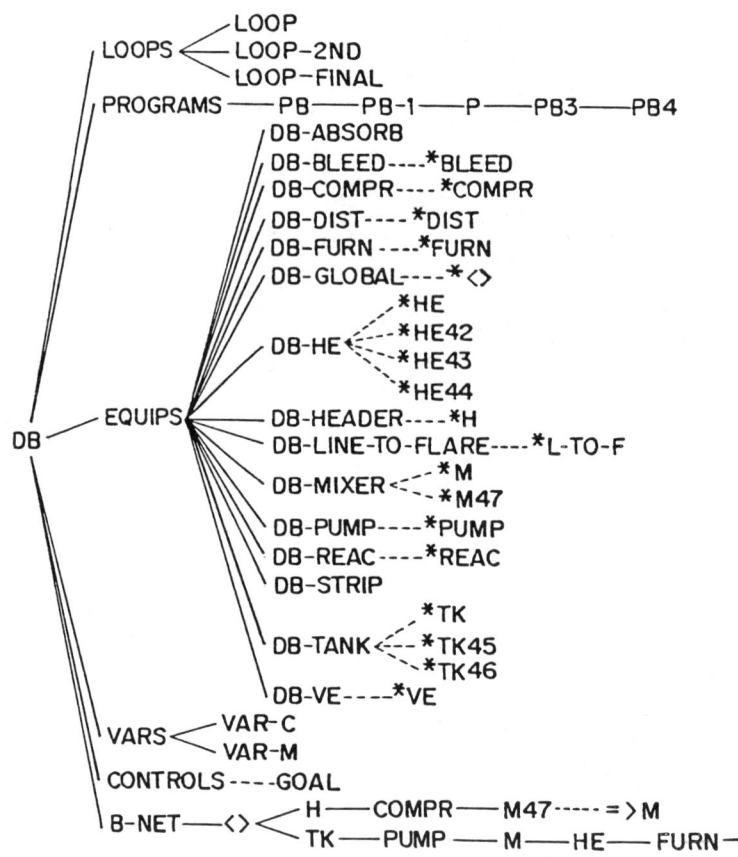

Fig. 4 Data base structure

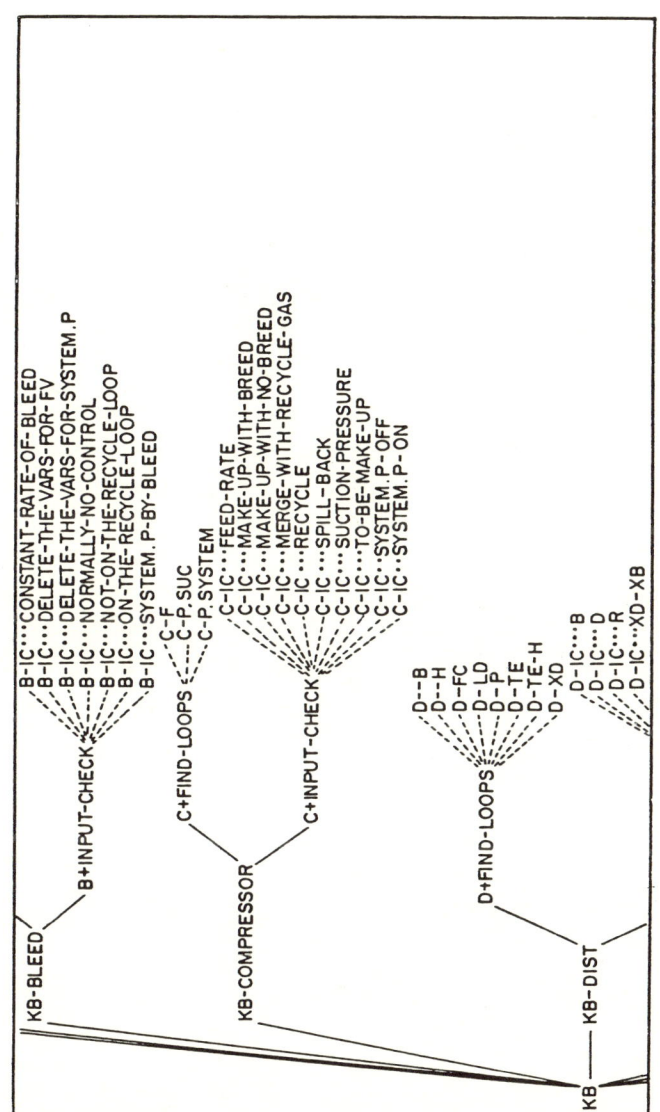

Fig.5 Part of knowledge base structure

```
(C-F   ; FIND FEED FLOW RATE CONTROL LOOP IN A COMPRESSOR

 (IF (AND (THE EQ OF GOAL IS ?EQ)                        ;<IF-PART>
          (THE TYPE OF ?EQ IS COMPRESSOR)                ;; EQ TYPE
          (AND (?V1 IS IN VAR-C)
               (A CONTR OF ?V1 IS F)                     ;; CONTROLLED VARIABLE
               (A EQ OF ?V1 IS ?EQ))                     ;;          F
          (AND (?V2 IS IN VAR-M)
               (A MANIP OF ?V2 IS F)                     ;; MANIPULATED VARIABLE
               (A EQ OF ?V IS ?EQ)))                     ;;          F

  DO                                                     ;<THEN-PART>
     (SETQ ?UNIT (INTERN (STRING-APPEND "C*"             ; MAKE ID NAME OF LOOP
                "F-F-----" (UNIT. NAME ?V1)
                "---"     (UNIT. NAME ?V2))))            ; LOOP SET UP

     (ASSERT '(,;?UNIT IS IN LOOP))
     (ASSERT '(THE EQ OF ,?UNIT IS ,?EQ))
     (ASSERT '(THE CV+ OF ,?UNIT IS F))
     (ASSERT '(THE MV+ OF ,?UNIT IS F))
     (UNITMSG 'PB 'ONE-UP-REPLACER ?V1 'LINKS)
     (UNITMSG 'PB 'ONE-UP-REPLACER ?V2 'LINKS)))


(C-IC** RECYCLE  ; FIND RECYCLE COMPRESSOR IN INPUT CHECK

 (IF (AND (THE EQ OF GOAL IS ?EQ)                        ;<IF-PART>
          (THE EQ-IN-EQUIP OF ALL ?EQ IS ?*EQ)
          (THE CHARACTER OF ?*EQ IS RECYCLE-COMPRESSOR)) ; RECYCLE

  DO                                                     ;<THEN-PART>
     (ADD.VALUE ?*EQ 'VARS-FOR-MANIP-NOT-RECOMMENDED 'SB) ; SET UP
     (ADD.VALUE ?*EQ 'VARS-FOR-CONTR-NOT-RECOMMENDED 'PS) ; NOT-PARMITTED VAR.
     (ADD.VALUE ?*EQ 'VARS-FOR-CONTR-NOT-RECOMMENDED 'PD)
     (ADD.VALUE ?*EQ  VARS-FOR-CONTR-NOT-RECOMMENDED 'F )));
```

Fig. 6 Some forward production rules in KEE

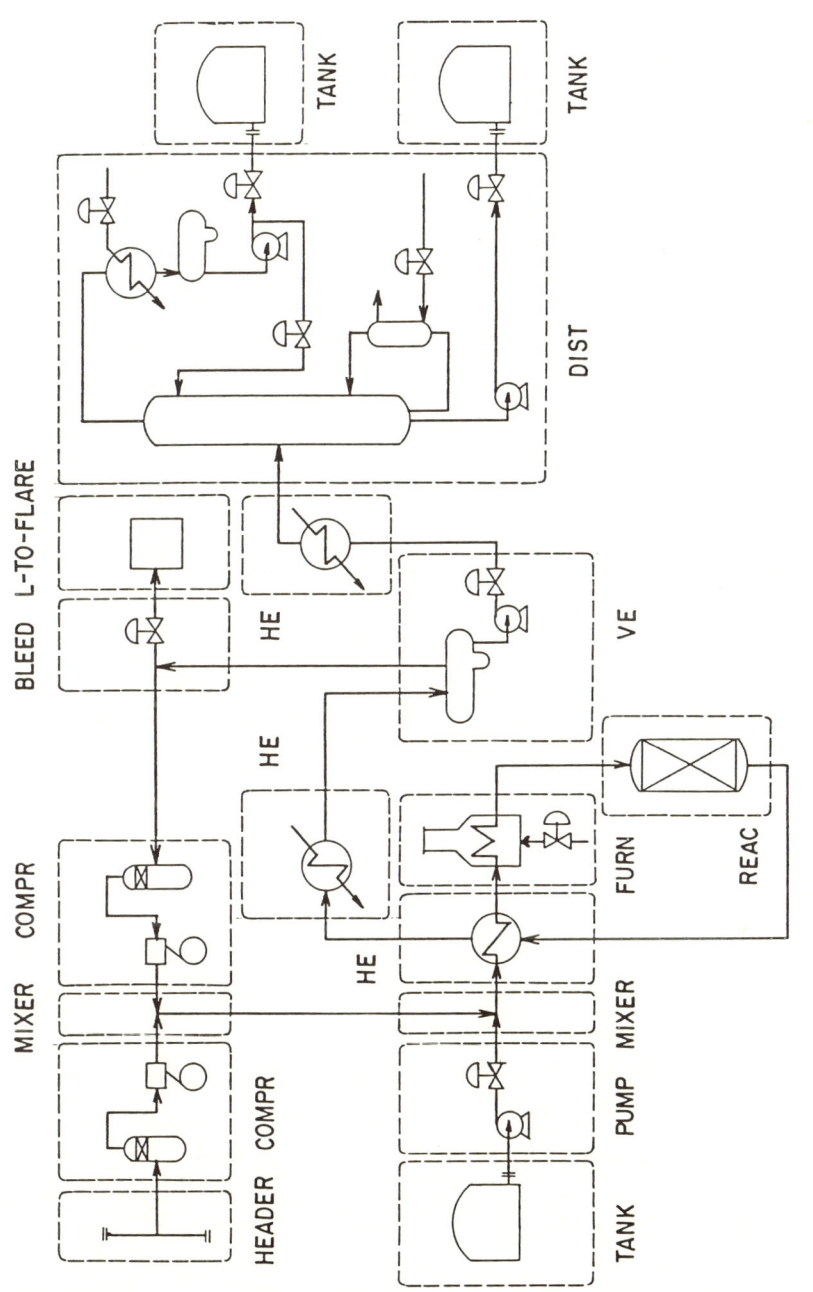

Fig. 7 Example: Definition of process network

CONTROLLED VARIABLE			MANIPULATED VARIABLE	
UNIT NAME	VAR. NAME		VAR. NAME	UNIT NAME
		< PRIMARY LOOPS >		
COMPR	F	←----	F	COMPR
COMPR	PI	←----	SB	COMPR
PUMP	F	←----	F	PUMP
FURN	T	←----	F	FURN
VE	SYS.P	←----	FB	BLEED
VE	L	←----	F1	VE
HE 44	T	←----	FI *	HE 44
DIST	LD	←----	D	DIST
DIST	R	←----	R	DIST
DIST	P	←----	C	DIST
DIST	TS	←----	H	DIST
DIST	LB	←----	B	DIST
		< SECONDARY LOOPS >		
DIST	P	←----	DV	DIST

Fig. 8 Intermediate solution for pairing between controlled and manipulated variables

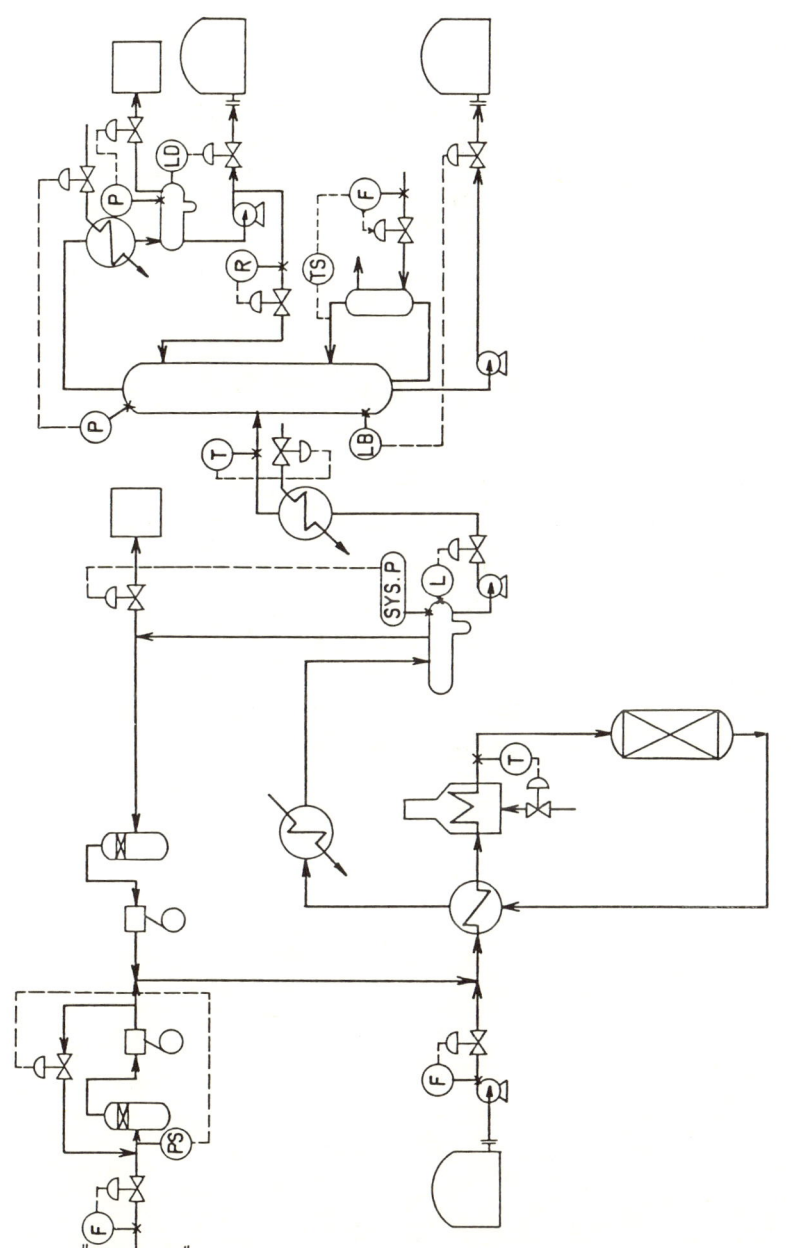

Fig. 9 Synthesized control loops

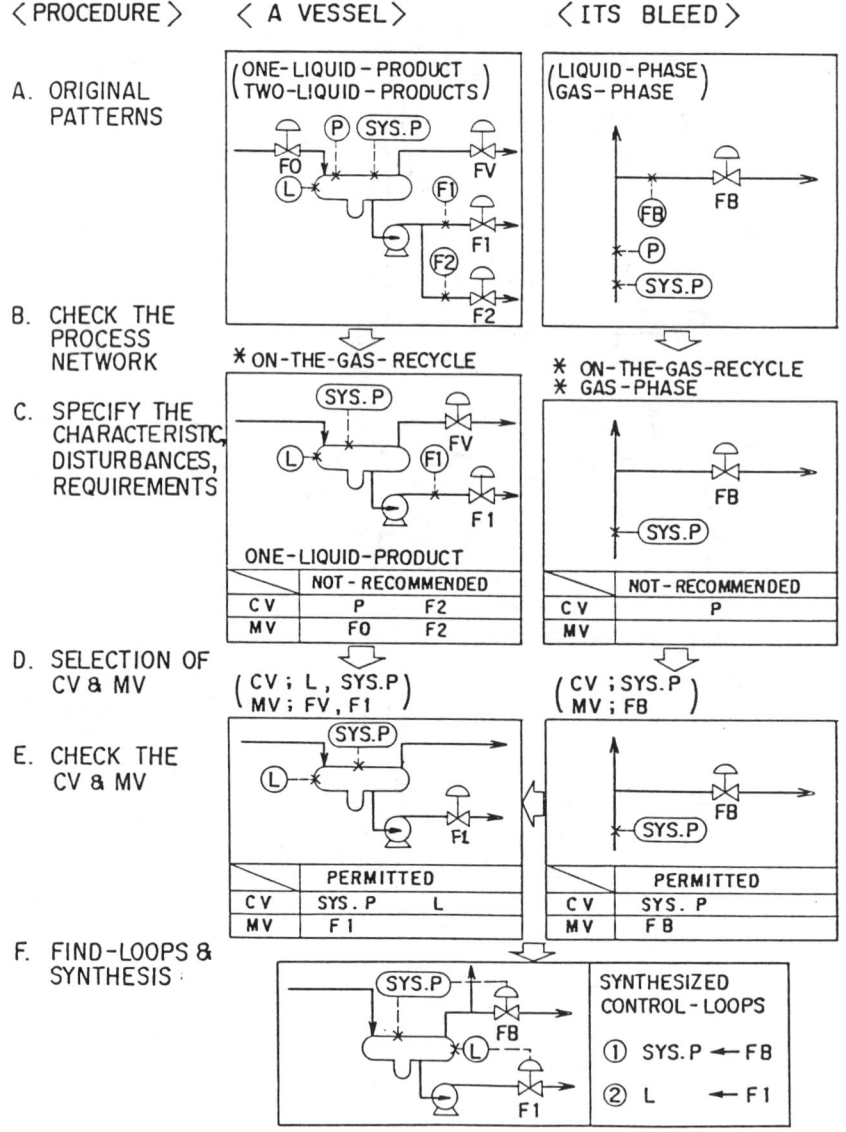

Fig. 10 Example : Synthesis around a vessel

AN EXPERT SYSTEM FOR THE DESIGN OF DISTILLATION CONTROLS

F. G. Shinskey
The Foxboro Company, Foxboro, MA 02035

Abstract. Distillation is the most common multivariable operation in the process industries; therefore its control has received much attention by both academics and practitioners. One approach has been the use of linear multivariable control systems. However, the process itself is fundamentally nonlinear, requiring multiplicative combinations of variables. It is possible to control a distillation column more effectively with a system structure that is far simpler and easier to operate and requires minimal tuning.

This paper describes an expert system for an IBM personal computer. This expert system selects the most effective control system for a distillation column based on relative gain analysis and estimates of integrated errors in response to load upsets. Graphic and numerical results are displayed and a schematic drawn with all control loops in place. Programs accommodate multicomponent two-product columns and single-sidestream columns.

Keywords. Distillation Control; Relative Gain Analysis; Expert Systems; Personal Computer; Disturbance Rejection.

INTRODUCTION

The profitability of most process operations, and particularly distillation, requires close control over product quality. But, this is not easy to accomplish, for distillation columns require

from four to six control loops which tend to interact severely with
each other.

The Multivariable Problem

Because distillation is the most common multivariable operation
in the process industries, its control has received considerable
attention from both academics and practitioners. Yet, there has
been much controversy among them over the best system
configuration. This is probably due to the sheer number of possible
configurations that could be applied.

If only single control loops are closed, the number of different
configurations varies factorially with the number of controlled
variables. Thus, 4 loops can be configured in 24 different ways, 5
loops in 120 ways, and 6 loops in 720 arrangements. But this is not
the end of the possibilities. We have found that composition
variables (such as analysis and temperature) are functions of flow
ratios, while inventory variables (such as level and pressure) are
functions of flow sums and differences. If we then extend our list
of manipulated variables to include all independent ratios of two
flows and sums and differences of two flows, the number of possible
configurations increases dramatically. For example, a system for
controlling four variables can be configured in over 37,000 ways;
for 5 variables there are over 3 million possibilities; and for 6
variables the possibilities exceed 1 billion.

Linear Multivariable Controllers

One approach to controlling multivariable processes is to use a
linear multivariable controller. Controlled variables are listed
against manipulated variables to form a matrix. At each
intersection, a weighting factor is applied to specify how much
influence each controlled variable will have on each manipulated
variable. Each weighting factor is essentially a proportional gain
element. Derivative can be added at each point if desired, but not
integral action, as there can be but one integrator for each
controlled variable.

For a process having n controlled variables, this approach would require tuning as many as n x n proportional gains and derivative time constants, along with n integrators. This totals to 36 coefficients for a 4-loop process, 55 for 5 loops, and 78 for 6 loops.

This approach has other limits as well. It assumes that the process to be controlled is essentially linear because the variables are combined only by adding and subtracting. But, as noted above, compositions are functions of flow ratios. These ratios need to change with changing stream compositions and enthalpies. Thus, a linear control system would require retuning as these conditions change.

DESIGN PHILOSOPHY

Our approach seeks the simplest control system structure that will minimize loop interaction and sensitivity to disturbances. This is the opposite philosophy to that of the linear multivariable system, which applies the same n x n structure to all processes having n controlled variables. However, we are faced with the difficult choice of selecting the single structure (out of thousands or millions) which best fits the process to be controlled.

Process Model

Except for very low-order systems, it is not feasible to determine detailed process characteristics by on-line testing. Furthermore, characterization often needs to be done before the plant is completed and put into operation. Therefore, analysis is best carried out on a process mathematical model, rather than on the process itself. Additionally, this approach can often provide a general solution to an entire class of problems.

We use shortcut analytical models that do not require iteration to be solved. While they are not sufficiently precise for process design, they are adequate to compare the relative effectiveness of different control system configurations. Steady-state relationships are the most important and are easiest to determine. Material and energy balances form the basis for the model, along with mass-transfer and heat-transfer relationships. From there, a steady-state solution for controlled variables in terms of manipulated variables (or <u>vice versa</u>) is found. Dynamic features are less important and may be considered qualitatively following the quantitative steady-state analysis.

Relative Gain Analysis

The next step in sorting requires an analysis of the steady-state characteristics of the process to evaluate the relative influence each of the manipulated variables has on each composition. The technique used is called Relative Gain Analysis (Shinskey, 1984).

Relative gain is defined as the response of a particular controlled variable to a particular manipulated variable with all loops open, divided by the response observed when all other loops are closed. If the two responses are identical, then there is no interaction with other loops, and the relative gain is 1.0.

Relative gain is a pure number, unaffected by process nonlinearities or common factors; therefore, it is quite useful and not difficult to calculate. Furthermore, an array of relative gains for a process will have each column and row sum to 1.0. For the 2 x 2 interaction between composition loops on the distillation column, we would have:

$$\begin{array}{cc} & \text{Manipulated Variables} \\ & \begin{array}{c|cc} & m_1 & m_2 \\ \hline y & \lambda_{y1} & 1-\lambda_{y1} \\ x & 1-\lambda_{y1} & \lambda_{y1} \end{array} \end{array}$$

Controlled Variables

For the 2 x 2 system, only one relative gain, λ_{y1}, need be calculated to specify the array. For a two-product distillation column, interaction between the 2 composition loops can be represented by 15 independent 2 x 2 arrays.

Each relative gain in the 0 to 1 range has a complement in the same range. Each number greater than 1.0, therefore, has a negative complement in its array. Loops should not be closed between variables having negative relative gains because the process steady-state gain changes sign whenever other loops are opened or closed. This changes the loop in question from negative to positive feedback, which destroys stability. A plant with this hazard cannot be operated safely.

Of the 30 relative gains for the distillation composition loops, 10 are negative, 10 are greater than 1, and 10 fall between 0 and 1. Those loops with relative gains less than 1 are dynamically slower than those whose relative gains exceed 1.

Disturbance Rejection

The integrated error sustained by well-tuned composition loops following a disturbance is as important as loop interaction. In a recent paper (Shinskey, 1985), the author described how manipulated variables differ in their sensitivity to certain disturbances. However, integrated error is a function of both the sensitivity and the relative gain, which determine the proportional band of the controller. In that paper, estimates of integrated error were verified by simulations performed on a dynamic column model.

Material Balances

Another factor bearing on the effectiveness of a control-system configuration is the relative sizes of streams manipulated by the controllers. For example, there are three streams affecting level in the base of a column: liquid entering from the trays, vapor leaving from the reboiler, and liquid leaving as bottom product. If the bottom product flow is very small, e.g., 1 percent of the other 2 flow rates, it will not be able to control base level. However, it should be very effective in controlling composition, while boilup is assigned to control base level.

The smallest stream in the column is generally most effective in controlling composition because it can be manipulated with the highest absolute accuracy.

EXPERT SYSTEMS

An expert system places the expertise of a skilled technologist at the disposal of someone with less experience or skill in that particular field. One such system, termed EXACT, is already being marketed by The Foxboro Company (Kraus and Myron, 1984). It is a PID controller which adjusts its own control modes with the skill and experience of several engineers programmed into it.

Still, it cannot correct for an ineffective control system structure. The task of structuring a control system, and it is a difficult one, requires a rare combination of skills: process understanding, control system knowledge, and the ability of mathematical modeling. This combination resides in relatively few individuals, who are in great demand. The offering of an expert system is one way to satisfy that demand.

Two-Product Columns

This expert system configures the control loops for a distillation column based on Relative Gain Analysis of the column model. One of its programs calculates relative gains and sensitivities to disturbances, and a second draws the column with its recommended controls.

Fig. 1 lists the data required to characterize a simple (two-product) distillation column. From this information, the program plots the operating curves shown in Fig. 2. Scales for the heavy impurity in the distillate (y_H) and the light impurity in the bottoms (x_L) are selected by the computer to position the operating point (where the curves cross) in the lower left quadrant of the plot. Each of the curves corresponds to one of the manipulated variables, e.g., distillate (D) being held constant.

DISTILLATION CONTROL CALCULATIONS

MOL. FR. LIGHT KEY IN TOP? 0.98   **AND IN THE BOTTOM?** 0.002
MOL. FR. HEAVY KEY IN TOP? 0.016   **AND IN THE BOTTOM?** 0.945
MOL. FR. LIGHT KEY IN FEED? 0.24   **FEED % VAPOR?** 20
NUMBER OF ACTUAL TRAYS? 50   **REFLUX/DISTILLATE RATIO?** 3.8

DO YOU KNOW TRAY EFFICIENCY (E) OR RELATIVE VOLATILITY (A)? E

TRAY EFFICIENCY, %? 65   **RELATIVE VOLATILITY=** 2.06

PLOT (P) OR CALCULATE RGA (R)?

Fig. 1. The Distillation Consultant requires answers to these questions to characterize a simple column.

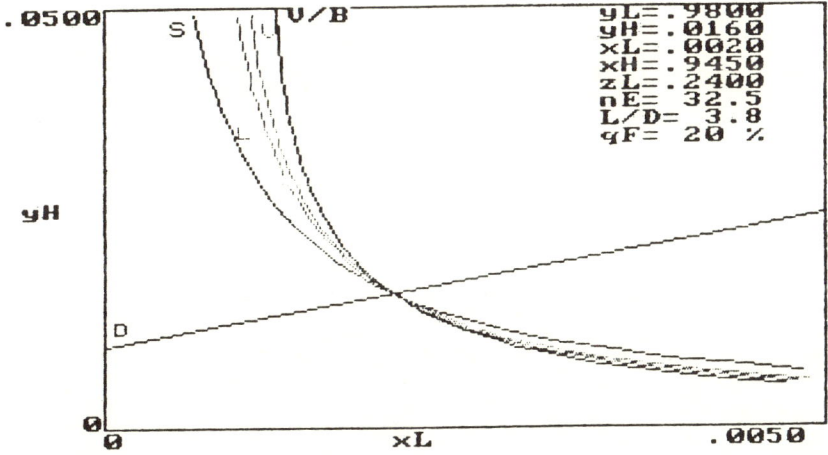

Fig. 2. Operating curves are plotted, relating the heavy key in the distillate (y_H) against the light key in the bottom (x_L) for conditions of constant distillate flow (D), reflux ratio (S), reflux flow (L), vapor flow (V), and reboil ratio (V/B). They cross at the operating point.

The slopes of these curves at the operating point are used to calculate the relative gains which are tabulated in Fig. 3. For example, the combination of reflux flow (L) manipulated for top-composition control (m_y) and vapor flow (V) for bottom-composition control (m_x) uses slopes of the L and V curves to calculate a relative gain of 15.2. The closer the curves lie, the higher will be the relative gain and, therefore, the greater the interaction. All material-balance control systems (manipulating either distillate D or bottom flow B) produce relative gains less than 1.0 in combination with any of the other curves. (Note that D and B have the same operating curve.)

DISTILLATION INTERACTIONS

	$m_y \rightarrow$ $m_x \downarrow$	D	L	S*
RELATIVE GAINS	D		0.237	0.282
	V	0.774	15.2	3.79
	R*	0.793	6.14	2.94
IE FOR UPSET IN ENERGY BALANCE	D		14	12
	V	4	100	12
	R	4	40	10
FEED COMP	D		56	84
	V	21	11	13
	R	26	11	13

$yL=0.9800$
$yH=0.0160$
$xL=0.0020$
$xH=0.9450$
$zL= 0.24$
$qF= 20\%$
$nE= 32.5$
$AL= 2.06$
$L/D=3.8$

← INTEGRATED ERROR (IE) IN RELATIVE UNITS.

* S=REFLUX RATIO; R=REBOIL RATIO(V/B)

Fig. 3. Candidates for controlling top composition are distillate flow (D), reflux flow (L), and reflux ratio (S); for controlling bottom composition, we can manipulate D, vapor (V), or reboil ratio (R). The best relative gains are obtained with the D-R and S-R combinations; if feed composition upsets are expected, S-R gives the best response.

There are actually 6 independent operating curves (only 5 are shown) which, taken 2 at a time, produce 15 independent 2 x 2 relative gain arrays. The total number of relative gains is therefore 30. However, 10 of these will be negative and are, therefore, excluded from consideration. Actually, the most favorable relative gains are developed from the outside curves S, V/B, and D. In the table of relative gains shown in Fig. 3, 0.793 is the highest of the numbers less than 1, and 2.94 is the lowest of the higher set. Because the ranges of the sets are so different (0 to 1 vs. 1 to ∞), algebraic comparison cannot be used.

To select the better of these two configurations, the integrated errors in response to upsets in the energy balance and feed composition are estimated for all configurations. These numbers are not absolute, but are relative to the integrated error of a reflux-boilup (L-V) configuration to an energy-balance upset at 100. If feed composition upsets are anticipated, then the S-R combination appears to be the best overall choice.

To draw the schematic diagram, the questionnaire must be filled in pertaining to the economic objective of the column and details about its auxiliaries. Note that the selection of loops made from the relative-gain and disturbance analysis is entered in reply to the second and third questions. The reply given to each of the questions determines the formulation of subsequent questions.

When the last question is answered, the computer draws the column and its auxiliaries (Fig. 4), then adds the control loops (Fig. 5). The drawing takes about 5 seconds. There are 24,780 different schematics that can be drawn, each an expert solution to a particular distillation control problem. (There are actually millions of possible configurations which are not available because they are deficient in one way or another.)

Whereas configuring such a control system once took days of study and effort, with no guarantee of successful operation, now it takes minutes, and the confidence level of excellent performance is very high.

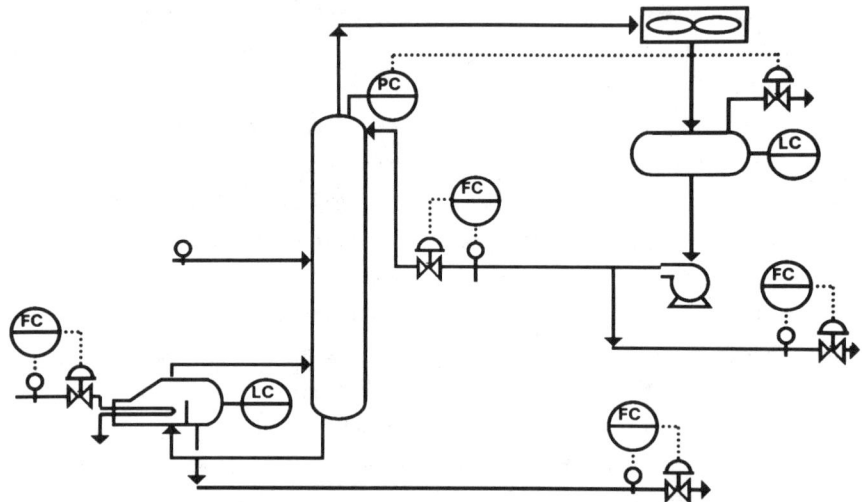

Fig. 4. This is the column described by the answers given to the questionnaire; it features an overhead air-cooled condenser, pressure controlled by releasing noncondensible vapor, and an oil-heated kettle reboiler.

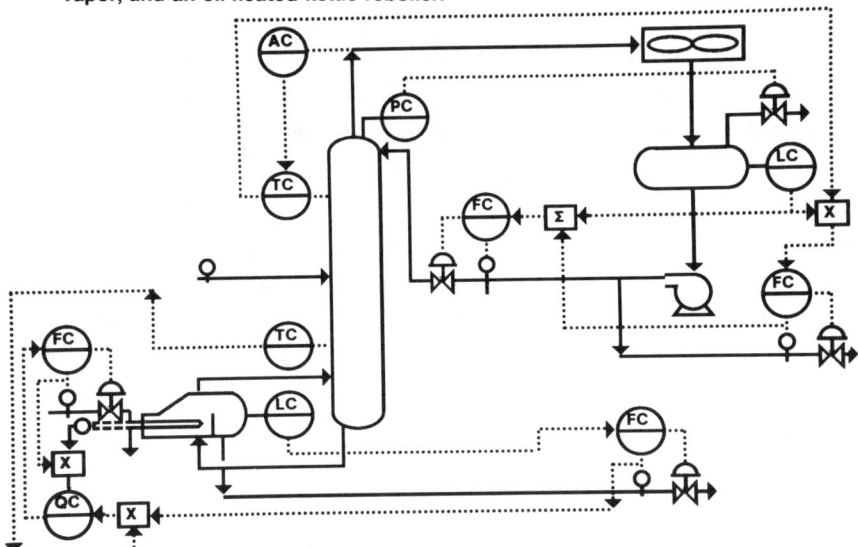

Fig. 5. The accumulator LC manipulates the sum of reflux and distillate, while top composition is cascaded through temperature to manipulate the D/V ratio. The lower TC manipulates the reboil ratio, setting heat flow calculated from oil flow and temperature drop.

Sidestream Columns

Extending the previous procedure to columns with a sidestream product is not an easy task. The additional stream brings with it a third composition loop, but the three controlled compositions are not always the same. The sidestream has two contaminants -- one light and one heavy -- either or both of which may be controlled. Furthermore, there are two sections to the column -- one above and one below the sidestream -- but separation can be manipulated independently in only one.

To use the procedure already developed for simple columns, sidestream columns are first classified into three groups. If one of the product streams is much smaller than the others, its flow must be manipulated to control the impurity it represents in the adjacent larger product. If none are particularly small, then separation is selected to be manipulated over the larger section of the column. In either case, classification arbitrarily assigns one composition loop and submits the other two to relative-gain analysis as before.

The column chosen to illustrate this program is an ethylene fractionator, designed to withdraw a pure ethylene product from a sidestream near the top. Ethane is removed from the bottom, and a small vapor stream, predominantly methane, is taken overhead.

Fig. 6 shows the data entered into the relative-gain program, and Fig. 7 gives the results. The flow of methane vapor is arbitrarily assigned to control its content in the sidestream, and interactions between the sidestream and bottom product are calculated and displayed. Because all relative gains above 1.0 are very high in this example, the best choice is the highest of the lower set, 0.590. The heavy key (ethane) in the sidestream should be controlled by sidestream flow, and the middle key (ethylene) in the bottom product by boilup ratio.

Column classification and other details are chained to the program which draws the schematic. Its questionnaire is similar to that for simple columns. After the last question is answered, the

RELATIVE GAINS FOR SIDESTREAM COLUMNS

THERE ARE 3 KEY COMPONENTS: LIGHT, MIDDLE, AND HEAVY.

MOL. FR. LIGHT KEY IN TOP? 0.63   IN THE SIDESTREAM? 0.0001   IN THE FEED? 0.005
MOL. FR. HEAVY KEY IN BOTTOM? 0.9993   IN THE SIDESTREAM? 0.0002   IN THE FEED? 0.17

COUNTING FROM THE BOTTOM,

TRAY NUMBER AT THE TOP? 121   AT THE SIDESTREAM? 110   AT THE FEED? 80
TRAY EFFICIENCY, %? 80
REFLUX/SIDESTREAM RATIO? 4.6
FEED % VAPOR? 0

IS SIDESTREAM LIQUID (L) OR VAPOR (V)? L

Fig. 6. The sidestream column needs this data for characterization.

DISTILLATION INTERACTIONS

CONTROL LIGHT KEY w_L IN P
USING DISTILLATE FLOW D.

$y_L = 0.6300$
$w_L = 0.0001$
$z_L = 0.0050$
$x_H = 0.9993$
$w_H = 0.0002$
$z_H = 0.1700$
$n = 121$
$n_P = 110$
$n_F = 80$
$E = 80$
$L/P = 4.6$
$D/F = 0.008$
$P/F = 0.822$
$B/F = 0.170$

RELATIVE GAINS

CONTROL $w_H \rightarrow$
CONTROL $x_M \downarrow$

	P	LL	SL
B	■	0.416	0.417
V	0.584	934	205
R	0.590	40.5	35.2

w_H = HEAVY KEY IN SIDESTREAM P
x_M = MIDDLE KEY IN BOTTOMS B
LL = REFLUX FLOW BELOW SIDESTREAM
SL = REFLUX RATIO BELOW SIDESTREAM
V = BOILUP
R = BOILUP RATIO V/B

Fig. 7. One arbitrary loop assignment is made, with the relative gains calculated for interactions between the other two.

column is drawn (Fig. 8), and the control loops added (Fig. 9).
Again, the program executes in only about 5 seconds, although it is
nearly twice as long (30,000 bytes) as the one for simple columns.
The number of possible diagrams is increased more than ten-fold, to
270,480. Note, for example, in Fig. 8 the need to identify which
analyzer controller (AC) controls the light key (L) in the
sidestream, and which controls the heavy key (H). There are also up
to six temperature points which appear when temperature is used to
control compositions.

BUILDING AN EXPERT SYSTEM

The methodology presently espoused for building expert systems
pools the skills of three individuals: the expert in his field, a
knowledge engineer, and a programmer. The duty of the knowledge
engineer is to "mine" the expert's information and judgements,
formulating them into models which can be solved and decision trees
which can be programmed. Then the programmer reduces this
information to code in whatever language is chosen -- some are said
to be more powerful for expert-system applications than others.

However, when tasks are divided among individuals with different
disciplines, results are rarely satisfactory. The expert tends to
be misrepresented, and the program may not operate like the expert
at all -- it may even be laced with pitfalls and serious mistakes.
If there are many thousands of possible solutions, as in the
diagrams described here, checkout and correction can be extremely
time-consuming.

To avoid these problems, the author assumed all three roles in
preparing this expert system. It was programmed entirely in BASICA,
which is sufficiently powerful to perform all the required
mathematical operations, curve plotting, logical operations, and
drawing the schematics. The programs are easily transportable, and
the results may be quickly recorded by any dot-matrix printer.

However, this approach has its limitations. The program for

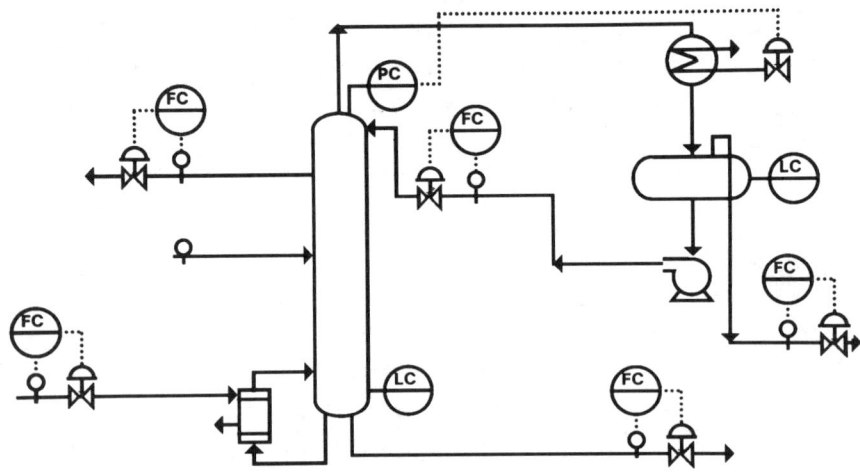

Fig. 8. Ethylene is removed as the upper sidestream, with methane as the overhead vapor and ethane as the bottom product.

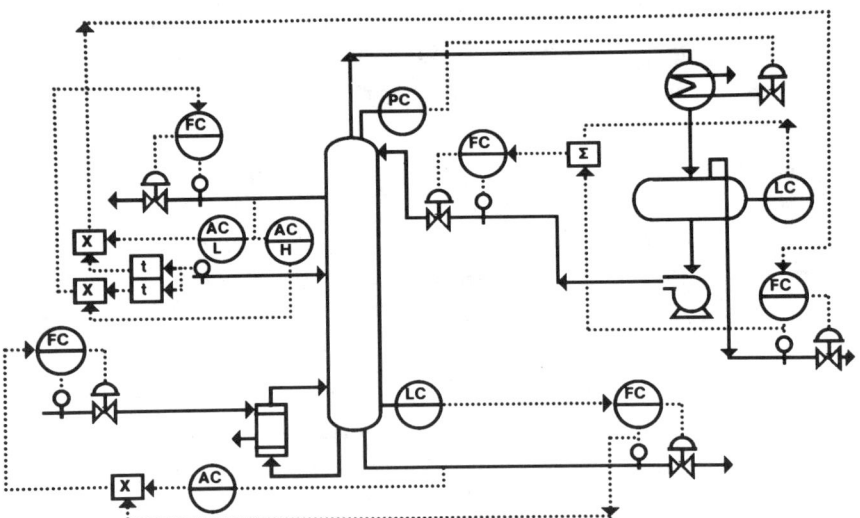

Fig. 9. The light key in the sidestream is controlled by the methane flow, and the heavy key by sidestream flow; bottom composition is controlled by boilup ratio.

generating the sidestream-column diagram is quite long and difficult to debug, yet there are still problems which it cannot accommodate, such as a second sidestream and differential temperature controls. Furthermore, the screen is quite crowded and resolution limited to 640 x 200 pixels.

CONCLUSIONS

Because the programs can run on a standard IBM or compatible personal computer, they can be sent to remote locations throughout the world without special hardware. They are intended to place the author's design expertise in the hands of Foxboro field engineers whenever distillation control systems need to be designed. These programs are the property of The Foxboro Company.

The Future of Expert Systems

Logical applications for an expert system would be in plant diagnostics and optimization. These will require detailed knowledge of the plant's operating characteristics, economic performance, and market demands. Rewards for the success of such a system are very great; but, it cannot be successful without an effective infrastructure supporting it. In other words, economic plant management depends on accurate and responsive quality control, which results from expertly configured systems of expertly tuned controllers.

REFERENCES

Shinskey, F. G. (1984). Distillation Control, 2nd Ed., McGraw-Hill Book Company, New York, Chap. 5.

Shinskey, F. G. (1985). Disturbance-Rejection Capabilities of Distillation Control Systems, Proceedings of the American Control Conference, Boston, MA, June 19-21.

Kraus, T. W. and T. J. Myron (1984). Self-Tuning PID Controller Uses Pattern-Recognition Approach. Control Engineering, June.

QUESTIONNAIRE FOR SIMPLE COLUMNS

Distillation Schematic Diagram

What is the economic objective of the column?   E
   * Maximize recovery of distillate (D)
   * Maximize recovery of bottoms (B)
   * Minimize consumption of energy (E)
   * Meet specifications for both products (P)

What variable controls top composition?   S
   * Distillate (D)
   * Reflux (L)
   * Reflux ratio (S)

What variable controls bottom composition?   R
   * Boilup (V)
   * Boilup-to-bottom ratio (R)
   * Bottom product (B)
   * Distillate (D)

What will be used for top-composition feedback?   C
   * Analyzer (A)
   * Temperature (T)
   * Analyzer-temperature cascade (C)
   * No measurement available (N)

And for bottom-composition feedback?   T
   * Analyzer (A)
   * Temperature (T)
   * Analyzer-temperature cascade (C)
   * No measurement available (N)

Is the distillate stream liquid (L) or vapor (V)?   L

Is the condenser submerged (S), or over (O) the accumulator?   O

What is the cooling medium?   A
   * Water (W)
   * Air (A)
   * Refrigeration (R)

How is heat removal controlled?   G
   * Gas release (G)
   * Condenser drain valve (D)
   * Coolant flow (C)
   * Liquid-filled accumulator (L)

What type of reboiler is used?   K
   * Vertical thermosyphon (V)
   * Horizontal thermosyphon (H)
   * Kettle (K)
   * Direct-fired (F)

What is the heating medium?   O
   * Steam or other vapor (S)
   * Hot oil or other liquid (O)

QUESTIONNAIRE FOR SIDESTREAM COLUMNS

Sidestream Column Diagram

What is the economic objective of the column?   E
   * Maximize recovery of bottoms (B)
   * Maximize recovery of sidestream (P)
   * Minimize consumption of energy (E)

What variable controls heavy key in sidestream?   P
   * Sidestream (P)
   * Reflux below sidestream (L)
   * Reflux ratio below sidestream (S)

What variable controls bottom composition?   R
   * Boilup (V)
   * Boilup-to-bottom ratio (R)

What will be used for sidestream-composition feedback --
light key?   A
   * Analyzer (A)
   * Temperature (T)
   * No measurement available (N)

What will be used for sidestream-composition feedback --
heavy key?   A
   * Analyzer (A)
   * Temperature (T)

What will be used for bottom-composition feedback?   A
   * Analyzer (A)
   * Temperature (T)
   * No measurement available (N)

Is the distillate stream liquid (L) or vapor (V)?   V

Is the condenser submerged (S), or over (O) the accumulator?   O

What is the cooling medium?   R
    * Water (W)
    * Air (A)
    * Refrigeration (R)

What type of reboiler is used?   V
    * Vertical thermosyphon (V)
    * Horizontal thermosyphon (H)
    * Kettle (K)
    * Direct-fired (F)

What is the heating medium?   S
    * Steam or other vapor (S)
    * Hot oil or other liquid (O)

REFLECTIONS ON CPC-III

The main feature of the final day of the conference was a presentation by F. G. Shinskey. It was followed by several hours of comments from the audience. The presentation and the subsequent discussions are reproduced below in abbreviated and edited form.

Greg Shinskey, Foxboro Company

What can research do that will advance process control? I think we ought to take a very, very serious look at this issue. I'm going to skip out of the chemical industry into the steel industry. Why? Well, more than the chemical industry, the steel industry in the U.S. is in a very depressed state and I think we have some lessons to learn from it; the chemical industry may follow suit.

There is a steel plant that not long ago employed 17,000 people. That plant is now shut down. Why? Foreign competition. The plant couldn't make steel from American ore to deliver to American sites as cheaply as the Japanese could use American ore to make steel to deliver to American users. Why not? We know that labor rates in the U.S. are higher than labor rates in Japan or Taiwan. In that respect, the labor unions have themselves to blame. However, recognize that in the steel industry, not all of the expense that goes into manufacturing a product is labor. A great deal of it is fuel, power, capital equipment, and products that have to be recycled because they have not met quality standards. And in a steel mill, it is common to heat and cool and reheat, etc.

Analyses in the steel industry or about the steel industry report: lack of innovation. The steel industry in the U.S. is outdated. It can't compete with the processes that have been developed and put into production in other places. If the steel industry doesn't have innovation, why? Where does innovation come from? Most of you will probably say - research. I'm going to examine that issue. If innovation comes from research, we must face up to the fact that this company has probably the most modern, best funded research facility in the entire steel industry. Unfortunately, it's an ivory tower and ivory towers tend to produce very little, not even ivory. I've been exposed to it enough to identify what I think are principle problem areas for great concern. It is characteristic of their research engineers, to have no deadlines nor budget limits, and they publish no papers.

What does the "no published papers" tell us? If there are important things that are going on in research, other parts of the steel industry don't know about them and people in control technology don't know about them. Now, they may say they're not going to tell because they don't want others to know their proprietary information. Mathematically, that may make sense, but there are many things in this world that do not follow the laws of mathematics. One of them is goodwill and another one is information transfer.

Let's look at "no deadlines" and "no budget limits". If you give an engineer or a scientist an unlimited length of time and an unlimited budget to solve a problem, he will consume all your money and all your time. How do the Japanese innovate? Innovation begins on a plant floor. They encourage everyone in the entire organization to participate in the design process. If it doesn't work, the managers want to know about it. If it can be made better, they want to know about it. Those responsible for design talk to the maintenance people, and those responsible for manufacturing talk to the operating people - it almost seems too democratic for the U.S. We have a problem in the U.S. that may not exist in Japan: people who belong to labor unions do not talk to people in management.

Why does innovation begin on the plant floor? Because problems exist in the plant and they need to be solved in the plant. You can spend your whole life working on the solutions to problems but if you have not addressed the problems that exist out in the plant, you're solving the wrong ones. If you come up with a solution that won't work because there is a missing link, because the hardware is not available, because the sensors don't exist, all of your theory will not reach fruition. Now, you can say to yourself "I was just born 50 years too soon", but that's not going to pay the bills, or make you a leader in world business.

If we don't want the chemical industry to fall into the same pit as the steel industry then we have to change our habits in such a way that we will maximize innovation.

Necessity is the mother of invention. If you want a clever engineer to come up with a novel product or a novel way of doing something, give him a deadline and expect results. Now of course, it has to be a reasonable deadline; but when you give him that deadline, stick with it and if he fails to meet the deadline, then give the job to somebody else. Limit the budget and don't extend it. If you're trying to make a product, barrel of petroleum or a ton of steel or whatever, and if it costs you more money to make it that you can get when you sell it, you're going to go out of business. That's competition, that is what made American industry strong and if you don't think American industry is strong, visit Poland, Russia, Czechoslovakia, China.

In order to come up with truly innovative products, put constraints on the engineers. Tell them that in contrast to today's technology or today's product, you want the next one to be smaller, lighter, faster. Demand new technology. One of the jobs that I benefited from as much as any other, from a standpoint of inducing creativity, stands out in my mind very clearly. We had a sales engineer in the oil fields of Louisiana who was kind of a nut, but he could sell anybody anything. Now the difficulty was that he would get the Foxboro Company in trouble because we couldn't always deliver what he sold. But in any case, he convinced people in the oil patch in 1962, that they needed a digital data logger that would stand out in the weather and record flows and pressures from a battery of wells being connected to a common separator (both production and test) with programming to select the order of wells tested and the duration of the testing. He wanted to printout three carbon copies with - get this - no electric power! I was given the assignment to build this monster. One problem was that Foxboro doesn't make any of those products. I had to buy a digital printer, and used a battery of Ni-Cad ceels to drive it. During printing, once an hour, it would draw about 4 amps but for only milliseconds - the average power consumption was around 1 milliwatt. This was all accomplished with latching type of circuitry. I built an A/D converter out of a Foxboro pressure transmitter, a cam, and a stepping motor. The product was a success, except that we couldn't print out three carbon copies. It was standing out in the salt and the sun and the heat and the sulfur down in the bayou, so we had problems with corrosion but it worked. Its most important contribution was innovation, which told me that if you ask for the impossible, you might get almost that. If you ask for what's easy, state of art, and you place no constraints on it, that's probably what you'll get. If you're out in the middle of the desert and your vehicle breaks down, you'll probably learn how to fix it with what you have available, and when you do, your solution is bound to be innovative.

Successful products are also simple. Einstein said, "Things should be kept as simple as possible, but no simpler". Inverting a 210 x 210 matrix in order to control four loops, is not "as simple as possible".

I have, in my engineering career, always invoked Parkinson's law" "The time required to do a job will equal the time allowed to do it". On a business trip, I have always booked in advance a return reservation, and I have always come home on that flight. There are many temptations to stay longer, and I refuse. I allow X days to complete a job, and that job will be completed in X days. Suppose you've allowed five days, Monday through Friday, to commission a control system on an evaporator, and you arrive Monday to find they don't have the steam line connected to the evaporator yet. I then announce I am leaving on Friday afternoon at the latest. If this isn't ready to be completely run and commissioned by Friday afternoon, then tell me now so I won't waste my time. They say we're not sure we can get that steam valve in place because we'll have to call in technicians to work overtime, etc. Well, you had better do it. If not, then I'll take the next plane home. You have to do that. If you don't do that, you're just extending deadlines - the time required to do the job will fill the time allowed to do it; if you allow an indefinite length of time, it will never be finished. The same applies to money.

Standards. Set standards and expect them to be maintained. I am suspicious about weighting factors applied to performance curves, which show performance to be acceptable, so that tests will comply with standards. This is revisionist history. The novel 1984, described a totalitarian country, where the principle character in the story had the job of retroactive new editor. His assignment was to correct old newspapers so that the stated government goals could be readjusted to the real production figures that were later attained.

I saw a bumper sticker recently that is a terrible tragedy: "If all else fails, lower your standards". Let's try to provide the control that's the best we can within the justification of economics.

the middle of the desert and your vehicle breaks down, you'll probably learn how to fix it with what you have available, and when you do, your solution is bound to be innovative.

Successful products are also simple. Einstein said, "Things should be kept as simple as possible, but no simpler". Inverting a 210 x 210 matrix in order to control four loops, is not "as simple as possible".

I have, in my engineering career, always invoked Parkinson's law" "The time required to do a job will equal the time allowed to do it". On a business trip, I have always booked in advance a return reservation, and I have always come home on that flight. There are many temptations to stay longer, and I refuse. I allow X days to complete a job, and that job will be completed in X days. Suppose you've allowed five days, Monday through Friday, to commission a control system on an evaporator, and you arrive Monday to find they don't have the steam line connected to the evaporator yet. I then announce I am leaving on Friday afternoon at the latest. If this isn't ready to be completely run and commissioned by Friday afternoon, then tell me now so I won't waste my time. They say we're not sure we can get that steam valve in place because we'll have to call in technicians to work overtime, etc. Well, you had better do it. If not, then I'll take the next plane home. You have to do that. If you don't do that, you're just extending deadlines - the time required to do the job will fill the time allowed to do it; if you allow an indefinite length of time, it will never be finished. The same applies to money.

Standards. Set standards and expect them to be maintained. I am suspicious about weighting factors applied to performance curves, which show performance to be acceptable, so that tests will comply with standards. This is revisionist history. The novel 1984, described a totalitarian country, where the principle character in the story had the job of retroactive new editor. His assignment was to correct old newspapers so that the stated government goals could be readjusted to the real production figures that were later attained.

I saw a bumper sticker recently that is a terrible tragedy: "If all else fails, lower your standards". Let's try to provide the control that's the best we can within the justification of economics.

Now let's take a look at the gap. We're here because there is a gap and if you think the gap is between industry and university, you're wrong. The gap is between theory and practice. I know this for a fact because there are some people in industry who are on the other side of the gap from me, and there are some people in academia, notably Tom McAvoy, Bill Luyben and some others, who are on the same side of the gap that I am. We have to identify the gap, ask why it is there, and what we can do about it.

This gap exists between research and production (Fig. 1).

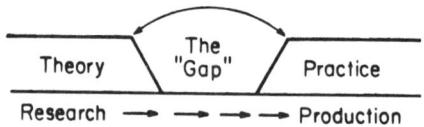

Figure 1

Research is supposed to contribute to production, but how is it going to do that when it's located physically and mentally, remote from production? Yet the ideas have to get there or else they're not going to bear fruit.

We have to bridge this gap, and the bridge is called communications. Research has to say, this is the right way to control this process. In order to do that, they have to be able to speak to the production people in words the production people understand. If you can't do that, you will not achieve your objective. But even if you can do that, you may not achieve your objective if the production people have not said "Here's what we need out in our plant, here's what our problems are, here is a process that needs to be controlled". And if we have not done that, those of us who are close to the process, then it is not surprising that the theorists work on processes that are represented by $\dot{x} = Ax + Bu$.

There is a complete spectrum from the theory on one side to the practice on the other side; but we must move toward practice because we must have production. We have different types of people whom I will represent by bell-shaped curves (Fig. 2).

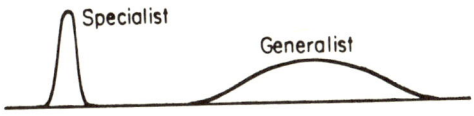

Figure 2

Consider the knowledge of the individual expressed as an area under the curve; the mean is his specialty and the standard deviation indicates the breadth of his experience and understanding. A generalist may not be particularly good at anything, but he knows much about a great number of things. The practitionist tends to be more generalist, and the theoreticians to be more specialists. I am a generalist I grow my own food at home and I fix my own car, as well as being a control engineer.

In a lifetime of experience, we can only accumulate a certain area under the curve. Someone who is a true specialist begins to look like an impulse function - a signal of an infinite height and zero width. In order for the specialist to contribute anything to production, which is beyond his immediate sphere of influence, he must convey his specialty to someone who can do that. If he doesn't, if he is unable to communicate his research findings in such a way that someone else can use them, then he accomplishes absolutely nothing. He can be the most brilliant person in the world and understand everything in his field of specialty, but if he can't tell anyone about it, he may as well not exist.

As examples of people who are on different sides of different gaps, I am only going to talk about my friends. I have plotted three people on this spectrum between theory on the left and practice on the right. On the left is Edgar Bristol. Ed and I have worked together for Foxboro Company so long that I can't remember when it began. Ed is definitely on the theory side more than I am, and much as I like Ed, there is much he says that I do not understand.

The area of overlap between us is fairly small but it does exist (Fig. 3). If it didn't exist, we might not have a word in our present glossary that is called "relative gain array". Ed Bristol developed his measure of interaction which he communicated in an IEEE journal (which I don't read because I can't read it). But Ed was kind enough to go through examples and to communicate to me an understanding of his new measure of interaction which I found very useful. I began to apply it, and it appeared under the name "Relative Gain" for the first time in my book "Process Control Systems" in 1967.

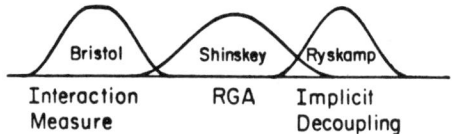

Figure 3

If it hadn't been for the communication between Edgar and myself, we would have both lost this opportunity. I believe that Ed developed this in response to a need that another engineer had to evaluate the interaction that exists in a boiler between variables such as steam pressure, steam temperature, and drum level. Now I don't know whether he ever successfully characterized the boiler at that early stage, but I think it gave Ed the idea of this interaction measure with the properties of being dimensionless with rows and columns adding up to 1.

In an area closer to production is Carrol Ryskamp. He is the type of individual who can go out to the process and understand what's going on without having to derive a series of differential equations. At CPC2, he presented a paper on implicit decoupling which he used successfully. He did not calculate the relative gains for that distillation column where he successfully applied the implicit decoupling, but he did report on it's superior dynamic responsiveness and it's reduced sensitivity to disturbances. Now after having previous difficulty with material-balance systems, I discovered the reasons why his idea worked so well. I followed his paper with another which analyzed his system. From that I was able to move back in the theoretical direction and develop a control system for the bottom of the column similar to what he applied to the top. It produces even less interaction on a relative gain basis, and that has been even more successful.

Notice that I have inputs from both directions. Edgar's theory impacts from the left, and Carrol's very valuable experience and insight and intuition moves from the right - this transfer of information is very important. We must work with all kinds of people having different mean positions on this axis and different standard deviations; if their curves don't overlap, we had better do something about it to close the gap.

If we want to do the best we possibly can, and make our processes run as perfectly and as efficiently as we can so, we would like to execute very high levels of control. We would like to optimize everything. However, before we can optimize anything, the process has to be operable, it has to run, the pumps have to go, the compressors, the pipes have to retain the fluids and so forth. Very, very elementary and basic things, but if they don't work, forget about the controls. If your car doesn't have any wheels, you can't drive it regardless of its electronic ignition and fuel injection and computer controlled whatever. If you can't measure process variables and manipulate flowrates, then you can't put in any higher level of controls.

We saw a slide that inferred that PID was obsolete. You had better not believe that; PID is a low-level control device which fits nicely between advanced control, and sensors and valves. It's cheap, reliable, robust, and does not require a process model. Examine your chemical plant, your petroleum refinery and you will discover that all flow controllers and all level controller and almost all pressure controllers perform very well their assigned tasks using PI controllers. If you add all those together in a typical chemical plant, it will total at least 85% of your controlled variables. If you want to replace them with model-based DMC, etc., it's expensive and it's time consuming; and therefore you apply it where you need it, but not everywhere. Also, you will discover that almost every advanced control system will execute it's function by sending a setpoint to a flow controller or pressure controller or a temperature controller which is PI of PID. If you take that item out and go directly to the valve, you are asking for trouble because of nonlinear valve characteristics, and sticky valves; and you will lose the information that the flow control loop provides.

Advanced control is used to hold product quality at setpoint in the face of upsets. Then we can operate at whatever optimum setpoints we select. Put in the grandest optimizer in the world, but if you don't have any regulation over your process, the controlled variable will never be at setpoint, so it doesn't matter whether the setpoint is optimum. Any optimizer must depend on the performance of everything under it (Fig. 4).

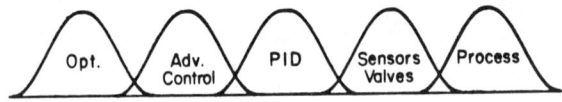

Figure 4

Furthermore, optimization is performed on models, not directly on the plant because the plant is too slow. You might optimize the model, but if your model doesn't represent the plant, you haven't optimized the plant.

For those of you who have a great amount of faith in optimization and you think that this can be done because we have the "technology" to do it, recognize that during this conference the slide projector jammed, and we can't seem to find the light switches to control the lights. Have you tried to control the temperature of the hot water in your shower? I am convinced that temperature is continuously variable; sometimes even with the hot water valve fully open, the water is not hot enough. We must expect to encounter constraints which are beyond our personal control. The function at this conference which has worked the best, bar none, is the kitchen. And that is the most manually controlled operation!

Those who are concerned with achieving high-level control, have to be able to communicate their findings to the control engineer who then must communicate them to the operator (Fig. 5).

Figure 5

That's not an easy task. I find it's difficult to communicate with operators, but I know I have to do it so I make a real effort. If it requires talking about last week's football game or the fishing in the bayou, that's what we have to do. If you're going to talk to him about pole placement and it's not a fishing pole, you will be unsuccessful. I place myself in the middle of this spectrum, for I have communicated with operators and am reasonably successful in getting my control systems used. However, I heard some papers delivered at this conference which communicated nothing to me. At least they did not communicate what the author intended.

There was a session on adaptive control that seemed like a different subject, because I didn't understand any of it. I had to go back and look at my credentials, and recall that I built a self-tuning controller in 1973 and it worked. I must have then known something about adaptive control; what have I forgotten? Why didn't I understand these presentations?

Communication is the key. Most of you have been in this business for a long time, not only designing, conceiving, theorizing, but also teaching. The business of explaining to your colleagues what you are trying to do, the business of trying to sell concepts to someone who may not be willing, who may not be interested, who may be defensive; whether you succeed depends on your communication skills. Make this an important facet of your research.

George Stephanopoulos, MIT

Let me start by saying that we should all recognize that control is a manmade engineering task. It is not like natural science where you try to find the mechanisms that govern catalytic reactions, the fluid mechanical aspects, the physical chemistry of interfaces, where you try to understand what nature is all about and model it appropriately. In process control we are dealing exclusively with manmade problems. We define our objectives, those may be economic or quality control, we define the constraints, we define our freedoms, we define everything and the quality of the solutions we get is directly proportional to the quality of the problem statement we had. If the solution is irrelevant it is because the problem formulation was irrelevant. So, we come to what has turned out to become a generic term, the gap. Greg said that the gap can only be bridged through communication and I am posing the following question; How are we going to bridge the gap by better communication if, as I have felt personally, there is very poor industrial articulation of the real problems and the real issues that would define the relevant problem? Allow me to say Greg that everything I know about control I have learned from your first book and I do not say that lightly. But it is an example of a poor articulation of problem statements and it is something that is very difficult to use for teaching someone the first time.

As far as the unified control theory is concerned, it is the biggest challenge that we are faced with and it is not a challenge in terms of numerical aspects, it is a challenge in terms of a thinking framework. Can we put within unified control theory all the issues that pertain to process control and how then are they interrelated to each other? That is all that unified control theory, the way I perceived it, is trying to do for all of us and in a certain way it is placing technological issues in front of us which are unachievable at present.

Roger Sargent, Imperial College, London

I was disappointed with industry. I guess they don't have any problems. Nobody gave me a new problem which they are waiting for solutions for. They obviously still have their old problems which we still keep talking about and still don't have solutions for but nothing new. My parting shot is to Greg Shinskey who talked a bit about PID. There is optimization on the one hand there is the process on the other and the two can only meet through the medium of PID controllers in the middle. He pointed out, of course, that about 90% of controllers in the process industries are PID controllers. One might wonder why that is? Of course the reason is because Foxboro only produces PID controllers and Honeywell and all the instrument manufacturers. You can't ask them for a self tuning regulator, because they haven't got a box that produces that. I'd like to put a challenge to these instrument manufacturers. If they put out some more flexible hardware into which you can program all these other things that we've been talking about and look at 20 years time and see how many PID controllers we are still using, that would be a better, more objective path than the utility of PID controllers. There is, perhaps, just one thought. There is a lot of fields of control other than process control. How is it that the process controllers are the only ones that have a thing about PID as the universal solution? Do they have different engineering laws than the rest of us? I'd like them to think about that a little bit.

Karl Astrom, Lund Institute of Technology, Sweden

I think that CPC3 has given me an interesting perspective on the field. If I compare it with other branches, I'm still coming away with the feeling that the chemical process control industry is somewhat backward. You are afraid to try new things. Maybe for good reasons.

I had a question for Greg: If you have about 85% PID regulators out there, wouldn't it be good to have automatic tuning devices on them since you have so many of them?

Jens Balchen, Norwegian Institute of Technology

I have a terrible feeling that the practitioners and particularly the American practitioners are so horrified by the thought that some progress should be made that is not coming all the way from the bottom that they're constantly wasting a lot of energy to prevent things going on, particularly in academia. I think that is a particular disease in this country and I think that Japan and Europe and maybe also China do not have that disease. I think that there are other fields of engineering, not process control engineering but say aerospace, microelectronics, you name it, where this country is great, where they have not had that disease to such an extent.

E. Ydstie, University of Massachusetts

My intention in coming to this conference was really to make friends and friends from industry mostly because we academics, we have a number of these conferences where we have the opportunity to meet. I'm not willing, however, to compromise my research to make friends with industry and you may wonder why don't I do research on PID control and RGA and these kind of things that maybe are more popular out in industry. I don't do that for the simple reason that I don't think I could contribute much. I could spend my whole lifetime on these things and I would never surpass what Greg Shinskey, Ed Bristol, Tom McAvoy and others have been doing. I think that is a waste of my time and effort to get into those kinds of problems. That does not mean that I don't think they're important and useful.

Industry does not believe convergence and stability proofs, so it's a waste of time to come and show them that kind of thing. Industry also is probably tired of simulations and we heard earlier that they are tired of seeing pilot plant experiments. If you take all those things out, there remains very few possibilities for academics to do any kind of useful research. What I would like to see is more industrial/university cooperation.

John MacGregor, McMaster University

One thing that has disturbed me about all these conferences and about the direction of chemical process control has been, in general, that we are really concerned with and dominated by the petrochemical petroleum industry. There is a lack of concern for quality and in particular quality in the specialty industries.

Bjorn Tyreus, E. I. Dupont Co.

I would like to defend academia and the research done in academia. I think a young control engineer does not have all the experience that Greg Shinskey and Page Buckley and others have. He will build it up, hopefully over the years but in the meantime he needs something to grab onto and to improve his understanding of the process control theory. I feel that a lot of the work that is being done in academia has been very useful in helping me become a better control engineer.

John Doyle, Honeywell and Caltech

There's a couple of comments I want to make on the technical side. One of them is that I noticed that there is a myth in chemical process control that is very similar to the one that exists in aerospace control and that is - what does a control engineer spend most of his time doing? It's not designing control systems but the most important role a control engineer can play in the process industry as a whole is to decide how to build plants. A lot of what you do is deciding not how to construct a controller but what are critical factors in a control problem, what are important things to measure, what are important things to control and how to configure plants. The reason for having a good theory is not so we can design hundreth order 10 x 10 multi-input multi-output control systems that squeeze the last bit of performance out of the system but rather that we can, in a systematic way, answer questions like - how good a model do I need in order to control this process to the level that I want to have it? In the end, maybe you'll decide that it's this level of model and it's okay to use a PID but it would be nice if you knew that rather than hooking a bunch of things up and if it works, great, and if it doesn't, well, you think it might be better to do something else.

Manfred Morari, Caltech

It is not for me to judge if this conference was a success. I feel that if everyone of you goes away from here, academics and industrials, questioning in some way what you have been doing - academics, if they are working on the right problems; industrials, if they look at the right methods to solve their problems - then I feel that this conference was a success.

AUTHOR AND DISCUSSOR INDEX

Arkun, Y.,
Dynamic Process Operability. Important Problems, Recent Results and New Challenges. 323

Astrom, K. J.,
Adaptation, Auto-Tuning and Smart Controls. 427
Discussor. 925

Balchen, J.,
Discussor. 925

Doyle, J. C.,
A Unifying Framework for Control System Design Under Uncertainty and its Implications for Chemical Process Control. 5
Discussor. 925

Dumont, G.,
On the Use of Adaptive Control in the Process Industries. 467

Edgar, T. F.,
On-Line Identification and Optimization. 513

Fjeld, M.,
Industrial Perspective of the Use of Adaptive Process Control. 501

Froisy, B.,
Industrial Applications of IDCOM. 233

Garcia, C. E.,
Advances in Industrial Model-Predictive Control. 245

Gilles, E. D.,
Some New Approaches for Controlling Complex Processes in Chemical Engineering. 689

Goodwin, G. C.,
Digital Adaptive Control. A Robust Approach. 517

Jensen, K. F.,
Control Problems in Microelectronics Processing. 623

Juba, M. R.,
Progress and Challenges in Batch Process Control. 139

Kantor, J. C.,
Control in the Presence of Model Uncertainty. 1

Lasdon, L.,
The Integration of Planning, Scheduling and Process Control. 579

MacGregor, J.,
Discussor. 929

Marlin, T. E.,
A Short-Cut Method for Process Control and Operability Analysis. 369

Moore, R. L.,
Expert Systems in On-Line Process Control. 839

Morari, M.,
Discussor. 929

Morshedi, A. M.,
Universal Dynamic Matrix Control. 547

Niida, K.,
Process Control System Synthesis by an Expert System. 869

Popeil, L.,
Coordinated Control. 295

Ray, W. H.,
Control of Chemical Reactors. 621

Reuss, M.,
Computer Control of Bioreactors—Present Limits and Challenges for the Future. 641

Rinard, I.,
Process Operability. 321

Roat, S.,
The Integration of Rigorous Dynamic Modeling and Control System Synthesis for Distillation Columns: An Industrial Approach. 99

Sargent, R.,
Discussor. 929

Schnelle, P. D.,
A Review of Industrial Reactor Control: Difficult Problems and Workable Solutions. 749

Seborg, D. E.,
Model Predictive Control. 231

Shinnar, R.,
Impact of Model Uncertainties and Nonlinearities on Modern Controller Design. Present Status and Future Goals. 53

Shinskey, F. G.,
An Expert System for the Design of Distillation Controls. 895
Reflections on CPC-3. 913

Stephanopoulos, G.,
Expert Systems in Process Control. 803
Discussor. 929

Taylor, J. H.,
Expert Systems for Computer-Aided Control Engineering. 807

Treiber, S.,
Multivariable Constraint Control Using a Frequency Domain Design Approach. 185

Tyreus, B.,
Discussor. 929

Wilhelm, Jr., R. G.,
An Industrial View of Advanced Process Control. 95

Ydstie, B. E.,
Adaptive Control. 421
Discussor. 929

KEYWORD INDEX

Adaptive Control, 427, 467, 501, 517
Alarm System, 689
ARIMA Noises, 53
Artificial Intelligence, 801, 869
Automatic Tuning, 427
Automation, 623

Batch Process Control, 139
Batch Reactor Control, 139
Batch-Reactor, 139
Biofiltrator, 641
Bioreactor Control, 641
Biotechnological Processes, 641

CAD Environments, 801
Calculus of Variation, 547
Chemical Industry, 467
Chemical Process, 869
CHIPS, 869
Computer Aided Design, 869
Computer Control, 233
Constraint Control, 185
Constraints, 245
Control Applications, 467
Control Survey, 749
Control System Design, 801, 323
Control System Synthesis, 869
Control Techniques, 749
Control Theory, 749
Controller Design, 53
Controller Tuning, 369
Coordinated Control, 295
Cross Machine Control, 501
Crystal Growth, 623

Dead-Time Compensator, 53
Decentralized Control, 323
Decoupling, 369
Delta Operators, 517
Diagnosis, 839
Digital Control, 233
Discrete Time Systems, 517
Distillation Column, 689

Distillation Control, 5, 99, 369, 895
Distillation with Reaction, 99
Distributed Parameter System, 697
Disturbance Rejection, 895
Dopant Diffusion, 623
Dynamic Distillation Simulation, 99
Dynamic Matrix Control, 254
Dynamic Matrix Control (DMC), 547
Dynamic Modelling, 369
Dynamic Process Operability, 323
Dynamic Simulation, 369

Engineering Design, 801
Euler-Lagrange Equations, 547
Expert Control, 427
Expert Systems, 801, 839, 895

Fault Detection, 839
Fed-Batch Fermentations, 627
Fed-Batch Reactor, 139
Feedback Control, 641
Feedforward Control, 641
Feedforward Controllers, 139
Fixed-Bed Reactor, 689
Frame Model, 869
Frequency Response, 53
Fundamental Limitations to Dynamic Operability, 323

Generalized Reduced Gradient (GRG), 547

H_∞-Optimal Control, 5
Heuristic Knowledge, 869
Higher Level Control, 579
Human Expertise, 869

IDCOM, 233
Identification, 139
Industrial Control, 467
Industrial Processes, 233
Inferential Control, 295
Integrated Circuits, 623
Interaction, 369

Interaction Measures, 99
Internal Model Control, 53, 233, 295, 323

Jacobian Matrix, 547

Kalman Filter, 689
KEE, 869
Knowledge Based Systems, 427

Lagrange Function, 547
Linear Programming (LP), 547
Linear Programming solution of Dynamic Matrix Control (LDMC), 547
LQG Adaptive Control, 501

Materials Handling, 623
Microelectronic Processing, 623
Model Based Control, 295
Model Error, 5
Model Predictive Control, 245
Model Reduction, 689
Model Sensitivity, 53
Model Space, 53
Model Uncertainty, 5, 53
Modern Control, 749
Multi-Input Single-Output Control, 295
Multivariable Control, 185, 295, 369

Nonlinear Controllers, 139
Nonlinear Models, 53, 139
Nonlinear Optimization, 547, 579
Nyquist Array, 185

Operability, 365
Optimal Control, 5, 139, 501
Optimization, 245

Paper Industry, 467
Paper Machines, 501
Parametrisations in Adaptive Control, 501
Performance, 245
Personal Computer, 895
Perturbation, 547

Petrochemical Control, 467
Piecewise Continuous, 547
Plasma Etching, 623
Polymerization Reactor, 689
Predictive Control, 233
Process Control, 139, 467, 839
Process Design, 323
Production Planning and Scheduling, 579
Production System, 869
Profile Control, 501
Pseudolinear Frequency Response, 53
Pulp Industry, 467

Quadratic Programming (QP), 547
Quadratic Programming Solution of Dynamic Matrix Control (QDMC), 547

Reactor Control, 749
Relative Gain Analysis, 895
Relative Gain Array, 99
Robust Estimation, 501
Robustness, 5, 53, 245, 427, 517
Robustness Measures, 323

Self Tuning Control, 517
Self-Tuning Regulators, 467
Semibatch Reactor, 139
Sensitivity Equation, 547
Simple Functions, 547
Singular Value, 5
Singular Value Decomposition, 99
Smith Predictor, 5
Stability Analysis, 53
State Estimation, 641
Stochastic Inputs, 53
Structured Singular Value, 5
Supervisory Control, 579, 839

Traditional Control, 749
Trajectory Optimization, 641

Uncertainty, 323
Unit Operation Control, 623